60 Years of CERN
Experiments and Discoveries

ADVANCED SERIES ON DIRECTIONS IN HIGH ENERGY PHYSICS

ISSN: 1793-1339

Advanced Series on
Directions in High Energy Physics — Vol. 23

60 Years of CERN
Experiments and Discoveries

Editors

Herwig Schopper
University of Hamburg and CERN

Luigi Di Lella
University of Pisa and CERN

W© World Scientific

NEW JERSEY · LONDON · SINGAPORE · BEIJING · SHANGHAI · HONG KONG · TAIPEI · CHENNAI

Published by

World Scientific Publishing Co. Pte. Ltd.
5 Toh Tuck Link, Singapore 596224
USA office: 27 Warren Street, Suite 401-402, Hackensack, NJ 07601
UK office: 57 Shelton Street, Covent Garden, London WC2H 9HE

Library of Congress Cataloging-in-Publication Data
60 years of CERN experiments and discoveries / edited by Herwig Schopper (CERN) and
Luigi Di Lella (CERN).
    pages cm -- (Advanced series on directions in high energy physics ; vol. 23)
  Includes bibliographical references and index.
   ISBN 978-9814644143 (hardcover : alk. paper) -- ISBN 978-9814663182 (pbk. : alk. paper)
  1. Particle accelerators--Experiments. 2. Particles (Nuclear physics)--Experiments. I. Schopper, Herwig F.,
editor. II. Di Lella, L., editor. III. Title: Sixty years of CERN experiments and discoveries. IV. Series:
Advanced series on directions in high energy physics ; v. 23.
  QC787.P3A16 2015
  539.7'309--dc23

          2015012852

British Library Cataloguing-in-Publication Data
A catalogue record for this book is available from the British Library.

Typeset by Stallion Press
Email: enquiries@stallionpress.com

Foreword

In 1949, Europe was in ruins. However, in that year a small group of visionary scientists and diplomats created a resonance that would change the way international science is done, and play an important role in transforming the continent. That resonance was the idea that became CERN, when Europe's laboratory for fundamental physics was established on 29 September 1954. Today, we can only marvel at the vision that drove those people, at their tenacity and determination, and at their remarkable foresight in creating a formula that would stand the test of time and act as a blueprint for long-term, international collaboration in science that has not been bettered to this day. Chronicling 60 years of scientific achievements at CERN is no easy task. Yet this book achieves it, and is fitting testament to the remarkable foresight of CERN's pioneers.

CERN's development has proceeded hand-in-hand with that of electroweak interaction physics, a fact that comes out loud and clear in this volume. From the early measurements of rare pion decays in the 60s to the detection of weak neutral currents in the 70s and subsequent discovery in the 80s of W and Z bosons, carriers of the weak interaction, CERN experiments have put in place many of the cornerstones of this vital element of the Standard Model. The 1990s saw the LEP experiments establish solid experimental foundations for the Standard Model, leaving just one ingredient missing to complete the picture. That, I hardly need to remind you, was the Higgs boson, herald of the Brout–Englert–Higgs mechanism, whose discovery was announced on 4 July 2012 by the ATLAS and CMS collaborations at the LHC.

It is no understatement to say that the Standard Model ranks among the greatest intellectual triumphs of humankind, bringing together theory and experiment, and people from around the world to achieve a common goal. This book delivers a compelling chronicle of CERN's role in that adventure, but it also goes much further. It explores other major contributions CERN has made in areas as diverse as pioneering measurements of the muon $g - 2$, and the achievements of a still unique low-energy antimatter facility.

Herwig Schopper and Luigi Di Lella have done a remarkable job. The table of contents is broad and inclusive, and the author list includes many of those who have left their mark on the history of particle physics at CERN. For any serious student of the development of particle physics from the middle of the 20th century, this book is a must-read.

Rolf-Dieter Heuer • Director-General

CERN • April 2015

Preface

In the year 2014 CERN celebrated its 60th Anniversary, an occasion to review the great success of this first European scientific organisation. The achievements are indeed manifold, scientific in the first place, joined by considerable technological triumphs, and not to forget, the promotion of international cooperation. This book is devoted to summarise the most important experimental accomplishments of CERN, covering also new instrumental technologies and the cooperation between experimentalists and theorists and accelerator development. The objective was not to reproduce more or less the original publications of the time, but we asked colleagues who were main protagonists to look back and describe from the present point of view their motivations and the different ways which led to success or sometimes to failure. It was unavoidable and even desirable that some of the reports reflect the personal view of the authors, each of them playing a major role in those activities. The book shows that scientific progress is based on new ideas combined sometimes with tedious hard work, open exchange of information and fair human collaboration between the laboratory staff and the many groups from external Institutes.

Because of the limited space in the book, we had the extremely difficult task to choose an appropriate selection of contributions of CERN experiments providing a reasonable summary of the overall programme. A part of the various articles shows how experiments done at CERN have provided a number of essential steps for the establishment of the Standard Model of particle physics. Some articles are devoted to the field of nuclear physics and in particular to the physics of nuclear matter. The book demonstrates the enormous progress particle physics has made during the past 60 years, from early beginnings to the final triumph of the detection of the Higgs particle. This development was accompanied by an incredible growth of technical equipment (accelerators-colliders and detectors) and from small groups of physicists to international collaborations composed of several thousands of scientists.

The book should be of interest to physicists, to students and teachers and also to historians of science.

We would like to thank warmly all authors for their efforts to write the contributions to this book.

Editors
Luigi Di Lella and Herwig Schopper
Geneva, March 2015

Contents

The Discovery of the Higgs Boson at the LHC

Peter Jenni[1] and Tejinder S. Virdee[2]

[1] *Albert Ludwigs University of Freiburg, 79085 Freiburg, Germany,*
and CERN, CH-1211 Geneva 23, Switzerland
peter.jenni@cern.ch
[2] *Blackett Laboratory, Imperial College, London SW7 2BW, UK*
t.virdee@imperial.ac.uk

The journey in search for the Higgs boson with the ATLAS and CMS experiments at the LHC started more than two decades ago. The discovery of a heavy scalar boson was announced on 4th July 2012, and subsequent data point strongly to the properties as expected for the boson associated with the Brout–Englert–Higgs mechanism.

1. Introduction

The Standard Model (SM) of particle physics is one of the most remarkable achievements of physics over the past 50 years. Its descriptive and predictive power has been experimentally demonstrated with unprecedented accuracy in many generations of experiments ranging from low to high energies. The SM comprises the fundamental building blocks of all visible matter, with the three fermion families of quarks and leptons, and their interactions via three out of the four fundamental interactions mediated by bosons, namely the massless photon for the electromagnetism, the heavy W and Z bosons for the weak force, these two interactions unified in the electroweak theory,[1-3] and the massless gluons for the strong interaction.

In order to solve the mystery of generation of mass, a spontaneous symmetry-breaking mechanism was proposed[4-9] introducing a complex scalar field that permeates the entire universe. This mechanism, labelled Brout–Englert–Higgs (BEH) mechanism, gives the W and Z their large masses and leaves the photon massless. Interaction with the scalar field imparts masses to the quarks and leptons in proportion to the strength of their couplings to it. This field leads to an additional massive scalar boson as its quantum, called the Higgs boson. After the discovery of the W and Z bosons in the early 1980s, the hunt for the Higgs boson, considered to be the keystone of the SM, became a central theme in particle physics, and also a primary motivation for the Large Hadron Collider (LHC). Finding the Higgs boson would establish the existence of the postulated BEH field, and thereby marking a crucial step in the understanding of Nature.

The great success in making the experimentally "clean" W and Z boson discoveries, despite the huge hadronic backgrounds, at the CERN SPS pbar–p Collider in early 1980s, described in the article by C. Rubbia of this book was crucial for the community to dare to even dream of a future powerful high-energy hadron collider in order to make a decisive search for the Higgs boson. The idea that the tunnel for the future Large Electron–Positron (LEP) machine should be able to house, at some future time, the LHC was already in the air in the late 1970s. Thankfully, those leading CERN at the time had the vision to plan for a tunnel with dimensions that could accommodate it. Enthusiasm for an LHC surfaced strongly in 1984 at a CERN-ECFA workshop in Lausanne entitled "LHC in the LEP Tunnel", which brought together working groups that comprised machine experts, theorists and experimentalists.

With the promise of great physics at the LHC, several motivating workshops and conferences followed, where the formidable experimental challenges started to appear manageable, provided that enough R&D work on detectors would be carried out. Highlights of these "LHC experiment preliminaries" were the 1987 Workshop in La Thuile of the so-called "Rubbia Long-Range Planning Committee" and the large Aachen ECFA LHC Workshop in 1990. Finally, in March 1992, the famous conference entitled "Towards the LHC Experimental Programme", took place in Evian-les-Bains, where several proto-collaborations presented their designs in "Expressions of Interest". Moreover, from the early 1990s, CERN's LHC Detector R&D Committee (DRDC), which reviewed and steered R&D collaborations, greatly stimulated innovative developments in detector technology.

The detection of the Higgs boson played a particularly important role in the design of the general-purpose experiments. In the region of low mass ($114 < m_{\rm H} < 150\,{\rm GeV}$), the two channels considered mostly suited for unambiguous discovery were the decay to two photons and the decay to two Z bosons, each decaying in turn into e^+e^- or $\mu^+\mu^-$, where one or both of the Z bosons could be virtual. As the natural width of the low-mass Higgs boson is $<10\,{\rm MeV}$, the width of any observed peak would be entirely dominated by the instrumental mass-resolution. This meant that in designing the general purpose detectors, considerable care was placed on the value of the magnetic field strength, on the precision tracking systems and on high-resolution electromagnetic (em) calorimeters. The high-mass region and signatures from supersymmetry drove the need for good resolution for jets and missing transverse energy ($E_{\rm T}^{\rm miss}$), as well as for almost full 4π calorimetry coverage.

In this short article there is no way to pay tribute to the two other absolutely essential ingredients in the Higgs boson discovery, besides the experiment (ATLAS and CMS). Indeed, the LHC project has to be seen as a global science project successfully combining the LHC accelerator complex, the experiments and the worldwide computing grid. The LHC has been developed alongside the experimental instruments, since the 1980s, whereas the plans for a powerful computing infrastructure started emerging in the late 1990s.

The LHC is a true 'Marvel of Technology'.[10] In the LHC, protons are accelerated in superconducting radio-frequency cavities and are guided around their circular orbits by powerful superconducting dipole magnets. The dipole magnets operate at 8.3 T and are cooled by superfluid helium to 1.9 K, a temperature lower than that found in inter-planetary space. The counter-rotating LHC beams are organised in 2808 bunches comprising $>10^{11}$ protons per bunch separated by 25 ns, leading to a bunch crossing rate of $\sim$40 MHz (up to now the LHC accelerator has operated at 50 ns bunch spacing with 1380 bunches). The main challenges for the accelerator were to build more than 1200 15-m long superconducting dipoles able to reach this magnetic field, the large distributed cryogenic plant to cool the magnets and some other accelerator structures, and the control of the beams, which will reach, at design operation, an impressive stored energy of 350 MJ, requiring extraordinary precautions.

After a technical incident in September 2008, the LHC started colliding protons in November 2009 at injection energy (450 GeV), followed by a very successful operation surpassing expectations, in 2010 and 2011 at 7 TeV, and then in 2012 at 8 TeV pp collision energy. The collider performed beyond its initial design parameters in almost all cases except energy, reaching peak luminosities of 7×10^{33} cm^{-2}s^{-1}.

The worldwide LHC computing grid (wLCG[a]) was developed to deal with the huge amounts of data generated by the experiments (tens of petabytes per year), requiring a fully distributed computing model. The wLCG provides universal access to the data within the collaborations, and consists of a hierarchical architecture of tiered centres, with one large Tier-0 centre at CERN, about 12 large Tier-1 centres at national/regional computing facilities, and more than 100 Tier-2 centres at various institutes.

The long duration of the LHC project so far is illustrated in Table 1, with a few selected milestones concerning the LHC and the general-purpose experiments. In the following, the article will concentrate on the ATLAS and CMS experiments, and their discovery of a scalar boson that has all characteristics of the SM Higgs boson of the BEH mechanism, within the accuracy of current measurements.

2. The ATLAS and CMS Experiments

To reach the ambitious physics goals, novel detector technologies had to be developed and most of the existing technologies had to be pushed to their limits. Several detector concepts were proposed, and finally two complementary ones, ATLAS[11a] and CMS,[12a] were selected by the LHC experiments committee (LHCC) as general-purpose detectors, to proceed to detailed design. These were fully developed, and their components prototyped and tested in beams, over numerous years before construction commenced in the second half of the 1990s.

---

[a]http://wlcg.web.cern.ch

Table 1   The LHC timeline.

1984	Workshop on a Large Hadron Collider in the LEP tunnel, Lausanne, Switzerland.
1987	Workshop on the Physics at Future Accelerators, La Thuile, Italy. The Rubbia "Long-Range Planning Committee" recommends the Large Hadron Collider as the right choice for CERN's future.
1990	LHC Workshop, Aachen, Germany (discussion of physics, technologies and detector design concepts).
1992	General Meeting on LHC Physics and Detectors, Evian-les-Bains, France (with four general-purpose experiment designs presented).
1993	Three Letters of Intent evaluated by the CERN peer review committee LHCC. ATLAS and CMS selected to proceed to a detailed technical proposal.
1994	The LHC accelerator approved for construction, initially in two stages.
1996	ATLAS and CMS Technical Proposals approved.
1997	Formal approval for ATLAS and CMS to move to construction (materials cost ceiling of 475 MCHF).
1997	Construction commences (after approval of detailed Technical Design Reports of detector subsystems).
2000	Assembly of experiments commences, LEP accelerator is closed down to make way for the LHC.
2008	LHC experiments ready for pp collisions. LHC starts operation. An incident stops LHC operation.
2009	LHC restarts operation, pp collisions recorded by LHC detectors.
2010	LHC collides protons at high energy (centre-of-mass energy of 7 TeV).
2012	LHC operates at 8 TeV: Discovery of a Higgs boson.

It cannot be stressed enough how important the many years of R&D were that preceded the final detector construction for both experiments. Technologies had to be taken far beyond their state-of-art of the late 1980s in terms of performance criteria in the anticipated harsh LHC environment, like granularity and speed of readout, radiation resistance, reliability, but also considering buildable sizes of the detector components and number of units, and very importantly at an affordable cost. For many detector subsystems there were initially a few parallel developments pursued as options, because it was not guaranteed from the onset that a given proposed technology would finally fulfil all the necessary requirements. Increasingly more realistic prototypes were developed, in a learning process for both the detector communities and the industries involved.

Some of the major technology decisions were taken by the Collaborations before the submission of the Technical Proposals[11b,12b] to the LHCC end of 1994, which were finally approved early in 1996. For other choices the R&D needed more time, and they could only be made in the subsequent years from 1996 to the early 2000s, thereby defining the timing for the final Technical Design Reports of the various detector components.

2.1.   *The ATLAS detector*

The design of the ATLAS detector[11c] is shown in Fig. 1 (top), and is based on a novel and challenging superconducting air-core toroid magnet system, containing about

80 km of superconductor cable in eight separate barrel coils (each $25 \times 5\,\text{m}^2$ in a 'racetrack' shape) and two matching endcap toroid systems. A field of $\sim$0.5 Tesla is generated over a large volume. The toroids are complemented with a thin solenoid (2.4 m diameter, 5.3 m length) at the centre which provides an axial magnetic field of 2 T.

The detector includes an electromagnetic (em) calorimeter complemented by a full coverage hadronic calorimeter for jet and E_T^{miss} measurements. The electromagnetic calorimeter is a cryogenic liquid argon–lead sampling calorimeter in a novel 'accordion' geometry allowing fine granularity, both laterally and in depth, and full coverage without any un-instrumented regions. A plastic scintillator — iron sampling hadronic calorimeter, also with a novel geometry, is used in the barrel part of the experiment. Liquid argon hadronic calorimeters are employed in the endcap regions near the beam axis. The em and hadronic calorimeters have almost 200000 and 20000 cells, respectively, and are in an almost field-free region between the toroids and the solenoid. They provide both fine lateral and longitudinal segmentation.

The momentum of the muons is precisely measured as they travel unperturbed by material for over $\sim$5 m in the air-core toroid field. About 1200 large muon chambers of various shapes, with a total area of $5000\,\text{m}^2$, measure the impact position with an accuracy of better than 0.1 mm. Another set of about 4200 fast chambers are used to provide the "trigger". The chambers were built in about 20 collaborating institutes on three continents. This was typical and also the case for other components of the experiment.

The reconstruction of all charged particles, including that of displaced vertices, is achieved in the inner detector, which combines highly granular pixel ($50 \times 400\,\mu\text{m}^2$ elements leading to 80 million channels) and microstrip (13 cm $\times$ 80 μm elements leading to 6 million channels) silicon semiconductor sensors placed close to the beam axis, and a 'straw tube' gaseous detector (350000 channels) which provides about 30–40 signal hits per track. The latter also helps in the identification of electrons using information from the effects of transition radiation.

The air-core magnet system allows a relatively lightweight overall structure leading to a detector weighing 7000 tons. The muon spectrometer defines the overall dimensions of the ATLAS detector: diameter of 25 m and length of 44 m.

2.2. The CMS detector

The design of the CMS detector,[12c] shown in Fig. 1 (bottom), is based on a super-conducting high-field solenoid, which first reached the design field of 4 T in 2006.

The solenoid generates a uniform magnetic field parallel to the direction of the LHC beams. The field is produced by a current of 20 kA flowing through a reinforced Nb-Ti superconducting coil built in four layers. Economic and transportation constraints limited the outer radius of the coil to 3 m and its length to 13 m. The field is returned through a 1.5 m thick iron yoke, which houses four muon

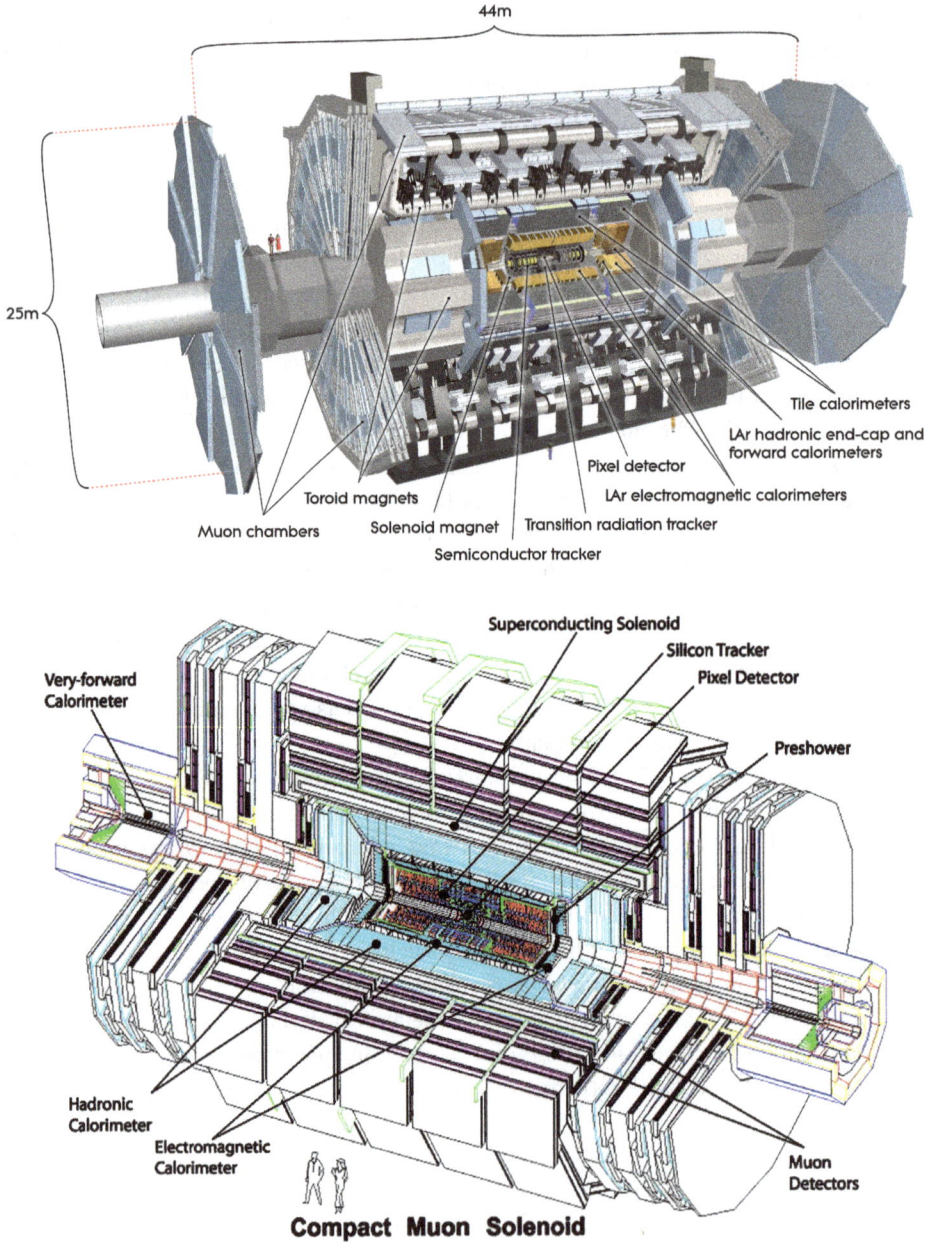

44m

25m

Tile calorimeters

LAr hadronic end-cap and
forward calorimeters

Pixel detector

LAr electromagnetic calorimeters

Toroid magnets

Transition radiation tracker

Muon chambers   Solenoid magnet

Semiconductor tracker

Superconducting Solenoid

Silicon Tracker

Pixel Detector

Very-forward
Calorimeter

Preshower

Hadronic
Calorimeter

Electromagnetic
Calorimeter

Muon
Detectors

Compact Muon Solenoid

Fig. 1.   Schematic longitudinal cut-away views of (top) the ATLAS and (bottom) the CMS
detectors, showing the different layers around the LHC beam axis, with the collision point in
the centre.

stations to ensure robustness of identification and measurement and full geometric coverage.

The CMS design was first optimised to cleanly identify, trigger and measure muons, over a wide range of momenta, e.g., arising from processes such as $H \rightarrow ZZ \rightarrow 4\mu$ and few TeV mass $Z' \rightarrow 2\mu$. In order to accomplish this, the region outside the inner tracker and calorimeters is surrounded with absorber material amounting to about 1.5–2 m of iron, to stop all the produced particles except for muons and neutrinos. The muons execute spiral trajectories in the magnetic field, and are identified and reconstructed in $\sim$3000 m^2 of gas chambers interleaved in the iron return yoke. Another set of about $\sim$500 fast chambers are used to provide a second system of detectors for the Level-1 muon trigger.

The next design priority was driven by the search for the decay of the SM Higgs boson into two photons. This motivated an em calorimeter with the best possible energy resolution. A new type of crystal was selected: lead tungstate (PbWO$_4$) scintillating crystal. Five years of research and development were necessary to improve the transparency and the radiation hardness of these crystals, and it then took over ten years (1998–2008) of round-the-clock production to manufacture the 75848 crystals, constituting the largest crystal calorimeter ever built.

The solution to charged particle tracking was to opt for a small number of precise position measurements of each charged track ($\sim$13 each with a position resolution of $\sim$15 μm per measurement) leading to a large number of cells distributed inside a cylindrical volume 5.8 m long and 2.5 m in diameter: 66 million 100 $\times$ 150 μm^2 silicon pixels and 9.3 million silicon microstrips ranging from $\sim$10 cm $\times$ 80 μm to $\sim$20 cm $\times$ 180 μm. With 198 m^2 of active silicon area the CMS tracker is by far the largest silicon tracker ever built.

Finally the hadron calorimeter, comprising $\sim$3000 small solid angle projective towers covering almost the full solid angle, is built from alternate plates of $\sim$5 cm brass absorber and $\sim$4 mm thick scintillator plates that sample the energy. The scintillation light is detected by photodetectors (hybrid photodiodes) that can operate in the strong magnetic field.

2.3. *Installation and commissioning*

The two very different and complementary detector concepts also had far-reaching consequences for the underground installation strategies in the two caverns on opposite locations on the LHC collider ring.

Given its size and its magnet structure, the ATLAS detector had to be assembled directly in the underground cavern. The installation process began in summer 2003 (after the completion of civil engineering work that started in 1998) and ended in summer 2008. Figure 2 shows one end of the cylindrical barrel detector after 3.5 years of installation work, 1.5 years before completion. The ends of four of the barrel toroid coils are visible, illustrating the eightfold symmetry of the structure.

Fig. 2.   Photograph of one end of the ATLAS detector barrel with the calorimeter end-cap still retracted before its insertion into the barrel toroid magnet structure (February 2007 during the installation phase).

The iron yoke of the CMS detector is sectioned into five barrel wheels and three endcap disks at each end, for a total weight of 12500 tons. The sectioning enabled the detector to be assembled and tested in a large surface hall while the underground cavern was being prepared. The sections, weighing between 350 tons and 2000 tons were then lowered sequentially (Fig. 3) between October 2006 and January 2008, using a dedicated gantry system equipped with strand jacks: a pioneering use of this technology to simplify the underground assembly of large experiments.

Individual detector components (e.g. chambers) were built and assembled in a distributed way all around the globe in the numerous participating institutes and were typically first tested at their production sites, then after delivery to CERN, and finally again after their installation in the underground caverns. The collaborations also invested enormous effort in testing representative samples of the detectors in test beams at CERN and other accelerator laboratories around the world. These test beam campaigns not only verified that performance criteria were met over the several years of production of detector components, but also were used to prepare the calibration and alignment data for LHC operation. Very important were the so-called large combined test beam set-ups, which represented whole 'slices' of the different detector layers of the final detectors.

During the progressing installation the experiments made extensive use of the constant flow of cosmic rays impinging on Earth providing a reasonable flux of muons even at a depth of 100 m underground, typically a few hundred per second

Fig. 3.   Photograph showing the lowering of the central barrel part and solenoid of the CMS detector during its installation in the cavern in 2007.

traversing the detectors. These muons were used to check the whole chain from hardware to analysis programs of the experiments, and to align the detector elements and calibrate their response prior to the pp collisions. In particular, after the LHC incident on 19th September 2008 the experiments used the 15 months LHC down time, before the first collisions on 23rd November 2009, to run the full detectors in very extensive cosmic ray campaigns, collecting many hundreds of millions of muon events. These runs allowed both ATLAS and CMS to be ready for physics operation, with pre-calibrated and pre-aligned detectors, by the time of the first pp collisions.

3.   Trigger, Computing, and Early Operation

3.1.   *Trigger and computing*

A particular challenge for ATLAS and CMS are the very high collision rates in the LHC, necessary for the Higgs search and studies, given its small production cross-section combined with the need to investigate final states with very small branching fractions. In the first three years of operation the LHC reached a peak instantaneous luminosity of 7×10^{33} cm^{-2}s^{-1} with a 50 ns bunch spacing, which meant that the detectors had to simultaneously cope with up to ∼50 overlapping (pile-up) events in a given bunch crossing. In the years ahead, the instantaneous luminosity is still expected to rise two- to three-fold.

It is technically not possible to store all data for all events, therefore a trigger system is used to reject large numbers of events and retain only the interesting ones from crossings with potential physics processes of interest. This is done in real time by sophisticated integrated trigger and data acquisition systems, involving custom made fast electronics in a first stage and large computing farms in subsequent stages before the data is transferred to mass storage for further analyses. The initial data rate from up to 40 MHz bunch crossings with multiple pile-up events is thereby reduced to a few hundreds of Hz for offline analysis. A description of these systems is far beyond the scope of this article, see Refs. 11 and 12 for details.

The ATLAS and CMS experiments generate huge amounts of data (tens of petabytes of data per year; $1\,\mathrm{PB} = 10^6\,\mathrm{GB}$) requiring a fully distributed computing model. The worldwide LHC Computing Grid allows any user anywhere access to any data recorded or produced in the analyses steps during the lifetime of the experiments. The centre at CERN receives the raw data, carries out prompt reconstruction, almost in real time, and exports the raw and reconstructed data to the Tier-1 centres and also to Tier-2 centres for physics analysis. The Tier-0 must keep pace with the event rate of several hundred Hz of typically 1 MB of raw data per event from each experiment. The large Tier-1 centres provide also long-term storage of raw data and reconstructed data outside of CERN (as a second copy). They carry out, for example, second-pass reconstruction, when better calibration constants are available. The large number of events simulated by Monte Carlo methods and necessary for quantifying the expectations are produced mainly in Tier-2 centres.

3.2.   *Standard model measurements to demonstrate the performance*

Observing, and measuring accurately, at the LHC collision energies, the production of known particles of the SM, was always considered to be a necessary stepping stone towards exploring the full potential of the LHC with its promise of new physics, firstly of the discovery of the Higgs boson. The SM processes, such as W and Z production, are often referred to as 'standard candles' for the experiments. An illustrative example of such a measurement is shown in Fig. 4, produced only after a few months or so after first high-energy collisions in spring 2010. ATLAS and CMS observed in such di-muon invariant mass distributions a 'summary' of decades of particle physics, with remarkable mass resolution.

However, there is much more value to measuring the SM processes than this: never before could the SM physics be studied at a hadron collider with such sophisticated and highly accurate detectors, allowing comparison with the predictions of the SM at an unprecedented precision and minimal instrumental systematic errors.

The data collected in the first three years of high-energy LHC operation have allowed ATLAS and CMS to make numerous precise measurements of SM processes, including production of bottom and top quarks, W and Z bosons, singly and in

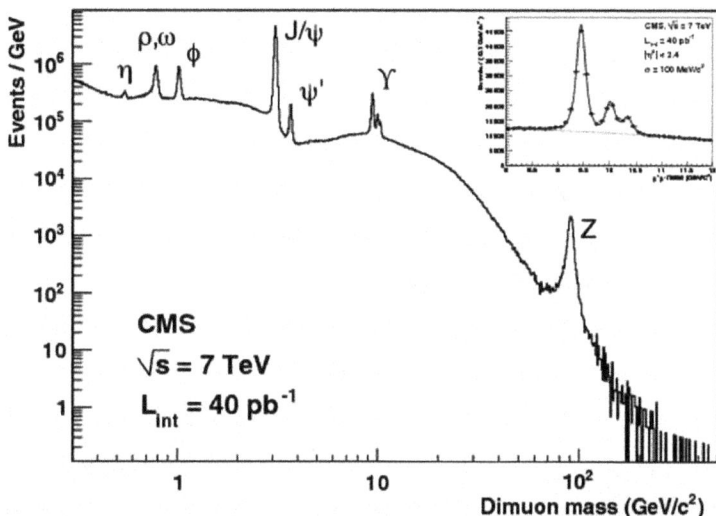

Fig. 4. The distribution of the invariant mass for di-muon events, shown here from CMS, displays the various well-known resonant states of the SM. The inset illustrates the excellent mass resolution for the three states of the Υ family.

pairs. In particular very detailed measurements of QCD processes have been made. A summary of examples of such studies is shown in Fig. 5 where measurements of cross-sections for various selected electroweak and QCD processes are compared with the SM predictions.[b] These very diverse measurements, probing cross-sections over a range of many orders of magnitude, confirm the predictions of the SM within the errors in all cases. Establishing this agreement is essential before any claims for discoveries can be made, i.e. to demonstrate on the one hand that the detector performance is well understood, and on the other hand that known SM processes are correctly observed in the experiments as they often constitute large backgrounds to signatures of new physics, such as those expected for the Higgs boson. The speed with which the wide range of measurements have shown that SM predictions for known physics have been essentially spot-on is a tribute to a large amount of work done by many particle physics theorists along with the results from the other collider experiments at LEP, Tevatron, HERA, and b-factories.

4. The Standard Model Higgs Boson and the LHC

The SM Higgs boson is a "special" particle: it is a unique fundamental spin-parity $J^P = 0^+$ particle, a quantum of an omnipresent fundamental scalar field that

---

[b]https://twiki.cern.ch/twiki/bin/view/AtlasPublic/StandardModelPublicResults
https://twiki.cern.ch/twiki/bin/view/CMSPublic/PhysicsResultsSMP

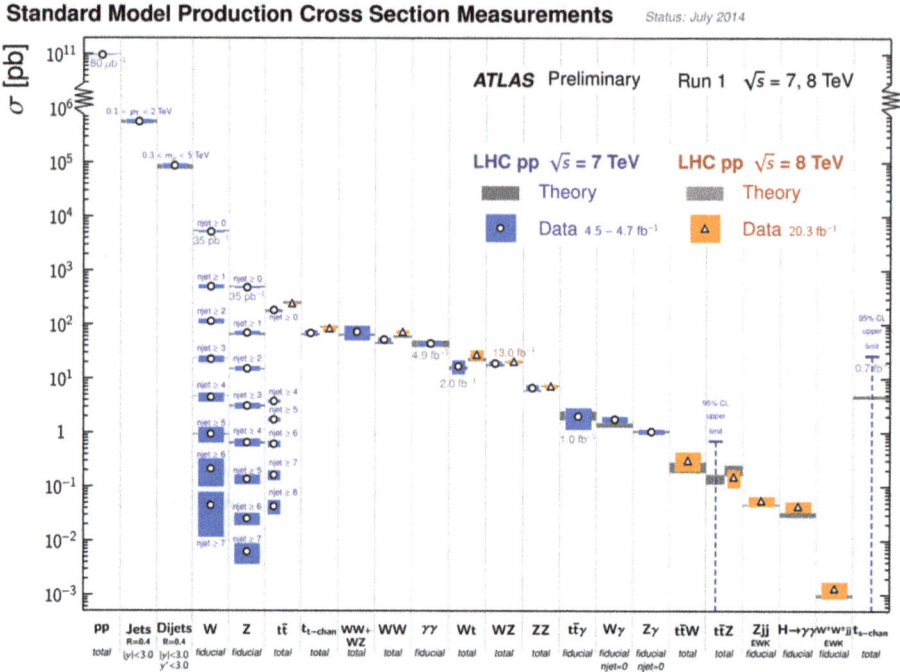

Fig. 5.   A comparison of cross-section measurements for electroweak and QCD processes with theoretical predictions from the SM, shown here as example from the ATLAS experiment.

interacts with, or couples to, other elementary particles with strengths related to their masses. The boson is short-lived (10^{-23} s) and hence experiments would only detect its decay products. However, the mass of the Higgs boson is not predicted by theory though once its mass is known all of its other properties are precisely predicted.

From general considerations $m_H < 1$ TeV whilst precision electroweak constraints imply that $m_H < 152$ GeV at 95% confidence level (CL).[13] The lower limit on the mass of the Higgs boson was established by the LEP experiments at 114.4 GeV.[14]

The production cross sections and the branching ratios into the various decay modes of the SM Higgs boson as a function of mass are illustrated in Figs. 6a and 6b, respectively.[15] Four main mechanisms are predicted for Higgs boson production in pp collisions: the gluon–gluon fusion mechanism, which has the largest cross-section, followed in turn by vector-boson fusion (VBF), associated WH and ZH production (VH), and production in association with top quark pairs.

The SM Higgs boson couples to the different pairs of particles in a proportion that is precisely predicted by the SM, i.e., for fermions (f) proportional to m_f^2 and for bosons (V) proportional m_V^4/v^2 where v is the vacuum expectation value of the scalar field ($v = 246$ GeV). A search had to be envisaged not only over a large range of masses but also many possible decay modes: into pairs of photons,

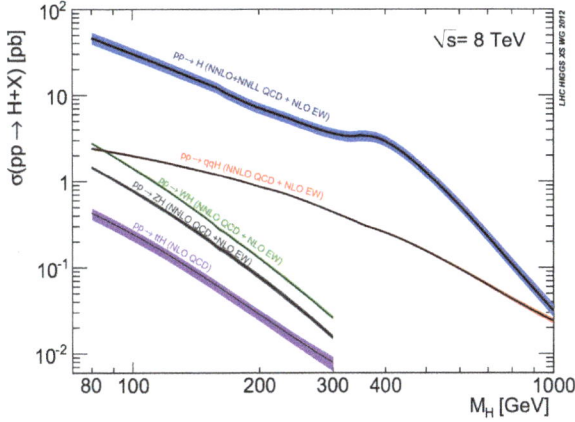

Fig. 6a.   The SM Higgs production cross-section at $\sqrt{s} = 8\,\text{TeV}$.

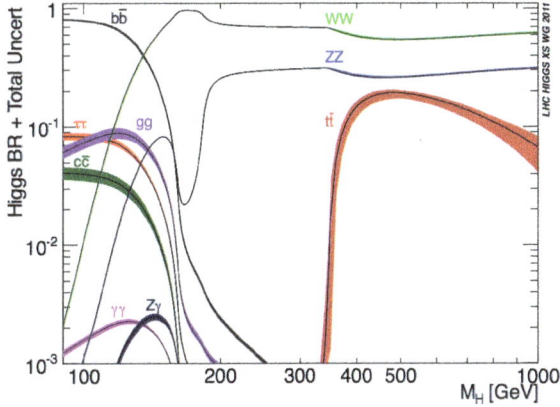

Fig. 6b.   The SM Higgs branching ratios as a function of the Higgs boson mass.

Z bosons, W bosons, τ leptons, and b quarks. For example, at $m_H = 125\,\text{GeV}$ the SM boson is predicted to decay into pairs of photons with BR $= 2.3 \times 10^{-3}$, into Z bosons and then four electrons or muons, or two muons and two electrons, with BR $- 1.25 \times 10^{-4}$, into a pair of W bosons and then into $ll\nu\nu$ with BR $\sim 1\%$, a pair of τ-leptons with BR $= 6.4\%$, and into a pair of b-quarks with BR $= 54\%$.

For simplicity, bb is used to denote b quark+anti-b quark, tt to denote t quark+anti-t quark, etc. Similarly, ZZ is used to denote ZZ$^{(*)}$ and WW to denote WW$^{(*)}$.

For a given Higgs boson mass hypothesis, the sensitivity of the search depends on the:

• mass of the Higgs boson,
• Higgs boson production cross-section (Fig. 6a),

- decay branching fraction into the selected final state (Fig. 6b),
- signal selection efficiency,
- observed Higgs boson mass resolution, and
- level of backgrounds with the same or a similar final state.

To improve the sensitivity of a given state, events are separated into categories with different signal/background ratios, and analysed independently. For many analyses all the relevant information on the discrimination between the signal and the background (aside from mass itself) is encoded into a single multivariate output that is, to first order, independent of mass.

4.1. *Discovery and properties of the Higgs boson*

The most striking result to emerge from the ATLAS[16] and CMS[17] experiments is the discovery of a new heavy boson with a mass of ~125 GeV.

In the 2011 data-taking run, the ATLAS and CMS experiments recorded data corresponding to an integrated luminosity of ~5 fb^{-1} at $\sqrt{s} = 7$ TeV. In December 2011, the very first "tantalising hints" of a new particle from both the ATLAS and CMS experiments were shown at CERN. The general conclusion was that both experiments were seeing an excess of unusual events at roughly the same place in mass (in the mass range 120–130 GeV) in two different decay channels. That set the stage for data taking in 2012.

In January 2012, it was decided to slightly increase the energy of the proton beams from 3.5 to 4 TeV, giving a centre-of-mass energy of 8 TeV, leading to ~20% increase in the production cross-section of an SM Higgs boson in the mass range 120–130 GeV. By June 2012, the number of high-energy collisions examined had doubled and both CMS and ATLAS had greatly improved their analyses. Each experiment decided to again look at the data, but only after all the algorithms and selection procedures had been agreed, in case a bias was inadvertently introduced. These data led to the discovery of a Higgs boson, independently in both the ATLAS and CMS experiments in July 2012 (see Section 4.2).

By the end of 2012 (LHC Run 1), the total amount of data that had been examined corresponded to ~5 fb^{-1} at $\sqrt{s} = 7$ TeV and ~20 fb^{-1} at $\sqrt{s} = 8$ TeV, equating to the examination of some 2000 trillion proton–proton collisions. Using these data first measurements of the properties of the new boson were also made (see Section 4.4).

4.2. *Results from the 2011 and partial 2012 datasets*

In this section we discuss the analyses that led to the discovery of a new heavy boson around a mass of 125 GeV using the data accumulated up to June 2012. The two channels that were particularly suited for unambiguous discovery are the decays to two photons and to two Z bosons, where one or both of the Z bosons could be virtual, subsequently decaying into four electrons, four muons or two electrons and two muons, as the observed mass resolution ($\sim 1\%$ of m_H) is the best and the backgrounds manageable or small.

4.2.1. *The H $\rightarrow \gamma\gamma$ decay mode*

In the H $\rightarrow \gamma\gamma$ analysis a search is made for a narrow peak in the diphoton invariant mass distribution in the mass range 110–150 GeV, on a large irreducible background from QCD production of two photons (via quark–antiquark annihilation and "box" diagrams). There is also a reducible background where one or more of the reconstructed photon candidates originate from misidentification of jet fragments, with the process of QCD Compton scattering dominating.

The event selection requires two photon candidates satisfying p_T and photon identification criteria. As an example, in CMS typically a p_T threshold of $m_{\gamma\gamma}/3$ ($m_{\gamma\gamma}/4$) is applied to the photon leading (sub-leading) in p_T, where $m_{\gamma\gamma}$ is the diphoton invariant mass. Scaling the p_T thresholds in this way avoids distortion of the shape of the $m_{\gamma\gamma}$ distribution. The background is estimated from data, without the use of MC simulation, by fitting the diphoton invariant mass distribution in a range ($100 < m_{\gamma\gamma} < 180$ GeV). Typically a polynomial function is used to describe the shape of the background.

The results from the CMS experiment are shown in Fig. 7a. A clear peak at a diphoton mass of around 125 GeV is seen.[17] A similar result was obtained in the ATLAS experiment.[16]

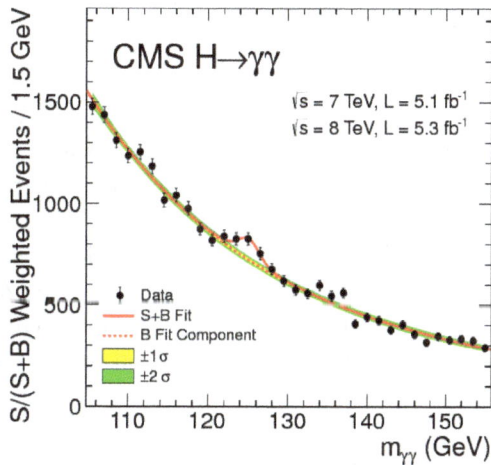

Fig. 7a. The two-photon invariant mass distribution of selected candidates in the CMS experiment, weighted by S/B of the category in which it falls. The lines represent the fitted background and the expected signal contribution ($m_H = 125$ GeV).

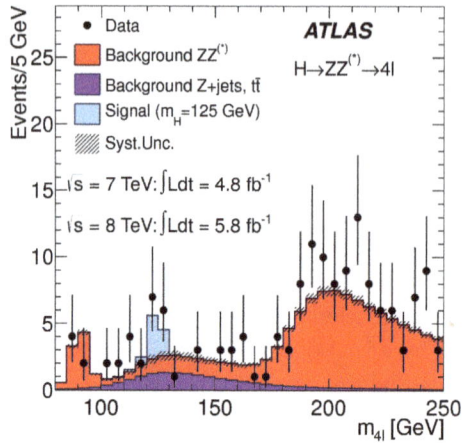

Fig. 7b.   The four-lepton invariant mass distribution in the ATLAS experiment for selected candidates relative to the background expectation. The expected signal contribution ($m_H = 125\,\mathrm{GeV}$) is also shown.

4.2.2.   *The $H \rightarrow ZZ \rightarrow 4l$ decay mode*

In the H→ZZ→4l decay mode a search is made for a narrow four-charged lepton mass peak in the presence of a small continuum background. The background sources include an irreducible four-lepton contribution from direct ZZ production via quark–antiquark and gluon–gluon processes. Reducible background contributions arise from Z + bb and tt production where the final states contain two isolated leptons and two b-quark jets producing secondary leptons.

The event selection requires two pairs of same-flavour, oppositely charged leptons. Since there are differences in the reducible background rates and mass resolutions between the sub-channels 4e, 4μ, and 2e2μ, they are analysed separately. Electrons are typically required to have $p_T > 7\,\mathrm{GeV}$. The corresponding requirements for muons are $p_T > 5\,\mathrm{GeV}$. Both electrons and muons are required to be isolated. The ZZ background, which is dominant, is evaluated from Monte Carlo simulation studies.

The m_{4l} distribution is shown in Fig. 7b for the ATLAS experiment.[16] A clear peak is observed at ∼125 GeV, in addition to the one at the Z mass. The latter is due to the conversion of an inner bremsstrahlung photon emitted simultaneously with the dilepton pair. A similar result was obtained by the CMS experiment.[17]

4.2.3.   *Combining the results*

A search was also made in other decay modes of a possible Higgs boson and combined to yield the final results published in August 2012 by ATLAS[16] and CMS.[17] Figure 8 presents the results in terms of local significance for a range of masses. Both ATLAS and CMS experiments independently discovered a new heavy

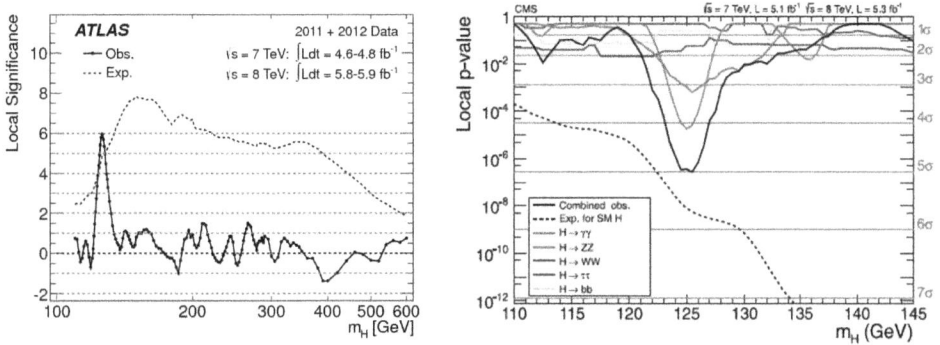

Fig. 8. The combined result of all searches in the ATLAS experiment (left) and the CMS experiment (right) for the observed and expected local significance as a function of mass. Note the vertical axes in the two plots are different. The probability for a background fluctuation to be at least as large as the observed maximum excess is termed the local p-value, and that for an excess *anywhere* in a specified mass range the global p-value.

boson at approximately the same mass, clearly evident in the two different decay modes, $\gamma\gamma$ and ZZ. The observed (expected) local significances were 6.0σ (5.0σ) and 5.0σ (5.8σ) in ATLAS and CMS respectively.

The decay into two bosons (two γ; two Z bosons; two W bosons) implied that the new particle is a boson with spin different from one and its decay into two photons that it carries either spin-0 or spin-2.

The results presented by both ATLAS and CMS collaborations were consistent, within uncertainties, with the expectations for an SM Higgs boson. Both noted that collection of more data would enable a more rigorous test of this conclusion and an investigation of whether the properties of the new particle imply physics beyond the SM.

4.3. *Results from the full 2011 and 2012 data set*

Now we present the results from the full 2011 and 2012 data sets corresponding to an integrated luminosity of $\sim 5\,\mathrm{fb}^{-1}$ at $\sqrt{s} = 7\,\mathrm{TeV}$ and $\sim 20\,\mathrm{fb}^{-1}$ at $\sqrt{s} = 8\,\mathrm{TeV}$. This larger dataset allowed confirmation of the discovery of the new boson, as well as a better examination of the decay channels other than the $H \rightarrow \gamma\gamma$ and the $H \rightarrow ZZ \rightarrow 4l$ decay modes and the first substantial investigations of the boson's properties.

4.3.1. *Decays to bosons: The $H \rightarrow \gamma\gamma$, the $H \rightarrow ZZ \rightarrow 4l$ and $H \rightarrow WW \rightarrow 2l2\nu$ decay modes*

The results from the ATLAS experiment are shown for the $H \rightarrow \gamma\gamma$ decay mode (Fig. 9a)[18] and those from the CMS experiment for the $H \rightarrow ZZ \rightarrow 4l$ mode (Fig. 9b).[19] The signal is unmistakable and the significances have increased

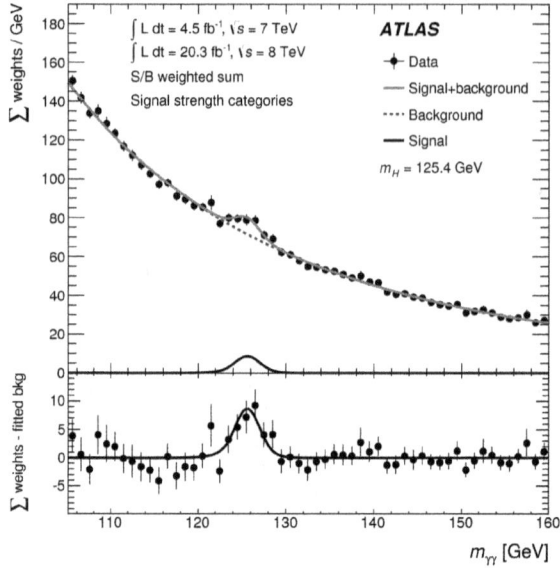

Fig. 9a.   Invariant mass distribution of di-photon candidates. The result of a fit to the background described by a polynomial and the sum of signal components is superimposed. The bottom inset displays the residuals of the data with respect to the fitted background component.

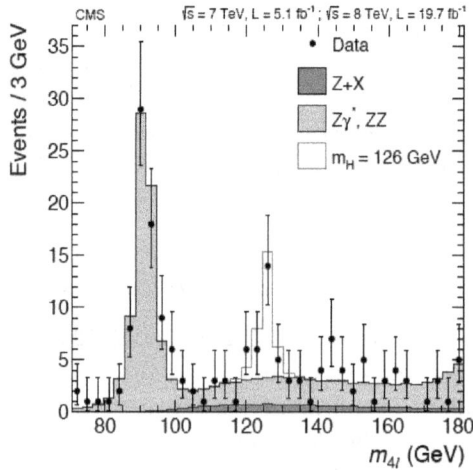

Fig. 9b.   The four-lepton invariant mass distribution in the CMS experiment for selected candidates relative to the background expectation. The expected signal contribution is also shown.

as can be seen from Table 2. The data show an even clearer excess of events above the expected background around 125 GeV. The complementary data from the two experiments can be found in Refs. 21 and 20.

The search for H → WW is primarily based on the study of the final state in which both W bosons decay leptonically, resulting in a signature with two

Table 2   The expected and observed local p-values in ATLAS and CMS,[30] expressed as the corresponding number of standard deviations of the observed excess from the background-only hypothesis, for $m_H = 125\,\text{GeV}$, for various decay modes.

Experiment Decay mode/combination	ATLAS		CMS	
	Expected (σ)	Observed (σ)	Expected (σ)	Observed (σ)
$\gamma\gamma$	4.6	5.2	5.3	5.6
ZZ	6.2	8.1	6.3	6.5
WW	5.8	6.1	5.4	4.7
bb	2.6	1.4	2.6	2.0
$\tau\tau$	3.4	4.5	3.9	3.8
$\tau\tau$+bb[26]	—	—	4.4	3.8

isolated, oppositely charged, high p_T leptons (electrons or muons) and large missing transverse momentum, E_T^{miss}, due to the undetected neutrinos. The signal sensitivity is improved by separating events according to lepton flavour; into e^+e^-, $\mu^+\mu^-$, and $e\mu$ samples, and according to jet multiplicity into 0-jet and 1-jet samples. The dominant background arises from irreducible non-resonant WW production. Any background arising from Z bosons, with same flavour but opposite sign leptons, is removed by a di-lepton mass cut $(m_Z - 15) < m_{ll} < (m_Z + 15)\,\text{GeV}$.

The m_{ll} distribution in the 0-jet and $e\mu$ final state is shown for CMS in Fig. 10a.[22] The expected contribution from a SM Higgs boson with $m_H = 125\,\text{GeV}$ is also shown. The transverse mass, m_T, distribution is shown in Fig. 10b from ATLAS, as well as the background-subtracted distribution.[23] Both show a clear excess of events compatible with a Higgs boson with mass $\sim 125\,\text{GeV}$. The observed (expected) significance of the excess with respect to the background only hypothesis is shown in Table 2.

4.3.2.   Decays to fermions: The H→ ττ and the H→ bb decay modes

It is important to establish whether this new particle also couples to fermions, and in particular to down-type fermions, since the measurements in Section 4.3.1 mainly constrain the couplings to the up-type top quark. Determination of the couplings to down-type fermions requires direct measurement of the Higgs boson decays to bottom quarks and τ leptons.

The $H \rightarrow \tau\tau$ search is typically performed using the final-state signatures $e\mu$, $\mu\mu$, $e\tau_h$, $\mu\tau_h$, $\tau_h\tau_h$, where electrons and muons arise from leptonic τ-decays and τ_h denotes a τ lepton decaying hadronically. Each of these categories is further divided into two exclusive sub-categories based on the number and the type of the jets in the event: (i) events with one forward and one backward jet, consistent with the VBF topology, (ii) events with at least one high p_T hadronic jet but not selected in the previous category. In each of these categories, a search is made for a broad excess in

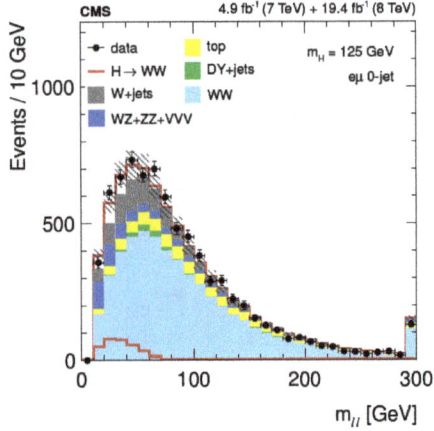

Fig. 10a.   Distribution of dilepton mass in the 0-jet, $e\mu$ final state in CMS for a $m_{\mathrm{H}} = 125\,\mathrm{GeV}$ SM Higgs boson decaying via H→WW→ $l\nu l\nu$ and for the main backgrounds.

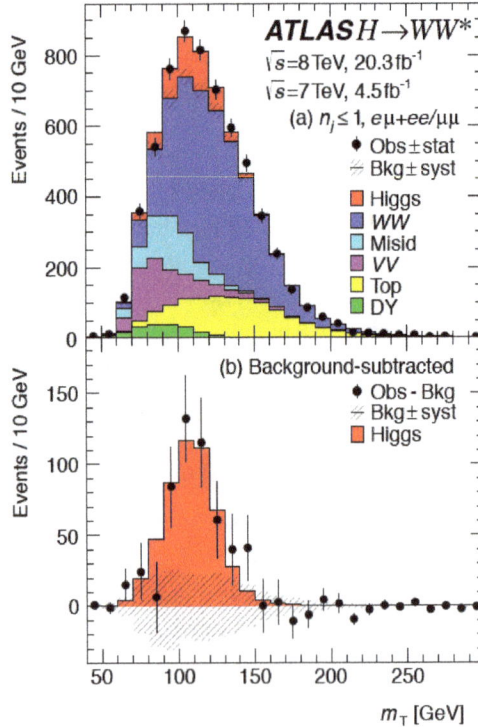

Fig. 10b.   The transverse mass distributions for events passing the full selection of the H→WW→ $l\nu l\nu$ analysis in ATLAS summed over all lepton flavours for final states with $N_{\mathrm{jet}} \leq 1$. In the lower part the residuals of the data with respect to the estimated background are shown, compared to the expected m_{T} distribution for a SM Higgs boson.

the reconstructed $\tau\tau$ mass distribution. The main irreducible background, $Z \to \tau\tau$ production, and the largest reducible backgrounds (W + jets, multijet production, $Z \to$ ee) are evaluated from various control samples in data.

The H $\to$ bb decay mode has by far the largest branching ratio ($\sim$54%). However since $\sigma_{bb}(QCD) \sim 10^7 \times \sigma(H \to bb)$ the search concentrates on Higgs boson production in association with a W or Z boson using the following decay modes: W $\to$ eν/$\mu\nu$ and Z $\to$ ee/$\mu\mu$/$\nu\nu$. The Z $\to \nu\nu$ decay is identified by the requirement of a large missing transverse energy. The Higgs boson candidate is reconstructed by requiring two b-tagged jets.

Evidence for a Higgs boson decaying to a $\tau\tau$ lepton pair is reported by the CMS[24] and ATLAS[27] Collaborations. The results are given in Table 2. CMS has updated its analysis and the results reported in Table 2 come from Ref. 30 where both the H $\to \tau\tau$ and H $\to$ WW contributions are considered as signal in the $\tau\tau$ decay-tag analysis. This treatment leads to an increased sensitivity to the presence of a Higgs boson that decays into both $\tau\tau$ and WW.

The CMS measurements in the H $\to \tau\tau$[27] and VH with H $\to$ bb[25] searches are mutually consistent, within the precision of the present data, and with the expectation for the production and decay of the SM Higgs boson. CMS has combined these two results, requiring the simultaneous analysis of the data selected by the two individual measurements.[26] Figure 11 shows the scan of the profile likelihood as a function of the signal strength relative to the expectation for the production and decay to fermions (bb and $\tau\tau$) of a standard model Higgs boson, m, for

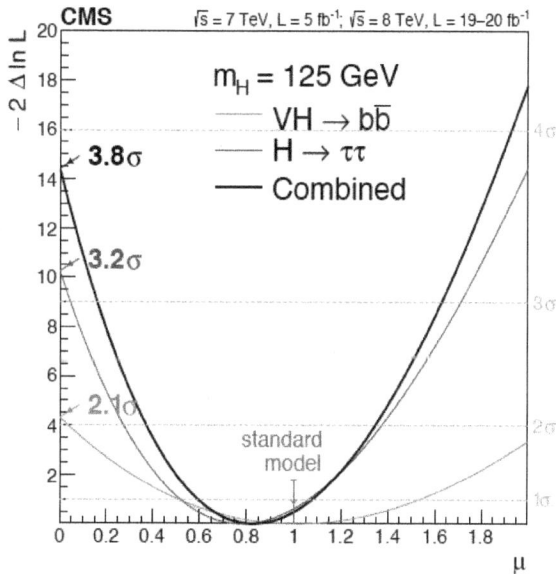

Fig. 11.   Scan of the profile likelihood as a function of the signal strength relative to the expectation for the production and decay of a standard model Higgs boson, m, for $m_H = 125$ GeV.

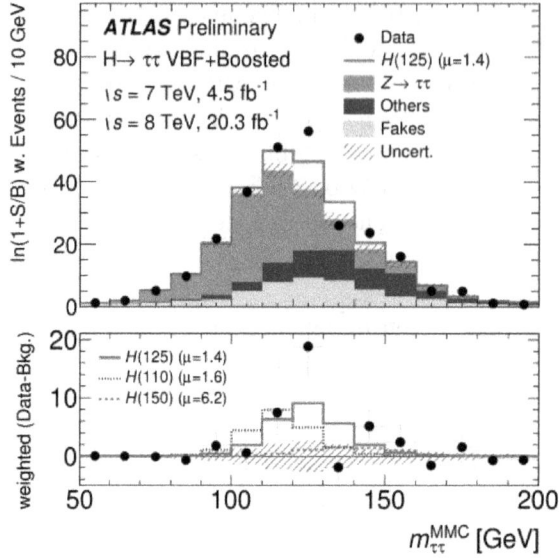

Fig. 12. Observed and expected weighted di-tau mass distributions in ATLAS. The bottom panel shows the difference between weighted data events and weighted background events (points) compared to signal events yields for various masses, with signal strengths set to their best-fit values.

$m_H = 125\,\text{GeV}$. The evidence against the background-only hypothesis is found to have a maximum of 3.8σ for $m_H = 125\,\text{GeV}$.

Figure 12 shows the observed and expected $\tau\tau$ mass distributions from the ATLAS experiment,[27] weighting all sub-distributions in each category of each channel by the ratio between the expected signal and background yields for the respective category in a di-tau mass interval containing 68% of the signal. The plot also shows the difference between the observed data and expected background distributions, together with the expected distribution for an SM Higgs boson signal with $m_H = 125\,\text{GeV}$. The observed (expected) significance of the excess with respect to the background only hypothesis at this mass is 4.5 (3.4) standard deviations in the ATLAS experiment.

The Tevatron experiments, CDF and D0, have also reported a combined observed significance of 3.0σ,[28] where the H$\to$ bb mode is the dominant one. All these results establish the existence of the fermionic decays of the new boson, consistent with the expectation from the SM.

4.4. *The ATLAS and CMS combinations of results from Run 1*

4.4.1. *The mass of the Higgs boson*

Both ATLAS and CMS experiments have separately combined their measurements of the mass of the Higgs bosons from the two channels that have the best mass resolution, namely H $\to \gamma\gamma$ and H $\to$ ZZ $\to 4l$. The signal in all channels is assumed to be due to a state with a unique mass, m_X. The obtained values are: ATLAS: $m_H = 125.36 \pm 0.37$ (stat) ± 0.18(syst) GeV;[29] CMS: $m_H = 125.02 \pm 0.27$ (stat) ± 0.14 (syst) GeV,[30] in excellent agreement.

4.4.2. *Significance of the observed excess*

Table 2 summarises the median expected and observed local significance for an SM Higgs boson mass hypothesis of 125 GeV from the individual decay modes in ATLAS and CMS.[30] Both experiments confirm the observation of a new particle with a mass near 125 GeV.

4.4.3. *Compatibility of the observed state with the SM Higgs boson hypothesis: Signal strength*

To establish whether or not the newly found state is the Higgs boson of the SM, we need to precisely measure its other properties and attributes. Several tests of compatibility of the observed excesses with those expected from a standard model Higgs boson have been made.

In one comparison labelled as the signal strength $\mu = \sigma/\sigma_{SM}$, the measured production $\times$ decay rate of the signal is compared with the SM expectation, determined for each decay mode individually and for the overall combination of all channels. A signal strength of one would be indicative of an SM Higgs boson.

Both the ATLAS and CMS experiments have measured μ values, by decay mode and by additional tags used to select preferentially events from a particular production mechanism. The best-fit value for the common signal strength μ, obtained in the different sub-combinations and the overall combination of all search channels in the ATLAS and CMS experiments is shown in Fig. 13. The observed μ value is 1.00 ± 0.09 (stat) ± 0.08 (theo) for CMS for a Higgs boson mass of 125.0 GeV[30] and 1.30 ± 0.20 in ATLAS for a Higgs boson mass of 125.5 GeV.[31] In both the experiments the μ-values are consistent with the value expected for the SM Higgs boson ($\mu = 1$). The Tevatron has also measured the value of this signal strength, primarily using the bb channel and find it to be 1.44 ± 0.59.[28]

4.4.4. *Couplings of the Higgs boson*

Figure 14 illustrates the dependence of the Higgs boson couplings on mass of the decay particles (τ, b-quark, W, Z and t-quark). The couplings are plotted in terms of λ or $\sqrt{(g/2v)}$. The line is the expectation from the SM. For the fermions, the values, λ, of the fitted Yukawa couplings Hff are shown, while for vector bosons the square-root of the coupling for the HVV vertex divided by twice the vacuum expectation value of the Higgs boson field $\sqrt{(g/2v)}$. For a Higgs boson with a mass of 125 GeV decaying to $\mu\mu$ CMS has found that the observed (expected) upper limit

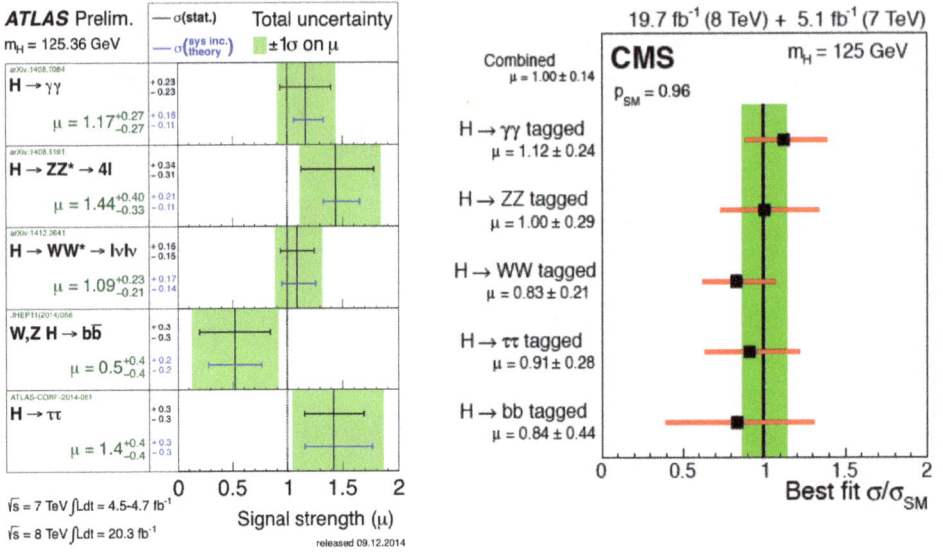

Fig. 13. Values of μ for sub-combinations by decay mode in (left) ATLAS and (right) in CMS.

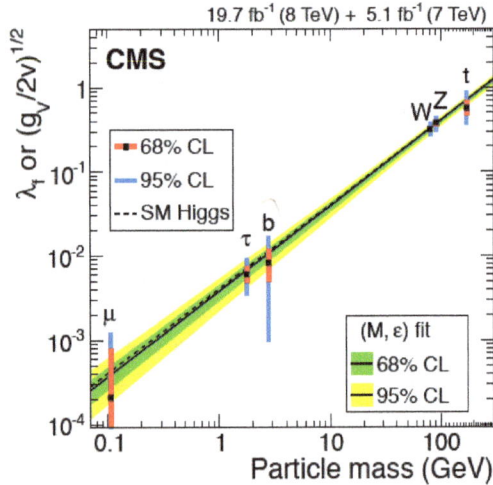

Fig. 14. Summary of the fits from the CMS experiment for deviations in the couplings λ or $\sqrt{(g/2v)}$ as function of particle mass for a Higgs boson with a mass of 125 GeV (see text).

on the production rate is 7.4 (6.5 +2.8, −1.9).[32] This corresponds to an upper limit on the branching fraction of 0.0016. The couplings are indeed proportional to mass, in the manner prescribed by the SM, over a broad mass range, from the τ-lepton mass (about 1.8 GeV) to that of the top quark (mass about one hundred times larger).

4.4.5. *Spin and parity*

Another key to the identity of the new boson is its quantum numbers amongst which is the spin-parity (J^P). The angular distributions of the decay particles can be used to test various spin hypotheses.

In the decay mode H $\rightarrow$ ZZ $\rightarrow$ $4l$ the full final state is reconstructed, including the angular variables sensitive to the spin-parity. The information from the five angles and the two di-lepton pair masses are combined to form boosted decision tree (BDT) discriminants. A decision tree is a set of cuts employed to classify events as "signal-like" or "background-like".

In the decay mode H $\rightarrow$ WW $\rightarrow$ $l\nu l\nu$, for example, in the ATLAS experiment the discriminants used in the fit are outputs of two different BDTs, trained separately against all backgrounds to identify 0^+ and 2^+ events, respectively. For the BDT the kinematic variables used are the transverse mass m_T, the azimuthal separation of the two leptons, $\Delta\varphi_{ll}, m_{ll}$ and dilepton p_{ll}^T.

A first study has been presented by CMS in the ZZ $\rightarrow$ $4l$ channel[33] with the data already disfavouring the pure pseudo-scalar hypothesis (Fig. 15). The CMS experiment has combined the ZZ $\rightarrow$ $4l$ and WW $\rightarrow$ $l\nu l\nu$ spin analyses.[34] Under the assumption that the observed boson has $J^P = 0^+$, the data disfavour the hypothesis of a graviton-like boson with minimal couplings produced in gluon fusion, $J^P = 2^+$, with a CLs value of 0.60%.

ATLAS has also presented a combined study of the spin of the Higgs boson candidate[35] using the H $\rightarrow$ $\gamma\gamma$, H $\rightarrow$ WW $\rightarrow$ $l\nu l\nu$ and H $\rightarrow$ ZZ $\rightarrow$ $4l$ decays to discriminate between the SM assignment of $J^P = 0^+$ and a specific model of

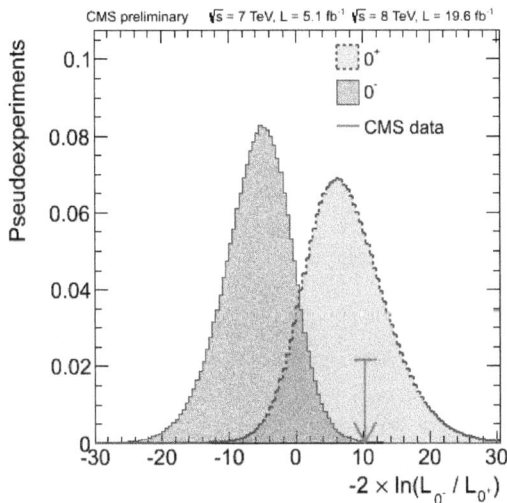

Fig. 15. Distribution of $q = -2\ln(L_{\rm JP}/L_{\rm SM})$ for two signal types, 0^+ (yellow/right histogram) and 0^- hypothesis (blue/left histogram) for $m_H = 126\,\text{GeV}$ for a large number of generated experiments. The arrow indicates the observed value.

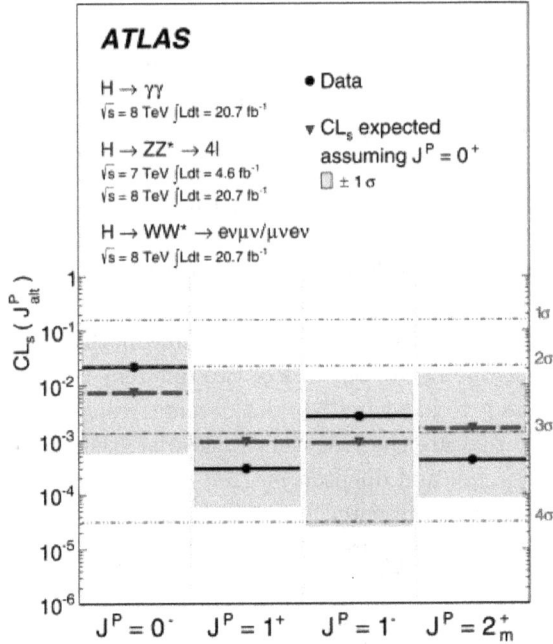

Fig. 16. Expected (blue triangles/dashed lines) and observed (black circles/solid lines) confidence level CL_S for alternative spin–parity hypotheses assuming a $J^P = 0^+$ signal. The green band represents the 68% $CL_S(J^P_{alt})$ expected exclusion range for a signal with assumed $J^P = 0^+$.

$J^P = 2^+$. The data strongly favour the $J^P = 0^+$ hypothesis (see Fig. 16). The specific $J^P = 2^+$ hypothesis is excluded with a confidence level above 99.9%, independently of the assumed contributions of gluon fusion and quark–antiquark annihilation processes in the production of the spin-2 particle.

The above-mentioned results show that the spin-parity $J^P = 0^+$ hypothesis is strongly favoured by both experiment, with the alternatives $J^P = 0^-$, 1^+, 1^-, 2^+ hypotheses rejected with confidence levels larger than 97.8%.

5. Conclusions and Outlook

The results from the two experiments show that a scalar Higgs boson has been discovered. It appears to be an elementary state with a spin-parity, and couplings to other SM particles, consistent with those predicted for the SM Higgs boson.

Although enormous progress has already been made by the ATLAS and CMS experiments to pin down the properties of this newly discovered particle, several outstanding questions remain. It is hoped that these questions will be addressed in the LHC Run 2, that will start in Spring 2015, at almost twice the energy of the LHC during Run 1, and later by the HL-LHC which has the goal of integrating some 3000 fb^{-1},[36,37] and by possible future colliders.

Improved measurements of the properties of the new particle, including the observation of rare decays such as H → $\mu\mu$, will provide more definitive information about its nature (e.g. whether it is elementary or composite). Physics beyond the SM is expected to modify the Higgs boson couplings to fermions and bosons by up to a few percent, depending on the energy scale of the new physics; hence experimental precision, from a few permil to a few percent, is required to detect significant deviations from the SM expectations. Any deviations will give clues for what should be new physics.

A definitive exploration of the electroweak symmetry breaking mechanism requires studies of WW, ZZ and WZ production at high masses of the boson pairs (m_{VV}). Such studies will also provide a powerful "closure test" of the SM. In the SM, without a Higgs boson, the cross-section for the scattering of two left-handed gauge bosons diverges with energy, becoming unphysical for $m_{VV} \geq 1$ TeV. It is therefore crucial to verify that the newly discovered particle restores the good behaviour of the theory, or else, reveal any additional dynamics contributing to electroweak symmetry breaking.

Higgs boson self-couplings, which would give access to the scalar potential in the SM Lagrangian, may be observed with the full luminosity of the upgraded LHC ($3000\,\text{fb}^{-1}$ per experiment). Around 100 million Higgs bosons would have been produced allowing also a search for exotic and rare decays of the new particle.

In parallel, searches for new physics may clarify whether or not the (light) Higgs boson mass is stabilised by a new symmetry. It is indeed a conundrum as to why the mass of an elementary scalar particle can be as low as $\sim$100 GeV. Quantum corrections make the mass of such fundamental particles float up to the next highest physical mass scale that, in the absence of extensions to the SM, is as high as 10^{15} GeV. One appealing hypothesis predicts a new symmetry labelled *supersymmetry*. In the simplest forms of supersymmetry five Higgs bosons are predicted to exist with one resembling the SM Higgs boson with a mass below $\sim$140 GeV. One of these would only be subtly different from the SM one. Much more data need to be collected to enable rigorous testing of the compatibility of the new boson with the SM and to establish whether precise measurements of its properties imply the existence of physics beyond the SM.

All data collected so far suggest that we have discovered a fundamental scalar field that pervades our universe. Astronomical and astrophysical measurements point to the following composition of energy–matter in the universe: $\sim$4% normal matter that "shines", $\sim$23% dark matter, and the rest forming "dark energy." Dark matter is weakly and gravitationally interacting matter with no electromagnetic or strong interactions. These are the properties carried by the lightest supersymmetic particle in models that conserve a quantum number called R-parity $R = (-1)^{3(B-L)+2s}$ where s, B and L are spin, baryon number and lepton number respectively. Hence arises the question: is dark matter supersymmetric in nature? Fundamental scalar fields could well have played a critical role in the conjectured inflation of our universe immediately after the Big Bang and in the recently observed accelerating expansion of the universe that, among other measurements, signals the presence of dark energy in our universe. Discovery of the Higgs boson now gives impetus to such conjectures.

After decades of superb theoretical and experimental efforts, and three years of LHC operation, the particle content of the Standard Model is now complete. However, the Standard Model is not the ultimate theory of particle physics, as many crucial questions remain unanswered. They include the composition of the universe, especially the identity of dark matter, the source of the asymmetry between matter and antimatter, the origin of neutrino masses, the motivation for the light mass of the Higgs boson, the extreme feebleness of gravity compared to the other forces. This discovery is widely expected to be a portal to physics beyond the SM. Physicists at the LHC are eagerly looking forward to the higher-energy running of the LHC and to establishing the true nature of the new boson, to find clues or answers to some of the other fundamental open questions in particle physics and cosmology. The physics exploitation of the LHC has just started and the expectations for other discoveries are high over the coming decades.

Acknowledgments

The construction, the operation, and exploitation of the large and complex ATLAS and CMS experiments have required the talents, the resources, and the dedication of thousands of scientists, engineers and technicians worldwide. This paper is dedicated to all who have worked on these experiments. The marvellous construction, and efficient operation of the LHC accelerator and the wLCG computing infrastructure are gratefully acknowledged.

Such a superb accomplishment is the result of the ingenuity, vision and perseverance of the high-energy physics community, and of more than twenty years of talented, dedicated work of those involved in the LHC projects.

References

1. S. L. Glashow, *Nucl. Phys.* **22**, 579 (1961).
2. S. Weinberg, *Phys. Rev. Lett.* **19**, 1264 (1967).
3. A. Salam, in *Proceedings of the Eighth Nobel Symposium*, ed. N. Svartholm (Almqvist & Wiskell, 1968), p. 367.
4. F. Englert and R. Brout, *Phys. Rev. Lett.* **13**, 321 (1964).
5. P. W. Higgs, *Phys. Lett.* **12**, 132 (1964).
6. P. W. Higgs, Phys. Rev. Lett. **13**, 508 (1964).
7. G. S. Guralnik, C. R. Hagen and T. W. B. Kibble, *Phys. Rev.* **155**, 1554 (1967).

8. P. W. Higgs, *Phys. Rev.* **145**, 1156 (1966).

9. T. W. B. Kibble, *Phys. Rev.* **155**, 1554 (1967).

10. L. Evans (ed.), *The Large Hadron Collider, a Marvel of Technology* (EPFL Press, 2009); L. Evans, P. Bryant (ed.) and LHC Machine, *JINST* **03**, S08001 (2008).

11. ATLAS Collaboration, a) Letter of Intent, CERN-LHCC-92-004 (1992); b) *Technical Proposal*, CERN-LHCC-1994-043 (1994); c) The ATLAS Experiment at the LHC, *JINST* **3**, S08003 (2008).

12. CMS Collaboration, a) Letter of Intent, CERN-LHCC-92-003 (1992); b) *Technical Proposal*, CERN-LHCC-1994-038 (1994); c) The CMS Experiment at the LHC, *JINST* **3**, S08004 (2008).

13. ALEPH, CDF, D0, DELPHI, L3, OPAL, SLD Collaborations, the LEP Electroweak Working Group, the Tevatron Electroweak Working Group, and the SLD Electroweak and Heavy Flavour Groups, Precision electroweak measurements and constraints on the standard model, CERN PH-EP-2010-095, http://lepewwg. web.cern.ch/LEPEWWG/plots/winter2012/, arXiv:1012.2367, 2010, http://cdsweb. cern.ch/record/1313716.

14. ALEPH, DELPHI, L3, OPAL Collaborations, and LEP Working Group for Higgs Boson Searches, *Phys. Lett. B* **565**, 61 (2003).

15. LHC Higgs Cross Section Working Group, *Handbook of LHC Cross Sections: 1. Inclusive Observables*, arXiv:1101.0593 (2011); *2. Differential Distributions*, arXiv:1201.3084 (2012); *3. Higgs Properties*, arXiv:1307.1347 (2013).

16. ATLAS Collaboration, Observation of a new particle in the search for the Standard Model Higgs boson with the ATLAS detector at the LHC, *Phys. Lett. B* **716**, 1 (2012).

17. CMS Collaboration, Observation of a new boson at a mass of 125 GeV with the CMS experiment at the LHC, *Phys. Lett. B* **716**, 30 (2012).

18. ATLAS Collaboration, Measurement of Higgs boson production in the diphoton channel in pp collisions at centre-of-mass energies of 7 and 8 TeV with the ATLAS detector, *Phys. Rev. D* **90**, 112015 (2014).

19. CMS Collaboration, Measurement of the properties of a Higgs boson in the four-lepton state, *Phys. Rev. D* **89**, 092007 (2014).

20. ATLAS Collaboration, Measurement of Higgs boson production and couplings in the four-lepton channel in pp collisions at center-of-mass energies of 7 and 8 TeV with the ATLAS detector, *Phys. Rev. D* **91**, 012006 (2015).

21. CMS Collaboration, Observation of the diphoton decay of the Higgs boson and measurement of its properties, *Eur. Phys.* **74**, 3076 (2014).

22. CMS Collaboration, Measurement of Higgs boson production and properties in the WW decay channel with leptonic states, *JHEP* **01**, 096 (2014).

23. ATLAS Collaboration, Observation and measurement of Higgs boson decays to WW* with the ATLAS detector, CERN-PH-EP-2-14-270, arXiv:1412.2641 (2014), submitted to *Phys. Rev. D* (2014).

24. CMS Collaboration, Evidence for the 125 GeV Higgs boson decaying into a pair of τ leptons, *JHEP* **05**, 104 (2014).

25. CMS Collaboration, Search for the standard model Higgs boson produced in association with a W or a Z boson and decaying to bottom quarks, *Phys. Rev. D* **89**, 012003 (2014).

26. CMS Collaboration, Evidence for the direct decay of the 125 GeV Higgs boson to fermions, *Nat. Phys.* **10**, 557 (2014).

27. ATLAS Collaboration, Evidence for Higgs Boson Yukawa coupling to tau leptons with the ATLAS Detector, arXiv: 1501.04943 (2015), submitted to *JHEP*.

28. CDF and D0 Collaborations, Higgs boson studies at the Tevatron, *Phys. Rev. D* **88**, 052014 (2013).

29. ATLAS Collaboration, Measurement of the Higgs boson mass from the H → γγ and H → ZZ*→ 4*l* with the ATLAS detector at the LHC, *Phys. Rev. D* **90**, 052004 (2014).

30. CMS Collaboration, Precise determination of the mass of the Higgs boson and test of compatibility of its couplings with the standard model predictions using proton collisions at 7 and 8 TeV, CMS-HIG-14-009, arXiv: 1412.8662 (December 2014), submitted to *EPJC*.

31. ATLAS Collaboration, Updated coupling measurements of the Higgs boson with the ATLAS detector using up to 25 fb^{-1} of proton-proton collision data, ATLAS-CONF-2014-009 (2014).

32. CMS Collaboration, Search for a standard model-like Higgs boson in the $\mu^+\mu^-$ and e$^+$e$^-$ decay channels at the LHC, arXiv: 1410.6679 (2014), submitted to *Phys. Lett. B*.

33. CMS Collaboration, Study of the mass and spin-parity of the Higgs boson candidate via its decays to Z boson pairs, *Phys. Rev. Lett.* **110**, 081803 (2013).

34. CMS Collaboration, Constraints on the spin-parity and anomalous HVV couplings of the Higgs boson in proton collisions at 7 and 8 TeV, CMS-HIG-14-018, arXiv: 1411.3441 (2014).

35. ATLAS Collaboration, Evidence for the spin-0 nature of the Higgs boson using ATLAS data, *Phys. Lett.* **726**, 120 (2014).

36. ATLAS Collaboration, Physics at a High-Luminosity LHC with ATLAS, ATL-PHYS-PUB-2013-007, arXiv:1307.7292.

37. CMS Collaboration, Projected Performance of an Upgraded CMS Detector at the LHC and HL-LHC, CMS Note-13-002, arXiv:1307.7135.

Precision Physics with Heavy-Flavoured Hadrons

Patrick Koppenburg[1] and Vincenzo Vagnoni[2]

[1] *Nikhef National Institute for Subatomic Physics,*
PO Box 41882, 1009 DB Amsterdam, The Netherlands

European Organization for Nuclear Research (CERN),
CH-1211 Geneva 23, Switzerland
patrick.koppenburg@cern.ch

[2] *Istituto Nazionale di Fisica Nucleare (INFN),*
Sezione di Bologna via Irnerio 46, 40126 Bologna, Italy
vincenzo.vagnoni@bo.infn.it

The understanding of flavour dynamics is one of the key aims of elementary particle physics. The last 15 years have witnessed the triumph of the Kobayashi–Maskawa mechanism, which describes all flavour changing transitions of quarks in the Standard Model. This important milestone has been reached owing to a series of experiments, in particular to those operating at the so-called B factories, at the Tevatron, and now at the LHC. We briefly review status and perspectives of flavour physics, highlighting the results where the LHC has given the most significant contributions, notably including the recent observation of the $B_s^0 \rightarrow \mu^+\mu^-$ decay.

1. Introduction

Flavour physics has played a central role in the development of the Standard Model (SM), which represents the state of the art of the fundamental theory of elementary physics interactions. The SM is able to describe with excellent accuracy all of the fundamental physics phenomena related to the electromagnetic, weak and strong forces, observed to date. Yet, it fails with some key aspects, notably including the fact that it does not provide an answer to one of the most fundamental questions: why is antimatter absent from the observed universe? Owing to the work of Andrei Sakharov in 1967,[1] the phenomenon of CP violation, i.e. the non-invariance of the laws of nature under the combined application of charge (C) and parity (P) transformations, is known to be one of the ingredients needed to dynamically generate a baryon asymmetry starting from an initially symmetric universe. However, it is also known that the size of CP violation in the SM is too small, by several orders of magnitude, to explain the observed baryon asymmetry of the universe.[2–4] As a consequence, other sources of CP violation beyond the SM (BSM), which should produce observable effects in the form of deviations from the

SM predictions of certain *CP*-violating quantities, must exist. Rare decays that are strongly suppressed in the SM are of particular interest, since BSM amplitudes could be relatively sizable with respect to those of the SM.

The first run of the Large Hadron Collider (LHC), with 7 and 8 TeV *pp* collisions, has led to the discovery of the Higgs boson,[5, 6] but no hint of the existence of other new particles has been found. Neither supersymmetry nor any other direct sign of BSM physics has popped out of the data. Besides the Higgs discovery, analyses from the first years of running have also firmly established the great impact of the ATLAS,[7] CMS[8] and LHCb[9] experiments in the field of *CP* violation and rare decays of heavy-flavoured hadrons. In particular, LHCb has produced a plethora of results on a broad range of flavour observables in the c- and b-quark sectors, and ATLAS and CMS have given significant contributions to the b-quark sector, mainly using final states containing muon pairs. Also these measurements do not provide hints of BSM physics.

Nevertheless, it is of fundamental importance for future developments of elementary particle physics to keep improving the theoretical and experimental knowledge of flavour physics. On the one hand, such improvements increase the reach of indirect searches for BSM physics, probing higher and higher mass scales in the event that no BSM effects were discovered by direct detection. On the other hand, they would enable the BSM Lagrangian to be precisely determined, if any new particle were detected in direct searches. Starting with a brief historical perspective on the development of heavy flavour physics, we review the present status, highlighting some of the results where the LHC has given the most significant contributions.

2. An Historical Perspective

2.1. *The origin of the Kobayashi–Maskawa mechanism*

As already mentioned, flavour physics has played a prominent role in the development of the SM. As an example, one of the most notable predictions made in this context was that of the existence of a third quark generation, in a famous paper of 1973 by Makoto Kobayashi and Toshihide Maskawa.[10] In that work, which won them the Nobel Price in Physics in 2008 "for the discovery of the origin of the broken symmetry which predicts the existence of at least three families of quarks in nature", Kobayashi and Maskawa extended the Cabibbo[11] (with only u, d, and s quarks) and the Glashow–Iliopoulos–Maiani[12] (GIM, including also the c quark) mechanisms, pointing out that *CP* violation could be incorporated into the emerging picture of the SM if six quarks were present. This is commonly referred to as the Kobayashi–Maskawa (KM) mechanism. It must be emphasised that at that time only hadrons made of the three lighter quarks had been observed. An experimental revolution took place in 1974, when a new state containing the c quark was discovered almost simultaneously at Brookhaven[13] and SLAC.[14] Then, the experimental observations of the b[15] and t[16] quarks were made at FNAL in 1977 and 1995, respectively.

The idea of Kobayashi and Maskawa, formalised in the so-called Cabibbo–Kobayashi–Maskawa (CKM) quark mixing matrix, was then included in the SM by the beginning of the 1980s. The phenomenon of *CP* violation, first revealed in 1964 using decays of neutral kaons,[17] was elegantly accounted for as an irreducible complex phase within the CKM matrix. The experimental proof of the validity of the KM mechanism and the precise measurement of the value of the *CP*-violating phase soon became questions of paramount importance.

2.2. *The rise of B physics*

Due to the nature of the CKM matrix, an accurate test of the KM mechanism required an extension of the physics programme to heavy-flavoured hadrons. Pioneering steps in the b-quark flavour sector were moved at the beginning of the 1980s by the CLEO experiment at CESR.[18] At the same time, Ikaros Bigi, Ashton Carter and Tony Sanda explored the possibility that large *CP*-violating effects could be present in the decay rates of B^0 mesons decaying to the $J/\psi K_S^0$ *CP* eigenstate.[19, 20] In addition, they pointed out that such a measurement could be interpreted in terms of the *CP*-violating phase without relevant theoretical uncertainties due to strong interaction effects. However, there were two formidable obstacles to overcome: first, an experimental observation required an enormous amount of B^0 mesons, well beyond what was conceivable to produce and collect at the time; second, a precise measurement of the decay time was required, together with the knowledge of the flavour of the B^0 meson at production.

Few years later, in 1987, the ARGUS experiment at DESY measured for the first time the mixing rate of B^0 and $\bar{B}^0$ mesons,[21] whose knowledge was an important ingredient to understand the feasibility of measuring *CP* violation with $B^0 \to J/\psi K_S^0$ decays. Another crucial ingredient came along due to tremendous developments in the performance of e^+e^- storage rings. By the late 1980s, many different possible designs for new machines were being explored. A novel idea was put forward by Pier Oddone in 1987: a high-luminosity asymmetrical e^+e^- circular collider operating at the centre-of-mass energy of the $\Upsilon(4S)$ meson.[22] Owing to the beam-energy asymmetry, B mesons would have been produced with a boost in the laboratory frame towards the direction of the most energetic beam. The consequent nonzero decay length, measured by means of state-of-the-art silicon vertex detectors, would have enabled precise measurements of the decay time to be achieved. Two machines based on Oddone's concept, so-called B factories, were eventually built: PEP-II at SLAC in the United States and KEKB at KEK in Japan. The associated detectors, BaBar[23] at PEP-II and Belle[24] at KEKB, were approved in 1993 and in 1994, respectively. If CESR was initially able to produce few tens of $b\bar{b}$ pairs per day, PEP-II and KEKB were capable of producing order of one million $b\bar{b}$ pairs per day.

Meanwhile, during the course of the 1990s, many b-physics measurements were being performed at the Z^0 factories, i.e. the LEP experiments[25–28] at CERN, and the SLD[29] experiment at SLAC. Despite the relatively small statistics, b hadrons produced in Z^0 decays were naturally characterised by a large boost, enabling measurements of lifetimes of all b-hadron species and of oscillation frequencies of neutral B mesons to be performed. In particular, for the first time it was possible to study samples of B_s^0-meson, b-baryon and even a handful of B_c^+-meson decays.[30, 31] Similar pioneering measurements were also made at the Tevatron with Run I data, using hadronic collisions as a source of b quarks.[32]

Soon after PEP-II and KEKB were turned on, the two machines broke any existing record of instantaneous and integrated luminosity of previous particle colliders. By the end of their research programmes, BaBar and Belle measured CP violation in $B^0 \rightarrow J/\psi K_S^0$ decays with a relative precision of about 3%.[33, 34] The large sample of B-meson decays collected at BaBar and Belle enabled a series of further measurements in the flavour sector to be performed, well beyond the initial expectations. In the same years, a major step forward in these topics was also made at the Tevatron with Run II data. Although with a somewhat limited scope if compared to B factories, the CDF and D0 experiments at FNAL collected large amounts of heavy-flavoured-hadron decays, performing some high precision measurements,[35] notably including the first observation of B_s^0-meson mixing in 2006.[36]

2.3. *The LHC era*

When the constructions of the BaBar and Belle detectors were being scrutinised for approval, three distinct proposals for a dedicated b-physics experiment at the LHC were put forward, so-called COBEX, GAJET, and LHB. GAJET and LHB were both based on fixed targets, the former working with a gas target placed inside the LHC beam pipe and the latter exploiting an extracted LHC beam. COBEX was instead proposed to work in proton–proton collider mode. The three groups of proponents were asked to join together and submit to the LHC Experiments Committee (LHCC) a proposal for a single collider-mode experiment, namely LHCb.[9] LHCb was then designed to exploit the potential for heavy-flavour physics at the LHC by instrumenting the forward region of proton–proton collisions, in order to take advantage of the large $b\bar{b}$ cross section in the forward (or backward) LHC beam direction. The LHCb experiment was approved in 1998, and started taking data with the start-up of LHC in 2009.

The LHCb detector,[9, 37] shown in Fig. 1, includes a high-precision tracking system consisting of a silicon-strip vertex detector surrounding the pp interaction region, a large-area silicon-strip detector located upstream of a dipole magnet with a bending power of about 4 Tm, and three stations of silicon-strip detectors and

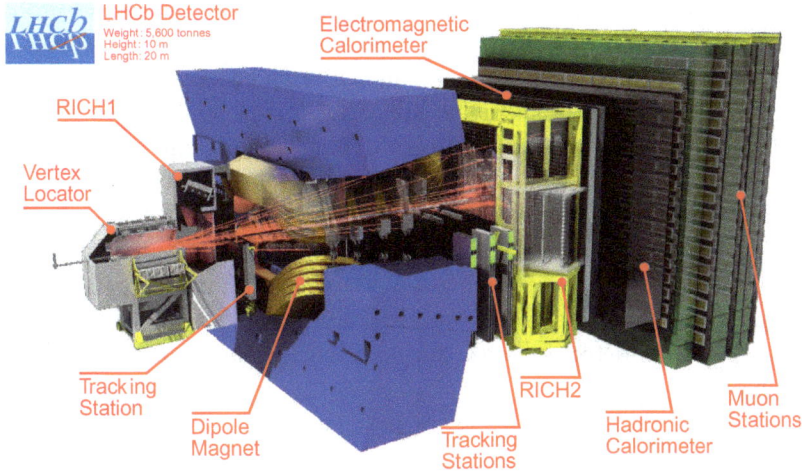

Fig. 1.   Sketch of the LHCb detector.

straw drift tubes placed downstream of the magnet. The tracking system provides a measurement of momentum of charged particles with a relative uncertainty that varies from 0.5% at low momentum to 1.0% at 200 GeV/c. The silicon sensors of the vertex detector come as close as 8 mm to the LHC beam. This allows for a very precise measurement of the track trajectory close to the interaction point, which is crucial to separate decays of beauty and charm hadrons, with typical flight distances of a few millimetres in the laboratory frame, from the background. The distance of a track to a primary vertex, the impact parameter, is measured with a resolution of 15–30 μm.

One distinctive feature of LHCb, when compared to the ATLAS and CMS detectors, is its particle identification capability for charged hadrons. This is mainly achieved by means of two ring-imaging Cherenkov ("RICH") detectors placed on either side of the tracking stations. Once particle momenta are measured, the two RICH detectors enable the identification of protons, kaons and pions to be obtained. An electromagnetic calorimeter, complemented with scintillating-pad and preshower detectors, provides energy and position of photons and electrons, and allow for their identification in conjunction with information from the tracking system. The electromagnetic calorimeter is followed by a hadronic calorimeter that also gives some information to identify hadrons. Finally, muons are identified by a system composed of alternating layers of iron and multiwire proportional chambers. The online event selection is performed by a trigger[38] which consists of a hardware stage, based on information from the calorimeter and muon systems, followed by a software stage, which applies full event reconstruction.

LHCb operates at a much lower instantaneous luminosity than the peak luminosity made available by the LHC. This is necessary to better distinguish charged particles resulting from b- and c-hadron decays from particles produced in other pp collisions, in the forward region covered by the detector acceptance. During the first run of the LHC, the average number of pp collisions per bunch crossing at the LHCb intersection was kept at about 1.7, corresponding to a luminosity of 4×10^{32} cm^{-2}s^{-1}. This was achieved by a dynamical adjustment of the transverse offset between the LHC beams during the fill, enabling constant luminosity to be kept throughout.

Besides LHCb, the general purpose ATLAS and CMS detectors were also designed with the aim of performing b-physics measurements, mainly using final states containing muon pairs due to constraints dictated by the trigger and to the absence of sub-detectors with strong particle identification capabilities for charged hadrons.

3.   The CKM Matrix

3.1.   Definition

In the SM, charged-current interactions of quarks are described by the Lagrangian

$$\mathcal{L}_{W^\pm} = -\frac{g}{\sqrt{2}}\overline{U}_i\gamma^\mu\frac{1-\gamma^5}{2}\,(V_{\text{CKM}})_{ij}\,D_jW_\mu^+ + \text{h.c.},$$

where g is the electroweak coupling constant and V_{CKM} is the CKM matrix

$$V_{\text{CKM}} = \begin{pmatrix} V_{ud} & V_{us} & V_{ub} \\ V_{cd} & V_{cs} & V_{cb} \\ V_{td} & V_{ts} & V_{tb} \end{pmatrix},$$

originating from the misalignment in flavour space of the up and down components of the $SU(2)_L$ quark doublet of the SM. The V_{ij} matrix elements represent the couplings between up-type quarks $U_i = (u, c, t)$ and down-type quarks $D_j = (d, s, b)$.

An important feature of the CKM matrix is its unitarity. Such a condition determines the number of free parameters of the matrix. A generic $N \times N$ unitary matrix depends on $N(N-1)/2$ mixing angles and $N(N+1)/2$ complex phases. In the CKM case, dealing with a mixing matrix between the quark flavour eigenstates, the Lagrangian enables the phase of each quark field to be redefined, such that $2N-1$ unphysical phases cancel out. As a consequence, any $N \times N$ complex matrix describing mixing between N generations of quarks has

$$\underbrace{\frac{1}{2}N(N-1)}_{\text{mixing angles}} + \underbrace{\frac{1}{2}(N-1)(N-2)}_{\text{physical complex phases}} = (N-1)^2$$

free parameters. The interesting case $N = 2$ leads to the GIM mixing matrix with only one free parameter, namely the Cabibbo angle θ_C[11]

$$V_{\text{GIM}} = \begin{pmatrix} \cos\theta_C & \sin\theta_C \\ -\sin\theta_C & \cos\theta_C \end{pmatrix}.$$

When formalised in 1970, the nature of V_{GIM} was invoked to explain the suppression of flavour changing neutral current (FCNC) processes, and put the basis for the discovery of the charm quark.[12-14] In the case $N = 3$, the resulting number of free parameters is four: three mixing angles and one complex phase. This phase alone is responsible for CP violation in the weak interactions of the SM.

3.2. Standard parametrisation

Among the various possible conventions, a standard choice to parametrise V_{CKM} is given by

$$
V_{\mathrm{CKM}} = \begin{pmatrix} c_{12}c_{13} & s_{12}c_{13} & s_{13}e^{-i\delta} \\ -s_{12}c_{23} - c_{12}s_{23}s_{13}e^{i\delta} & c_{12}c_{23} - s_{12}s_{23}s_{13}e^{i\delta} & s_{23}c_{13} \\ s_{12}s_{23} - c_{12}c_{23}s_{13}e^{i\delta} & -c_{12}s_{23} - s_{12}c_{23}s_{13}e^{i\delta} & c_{23}c_{13} \end{pmatrix},
$$

where $s_{ij} = \sin\theta_{ij}$, $c_{ij} = \cos\theta_{ij}$ and δ is the CP-violating phase. All the θ_{ij} angles can be chosen to lie in the first quadrant, thus $s_{ij}, c_{ij} \geq 0$. The coupling between quark generations i and j vanishes if the corresponding θ_{ij} is equal to zero. In the case where $\theta_{13} = \theta_{23} = 0$, the third generation would decouple and the CKM matrix would take the form of V_{GIM}. The presence of a complex phase in the mixing matrix is a necessary condition for CP violation, although not sufficient. As pointed out in Ref. 39, a key condition is

$$
(m_t^2 - m_c^2)(m_t^2 - m_u^2)(m_c^2 - m_u^2)(m_b^2 - m_s^2)(m_b^2 - m_d^2)(m_s^2 - m_d^2) \times J_{CP} \neq 0,
$$

where

$$
J_{CP} = |\Im(V_{i\alpha}V_{j\beta}V_{i\beta}^*V_{j\alpha}^*)| \quad (i \neq j, \alpha \neq \beta)
$$

is the so-called Jarlskog parameter. This condition is related to the fact that the CKM phase could be eliminated if any of two quarks with the same charge were degenerate in mass. As a consequence, the origin of CP violation in the SM is deeply connected to the origin of the quark mass hierarchy and to the number of fermion generations.

The Jarlskog parameter can be interpreted as a measure of the size of CP violation in the SM. Its value does not depend on the phase convention of the quark fields, and adopting the standard parametrisation it can be written as

$$
J_{CP} = s_{12}s_{13}s_{23}c_{12}c_{23}c_{13}^2 \sin\delta.
$$

Experimentally one has $J_{CP} = \mathcal{O}(10^{-5})$, which quantifies how small CP violation is in the SM.

3.3. Wolfenstein parametrisation

Experimental information leads to the evidence that transitions within the same generation are characterised by V_{CKM} elements of $\mathcal{O}(1)$. Instead, those between the first and second generations are suppressed by a factor $\mathcal{O}(10^{-1})$; those between the second and third generations by a factor $\mathcal{O}(10^{-2})$; and those between the first and third generations by a factor $\mathcal{O}(10^{-3})$. It can be stated that

$$s_{12} \simeq 0.22 \gg s_{23} = \mathcal{O}(10^{-2}) \gg s_{13} = \mathcal{O}(10^{-3}).$$

It is useful to introduce a parametrisation of the CKM matrix, whose original formulation was due to Wolfenstein,[40] defining

$$s_{12} = \lambda = \frac{|V_{us}|}{\sqrt{|V_{ud}|^2 + |V_{us}|^2}},$$

$$s_{23} = A\lambda^2 = \lambda \left| \frac{V_{cb}}{V_{us}} \right|,$$

$$s_{13} e^{-i\delta} = A\lambda^3 (\rho - i\eta) = V_{ub}.$$

The CKM matrix can be re-written as a power expansion of the parameter λ (which corresponds to $\sin \theta_C$)

$$V_{\mathrm{CKM}} = \begin{pmatrix} 1 - \frac{1}{2}\lambda^2 - \frac{1}{8}\lambda^4 & \lambda & A\lambda^3(\rho - i\eta) \\ -\lambda + \frac{1}{2}A^2\lambda^5[1 - 2(\rho + i\eta)] & 1 - \frac{1}{2}\lambda^2 - \frac{1}{8}\lambda^4(1 + 4A^2) & A\lambda^2 \\ A\lambda^3\left[1 - (\rho + i\eta)\left(1 - \frac{1}{2}\lambda^2\right)\right] & -A\lambda^2 + \frac{1}{2}A\lambda^4[1 - 2(\rho + i\eta)] & 1 - \frac{1}{2}A^2\lambda^4 \end{pmatrix},$$

which is valid up to $\mathcal{O}(\lambda^6)$. With this parametrisation, the CKM matrix is complex, and hence *CP* violation is allowed for, if and only if η differs from zero. To lowest order the Jarlskog parameter becomes

$$J_{CP} = \lambda^6 A^2 \eta,$$

and, as expected, is directly related to the *CP*-violating parameter η.

3.4. *The unitarity triangle*

The unitarity condition of the CKM matrix, $V_{\mathrm{CKM}} V_{\mathrm{CKM}}^\dagger = V_{\mathrm{CKM}}^\dagger V_{\mathrm{CKM}} = \mathbb{I}$, leads to a set of 12 equations: 6 for diagonal terms and 6 for off-diagonal terms. In particular, the equations for the off-diagonal terms can be represented as triangles in the complex plane, all characterised by the same area $J_{CP}/2$

$$\underbrace{V_{ud}V_{us}^*}_{\mathcal{O}(\lambda)} + \underbrace{V_{cd}V_{cs}^*}_{\mathcal{O}(\lambda)} + \underbrace{V_{td}V_{ts}^*}_{\mathcal{O}(\lambda^5)} = 0,$$

$$\underbrace{V_{us}V_{ub}^*}_{\mathcal{O}(\lambda^4)} + \underbrace{V_{cs}V_{cb}^*}_{\mathcal{O}(\lambda^2)} + \underbrace{V_{ts}V_{tb}^*}_{\mathcal{O}(\lambda^2)} = 0,$$

$$V_{ud}V_{ub}^* + V_{cd}V_{cb}^* + V_{td}V_{tb}^* = 0,$$
$$\underbrace{\phantom{V_{ud}V_{ub}^*}}_{\mathcal{O}(\lambda^3)} \quad \underbrace{\phantom{V_{cd}V_{cb}^*}}_{\mathcal{O}(\lambda^3)} \quad \underbrace{\phantom{V_{td}V_{tb}^*}}_{\mathcal{O}(\lambda^3)}$$

$$V_{ud}V_{cd}^* + V_{us}V_{cs}^* + V_{ub}V_{cb}^* = 0,$$
$$\underbrace{\phantom{V_{ud}V_{cd}^*}}_{\mathcal{O}(\lambda)} \quad \underbrace{\phantom{V_{us}V_{cs}^*}}_{\mathcal{O}(\lambda)} \quad \underbrace{\phantom{V_{ub}V_{cb}^*}}_{\mathcal{O}(\lambda^5)}$$

$$V_{cd}V_{td}^* + V_{cs}V_{ts}^* + V_{cb}V_{tb}^* = 0,$$
$$\underbrace{\phantom{V_{cd}V_{td}^*}}_{\mathcal{O}(\lambda^4)} \quad \underbrace{\phantom{V_{cs}V_{ts}^*}}_{\mathcal{O}(\lambda^2)} \quad \underbrace{\phantom{V_{cb}V_{tb}^*}}_{\mathcal{O}(\lambda^2)}$$

$$V_{ud}V_{td}^* + V_{us}V_{ts}^* + V_{ub}V_{tb}^* = 0.$$
$$\underbrace{\phantom{V_{ud}V_{td}^*}}_{\mathcal{O}(\lambda^3)} \quad \underbrace{\phantom{V_{us}V_{ts}^*}}_{\mathcal{O}(\lambda^3)} \quad \underbrace{\phantom{V_{ub}V_{tb}^*}}_{\mathcal{O}(\lambda^3)}$$

Only two out of these six triangles have sides of the same order of magnitude, $\mathcal{O}(\lambda^3)$. In terms of the Wolfenstein parametrisation, up to $\mathcal{O}(\lambda^7)$ the corresponding equations can be written as

$$A\lambda^3\{(1 - \lambda^2/2)(\rho + i\eta) + [1 - (1 - \lambda^2/2)(\rho + i\eta)] + (-1)\} = 0,$$
$$A\lambda^3\{(\rho + i\eta) + [1 - \rho - i\eta - \lambda^2(1/2 - \rho - i\eta)] + [-1 + \lambda^2(1/2 - \rho - i\eta)]\} = 0.$$

Eliminating the common factor $A\lambda^3$ from both equations, the two triangles in the complex plane represented in Fig. 2 are obtained. In particular, the triangle defined by the former equation is commonly referred to as the unitarity triangle (UT). The sides of the UT are given by

$$R_u \equiv \left| \frac{V_{ud}V_{ub}^*}{V_{cd}V_{cb}^*} \right| = \sqrt{\bar{\rho}^2 + \bar{\eta}^2},$$

$$R_t \equiv \left| \frac{V_{td}V_{tb}^*}{V_{cd}V_{cb}^*} \right| = \sqrt{(1 - \bar{\rho})^2 + \bar{\eta}^2},$$

where to simplify the notation the parameters $\bar{\rho}$ and $\bar{\eta}$, namely the coordinates in the complex plane of the only non-trivial apex of the UT, the others being $(0,0)$ and $(1,0)$, have been introduced. The exact relation between $\bar{\rho}$ and $\bar{\eta}$ and the Wolfenstein

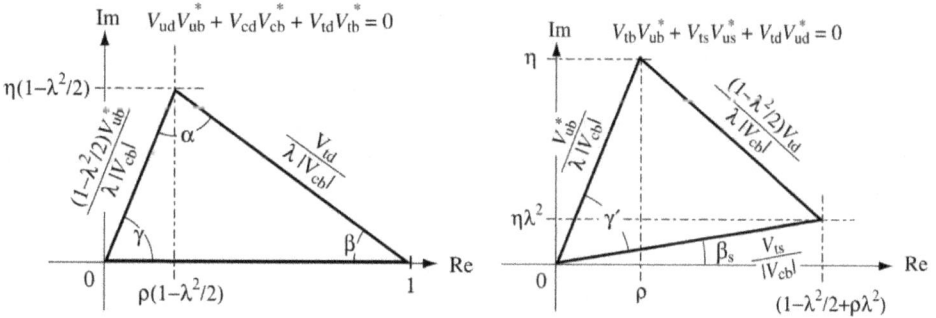

Fig. 2. Representation in the complex plane of the non-squashed triangles obtained from the off-diagonal unitarity relations of the CKM matrix.

parameters is defined by

$$\rho + i\eta = \sqrt{\frac{1 - A^2\lambda^4}{1 - \lambda^2}} \frac{\bar{\rho} + i\bar{\eta}}{1 - A^2\lambda^4(\bar{\rho} + i\bar{\eta})},$$

which, at the lowest non-trivial order in λ, yields

$$\rho = \left(1 + \frac{\lambda^2}{2}\right)\bar{\rho} + \mathcal{O}(\lambda^4), \quad \eta = \left(1 + \frac{\lambda^2}{2}\right)\bar{\eta} + \mathcal{O}(\lambda^4).$$

The angles of the UT are related to the CKM matrix elements as

$$\alpha \equiv \arg\left(-\frac{V_{td}V_{tb}^*}{V_{ud}V_{ub}^*}\right) = \arg\left(-\frac{1 - \bar{\rho} - i\bar{\eta}}{\bar{\rho} + i\bar{\eta}}\right),$$

$$\beta \equiv \arg\left(-\frac{V_{cd}V_{cb}^*}{V_{td}V_{tb}^*}\right) = \arg\left(\frac{1}{1 - \bar{\rho} - i\bar{\eta}}\right),$$

$$\gamma \equiv \arg\left(-\frac{V_{ud}V_{ub}^*}{V_{cd}V_{cb}^*}\right) = \arg\left(\bar{\rho} + i\bar{\eta}\right).$$

The second non-squashed triangle has similar characteristics with respect to the UT. The apex is placed in the point (ρ, η) and is tilted by an angle

$$\beta_s \equiv \arg\left(-\frac{V_{ts}V_{tb}^*}{V_{cs}V_{cb}^*}\right) = \lambda^2\eta + \mathcal{O}(\lambda^4).$$

3.5.   *Phenomenology of CP violation*

The phenomenon of *CP* violation has been observed at a level above five standard deviations in a dozen of processes involving charged and neutral *B*-meson decays, as well as in a couple of neutral kaon decays.[41] In this section we focus in particular on the phenomenology of *CP* violation in the *b*-quark sector.

Three types of *CP* violation can occur in the quark sector: *CP* violation in the decay (also known as direct *CP* violation), *CP* violation in the mixing of neutral mesons, and *CP* violation in the interference between mixing and decay.

Defining the amplitude of a *B* meson decaying to the final state f as A_f, and that of its *CP* conjugate $\bar{B}$ to the *CP* conjugate final state $\bar{f}$ as $\bar{A}_{\bar{f}}$, direct *CP* violation occurs when $|A_f| \neq |\bar{A}_{\bar{f}}|$. This is the only possible type of *CP* violation that can be observed in the decays of charged mesons and baryons, where mixing is not allowed for. If for example there are two distinct processes contributing to the decay amplitude, one can write

$$A_f = e^{i\varphi_1}|A_1|e^{i\delta_1} + e^{i\varphi_2}|A_2|e^{i\delta_2},$$

$$\bar{A}_{\bar{f}} = e^{-i\varphi_1}|A_1|e^{i\delta_1} + e^{-i\varphi_2}|A_2|e^{i\delta_2},$$

where $\varphi_{1,2}$ denotes *CP*-violating weak phases and $|A_{1,2}|e^{i\delta_{1,2}}$ *CP*-conserving strong amplitudes of the two processes, labelled by the subscripts 1 and 2. The *CP*-violating

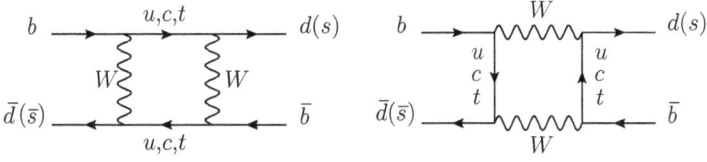

Fig. 3. Box diagrams contributing to $\bar{B}^0-B^0$ and $\bar{B}^0_s-B^0_s$ mixing.

asymmetry is given by

$$A_{CP} \equiv \frac{\Gamma_{\bar{B}\to\bar{f}} - \Gamma_{B\to f}}{\Gamma_{\bar{B}\to\bar{f}} + \Gamma_{B\to f}} = \frac{|\bar{A}_{\bar{f}}|^2 - |A_f|^2}{|\bar{A}_{\bar{f}}|^2 + |A_f|^2}$$

$$= \frac{2|A_1||A_2|\sin(\delta_2 - \delta_1)\sin(\varphi_2 - \varphi_1)}{|A_1|^2 + 2|A_1||A_2|\cos(\delta_2 - \delta_1)\cos(\varphi_2 - \varphi_1) + |A_2|^2}. \tag{1}$$

A nonzero value of the asymmetry A_{CP} arises from the interference between the two processes, and requires both a nonzero weak phase difference $\varphi_2 - \varphi_1$ and a nonzero strong phase difference $\delta_2 - \delta_1$. The presence of (at least) two interfering processes is a distinctive feature to observe CP violation.

When neutral heavy mesons are involved, the phenomenon of mixing between opposite flavours takes place. In the case of B^0 and B^0_s mesons, due to the diagrams sketched in Fig. 3, an initial $|B\rangle$ state will evolve as a superposition of $|B\rangle$ and $|\bar{B}\rangle$ states. Hence the mass eigenstates do not coincide with the flavour eigenstates, but are related to them by

$$|B_{\mathrm{H}}\rangle = \frac{p|B\rangle + q|\bar{B}\rangle}{\sqrt{|p|^2 + |q|^2}}, \quad |B_{\mathrm{L}}\rangle = \frac{p|B\rangle - q|\bar{B}\rangle}{\sqrt{|p|^2 + |q|^2}},$$

where p and q are two complex parameters, and $|B_{\mathrm{H}}\rangle$ and $|B_{\mathrm{L}}\rangle$ denote the two eigenstates of the $B^0_{(s)}-\bar{B}^0_{(s)}$ system. These two eigenstates are split in mass and lifetime, and we can define the mass and width differences $\Delta m_{d(s)} \equiv m_{d(s),\mathrm{H}} - m_{d(s),\mathrm{L}}$ and $\Delta\Gamma_{d(s)} \equiv \Gamma_{d(s),\mathrm{L}} - \Gamma_{d(s),\mathrm{H}}$. The subscripts H and L denote the heavy and light eigenstates. With this convention, the values of Δm_d and Δm_s are positive by definition. The present knowledge of B^0- and B^0_s-mixing processes is obtained from flavour-tagged time-dependent studies of semileptonic decays and of other decays involving flavour-specific final states, such as $B^0_s \to D^-_s\pi^+$. The world averages of the mass differences are $\Delta m_d = 0.510 \pm 0.003\,\mathrm{ps}^{-1}$ and $\Delta m_s = 17.757 \pm 0.021\,\mathrm{ps}^{-1}$.[42] The value of $\Delta\Gamma_s$ is measured to be positive,[43, 44] $\Delta\Gamma_s = 0.106 \pm 0.011\,(\mathrm{stat}) \pm 0.007\,(\mathrm{syst})\,\mathrm{ps}^{-1}$.[44] The value of $\Delta\Gamma_d$ is also positive in the SM and is expected to be much smaller than that of $\Delta\Gamma_s$, $\Delta\Gamma_d \simeq 3\times10^{-3}\,\mathrm{ps}^{-1}$.[45]

CP violation in the mixing of neutral mesons arises when the rate of, e.g. $B^0_{(s)}$ mesons transforming into $\overline{B}^0_{(s)}$ mesons differs from the rate of $\overline{B}^0_{(s)}$ mesons transforming into $B^0_{(s)}$ mesons. The condition to have CP violation in the mixing is given by $|q/p| \neq 1$. However, to a very good approximation the SM predicts $|q/p| \simeq 1$, i.e. CP violation in the mixing is very small, as also confirmed by experimental determinations.[42, 46] For neutral B^0 and B_s^0 mesons, the value of $|q/p|$ can be measured by means of the so-called semileptonic asymmetry

$$A_{\rm sl}^{d(s)} \equiv \frac{\Gamma_{\overline{B}^0_{(s)} \to \ell^+ X}(t) - \Gamma_{B^0_{(s)} \to \ell^- X}(t)}{\Gamma_{\overline{B}^0_{(s)} \to \ell^+ X}(t) + \Gamma_{B^0_{(s)} \to \ell^- X}(t)} = \frac{1 - |q/p|_{d(s)}^4}{1 + |q/p|_{d(s)}^4},$$

which actually turns out to be not dependent on time.

Finally, CP violation may arise in the interference between decay and mixing processes. Assuming CPT invariance, the CP asymmetry as a function of decay time for a neutral B^0 or B_s^0 meson decaying to a self-conjugate final state f, is given by

$$A(t) \equiv \frac{\Gamma_{\overline{B}^0_{(s)} \to f}(t) - \Gamma_{B^0_{(s)} \to f}(t)}{\Gamma_{\overline{B}^0_{(s)} \to f}(t) + \Gamma_{B^0_{(s)} \to f}(t)} = \frac{-C_f \cos(\Delta m_{d(s)} t) + S_f \sin(\Delta m_{d(s)} t)}{\cosh\left(\frac{\Delta \Gamma_{d(s)}}{2} t\right) + A_f^{\Delta \Gamma} \sinh\left(\frac{\Delta \Gamma_{d(s)}}{2} t\right)}.$$

The quantities C_f, S_f and $A_f^{\Delta \Gamma}$ are

$$C_f \equiv \frac{1 - |\lambda_f|^2}{1 + |\lambda_f|^2}, \quad S_f \equiv \frac{2 \Im \lambda_f}{1 + |\lambda_f|^2} \quad \text{and} \quad A_f^{\Delta \Gamma} \equiv -\frac{2 \Re \lambda_f}{1 + |\lambda_f|^2},$$

where λ_f is given by

$$\lambda_f \equiv \frac{q}{p} \frac{\bar{A}_f}{A_f}.$$

The parameter λ_f is thus related to $B^0_{(s)}$–$\overline{B}^0_{(s)}$ mixing (via q/p) and to the decay amplitudes of the $B^0_{(s)} \to f$ decay (A_f) and of the $\overline{B}^0_{(s)} \to f$ decay ($\bar{A}_f$). With negligible CP violation in the mixing ($|q/p| = 1$), the terms C_f and S_f parametrise CP violation in the decay and in the interference between mixing and decay, respectively. The following relation between C_f, S_f and $A_f^{\Delta \Gamma}$ holds

$$(C_f)^2 + (S_f)^2 + (A_f^{\Delta \Gamma})^2 = 1.$$

Notably, one has direct CP violation ($\bar{A}_f \neq A_f$) when $C_f \neq 0$. But even in the case of suppressed CP violation in the decay, it is still possible to observe CP violation if a relative phase between q/p and $\bar{A}_f/A_f$ exists. In such a case, one has $S_f = \Im(q/p \times \bar{A}_f/A_f)$. For example, for the $B^0 \to J/\psi K_S^0$ decay, to an approximation that is valid in the SM at the percent level or below, one has $S_{B^0 \to J/\psi K_S^0} = \sin(\phi_d)$ and $C_{B^0 \to J/\psi K_S^0} = 0$, with $\phi_d = 2\beta$. Similarly, for the $B_s^0 \to J/\psi \phi$ decay, CP violation in the interference between mixing and decay gives access to $\phi_s = -2\beta_s$.

3.6. Experimental determination of the unitarity triangle

The experimental determination of the UT is here presented in brief. Many excellent reviews are available in the literature,[42, 47–50] where detailed discussions on the various topics can be found. Several pieces of information must be combined by means of sophisticated fits in order to determine the apex of the UT with the highest possible precision. Relevant inputs to the fits are

- $|\varepsilon_K|$: This parameter is determined by measuring indirect CP violation in the neutral kaon mixing, using $K \to \pi\pi$, $K \to \pi l \nu$, and $K^0_L \to \pi^+\pi^- e^+ e^-$ decays, and provides a very important constraint on the position of the UT apex.
- $|V_{ub}|/|V_{cb}|$: The measurements of branching fractions of semileptonic decays governed by $b \to u l \bar\nu$ and $b \to c l \bar\nu$ transitions give information about the magnitudes of V_{ub} and V_{cb}, respectively. The ratio between these two quantities constrains the side of the UT between the γ and α angles.
- Δm_d: This parameter represents the frequency of $B^0 - \bar{B}^0$ mixing. It is proportional to the magnitude of V_{td} and thus constrains the side of the UT between the β and α angles.
- $\Delta m_s/\Delta m_d$: Δm_s is the analogue of Δm_d in the case of $B^0_s - \bar{B}^0_s$ mixing and its value is proportional to the magnitude V_{ts}. However, in order to reduce theoretical uncertainties on hadronic parameters determined using Lattice QCD calculations, the use of the ratio $\Delta m_s/\Delta m_d$ is more effective. This also provides a constraint on the side of the UT between the β and α angles.
- $\sin(2\beta)$: This quantity is mainly determined from time-dependent CP-violation measurements of $B^0 \to J/\psi K^0_S$ decays, and provides a powerful constraint on the angle β of the UT.
- α: This UT angle is determined from the measurements of CP-violating asymmetries and branching fractions in $B \to \pi\pi$, $B \to \rho\rho$ and $B \to \rho\pi$ decays.
- γ: The determination of this angle is performed by measuring time-integrated CP-violating asymmetries and branching fractions of $B \to D^{(*)}K^{(*)}$ decays, and time-dependent CP violation in $B^0_s \to D^\pm_s K^\mp$ decays.

World averages of the various experimental measurements are kept up to date by the Heavy Flavour Averaging Group.[42] Each of these measurements yields a constraint on the position of the UT apex, i.e. on the values of the $\bar\rho$ and $\bar\eta$ parameters. Two independent groups, namely CKMfitter[51] and UTfit,[45] regularly perform global CKM fits starting from the same set of experimental and theoretical inputs, but using different statistical approaches. In particular, CKMfitter performs a frequentist analysis, whereas UTfit follows a Bayesian method. Their latest results are displayed in Fig. 4. Each of the experimental constraints is represented as a 95% probability region by a filled area of different colour. The intersection of all regions identifies the position of the UT apex. The level of agreement amongst the various regions in pinpointing the UT apex is also a measure of the level of consistency of the KM mechanism with data. If (at least) one of the areas were not in agreement with the others, that would be an indication of the existence of BSM physics. At present, no striking evidence of any disagreement is found. The latest values of A, λ, $\bar\rho$ and $\bar\eta$ obtained by the CKMfitter and UTfit groups are reported in Table 1. Besides small fluctuations, also stemming from slightly different theoretical inputs and statistical procedures, the two sets of results are in good agreement.

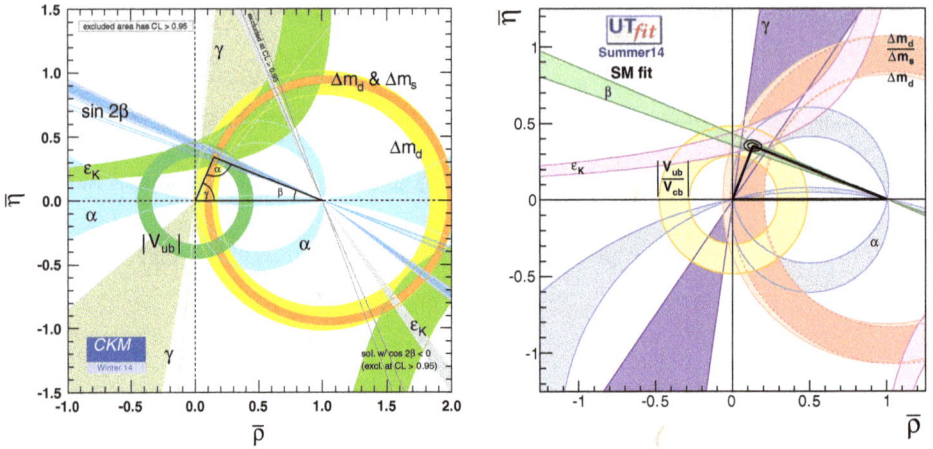

Fig. 4. Results of global CKM fits performed by the (left) CKMfitter[51] and (right) UTfit[45] groups. The 95% probability regions corresponding to each of the experimental measurements are indicated by filled areas of different colours. The various areas intersect in the position of the UT apex.

Table 1   Results of global CKM fits by the CKMfitter[51] and UTfit[45] groups.

Group	A	λ	$\bar{\rho}$	$\bar{\eta}$
CKMfitter	$0.810^{+0.018}_{-0.024}$	$0.22548^{+0.00068}_{-0.00034}$	$0.1453^{+0.0133}_{-0.0073}$	$0.343^{+0.011}_{-0.012}$
UTfit	0.821 ± 0.012	0.22534 ± 0.00065	0.132 ± 0.023	0.352 ± 0.014

4.   Overview of Beauty Physics at the LHC

4.1.   *CP violation*

Owing to the legacy of the B factories,[52] we have entered the era of precision tests of CKM physics, where ultimate sensitivity is needed to search for new sources of *CP* violation beyond the single phase of the CKM matrix.

The LHC is often considered as a B_s^0-meson factory, due to the large production cross-sections and to the excellent capabilities of the LHC experiments to precisely resolve B_s^0 oscillations. This has opened the door to precision measurements of the *CP*-violating phase $\varphi_s^{c\bar{c}s}$, which is equal to $-2\beta_s$ in the SM, neglecting sub-leading penguin contributions. It has been measured at the LHC by ATLAS, CMS and LHCb using the flavour eigenstate decays $B_s^0 \to J/\psi K^+ K^-$[53–55] and $B_s^0 \to J/\psi \pi^+ \pi^-$.[56] Recently LHCb has used the decay $B_s^0 \to J/\psi K^+ K^-$ for the first time in a polarisation-dependent way.[57] The quantity $\varphi_s^{c\bar{c}s}$ has also been measured with a fully hadronic final state using the decay $B_s^0 \to D_s^+ D_s^-$ with $D_s^\pm \to K^+ K^- \pi^\pm$.[58] Combining all determinations, LHCb obtains $\varphi_s^{c\bar{c}s} = -0.010 \pm 0.039$ rad. Including also other results, and in particular those from ATLAS and CMS, the uncertainty

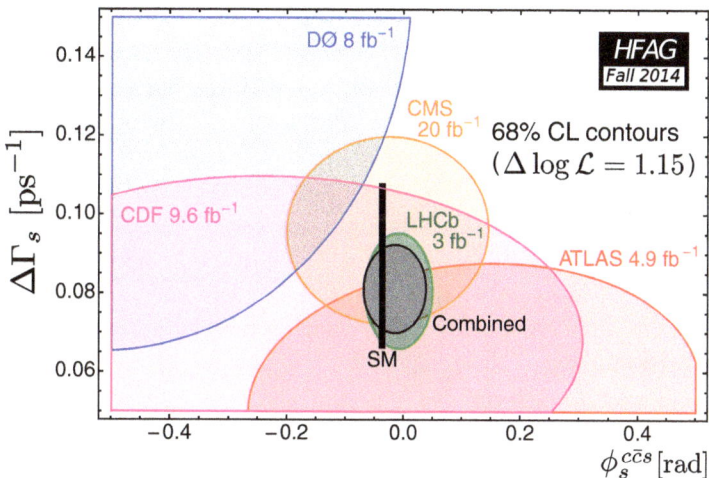

Fig. 5.   Constraints on $\Delta\Gamma_s$ and $\varphi_s^{c\bar{c}s}$ from various experiments.

is slightly reduced, obtaining $\varphi_s^{c\bar{c}s} = -0.015 \pm 0.035$ rad. The constraints on $\varphi_s^{c\bar{c}s}$ and on the decay width difference $\Delta\Gamma_s$ are shown in Fig. 5, together with the corresponding SM expectations.

With the precision reaching the degree level, the effects of suppressed penguin topologies cannot be neglected anymore.[59–64] Such effects may lead to a shift $\delta\varphi_s$ in the measured value of $\varphi_s^{c\bar{c}s}$, which can be constrained using Cabibbo-suppressed decay modes, where penguin topologies are relatively more prominent. This programme has started with studies of the decays $B_s^0 \rightarrow J/\psi K_S^0$,[65, 66] $B_s^0 \rightarrow J/\psi \bar{K}^{*0}$[67] and more recently $B^0 \rightarrow J/\psi \pi^+ \pi^-$.[68] These studies enable a constraint on $\delta\varphi_s$ to be placed in the range $[-0.018, 0.021]$ rad at 68% CL. Considering the present uncertainty of 0.039 rad, such a shift needs to be to constrained further.

Another interesting test of the SM is provided by the measurement of the mixing phase $\varphi_s^{s\bar{s}s}$ with a penguin-dominated mode as $B_s^0 \rightarrow \phi\phi$. In this case the measured value is $-0.17 \pm 0.15 \pm 0.03$ rad,[69] which is compatible with the SM expectation.

Similarly, the decays $B \rightarrow hh$ with $h = \pi, K$ receive large contributions from penguin topologies, and are sensitive to γ and β_s. LHCb for the first time measured time-dependent *CP*-violating observables in the decay $B_s^0 \rightarrow K^+K^-$.[70] Using methods outlined in Refs. 64, 71 and 72, a combination of this and other results from $B \rightarrow \pi\pi$ modes enables the determination $-2\beta_s = -0.12\,^{+0.14}_{-0.16}$ rad using as input the angle γ from tree-level decays (see below), or $\gamma = (63.5\,^{+7.2}_{-6.7})°$ relating $-2\beta_s$ to the SM expectation.[73] These values are in principle sensitive to the amount of U-spin (a subgroup of $SU(3)$ analogous to isospin, but involving d and s quarks instead of d and u quarks) breaking in the involved decay amplitudes and are given here for a maximum allowed breaking of 50%. This value of γ can be compared to that obtained from tree-dominated $B \rightarrow DK$ decays, where the *CP*-violating phase appears in the interference of the $b \rightarrow c$ and $b \rightarrow u$ topologies. The determination of γ from tree decays is considered free from contributions beyond the SM and unaffected by hadronic uncertainties. Yet, its precise measurement is important to test the consistency of the KM mechanism, also allowing for comparisons with measurements from modes dominated by penguin topologies.

The most precise determination of γ from a single tree-level decay mode is achieved with the decay $B^+ \rightarrow DK^+$ followed by $D \rightarrow K_s^0 h^+ h^-$ with $h = \pi, K$,[74] yielding $\gamma = (62 {}^{+15}_{-14})°$. Here, the interference of the D^0 and $\bar{D}^0$ decays to $K_s^0 h^+ h^-$ is exploited to measure CP asymmetries.[75] The method needs external input in the form of a measurement of the strong phase over the Dalitz plane of the D decay, coming from CLEO-c data.[76] The same decay mode is also used in a model-dependent measurement.[77]

A different way for determining γ is provided by the decay $B_s^0 \rightarrow D_s^{\pm} K^{\mp}$.[78–81] In this case the phase is measured in a time-dependent tagged CP-violation analysis. Using a dataset corresponding to 1 fb^{-1}, LHCb determines $\gamma = (115 {}^{+28}_{-43})°$, which is not yet competitive with other methods but will provide important cross-checks with more data.

The γ measurements of Refs. 74, 81–85 are then combined in an LHCb average.[86] Using all $B \rightarrow DK$ decay modes one finds $\gamma = (73 {}^{+9}_{-10})°$, which is more precise than the corresponding combination of measurements from the B factories.[52] The LHCb likelihood profile is shown in Fig. 6.

The same-sign dimuon asymmetry measured by the D0 collaboration[87] and interpreted as a combination of the semileptonic asymmetries A_{sl}^d and A_{sl}^s in B^0 and B_s^0 decays, respectively, differs from the SM expectation by 3σ. So far LHCb has not been able to confirm or disprove this result. The measurement from LHCb looks at the CP asymmetry between partially reconstructed $B \rightarrow D\mu\nu$ decays, where the flavour of the D meson identifies that of the B. The measured value of A_{sl}^{s}[46] and the newly reported A_{sl}^{d}[88] are both consistent with the SM and the D0 values.

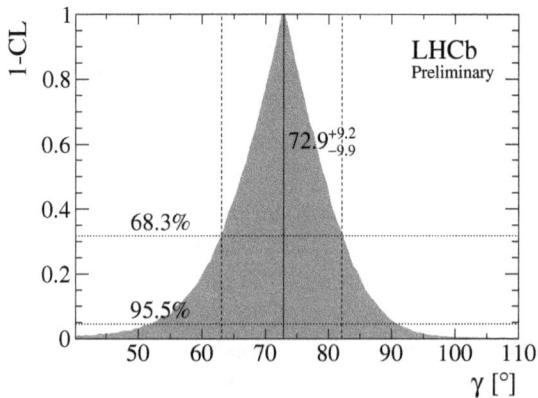

Fig. 6.   LHCb combination of $B \rightarrow DK$ decays measuring γ.[86]

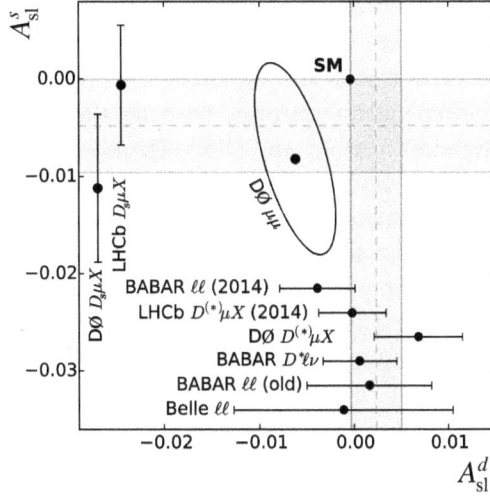

Fig. 7.   Experimental constraints on A_{sl}^d and A_{sl}^s from various experiments. The symbol ℓ stands for electron or muon. The horizontal and vertical bands represent the uncertainties on the averages of the experimental measurements. The elliptic contour represents the measurement of the same-sign dimuon asymmetry by the D0 experiment.

The world average including measurements from the B factories and D0 is not more conclusive, as shown in Fig. 7.

Large CP violation has been found in charmless b-hadron decays like $B^+ \to h^+h^-h^{+89}$ ($h = \pi, K$) and $B^+ \to p\overline{p}h^{+.90}$ Particularly striking features of these decays are the very large asymmetries observed in small regions of the phase-space not related to any resonance, which are opposite in sign for $B^\pm \to h^\pm K^+K^-$ and $B^\pm \to h^\pm\pi^+\pi^-$ decays. This observation could be a sign of long-distance $\pi^+\pi^- \leftrightarrow K^+K^-$ rescattering.

Another important field is the study of CP violation in beauty baryons. The probability that a b quark hadronises to a Λ_b^0 baryon is measured to be surprisingly large at the LHC in the forward region,[91] almost half of that to a B^0 meson. These baryons can be used for measurements of CP violation with better precision than B^0 mesons. Searches have been performed by LHCb with the decays $\Lambda_b^0 \to J/\psi p\pi^-$,[92] $\Lambda_b^0 \to K^0 p\pi^-$,[93] and by CDF with $\Lambda_b^0 \to p\pi^-$ and $\Lambda_b^0 \to pK^-$.[94] It is worth noting that no evidence for CP violation in any decay of a baryon has ever been reported.

4.2. *Rare electroweak decays*

The family of decays $b \to s\ell^+\ell^-$ is a laboratory for BSM-physics searches on its own. In particular the exclusive decay $B^0 \to K^{*0}\ell^+\ell^-$ ($\ell = e, \mu$) provides a very rich set of observables with different sensitivities to BSM physics and for which the available SM predictions are affected by varying levels of hadronic uncertainties. For some ratios of such observables, most of the theoretical uncertainties cancel out, thus providing a clean test of the SM.[95–100]

The differential decay width with respect to the dilepton mass squared q^2, the forward-backward asymmetry A_{FB}, and the longitudinal polarisation fraction F_L of the K^* resonance have been measured by many experiments[101–106] with no significant sign of deviations from the SM expectation.

In a second analysis of the already published 2011 data,[106] LHCb has published another set of angular observables[107] suggested in Ref. 100. In particular a 3.7σ local deviation from the SM expectation of one of these observables has been found in one bin of q^2. This measurement has triggered a lot of interest in the theoretical community, with interpretation articles being submitted very quickly to journals. See Refs. 108–112 for a small subset. It is not clear whether this discrepancy is an experimental fluctuation, is due to under-estimated form factor uncertainties (see Ref. 113), or is the manifestation of a heavy Z' boson, among many other suggested explanations. The contribution of $c\bar{c}$ resonances is also being questioned[114] after the LHCb observation of $B^+ \to \psi(4160)K^+$ with $\psi(4160) \to \mu^+\mu^-$,[115] where the $\psi(4160)$ and its interference with the non-resonant component account for 20% of the rate for dimuon masses above 3770 MeV/c^2. Such a large contribution was not expected.

Given a hint of abnormal angular distributions, LHCb tried to look for other deviations in several asymmetry measurements. The CP asymmetry in $B^0 \to K^{(*)0}\mu^+\mu^-$ and $B^\pm \to K^\pm\mu^+\mu^-$ decays turns out to be compatible with zero as expected,[116] as does the isospin asymmetry between $B^0 \to K^{(*)0}\mu^+\mu^-$ and $B^+ \to K^{(*)+}\mu^+\mu^-$ decays.[117] The lepton universality factor $R_K = \frac{\mathcal{B}(B^+ \to K^+\mu^+\mu^-)}{\mathcal{B}(B^+ \to K^+e^+e^-)}$ is measured to be $0.745\,^{+0.090}_{-0.074} \pm 0.036$[118] in the $1 < q^2 < 6$ GeV^2/c^4 range, which indicates a 2.6σ deviation from unity. This result can be interpreted as a possible indication of a new vector particle that would couple more strongly to muons and interfere destructively with the SM vector current.[119–123]

4.3. *Observation of the $B_s^0 \to \mu^+\mu^-$ decay*

The measurement of the branching fractions of the rare $B^0 \to \mu^+\mu^-$ and $B_s^0 \to \mu^+\mu^-$ decays is considered amongst the most promising avenues to search for BSM effects at the LHC. These decays proceed via FCNC processes and are highly suppressed in the SM. Moreover, the helicity suppression of axial vector terms makes them sensitive to scalar and pseudoscalar BSM contributions that can alter their branching fractions with respect to SM expectations. The untagged time-integrated SM predictions for the branching fractions of these decays are[124]

$$\mathcal{B}(B_s^0 \to \mu^+\mu^-)_{SM} = (3.66 \pm 0.23) \times 10^{-9},$$

$$\mathcal{B}(B^0 \to \mu^+\mu^-)_{SM} = (1.06 \pm 0.09) \times 10^{-10},$$

which are obtained using the latest combination of values for the top-quark mass from LHC and Tevatron experiments.[125] The ratio $\mathcal{R}$ between these two branching fractions is also a powerful tool to discriminate amongst BSM models. In the SM it is predicted to be

$$\mathcal{R} = \frac{\mathcal{B}(B^0 \to \mu^+\mu^-)}{\mathcal{B}(B^0_s \to \mu^+\mu^-)} = \frac{\tau_{B^0}}{1/\Gamma^s_H} \left(\frac{f_{B^0}}{f_{B^0_s}}\right)^2 \left|\frac{V_{td}}{V_{ts}}\right|^2 \frac{M_{B^0}\sqrt{1 - \frac{4m^2_\mu}{M^2_{B^0}}}}{M_{B^0_s}\sqrt{1 - \frac{4m^2_\mu}{M^2_{B^0_s}}}} = 0.0295 \, {}^{+\,0.0028}_{-\,0.0025} \,,$$

where τ_{B^0} and $1/\Gamma^s_H$ are the lifetimes of the B^0 and of the heavy mass eigenstate of the B^0_s–$\overline{B}^0_s$ system, $M_{B^0_{(s)}}$ is the mass and $f_{B^0_{(s)}}$ is the decay constant of the $B^0_{(s)}$ meson, V_{td} and V_{ts} are the elements of the CKM matrix and m_μ is the mass of the muon. In minimal flavour-violating BSM scenarios, the branching fractions of both decays can change, but their ratio is predicted to be equal to that of the SM.

The LHCb collaboration reported the first evidence of the $B^0_s \to \mu^+\mu^-$ decay with a 3.5σ significance in 2012 using $2\,\text{fb}^{-1}$ of data.[126] One year later, CMS and LHCb presented their updated results based on $25\,\text{fb}^{-1}$ and $3\,\text{fb}^{-1}$, respectively.[127, 128] The two measurements resulted in good agreement with comparable precisions. However, none of them was precise enough to claim the first observation of the $B^0_s \to \mu^+\mu^-$ decay. A naive combination of CMS and LHCb results was presented in 2013,[129] but no attempt was made to take into account all correlations stemming from common physical quantities, and no statistical significance for the existence of the signals was provided.

More recently, a combination of the CMS and LHCb results based on a simultaneous fit to the two datasets has been performed. This fit correctly takes into account correlations between the input parameters. The CMS and LHCb experiments have very similar analysis strategies. $B^0_{(s)} \to \mu^+\mu^-$ candidates are selected as two oppositely charged tracks. A soft first selection is applied in order to reduce the background while keeping high the efficiency on the signal. After this selection, the remaining backgrounds are mainly due to random combinations of muons from semileptonic B decays (combinatorial background), semileptonic decays, such as $B \to h\mu\nu$, $B \to h\mu\mu$ and $\Lambda^0_b \to p\mu^-\bar{\nu}$, and $B^0_{(s)} \to h^+h'^-$ decays (peaking background) where hadrons are misidentified as muons. Further separation between signal and background is achieved exploiting the power of a multivariate classifier. The classification of the events is done using the dimuon invariant mass $m_{\mu\mu}$ and the multivariate classifier output. The multivariate classifier is trained using kinematic and geometrical variables. The calibration of the dimuon mass $m_{\mu\mu}$ is performed using the dimuon resonances and, for LHCb, also by using $B^0_{(s)} \to h^+h'^-$ decays. For both analyses the $B^0_{(s)} \to \mu^+\mu^-$ yield is normalised with respect to the $B^+ \to J/\psi K^+$ yield, taking into account the hadronisation fractions of a b quark to B^0_s and B^0 mesons measured by the LHCb experiment.[130–132] LHCb also uses the $B^0 \to K^+\pi^-$ decay as a normalisation channel.

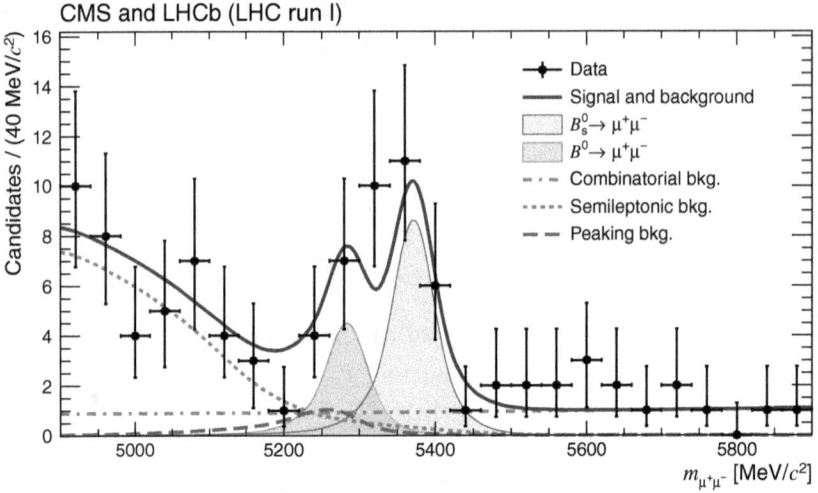

Fig. 8.   Dimuon mass distribution for the six multivariate-classifier categories with highest B_s^0 signal purity. The result of the simultaneous fit is overlaid.

A simultaneous fit is performed to evaluate the branching fractions of the $B_s^0 \to \mu^+\mu^-$ and $B^0 \to \mu^+\mu^-$ decays. The CMS and LHCb datasets are used together as in a single combined experiment. A simultaneous unbinned extended maximum likelihood fit is performed to the invariant mass spectrum in 20 categories of multivariate classifier output for the two experiments: 8 categories for LHCb and 12 categories for CMS. The various categories are characterised by construction by different values of signal purity. In each category the mass spectrum is described as the sum of each background source and the two signals. The parameters shared between the two experiments are the branching fractions of the two signal decays being looked for, $\mathcal{B}(B_s^0 \to \mu^+\mu^-)$ and $\mathcal{B}(B^0 \to \mu^+\mu^-)$, the already measured branching fraction of the common normalisation channel $\mathcal{B}(B^+ \to J/\psi K^+)$, and the ratio of the hadronisation fractions f_s/f_d. Assuming the SM, 94 ± 7 $B_s^0 \to \mu^+\mu^-$ events and 10.5 ± 0.6 $B^0 \to \mu^+\mu^-$ events are expected in the full dataset. For illustrative purposes, Fig. 8 shows the dimuon mass distribution for the events corresponding to the six multivariate categories with highest B_s^0 signal purity. The results of the simultaneous fit are[133]

$$\mathcal{B}(B_s^0 \to \mu^+\mu^-) = \left(2.8^{+0.7}_{-0.6}\right) \times 10^{-9},$$

$$\mathcal{B}(B^0 \to \mu^+\mu^-) = \left(3.9^{+1.6}_{-1.4}\right) \times 10^{-10}.$$

The statistical significances, evaluated using the Wilks' theorem,[134] are 6.2σ and 3.2σ for $B_s^0 \to \mu^+\mu^-$ and $B^0 \to \mu^+\mu^-$, respectively. The expected significances assuming the SM branching fractions are 7.4σ and 0.8σ for B_s^0 and B^0 channels, respectively. Since the Wilks' theorem shows a $B^0 \to \mu^+\mu^-$ signal significance slightly above the 3σ level, a more refined method based on the Feldman–Cousins

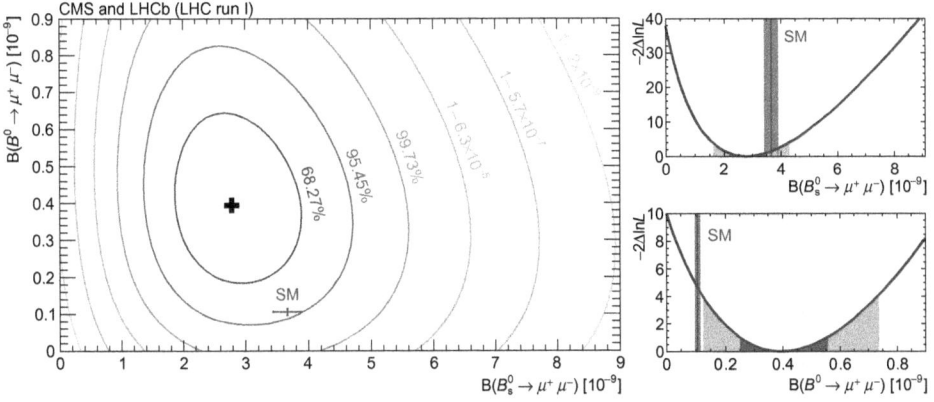

Fig. 9. (Left) Likelihood contours in the $\mathcal{B}(B_s^0 \to \mu^+\mu^-)$–$\mathcal{B}(B^0 \to \mu^+\mu^-)$ plane. Likelihood profile for (top-right) $\mathcal{B}(B_s^0 \to \mu^+\mu^-)$ and (bottom-right) $\mathcal{B}(B^0 \to \mu^+\mu^-)$. The dark and light areas define the $\pm 1\sigma$ and $\pm 2\sigma$ confidence intervals, respectively. The SM expectations are indicated with vertical bands.

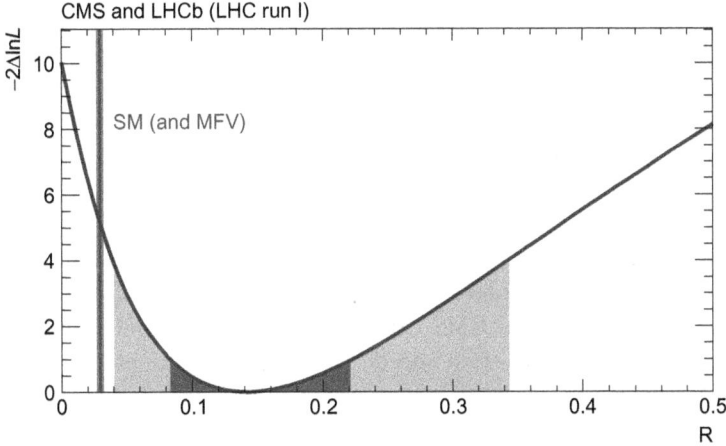

Fig. 10. Likelihood profile for $\mathcal{R}$. The dark and light areas define the $\pm 1\sigma$ and $\pm 2\sigma$ confidence intervals, respectively. The SM expectation is indicated with a vertical band.

construction[135] is also used for the $B^0 \to \mu^+\mu^-$ mode. A statistical significance of 3.0σ is obtained in this case. The Feldman–Cousins confidence intervals at $\pm 1\sigma$ and $\pm 2\sigma$ are $[2.5, 5.6] \times 10^{-10}$ and $[1.4, 7.4] \times 10^{-10}$, respectively. In Fig. 9 the likelihood contours in the $\mathcal{B}(B_s^0 \to \mu^+\mu^-)$–$\mathcal{B}(B^0 \to \mu^+\mu^-)$ plane are shown. In the same figure, the likelihood profile for each signal mode is displayed. The compatibilities of $\mathcal{B}(B_s^0 \to \mu^+\mu^-)$ and $\mathcal{B}(B^0 \to \mu^+\mu^-)$ with the SM are evaluated at the 1.2σ and 2.2σ levels, respectively. A separate fit to the ratio of B^0 to B_s^0 gives $\mathcal{R} = 0.14^{+0.08}_{-0.06}$, which is compatible with the SM at the 2.3σ level. The likelihood profile for $\mathcal{R}$ is shown in Fig. 10.

5. Conclusions

The LHC is the new b-hadron factory. The ATLAS, CMS and LHCb experiments will be dominating heavy-flavour physics for the next decade, together with the forthcoming Belle II experiment, and even beyond with the high luminosity LHC phase. During Run I, ATLAS, CMS and LHCb have performed fundamental measurements in the field of CP violation and rare decays of B mesons. In this paper we have discussed some of these measurements, notably including that of the B_s^0-meson mixing phase from $b \rightarrow c\bar{c}s$ decays, $\varphi_s^{c\bar{c}s}$; of the UT angle γ from tree-level decays; of B^0 and B_s^0 semileptonic asymmetries; of angular observables in $b \rightarrow s\ell^+\ell^-$ transitions; and of the $B_s^0 \rightarrow \mu^+\mu^-$ branching fraction, with the first observation of this decay at more than five standard deviations. No striking evidences of deviations from SM expectations have emerged from any of these measurements so far. However, they have enabled strong constraints to be set on many BSM models. The upcoming Run II, with its higher centre-of-mass energy translating into a higher $b\bar{b}$ cross-sections, will witness substantial improvements in the study of B physics, and will hopefully lead to the observation of new physics phenomena not accounted for in the SM.

  Heavy flavour physics in the quark sector is not limited to beauty hadrons alone. The LHC is also an abundant source of charmed hadrons, which provide another interesting laboratory for BSM-physics searches. The recent experimental improvements in the measurement of mixing-related observables of D^0 mesons at LHCb have raised the interest for CP-violation measurements in this sector. Belle II and LHCb, with its upgraded detector that will be operational in the third LHC run, will probe CP violation in charm mixing with ultimate precision. The top quark is also another excellent tool for seeking BSM physics. The unprecedented samples collected by the ATLAS and CMS experiments will enable relevant studies of CP violation in top-quark production and decays to be carried out.

References

1. A. D. Sakharov, Violation of CP invariance, C asymmetry, and baryon asymmetry of the universe, *Pisma Zh. Eksp. Teor. Fiz.* **5**, 32–35 (1967).
2. A. G. Cohen, D. Kaplan and A. Nelson, Progress in electroweak baryogenesis, *Ann. Rev. Nucl. Part. Sci.* **43**, 27–70 (1993). doi: 10.1146/annurev.ns.43.120193.000331.
3. A. Riotto and M. Trodden, Recent progress in baryogenesis, *Ann. Rev. Nucl. Part. Sci.* **49**, 35 (1999). doi: 10.1146/annurev.nucl.49.1.35.
4. W.-S. Hou, Source of CP violation for the baryon asymmetry of the universe, *Chin. J. Phys.* **47**, 134 (2009).
5. G. Aad *et al.*, Observation of a new particle in the search for the Standard Model Higgs boson with the ATLAS detector at the LHC, *Phys. Lett. B* **716**, 1–29 (2012). doi: 10.1016/j.physletb.2012.08.020.

6. S. Chatrchyan *et al.*, Observation of a new boson at a mass of 125 GeV with the CMS experiment at the LHC, *Phys. Lett. B* **716**, 30–61 (2012). doi: 10.1016/j.physletb.2012.08.021.

7. G. Aad *et al.*, The ATLAS Experiment at the CERN Large Hadron Collider, *JINST* **3**, S08003, (2008). doi: 10.1088/1748-0221/3/08/S08003.

8. S. Chatrchyan *et al.*, The CMS experiment at the CERN LHC, *JINST* **3**, S08004 (2008). doi: 10.1088/1748-0221/3/08/S08004.

9. A. A. Alves Jr. *et al.*, The LHCb detector at the LHC, *JINST* **3**, S08005 (2008). doi: 10.1088/1748-0221/3/08/S08005.

10. M. Kobayashi and T. Maskawa, *CP* violation in the renormalizable theory of weak interaction, *Prog. Theor. Phys.* **49**, 652 (1973).

11. N. Cabibbo, Unitary symmetry and leptonic decays, *Phys. Rev. Lett.* **10**, 531–532 (1963).

12. S. L. Glashow, J. Iliopoulos and L. Maiani, Weak interactions with lepton-hadron symmetry, *Phys. Rev. D* **2**, 1285–1292 (1970).

13. J. J. Aubert *et al.*, Experimental observation of a heavy particle J, *Phys. Rev. Lett.* **33**, 1404–1406 (1974).

14. J. E. Augustin *et al.*, Discovery of a narrow resonance in e^+e^- annihilation, *Phys. Rev. Lett.* **33**, 1406–1408 (1974).

15. S. Herb *et al.*, Observation of a Dimuon Resonance at 9.5-GeV in 400-GeV Proton-Nucleus Collisions, *Phys. Rev. Lett.* **39**, 252–255 (1977). doi: 10.1103/PhysRevLett.39.252.

16. F. Abe *et al.*, Observation of top quark production in $\bar{p}p$ collisions, *Phys. Rev. Lett.* **74**, 2626–2631 (1995). doi: 10.1103/PhysRevLett.74.2626.

17. J. H. Christenson, J. W. Cronin, V. L. Fitch, and R. Turlay, Evidence for the 2π Decay of the K_2^0 Meson, *Phys. Rev. Lett.* **13**, 138–140 (1964).

18. D. Andrews *et al.*, The CLEO detector, *Nucl. Instrum. Meth.* **211**, 47 (1983). doi: 10.1016/0167-5087(83)90556-2.

19. A. B. Carter and A. Sanda, *CP* violation in *B* meson decays, *Phys. Rev. D* **23**, 1567 (1981). doi: 10.1103/PhysRevD.23.1567.

20. I. I. Bigi and A. Sanda, On B^0–$\bar{B}^0$ mixing and violations of *CP* symmetry, *Phys. Rev. D* **29**, 1393 (1984). doi: 10.1103/PhysRevD.29.1393.

21. H. Albrecht *et al.*, ARGUS: A universal detector at DORIS-II, *Nucl. Instrum. Meth. A* **275**, 1–48 (1989). doi: 10.1016/0168-9002(89)90334-3.

22. P. Oddone, Detector considerations, *eConf C.* **870126**, 423–446 (1987).

23. B. Aubert *et al.*, The BaBar detector, *Nucl. Instrum. Meth. A* **479**, 1–116, (2002). doi: 10.1016/S0168-9002(01)02012-5.

24. A. Abashian *et al.*, The Belle detector, *Nucl. Instrum. Meth. A* **479**, 117–232 (2002). doi: 10.1016/S0168-9002(01)02013-7.

25. D. Decamp *et al.*, ALEPH: A detector for electron-positron annihilations at LEP, *Nucl. Instrum. Meth. A* **294**, 121–178 (1990). doi: 10.1016/0168-9002(90)91831-U.

26. P. Aarnio *et al.*, The DELPHI detector at LEP, *Nucl. Instrum. Meth. A* **303**, 233–276 (1991). doi: 10.1016/0168-9002(91)90793-P.

27. L3 collaboration, The Construction of the L3 Experiment, *Nucl. Instrum. Meth. A* **289**, 35–102 (1990). doi: 10.1016/0168-9002(90)90250-A.

28. K. Ahmet *et al.*, The OPAL detector at LEP, *Nucl. Instrum. Meth. A* **305**, 275–319 (1991). doi: 10.1016/0168-9002(91)90547-4.

29. SLD collaboration, SLD design report. (1984). SLAC Report-273.

30. G. J. Barker, b-physics at LEP, *Springer Tracts Mod. Phys.* **236**, 1–21 (2010). doi: 10.1007/978-3-642-05279-8_1.

31. P. Rowson, D. Su, and S. Willocq, Highlights of the SLD physics program at the SLAC linear collider, *Ann. Rev. Nucl. Part. Sci.* **51**, 345–412 (2001). doi: 10.1146/annurev.nucl.51.101701.132413.

32. M. Paulini, B lifetimes, mixing and CP violation at CDF, *Int. J. Mod. Phys. A* **14**, 2791–2886 (1999). doi: 10.1142/S0217751X99001391.

33. B. Aubert *et al.*, Measurement of time-dependent CP asymmetry in $B^0 \to c\bar{c}K^{(*)0}$ decays, *Phys. Rev. D* **79**, 072009 (2009). doi: 10.1103/PhysRevD.79.072009.

34. I. Adachi *et al.*, Precise measurement of the CP violation parameter $\sin 2\phi_1$ in $B^0 \to (c\bar{c})K^0$ decays, *Phys. Rev. Lett.* **108**, 171802 (2012). doi: 10.1103/PhysRevLett.108.171802.

35. T. Kuhr, Flavor physics at the Tevatron, *Springer Tracts Mod. Phys.* **249** 1–161 (2013). doi: 10.1007/978-3-642-10300-1.

36. A. Abulencia *et al.*, Observation of B_s^0–$\bar{B}_s^0$ Oscillations, *Phys. Rev. Lett.* **97**, 242003 (2006). doi: 10.1103/PhysRevLett.97.242003.

37. R. Aaij *et al.*, LHCb detector performance, *Int. J. Mod. Phys. A* **30**, 1530022 (2015).

38. R. Aaij *et al.*, The LHCb trigger and its performance in 2011, *JINST* **8**, P04022 (2013). doi: 10.1088/1748-0221/8/04/P04022.

39. C. Jarlskog, Commutator of the quark mass matrices in the Standard Electroweak Model and a measure of maximal CP violation, *Phys. Rev. Lett.* **55**, 1039 (1985).

40. L. Wolfenstein, Parametrization of the Kobayashi-Maskawa matrix, *Phys. Rev. Lett.* **51**, 1945 (1983).

41. K. A. Olive *et al.*, Review of particle physics, *Chin. Phys. C* **38**, 090001 (2014). doi: 10.1088/1674-1137/38/9/090001.

42. Y. Amhis *et al.*, Averages of b-hadron, c-hadron, and τ-lepton properties as of summer 2014 (2014). updated results and plots available at http://www.slac.stanford.edu/xorg/hfag/.

43. Y. Xie, P. Clarke, G. Cowan, and F. Muheim, Determination of $2\beta_s$ in $B_s^0 \to J/\psi K^+K^-$ decays in the presence of a K^+K^- S-wave contribution, *JHEP* **0909**, 074 (2009). doi: 10.1088/1126-6708/2009/09/074.

44. R. Aaij *et al.*, Determination of the sign of the decay width difference in the B_s^0 system, *Phys. Rev. Lett.* **108**, 241801 (2012). doi: 10.1103/PhysRevLett.108.241801.

45. M. Bona *et al.*, The unitarity triangle fit in the Standard Model and hadronic parameters from Lattice QCD: a reappraisal after the measurements of Δm_s and BR($B \to \tau\nu$), *JHEP* **0610**, 081 (2006). doi: 10.1088/1126-6708/2006/10/081.

46. R. Aaij *et al.*, Measurement of the flavour-specific CP-violating asymmetry a_{sl}^s in B_s^0 decays, *Phys. Lett. B* **728**, 607 (2014). doi: 10.1016/j.physletb.2013.12.030.

47. M. Ciuchini and A. Stocchi, Physics Opportunities at the Next Generation of Precision Flavuor Physics, *Ann. Rev. Nucl. Part. Sci.* **61**, 491–517 (2011). doi: 10.1146/annurev-nucl-102010-130424.

48. K. Olive *et al.*, Review of Particle Physics, *Chin. Phys. C* **38**, 090001 (2014). doi: 10.1088/1674-1137/38/9/090001.

49. R. Aaij *et al.*, Implications of LHCb measurements and future prospects, *Eur. Phys. J. C* **73**(4), 2373 (2013). doi: 10.1140/epjc/s10052-013-2373-2.
50. R. Fleischer, *CP* violation in the *B* system and relations to $K \to \pi\nu\bar{\nu}$ decays, *Phys. Rept.* **370**, 537–680 (2002). doi: 10.1016/S0370-1573(02)00274-0.
51. J. Charles *et al.*, Current status of the Standard Model CKM fit and constraints on $\Delta F = 2$ new physics (2015).
52. A. Bevan *et al.*, The physics of the *B* factories (2014).
53. G. Aad *et al.*, Flavour tagged time dependent angular analysis of the $B_s^0 \to J/\psi\phi$ decay and extraction of $\Delta\Gamma$ and the weak phase ϕ_s in ATLAS, *Phys. Rev. D* **90**, 052007 (2014). doi: 10.1103/PhysRevD.90.052007.
54. CMS collaboration, Measurement of the *CP*-violating weak phase ϕ_s and the decay width difference $\Delta\Gamma_s$ using the $B_s^0 \to J/\psi\phi(1020)$ decay channel (2014), CMS-PAS-BPH-13-012.
55. R. Aaij *et al.*, Measurement of *CP* violation and the B_s^0 meson decay width difference with $B_s^0 \to J/\psi K^+ K^-$ and $B_s^0 \to J/\psi\pi^+\pi^-$ decays, *Phys. Rev. D* **87**, 112010 (2013). doi: 10.1103/PhysRevD.87.112010.
56. R. Aaij *et al.*, Measurement of the *CP*-violating phase ϕ_s in $\bar{B}_s^0 \to J/\psi\pi^+\pi^-$ decays, *Phys. Lett. B* **736**, 186 (2014). doi: 10.1016/j.physletb.2014.06.079.
57. R. Aaij *et al.*, Precision measurement of *CP* violation in $B_s^0 \to J/\psi K^+ K^-$ decays, *Phys. Rev. Lett.* **114**, 041801 (2015). doi: 10.1103/PhysRevLett.114.041801.
58. R. Aaij *et al.*, Measurement of the *CP*-violating phase ϕ_s in $\bar{B}_s^0 \to D_s^+ D_s^-$ decays, *Phys. Rev. Lett.* **113**, 211801 (2014). doi: 10.1103/PhysRevLett.113.211801.
59. R. Fleischer, Extracting γ from $B_{s(d)} \to J/\psi K_S$ and $B_{d(s)} \to D_{(d/s)}^+ D_{(d/s)}^-$, *Eur. Phys. J. C* **10**, 299–306 (1999). doi: 10.1007/s100529900099.
60. R. Fleischer, Extracting CKM phases from angular distributions of $B_{(d,s)}$ decays into admixtures of *CP* eigenstates, *Phys. Rev. D* **60**, 073008 (1999). doi: 10.1103/PhysRevD.60.073008.
61. R. Fleischer, Recent theoretical developments in *CP* violation in the *B* system, *Nucl. Instrum. Meth. A* **446**, 1–17 (2000). doi: 10.1016/S0168-9002(00)00003-6.
62. S. Faller, M. Jung, R. Fleischer, and T. Mannel, The golden modes $B^0 \to J/\psi K_{(S,L)}$ in the era of precision flavour physics, *Phys. Rev. D* **79**, 014030 (2009). doi: 10.1103/PhysRevD.79.014030.
63. K. De Bruyn, R. Fleischer, and P. Koppenburg, Extracting γ and penguin topologies through *CP* violation in $B_s^0 \to J/\psi K_S^0$, *Eur. Phys. J. C* **70**, 1025–1035 (2010). doi: 10.1140/epjc/s10052-010-1495-z.
64. M. Ciuchini, M. Pierini, and L. Silvestrini, The effect of penguins in the $B^0 \to J/\psi K^0$ *CP* asymmetry, *Phys. Rev. Lett.* **95**, 221804 (2005). doi: 10.1103/PhysRevLett.95.221804.
65. R. Aaij *et al.*, Measurement of the effective $B_s^0 \to J/\psi K_S^0$ lifetime, *Nucl. Phys. B* **873**, 275–292 (2013). doi: 10.1016/j.nuclphysb.2013.04.021.
66. R. Aaij *et al.*, Search for *CP* violation in the decay $B_s^0 \to J/\psi K_S^0$ (2014), in preparation.
67. R. Aaij *et al.*, Measurement of the $B_s^0 \to J/\psi \bar{K}^{*0}$ branching fraction and angular amplitudes, *Phys. Rev. D* **86**, 071102(R) (2012). doi: 10.1103/PhysRevD.86.071102.
68. R. Aaij *et al.*, Measurement of the *CP*-violating phase β in $\bar{B}^0 \to J/\psi\pi^+\pi^-$ decays and limits on penguin effects (2014), to appear in *Phys. Lett. B*.

69. R. Aaij *et al.*, Measurement of *CP* violation in $B_s^0 \to \phi\phi$ decays, *Phys. Rev. D* **90**, 052011 (2014). doi: 10.1103/PhysRevD.90.052011.

70. R. Aaij *et al.*, First measurement of time-dependent *CP* violation in $B_s^0 \to K^+K^-$ decays, *JHEP* **10**, 183 (2013). doi: 10.1007/JHEP10(2013)183.

71. R. Fleischer, New strategies to extract β and γ from $B^0 \to \pi^+\pi^-$ and $B_s^0 \to K^+K^-$, *Phys. Lett. B* **459**, 306–320 (1999). doi: 10.1016/S0370-2693(99)00640-1.

72. R. Fleischer, $B_{s,d}^0 \to \pi\pi, \pi K, KK$: Status and Prospects, *Eur. Phys. J. C* **52**, 267–281 (2007). doi: 10.1140/epjc/s10052-007-0391-7.

73. R. Aaij *et al.*, First observation of $\bar{B}^0 \to J/\psi K^+K^-$ and search for $\bar{B}^0 \to J/\psi\phi$ decays, *Phys. Rev. D* **88**, 072005 (2013). doi: 10.1103/PhysRevD.88.072005.

74. R. Aaij *et al.*, Measurement of the CKM angle γ using $B^\pm \to DK^\pm$ with $D \to K_S\pi^+\pi^-, K_SK^+K^-$ decays, *JHEP* **10**, 097 (2014). doi: 10.1007/JHEP10(2014)097.

75. A. Giri, Y. Grossman, A. Soffer, and J. Zupan, Determining γ using $B^\pm \to DK^\pm$ with multibody D decays, *Phys. Rev. D* **68**, 054018 (2003). doi: 10.1103/PhysRevD.68.054018.

76. J. Libby *et al.*, Model-independent determination of the strong-phase difference between D^0 and $\bar{D}^0 \to K_{S,L}^0 h^+h^-$ ($h = \pi, K$) and its impact on the measurement of the CKM angle γ/ϕ_3, *Phys. Rev. D* **82**, 112006 (2010). doi: 10.1103/PhysRevD.82.112006.

77. R. Aaij *et al.*, Measurement of *CP* violation and constraints on the CKM angle γ in $B^\pm \to DK^\pm$ with $D \to K_S^0\pi^+\pi^-$ decays, *Nucl. Phys. B* **888**, 169 (2014). doi: 10.1016/j.nuclphysb.2014.09.015.

78. I. Dunietz and R. G. Sachs, Asymmetry between inclusive charmed and anticharmed modes in B^0, $\bar{B}^0$ decay as a measure of *CP* violation, *Phys. Rev. D* **37**, 3186 (1988). doi: 10.1103/PhysRevD.37.3186, 10.1103/PhysRevD.39.3515.

79. R. Aleksan, I. Dunietz, and B. Kayser, Determining the *CP* violating phase gamma, *Z. Phys. C* **54**, 653–660 (1992). doi: 10.1007/BF01559494.

80. R. Fleischer, New strategies to obtain insights into *CP* violation through $B_s^0 \to D_s^\pm K^\mp, D_s^{*\pm}K^\mp, \ldots$ and $B^0 \to D^\pm\pi^\mp, D^{*\pm}\pi^\mp, \ldots$ decays, *Nucl. Phys. B* **671**, 459–482 (2003). doi: 10.1016/j.nuclphysb.2003.08.010.

81. R. Aaij *et al.*, Measurement of *CP* asymmetry in $B_s^0 \to D_s^\mp K^\pm$ decays, *JHEP* **11**, 060 (2014). doi: 10.1007/JHEP11(2014)060.

82. R. Aaij *et al.*, Observation of *CP* violation in $B^\pm \to DK^\pm$ decays, *Phys. Lett. B* **712**, 203 (2012). doi: 10.1016/j.physletb.2012.04.060.

83. R. Aaij *et al.*, Observation of the suppressed ADS modes $B^\pm \to [\pi^\pm K^\mp\pi^+\pi^-]_D K^\pm$ and $B^\pm \to [\pi^\pm K^\mp\pi^+\pi^-]_D \pi^\pm$, *Phys. Lett. B* **723**, 44 (2013). doi: 10.1016/j.physletb.2013.05.009.

84. R. Aaij *et al.*, A study of *CP* violation in $B^\pm \to DK^\pm$ and $B^\pm \to D\pi^\pm$ decays with $D \to K_S^0 K^\pm\pi^\mp$ final states, *Phys. Lett. B* **733**, 36 (2014). doi: 10.1016/j.physletb.2014.03.051.

85. R. Aaij *et al.*, Measurement of *CP* violation parameters in $B^0 \to DK^{*0}$ decays, *Phys. Rev. D* **90**, 112002 (2014). doi: 10.1103/PhysRevD.90.112002.

86. LHCb collaboration. Improved constraints on γ: CKM2014 update (Sept 2014), LHCb-CONF-2014-004.

87. V. M. Abazov *et al.*, Study of *CP*-violating charge asymmetries of single muons and like-sign dimuons in $p\bar{p}$ collisions, *Phys. Rev. D* **89**(1), 012002 (2014). doi: 10.1103/PhysRevD.89.012002.

88. R. Aaij *et al.*, Measurement of the semileptonic *CP* asymmetry in B^0–$\bar{B}^0$ mixing, *Phys. Rev. Lett.* **114**, 041601 (2015). doi: 10.1103/PhysRevLett.114.041601.

89. R. Aaij *et al.*, Measurement of *CP* violation in the three-body phase space of charmless $B^{\pm}$ decays, *Phys. Rev. D* **90**, 112004 (2014). doi: 10.1103/PhysRevD.90.112004.

90. R. Aaij *et al.*, Evidence for *CP* violation in $B^+ \to p\bar{p}K^+$ decays, *Phys. Rev. Lett.* **113**, 141801 (2014). doi: 10.1103/PhysRevLett.113.141801.

91. R. Aaij *et al.*, Study of the kinematic dependences of Λ_b^0 production in pp collisions and a measurement of the $\Lambda_b^0 \to \Lambda_c^+ \pi^-$ branching fraction, *JHEP* **08**, 143 (2014). doi: 10.1007/JHEP08(2014)143.

92. R. Aaij *et al.*, Observation of the $\Lambda_b^0 \to J/\psi p \pi^-$ decay, *JHEP* **07**, 103 (2014). doi: 10.1007/JHEP07(2014)103.

93. R. Aaij *et al.*, Searches for Λ_b^0 and Ξ_b^0 decays to $K_S^0 p \pi^-$ and $K_S^0 p K^-$ final states with first observation of the $\Lambda_b^0 \to K_S^0 p \pi^-$ decay, *JHEP* **04**, 087 (2014). doi: 10.1007/JHEP04(2014)087.

94. T. Aaltonen *et al.*, Measurements of direct *CP* violating asymmetries in charmless decays of strange bottom mesons and bottom baryons, *Phys. Rev. Lett.* **106**, 181802 (2011). doi: 10.1103/PhysRevLett.106.181802.

95. A. Ali, P. Ball, L. Handoko, and G. Hiller, A comparative study of the decays $B \to (K, K^*)\ell^+\ell^-$ in the Standard Model and supersymmetric theories, *Phys. Rev. D* **61**, 074024 (Mar 2000).

96. Frank Krüger and Joaquim Matias, Probing new physics via the transverse amplitudes of $B \to K^*(K\pi^+)\ell\ell$ at large recoil, *Phys. Rev. D* **71**, 094009 (2005).

97. W. Altmannshofer *et al.*, Symmetries and asymmetries of $B \to K^*\mu^+\mu^-$ decays in the Standard Model and beyond, *JHEP* **0901**, 019 (2009). doi: 10.1088/1126-6708/2009/01/019.

98. U. Egede, T. Hurth, J. Matias, M. Ramon, and W. Reece, New observables in the decay mode $\bar{B} \to \bar{K}^*\ell^+\ell^-$, *JHEP* **0811**, 032 (2008). doi: 10.1088/1126-6708/2008/11/032.

99. C. Bobeth, G. Hiller, and G. Piranishvili, *CP* asymmetries in $\bar{B} \to \bar{K}^*(\to \bar{K}\pi)\ell\ell$ and untagged $\bar{B}_s, B_s \to \phi(\to K^+K^-)\ell\ell$ decays at NLO, *JHEP* **0807**, 106 (2008). doi: 10.1088/1126-6708/2008/07/106.

100. S. Descotes-Genon, T. Hurth, J. Matias, and J. Virto, Optimizing the basis of $B \to K^*\ell^+\ell^-$ observables in the full kinematic range, *JHEP* **1305**, 137 (2013). doi: 10.1007/JHEP05(2013)137.

101. B. Aubert *et al.*, Measurements of branching fractions, rate asymmetries, and angular distributions in the rare decays $B \to K\ell^+\ell^-$ and $B \to K^*\ell^+\ell^-$, *Phys. Rev. D* **73**, 092001 (2006). doi: 10.1103/PhysRevD.73.092001.

102. J.-T. Wei *et al.*, Measurement of the Differential Branching Fraction and Forward-Backward Asymmetry for $B \to K^*\ell^+\ell^-$, *Phys. Rev. Lett.* **103**, 171801 (2009). doi: 10.1103/PhysRevLett.103.171801.

103. T. Aaltonen *et al.*, Measurements of the Angular Distributions in the Decays $B \to K^{(*)}\mu^+\mu^-$ at CDF, *Phys. Rev. Lett.* **108**, 081807 (2012). doi: 10.1103/PhysRevLett.108.081807.

104. S. Chatrchyan *et al.*, Angular analysis and branching fraction measurement of the decay $B^0 \to K^{*0}\mu^+\mu^-$, *Phys. Lett. B* **727**, 77–100 (2013). doi: 10.1016/j.physletb.2013.10.017.

105. ATLAS collaboration, Angular analysis of $B_d \to K^{*0}\mu^+\mu^-$ with the ATLAS experiment (Apr, 2013), ATLAS-CONF-2013-038.

106. R. Aaij *et al.*, Differential branching fraction and angular analysis of the decay $B^0 \to K^{*0}\mu^+\mu^-$, *JHEP* **08**, 131 (2013). doi: 10.1007/JHEP08(2013)131.

107. R. Aaij *et al.*, Measurement of form-factor-independent observables in the decay $B^0 \to K^{*0}\mu^+\mu^-$, *Phys. Rev. Lett.* **111**, 191801 (2013). doi: 10.1103/PhysRevLett.111.191801.

108. R. Gauld, F. Goertz, and U. Haisch, An explicit Z'-boson explanation of the $B \to K^*\mu^+\mu^-$ anomaly, *JHEP* **1401**, 069 (2014). doi: 10.1007/JHEP01(2014)069.

109. S. Descotes-Genon, J. Matias, and J. Virto, Understanding the $B \to K^*\mu^+\mu^-$ anomaly, *Phys. Rev. D* **88**(7), 074002 (2013). doi: 10.1103/PhysRevD.88.074002.

110. W. Altmannshofer and D. M. Straub, New physics in $B \to K^*\mu\mu$?, *Eur. Phys. J. C* **73**, 2646 (2013). doi: 10.1140/epjc/s10052-013-2646-9.

111. A. Datta, M. Duraisamy, and D. Ghosh, Explaining the $B \to K^*\mu^+\mu^-$ data with scalar interactions, *Phys. Rev. D* **89**(7), 071501 (2014). doi: 10.1103/PhysRevD.89.071501.

112. F. Mahmoudi, S. Neshatpour, and J. Virto, $B \to K^*\mu^+\mu^-$ optimised observables in the MSSM, *Eur. Phys. J. C* **74**, 2927 (2014). doi: 10.1140/epjc/s10052-014-2927-y.

113. F. Beaujean, C. Bobeth, and D. van Dyk, Comprehensive Bayesian analysis of rare (semi)leptonic and radiative B decays, *Eur. Phys. J. C* **74**, 2897 (2014). doi: 10.1140/epjc/s10052-014-2897-0.

114. J. Lyon and R. Zwicky, Resonances gone topsy turvy — the charm of QCD or new physics in $b \to s\ell^+\ell^-$? (2014).

115. R. Aaij *et al.*, Observation of a resonance in $B^+ \to K^+\mu^+\mu^-$ decays at low recoil, *Phys. Rev. Lett.* **111**, 112003 (2013). doi: 10.1103/PhysRevLett.111.112003.

116. R. Aaij *et al.*, Measurement of CP asymmetries in the decays $B^0 \to K^{*0}\mu^+\mu^-$ and $B^+ \to K^+\mu^+\mu^-$, *JHEP* **09**, 177 (2014). doi: 10.1007/JHEP09(2014)177.

117. R. Aaij *et al.*, Differential branching fractions and isospin asymmetries of $B \to K^{(*)}\mu^+\mu^-$ decays, *JHEP* **06**, 133 (2014). doi: 10.1007/JHEP06(2014)133.

118. R. Aaij *et al.*, Test of lepton universality using $B^+ \to K^+\ell^+\ell^-$ decays, *Phys. Rev. Lett.* **113**, 151601 (2014). doi: 10.1103/PhysRevLett.113.151601.

119. G. Hiller and M. Schmaltz, R_K and future $b \to s\ell\ell$ BSM opportunities (2014).

120. D. Ghosh, M. Nardecchia, and S. Renner, Hint of lepton flavour non-universality in B meson decays, *JHEP* **1412**, 131 (2014). doi: 10.1007/JHEP12(2014)131.

121. W. Altmannshofer and D. M. Straub, State of new physics in $b \to s$ transitions (2014).

122. A. Crivellin, G. D'Ambrosio, and J. Heeck, Explaining $h \to \mu^\pm\tau^\mp$, $B \to K^*\mu^+\mu^-$ and $B \to K\mu^+\mu^-/B \to Ke^+e^-$ in a two-Higgs-doublet model with gauged $L_\mu - L_\tau$ (2015).

123. S. L. Glashow, D. Guadagnoli, and K. Lane, Lepton flavor violation in B decays? (2014).

124. C. Bobeth *et al.*, $B_{s,d} \to l^+l^-$ in the Standard Model with reduced theoretical uncertainty, *Phys. Rev. Lett.* **112**, 101801 (2014). doi: 10.1103/PhysRevLett.112.101801.

125. ATLAS collaboration, CDF collaboration, CMS collaboration and D0 collaboration, First combination of Tevatron and LHC measurements of the top-quark mass (2014).

126. R. Aaij *et al.*, First evidence for the decay $B_s^0 \to \mu^+\mu^-$, *Phys. Rev. Lett.* **110**, 021801 (2013). doi: 10.1103/PhysRevLett.110.021801.

127. R. Aaij *et al.*, Measurement of the $B_s^0 \to \mu^+\mu^-$ branching fraction and search for $B^0 \to \mu^+\mu^-$ decays at the LHCb experiment, *Phys. Rev. Lett.* **111**, 101805 (2013). doi: 10.1103/PhysRevLett.111.101805.

128. S. Chatrchyan *et al.*, Measurement of the $B_s^0 \to \mu^+\mu^-$ branching fraction and search for $B^0 \to \mu^+\mu^-$ with the CMS Experiment, *Phys. Rev. Lett.* **111**, 101804 (2013). doi: 10.1103/PhysRevLett.111.101804.

129. CMS and LHCb collaborations. Combination of results on the rare decays $B_{(s)}^0 \to \mu^+\mu^-$ from the CMS and LHCb experiments (Jul 2013), CMS-PAS-BPH-13-007, LHCb-CONF-2013-012.

130. R. Aaij *et al.*, Determination of f_s/f_d for 7 TeV pp collisions and measurement of the $B^0 \to D^-K^+$ branching fraction, *Phys. Rev. Lett.* **107**, 211801 (2011). doi: 10.1103/PhysRevLett.107.211801.

131. R. Aaij *et al.*, Measurement of b hadron production fractions in 7 TeV pp collisions, *Phys. Rev. D* **85**, 032008 (2012). doi: 10.1103/PhysRevD.85.032008.

132. LHCb collaboration. Updated average f_s/f_d b-hadron production fraction ratio for 7 TeV pp collisions (Jul 2013), LHCb-CONF-2013-011.

133. V. Khachatryan *et al.*, Observation of the rare $B_s^0 \to \mu^+\mu^-$ decay from the combined analysis of CMS and LHCb data (2014), submitted to *Nature*.

134. S. Wilks, The large-sample distribution of the likelihood ratio for testing composite hypotheses, *Annals Math. Statist.* **9**(1), 60–62 (1938). doi: 10.1214/aoms/1177732360.

135. G. Feldman and R. Cousins, Unified approach to the classical statistical analysis of small signals, *Phys. Rev. D* **57**, 3873–3889 (Apr 1998).

Toward the Limits of Matter:
Ultra-relativistic Nuclear Collisions at CERN

Jurgen Schukraft[1] and Reinhard Stock[2]

[1] *CERN, CH-1211 Geneva 23, Switzerland*
schukraft@cern.ch
[2] *FIAS, 60438 Frankfurt am Main, Germany*
stock@ikf.uni-frankfurt.de

Strongly interacting matter as described by the thermodynamics of QCD undergoes a phase transition, from a low temperature hadronic medium, to a high temperature quark–gluon plasma state. In the early universe this transition occurred during the early microsecond era. It can be investigated in the laboratory, in collisions of nuclei at relativistic energy, which create "fireballs" of sufficient energy density to cross the QCD phase boundary. We describe three decades of work at CERN, devoted to the study of the QCD plasma and the phase transition. From modest beginnings at the SPS, ultra-relativistic heavy ion physics is today a central pillar of contemporary nuclear physics and forms a significant part of the LHC programme.

1. Strongly Interacting Matter

We recall here the development of a novel research field at CERN, devoted to the phases, and phase structure of matter governed by the strong fundamental force. Its proper field theory was discovered around 1970: Quantum Chromodynamics (QCD) addresses the fundamental interactions of elementary quarks, as mediated by gluons. Importantly, the gluons carry strong charges themselves, unlike the uncharged photons that mediate the QED interaction. Thus QCD is a much more complicated theory, mathematically. Both these field theories constitute a part of the modern Standard Model of elementary interactions. They thus enter electrons, photons, quarks and gluons into our inventory of elementary particles. Their predicted properties have been meticulously studied and confirmed by decades of particle physics research.

Now, our interest here is not the study of elementary QCD collisions, but of extended matter with architecture governed by the strong interaction, from a partonic plasma state to protons and nuclei, and to neutron stars. Matter has arisen at an extremely early stage of the cosmological evolution, and is thus a part of the Cosmological Standard Model (developed in parallel to the Standard Model of elementary interactions). The universe has gone through successive, distinct stages of matter composition. From attoseconds to early microseconds, the expanding cosmos was governed by a plasma 'fireball' era, and composed of elementary quarks, gluons, electrons, etc. The Big Bang matter was conductive toward, both, electric and colour charge currents, but structureless otherwise. Our research field of QCD matter begins with this era.

Cooling provokes structure formation via phase transitions. Note that they occur in macroscopic volumes only: we talk about phase transitions of extended matter. More precisely: although the transformations occur due to inter-quark neutralisation and binding effects taking place at the microscopic scale, it is their collective synchronisation by macroscopic thermodynamic conditions such as density and temperature that leads to the emergence of states and phases of matter. The laws of thermodynamics, plus the microscopic intrinsic features of the carriers of degrees of freedom, and their interaction, constitute a characteristic phase diagram, with boundary lines between phases, in a plane of temperature and density.

What is the phase diagram of a macroscopic volume of QCD matter consisting of quarks and gluons?

This question represents the goal of the research field, addressed here. In other words, we thus ask for the thermodynamics of QCD. Thus, in marked contrast to particle physics focus on the elementary properties of partons, we wish to know what will go on in, say, a cubic metre of quark–gluon plasma, of primordial cosmological vintage, once it cools down to nucleons, or even recompresses in the interior of neutron stars. This is a deeply non-trivial question, concerning partons and hadrons with non-perturbative, in-medium modified interactions.

In the Big Bang cosmological evolution expansive cooling descends, from an initially arbitrarily high energy density of partons without structure because QCD bound states (hadrons) could not form, down to a much lower temperature where, in turn, no free partons could exist, but hadrons. The partons exhibit a 'horror vacui' according to QCD that results from the fact that the force-mediating gluons carry strong colour charges themselves; emitted from a quark they have to be absorbed by another quark. QCD tells us that the quarks have to stay 'confined', at low energy density, within colour-neutral bound states, the hadrons. Indeed, the cosmic evolution has left us with 10^{78} protons and neutrons but free quarks have never been observed. Thus, there must have occurred a phase transition from partonic to hadronic matter, at some intermediary density. We cannot read the corresponding critical temperature directly off the present cosmological matter.

Two considerations might give insight before starting with experiments. As hadrons must form in a phase transition from quark matter plasma to bound quark clusters, i.e. in some form of condensation, we should expect that the protons interior energy density should closely resemble that of the medium they condense from. Now we know that the proton interior energy density amounts to about 1 GeV per cubic Fermi. Second, the CERN physicist R. Hagedorn had demonstrated long before the advent of QCD that the hadronic world must have a temperature limit, of about 165 MeV. Remarkably, in a gas of partons one could estimate that the energy density at this temperature roughly corresponds to the above $1\,\mathrm{GeV/fm^3}$! Now one infers from the Einstein–Friedman equations, long known to describe the space–time development, that this density corresponds to the cosmological period in the early microsecond time domain. One cubic metre of matter at this time had ten times more mass than the entire Nanga Parbat mountain range.

2. QCD Matter Research: Gaining Confidence

The perspective of a phase transition in QCD matter, sketched above, reflects the state of theoretical physics at around 1976 when Steven Weinberg[1] wrote his famous book 'The First Three Minutes'. Its last chapter addresses the early stages of the cosmological expansion, and it is his point that the universe was a thermal fireball evolving in thermodynamic equilibrium. Hot, thermal QCD had just been developed by Shuryak,[2] Baym[3] and Kapusta,[4] introducing the name 'Quark–Gluon Plasma' (QGP) for a free parton 'gas', following a first suggestion by Collins and Perry.[5] They had referred to the just-discovered[6] property of QCD to become a weak interaction at extremely high energy densities (a property called 'asymptotic freedom'; we shall return to this later on), which could not hold together the bound hadrons. This idea turned out later to miss the real point of the QCD phase transition. But the first sketches of a QCD matter phase diagram, in terms of the variables temperature and density, had already been given.[4] And, most importantly, there also appeared to be an experimental avenue to investigate hot, dense nuclear matter, perhaps up to, and beyond the hypothetical phase transition:

Collisions of heavy nuclei (so-called 'Heavy Ion' collisions) at relativistic energy could compress and heat the initial nucleonic matter, up to, and beyond the critical energy density of QCD, thus estimated.

In fact, concurrent experiments at the Berkeley Bevalac (a linac injecting nuclei into the Bevatron synchrotron) had shown by 1980 that, at the prevailing modest energies in the low GeV per nucleon range, the two intersecting nuclear matter spheres stopped each other down creating a fireball at the centre-of-mass coordinates. The major part of the initially longitudinal beam energy was trapped in it, leading to heating and compression. Of course we are not in cosmology here, nuclei are small and the fireball will disintegrate fast, so the crucial issue of thermal equilibrium attainment needed attention. Can we apply QCD thermodynamics, or hydrodynamics to the fireball? Phase transitions require a certain minimum ('relaxation') time to be completed. How fast does the fireball re-expand? Will thermodynamic processes leave a trace in the eventually emitted hadrons, or will these look like a trivial superposition of elementary nucleon collisions?

The main question: could one define observable properties that should reveal characteristic stages or processes such as a typical plasma radiation, or characteristic changes of the initial 'partonic inventory' unambiguously due to cooking through a partonic plasma fireball? Toward the early 1980s some of these questions had found first affirmative answers, however tentative, from extrapolating the results obtained at the Bevalac were collective processes such as hydrodynamic flow of hadronic matter had been clearly observed. The most important step, however, came from the progress of thermal QCD theory solving the unsurmountable mathematical problems of low energy, 'non-perturbative' QCD numerically, in a lattice approximation. The existence of a phase change between hadrons and partons was shown, for the first time, with a plot by the US lattice group[7] showing the specific heat capacity of QCD matter as a function of the temperature. It showed a dramatic, steep upward jump at some critical temperature T_c. This signals the massive increase in the number of degrees of freedom, that should indeed take place when nucleons decompose into their constituent quarks at the phase boundary. Recall that a nucleon consists of three bound quarks which, moreover, each carry one out of three different colour charge units. They become the new set of degrees of freedom in a deconfined QCD plasma state. And, equally sensational, the critical temperature was found to be about 170 MeV (initially with considerable uncertainties), essentially equal to Hagedorns former upper boundary of matter composed of hadrons!

After these and other theoretical lessons learned, scientists got encouraged to sketch a universal phase diagram of strongly interacting matter,[8] as shown in Fig. 1. Looking at this plot one is reminded of the fact that a high energy density can be achieved, not only by heating but also by compressing. Both can, alone or in conjunction, drive QCD matter to the phase boundary. Thus the left hand part of the diagram refers to hot Big Bang dynamics and Quark–Gluon Plasma, the right hand side to gravitationally recompressed cold matter in supernovae, neutron stars, perhaps black holes. Are there quark stars?

3. Hot QCD Matter Research at CERN

At such heights of fascinating theory and speculation one MUST turn to experiments! Extrapolating from the Bevalac experiences it became clear that nuclear

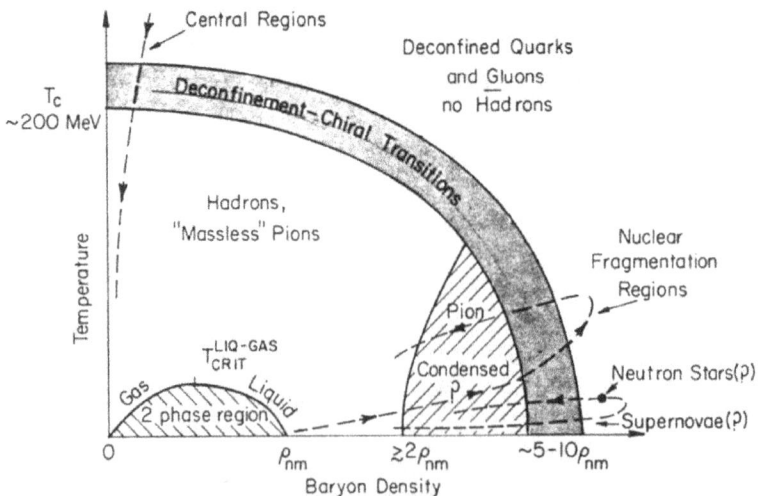

Fig. 1. An early sketch of the QCD phase diagram.[8]

projectile energies in the range of 10 to 100 GeV per nucleon would be required in fixed target experiments at synchrotrons, in order to reach, and surpass the predicted QCD phase boundary. Clearly a task for synchrotron facilities such as the PS and SPS of CERN, and for the AGS at Brookhaven National Laboratory. First research proposals were thus directed at these laboratories in the period from 1982 to 1986. Concurrently, however, more far-reaching concepts were first formulated, aiming at nucleus–nucleus collisions at much higher centre-of-mass (CM) energies that could be reached by colliders. Thus the idea was born to finish the construction of the temporarily abandoned superconducting ISABELLE collider project at Brookhaven[9] which would reach up to ten times the cm energy per nucleon attainable at the CERN SPS. Also the Large Hadron Collider (LHC) at CERN appeared already on the horizon which, equipped with nuclear projectiles would reach up in energy by yet a further factor of 25. Unlimited future research opportunities were coming into view, and there occurred another phase transition as the number of interested scientists jumped up by more than an order of magnitude, from the modest Bevalac beginnings. And, in fact, ALL these projects did materialize in the following 30 years! Nuclear collision experiments started taking data in 1986 at the CERN SPS, in 1994 really heavy nuclear projectiles like Lead (^{208}Pb) became available at the SPS, in 2001 at the newly built Relativistic Heavy Ion Collider, RHIC at Brookhaven, and in 2010 at the CERN LHC collider.

Ever higher collisional energies, and the accompanying relevant new physics observables have resulted in an overwhelming increase of experiment size, and complexity. We will turn below to a sketchy history of the corresponding physics ideas but wish, for now, to catch a typical glimpse at the developing dimensions of the experiments at CERN. To this end we show in Fig. 2 the layout of NA35, one of the first SPS experiments, and confront it in Fig. 3 with a sketch of the contemporary ALICE experiment at the LHC. The size of the international collaborations that have constructed these experiments, with significant help of CERN work force and funding, went up from about a hundred to more than a thousand physicists. Take the two experiments, illustrated above, for a typical example. NA35 was planned from 1981 onwards, and took SPS data from 1986 to 1992. It was constructed by about 80 physicists. The plans for the ALICE LHC experiment where first drafted in 1991, to be continually improved within the course of a long construction period that significantly changed its initially proposed instrumental techniques, to take first data at the onset of LHC experiments at the end of 2009. It united more than thousand physicists, and is planned to stay data taking for another decade, not to mention the to-be-expected final evaluation period. Such experiments require an unprecedented span of continuous preservation of instrumental expertise, information storage integrity, and a stability, over decades, of the intellectual pursuit of the scientific goals. This latter aspect of 'big science' experiments is, perhaps, one of the most interesting sociological examples of human intellectual cooperation.

The most immediate consequence of the increase in collisional cm energy, from 17.3 GeV at the SPS to (presently) 2.76 TeV at the LHC, is a rising number of

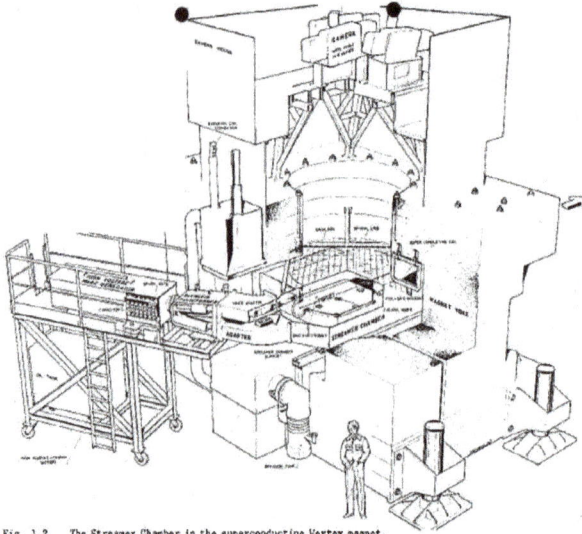

Fig. 1.2   The Streamer Chamber in the superconducting Vertex magnet.

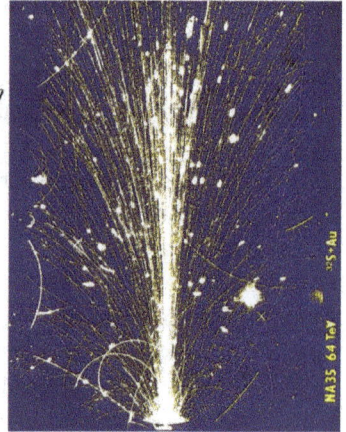

Fig. 2.   NA35 experiment (left) and event display (right).

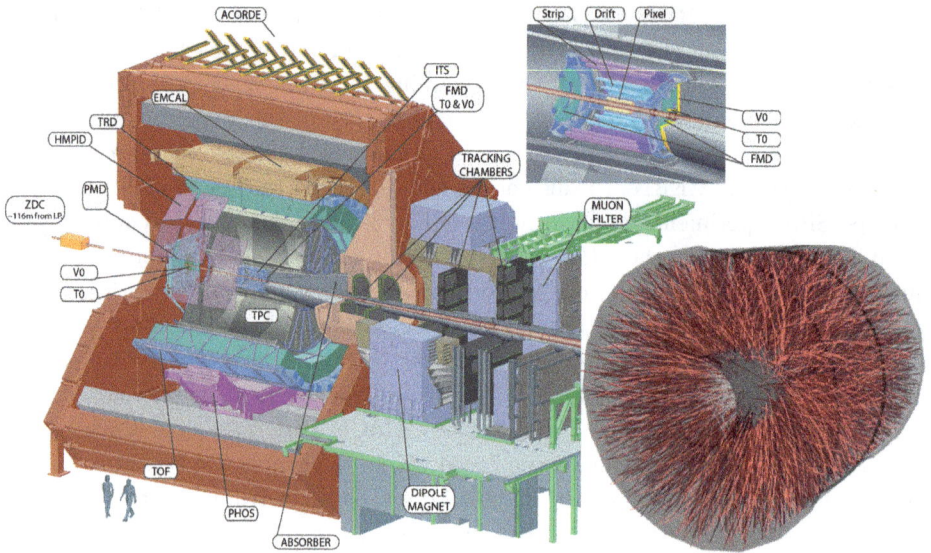

Fig. 3.   ALICE experiment (left) and event display (right).

created charged particles, from about 2000 to about 20000. These need to be resolved by tracking in a magnetic field, thus measuring their momenta, by track curvature, and their ionisation intensity in the gas, along the track. The modern detector of choice is the Time Projection Chamber (TPC), employed in SPS NA49, in RHIC STAR, and LHC ALICE. Particle identification (there are pions, kaons, protons, electrons, etc.) is completed by an outer shell of detectors that measure each particle's velocity. The tenfold rise in the number of charged particles produced in head-on collisions of Lead nuclei requires a larger tracking volume, and a much higher readout granularity, in the ALICE experiment. Moreover, the physics focus has shifted from SPS to LHC where the so-called 'hard QCD processes', like jet production, came into the centre of attention. The involved high momentum hadrons, from tens to hundred of GeV/c, require higher tracking accuracy than the 'thermal' hadrons in the former SPS experiments. This resolution is proportional to BL^2 with B the magnetic field strength and L the tracking length inside the field. The incredible complexity of the tracking tasks is also illustrated in Figs. 2 and 3. The former shows the track information in a single ^{32}S+Au collision at the top SPS energy, 200 GeV per projectile nucleon, which is about 20 GeV per nucleon pair in the cm frame. The figure shows the tracks in the NA35 Streamer Chamber, a 3D photographically recorded gas detector. Its somewhat archaic tracking technique was replaced 1994 by the Time Projection Chamber (TPC) in the NA49 experiment succeeding NA35. The yet far higher tracking effort at ALICE (with 2.76 TeV per nucleon pair cm energy collisions) is illustrated in Fig. 3; this TPC has over 600000 electronic channels, producing a raw data flux of about 20 Gbyte per collision event. At 1 kHz event frequency the TPC yields 20 Tbytes per second of raw data, clearly at the upper end of todays digital electronics capacity.

Let us note right here that a multiplicity of several thousand hadrons does not imply a dull accumulation of 'more of the same': there are subtle forms of global structures in each event, caused by a hydrodynamic collective expansion of partonic and hadronic QCD matter. This gives specific momentum kicks to all particles which are small individually but can be well quantified if all particles of an event are recorded, exhaustively. Furthermore, the very high multiplicity of charged particles, produced in a single event, results in a completely new phenomenon: the event originates from a single quantum mechanically coherent process but it can be individually analyzed with statistical significance. Single events become self-analysing.

3.1. *The acceleration of heavy nuclei at CERN*

Synchrotrons normally accelerate protons or electrons, as well as their antiparticles. The protons have a charge to mass ratio of $Q = 1$. Stable nuclei have $Q = 0.5$ up to Calcium (charge 20, mass 40), but exhibit an excess of neutrons over protons from there on such that $Q = 0.39$ for ^{208}Pb. The acceleration rate diminishes with lower Q; a proton synchrotron with 450 GeV top proton energy (the SPS) delivers Pb beams of up to 175 GeV per nucleon only, but stable acceleration modes can be accommodated. The problem is that the projectiles travel for about a million kilometres during acceleration, through the finite vacuum in the acceleration cavity. They will change their ionic charge state by stripping in the dilute gas, thus getting lost unless they are totally ionised already at injection, or if the vacuum is of extraordinary perfection. For proton acceleration this is of no concern, and thus synchrotrons tend to have a modest vacuum. They accept completely stripped ions only. The name 'Heavy Ion Physics' initially given to the field reflects the fact that fully stripped ions are needed, which can only be achieved in a complicated, multi-step pre-acceleration system. At its beginning a special ion source is required which delivers partially stripped ions out of an atomic plasma generated by strong electric and magnetic fields. It should give as high charge states as possible, and yet sufficient ionic current to be able to maintain beam stability in the subsequent synchrotrons, and to do experiments with sufficient event rate. During the early planning stage of CERN heavy ion research, 1982 to 1985, the preferred source type was the Electron Cyclotron Resonance (ECR) source.[10] The first nuclear beams were Oxygen in 1986, a light nucleus, swiftly followed by Sulphur in 1988. A much more elaborate acceleration scheme was employed for Lead ions in the SPS that came into operation in 1994. With some modifications it is still used today at the LHC.

A new high power ECR source produced Pb ions with charge states up to 20+. After charge state analysis this beam was accepted by the tanks of a newly built Linac which accelerated up to about 5 MeV per nucleon. Passing the beam through a thin stripper foil at the corresponding velocity produces a broad distribution of charge states around 50+. and the selection of a single charge state results in a 90% loss of beam intensity at this stage. Now injecting into the first synchrotron, the so-called PS-Booster, one loses another large fraction of the precious beam because the synchrotron gets injected (at bottom field) for a few tens of microseconds only, and then again at the next acceleration cycle. The Booster vacuum was improved to 10^{-9} Torr to minimise charge state changes. At the energy of 150 MeV per nucleon reached at extraction the Pb ion still cannot be fully stripped, and a further grave intensity loss would result from spreading over several charge states. Thus the vacuum of the Proton Synchrotron (PS), the next element in line, had also to be improved to 10^{-9} Torr, while still loosing 50% of the beam. The PS then extracts Pb(50+) at about 7 GeV per nucleon; stripping now produces fully ionised Pb(82+) nuclei for SPS injection. The PS has an acceleration cycle of one second duration whereas the SPS requires more than 15 seconds at top energy. The accelerator scientists thus can employ a complicated multi-turn injection technique at this stage, the SPS accommodating four successive PS extractions before its acceleration cavity is full of precisely positioned beam bunches; then acceleration begins. The fourfold intensity gain far outweighs the slightly lengthened overall cycle duration. A final SPS energy of 158 GeV per nucleon was adopted for Lead projectiles. This outstanding facility could also accelerate all lighter elements compatible with the given ECR source technique, and operate over a wide range of energy, from about 15 to 160 GeV/A. Completed in 1994, it faced no competition worldwide until the turn-on of the Brookhaven RHIC collider, which offered about ten times higher energy in 2000.

3.2. The CERN SPS experiments and their physics

The first proposal[11] was submitted to CERN in 1982, by a GSI–LBL–Heidelberg–Marburg–Warsaw Collaboration, of research groups established in nuclear physics, and concurrent Bevalac or Dubna Synchrophasotron engagement. It called for the establishment of an extracted CERN PS beam of Oxygen where GSI would purchase an ECR source from the Grenoble group of R. Geller, and LBL would construct an RFQ micro-linac, then to inject the existing Linac1 of CERN, followed by Booster and PS, extracting at $13\,GeV/A$. It was proposed to perform two parallel experiments, based on experience with two concurrent Bevalac experiments, the multi-segmented scintillator Plastic Ball[12] and the visual tracking Streamer Chamber spectrometer.[13] The former would investigate hydrodynamic nuclear matter flow, the latter meson production with a look at phase transition signals.

This initial proposal was accepted by CERN, but the accommodation of further experiments in the East Hall PS extraction area met with substantial difficulty. Of seminal consequence was then the suggestion[14] of CERN management to transport the PS beam to the SPS and distribute the resulting beams, at $200\,GeV/A$, via the external SPS beam line system to the then little used, huge experimental halls in the North and West of the SPS, where the former SPS proton beam experiments had been conducted. This idea catalysed the much more forward-looking idea of a full-fledged SPS heavy ion acceleration program, with beam energies ranging up to $200\,GeV/A$ which was enthusiastically welcomed as it also met with the intentions of several CERN experiment groups to establish a continuation of the formerly abandoned initiatives toward heavy ion experiments at the CERN ISR collider. A wealth of established experimental infrastructure was available here and, most importantly, three intact experiments, with still existing physics collaborations, as well as technicians: the huge magnetic hadron spectrometer OMEGA, the dilepton spectrometer from experiment NA10, and an almost complete Streamer Chamber plus calorimeter experiment, with a 400 ton superconducting dipole magnet, from experiments NA5 and NA24. The turn to the SPS really was a strike of genius!

The anticipated SPS research programme of CERN attracted further groups from nuclear physics but also a fraction of the particle physics groups working at CERN already. The Omega spectrometer group reshaped as WA85, the dilepton spectrometer as NA38, the Streamer Chamber experiment as NA35, the large calorimeter experiment NA34 (formerly engaged in a CERN ISR study of ^{4}He collisions — a precursor of the SPS programme[15]) became NA34-2. The Plastic Ball spectrometer moved from LBL and was amended with Lead glass electromagnetic calorimetry, to become WA80. Initially, only experiment NA45 was completely newly constructed, a double Cherenkov (RICH) magnetic spectrometer for dielectron spectroscopy. The culminating part of the programme was carried out from 1994 to 2002, with Lead nuclei, also including lower energies, 20, 30, 40 and $80\,GeV/A$. This setup was reactivated in 2005 with Indium (^{115}In) beams for NA60, a high precision charmonium and di-muon spectrometer constructed on the basis of former NA38 and NA50, and with Lead beams for NA61, a large acceptance hadron spectrometer based on NA49. We list below the experiments from the main Lead beam programme:

- NA44: Small angle focusing magnetic spectrometer for antiprotons and kaons.
- NA45: Double Cherenkov ring imaging magnetic spectrometer for di-electrons.
- NA49: Large acceptance TPC and calorimeter spectrometer for all hadrons.
- NA50: Magnetic di-muon spectrometer, EM calorimeter, for vector Mesons.
- NA52: 'NEW MASS' beam line spectrometer looking for strangelets.
- WA97/NA57: Hyperon and antihyperon spectrometer with Si pixel technique, WA85/WA97/NA57.
- WA98: Large acceptance hadron and photon spectrometer for direct photons.
- NA60 and NA61 followed later, as stated above.

4.   Results at the Millenium

We shall briefly sketch below a (subjective) selection of seven physics observables, emerging from the SPS programme and representing the increasing understanding of QCD matter and QCD phases. In this section we have to ask the reader for some patience because we have to turn to a more detailed physics argumentation. The experiments addressed mostly 'to be or not to be' questions. A quantitative description of the QCD plasma has resulted from the next following research era, at the colliders RHIC at Brookhaven and LHC at CERN, as we will show at the end of this article. Now let us take a closer look at the main topics of SPS research.

4.1.   *Fireball energy density*

Calorimeter experiments (NA34, NA49, WA98) measured the total transversal energy produced in head-on collisions of two ^{208}Pb nuclei.[16] Central, head-on collisions fall into the tail of the distribution. Here one has, on average, about 190 participating nucleon pairs from the initial target-projectile nuclear density distributions. So we know the collision geometry, the total initial energy in the cm system, and the newly created transversal energy. A formula derived by Bjorken[17] provides for an estimate of the corresponding energy density in the primordial fireball volume. In the present case it results in 3.0 ± 0.6 GeV per cubic Fermi. Comparing to a 'year 2000' view of the parton–hadron phase transition from Lattice QCD[18] one finds that we are just above the phase transition at this energy density, and that the critical QCD energy density is about 1 GeV per cubic Fermi.

4.2.   *Fireball temperature*

Lattice QCD also gives a plasma temperature estimate, prevailing at 3 GeV/fm cubed energy density, of $T = 210$ MeV. The photons from thermal plasma radiation should escape from the fireball unaffected because they lack strong interaction. A first measurement[19] was undertaken by WA98 of the so-called 'direct photons', which result from a meticulous subtraction of the trivial photon fraction resulting from electromagnetic decay of neutral mesons, that occurs much later. These data can be described[20] with a plasma temperature in the 200–250 MeV domain, in accord with the Lattice QCD temperature estimate given above.

4.3. *Hadrons form at $T = 160 \pm 10$ MeV: Close to lattice QCD prediction*

If a primordial QCD parton plasma is formed in central Pb+Pb collisions at top SPS energy, subsequent expansive cooling will bring the fireball volume back down to the QCD parton–hadron phase boundary, where confinement enforces hadron formation. If a critical QCD hadronisation temperature uniformly governs the fireball volume the various hadronic species will be simultaneously produced in proportion to their so-called statistical weights, a universal law that was discovered by E. Fermi and carries the name 'Fermis golden rule'. This is articulated in the Statistical Hadronisation Model (SHM) which predicts a universal yield order among the produced hadrons, with just only one essential parameter: the temperature T prevailing at birth of the hadronic final state.[21] Hadron multiplicity distributions were systematically measured by NA44, NA49 and WA57. Figure 4 shows an example[22] of an SHM fit to hadron multiplicities per collision event, observed at top SPS energy by NA49 in central Pb+Pb collisions. It gives a hadronisation temperature of $T = 158 \pm 5$ MeV, in close agreement with the critical temperature

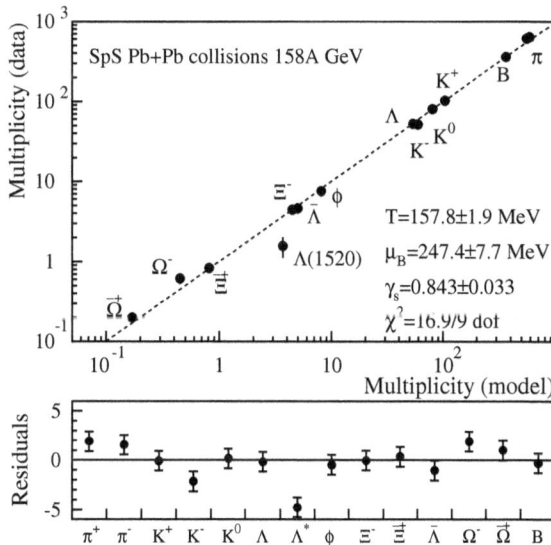

Fig. 4.   Hadron multiplicities vs. prediction of the Statistical Model.

T_c predicted by Lattice QCD. At this low temperature, we are very far away from QCD asymptotic freedom.[5, 6] The transition from deconfined to confined QCD matter must be driven by other, genuinely non-perturbative QCD mechanisms.

4.4. *Strange baryon and antibaryon production is enhanced*

The 'cooking' in an extended, thermal QCD matter state enhances the production rate of strange hadrons, in particular of the strangeness 1, 2 and 3 carrying so-called hyperons (Lambda, Xi and Omega as well as their antiparticles). This observation, made in nucleus–nucleus collisions, was called strangeness enhancement.

Figure 5 shows a measurement from NA57.[23] Pb+Pb collisions at top SPS energy are considered here, and the various hyperon and anti-hyperon multiplicities are observed in different, successive windows of collisional impact geometry, as monitored by the (simultaneously determined) number of nucleons participating in the collision, that increases toward central, head-on collision geometry. The fireball volume is growing with N(part). The yields are shown here as multiplicity per participant nucleon number, N(part). Moreover, they are normalised relative to the yields per participant observed in proton–nucleus collisions where no QCD plasma is expected to form. If the fireball outcome was merely a trivial superposition of elementary nucleon–nucleus collision multiplicities (here represented by p+Be and p+Pb results at similar energy) the hyperon production rate should simply

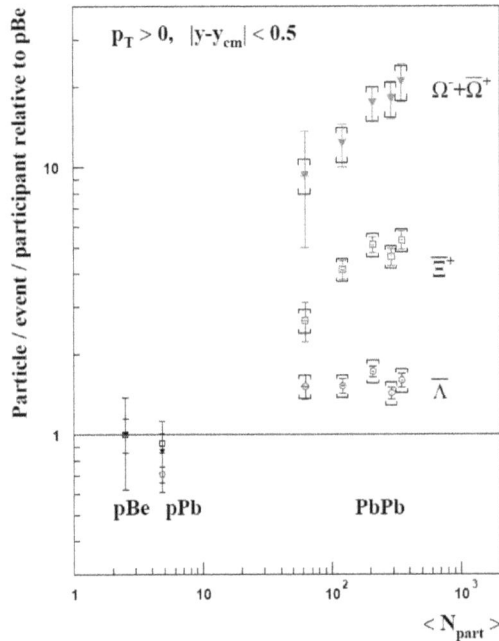

Fig. 5.   Strangeness enhancement in Pb+Pb hyperon yields.

stay constant. However, on the contrary, we observe an increase of the relative yields with Pb+Pb centrality, by up to above a factor of 10, for Omega and anti-Omega hyperons. Pb+Pb collisions are not a trivial superposition of p+A collisions, and strangeness enhancement grows with increasing fireball volume, but saturates toward central collisions. The fireball volume is seen to act collectively, as implied by the grand canonical (large volume limit) Statistical Hadronisation Model (Fig. 4).

4.5. *Charmonium (J/Ψ) suppression reveals QCD plasma formation*

J/Ψ vector mesons are called charmonia. They arise from ccbar, charm–anticharm quark pairs that are produced by 'hard' parton collisions in the initial phase of an A+A collision. If, subsequently, a deconfined QCD plasma phase governs the dynamical evolution, the primordial c–cbar pairs co-travel with the medium and can break up by QCD Debye screening of the interquark colour exchange force,[24] instead of evolving into the final charmonium states. As a result, J/Ψ production in A+A collisions will be suppressed by QCD plasma formation. Figure 6 shows the experimental verification by experiments NA38 and NA50. J/Ψ production is illustrated for a number of different reactions at various centralities.[25] The data can be represented on a common scale, the fireball energy density introduced previously, which increases monotonically with increasing number of target and projectile participants. We would now go on simply plotting the respective yields

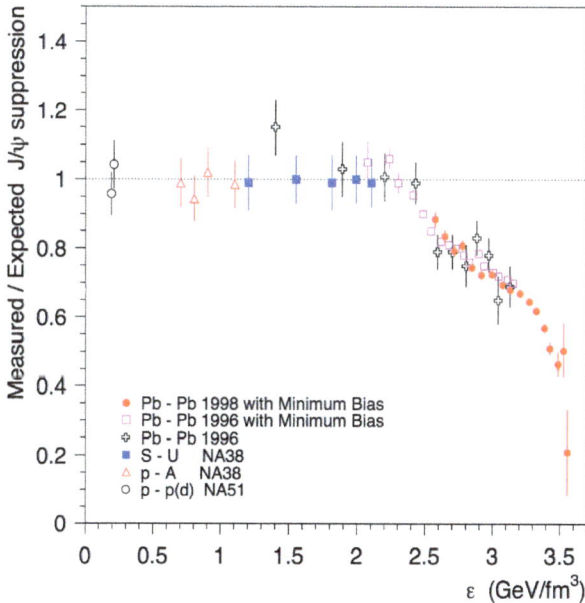

Fig. 6. J/Ψ suppression vs. energy density at the SPS.

per participant, to find a suppression uniformly increasing with energy density: J/Ψ production gets suppressed toward central collisions, the deconfinement signal we wanted to quantify. However, there also occurs an absorption of ccbar pairs in ordinary, cold nuclear matter. It gets quantified by measuring p+A collisions at similar energy, and a complicated method is employed to simulate the effect of this 'normal' absorption cross-section in an A+A collision geometry (as if it were cold). Figure 6 then, finally, shows the real J/Ψ yield relative to a constructed yield based on the normal cold absorption only. One calls this ratio the 'anomalous suppression of J/Ψ'. Figure 6 shows that this suppression gets stronger with increasing energy density: an ever smaller fraction of the initially produced c–cbar pairs find their way to become final hadrons because they dissolve in the plasma.

4.6.　QCD chiral symmetry restoration: Hadrons melt near T_c

What becomes of the large hadron mass, and of their quark 'wave functions', once hadronic matter approaches the QCD phase boundary? The plasma quarks are massless in QCD, their small masses, in the low MeV range for the up and down quarks out of which hadronic matter consists, stem from the Higgs meachnism QCD has the property of chiral symmetry, a pecularity of massless particles propagating at the speed of light: if their spin turns them left or right handed with respect to the direction of their momentum they will stay that way forever. This symmetry is completely lost ('broken') in hadrons where quarks dress up with massive vacuum polarisation clouds. This transition can be observed. Does it coincide with the QCD deconfinement process occurring at the critical temperature T_c? That is, will we see 'dissolving' hadron wave functions near T_c? All neutral, non-strange mesons decay to electron and/or muon pairs. Their intensity and invariant mass should reveal a melting of the mesonic wave functions. The spectroscopy of lepton pairs was pioneered by NA45, and perfectioned (for muon pairs) by NA60. One constructs an 'invariant mass' spectrum from the observed momentum spectra of both leptons, in which each meson species creates a peak at its proper rest mass, of characteristic height, and width. A simple superposition of the known in vacuum dilepton decays (called the 'cocktail') should fail to reproduce the observed invariant mass spectrum if the fraction of overall dilepton decays of mesons that is created in the immediate vicinity of the deconfinement phase transition temperature exhibits effects of the concurrent chiral phase transition of QCD. We show the excess of the data over the cocktail prediction in the $\mu^+\mu^-$ invariant mass spectrum obtained by NA60[26, 30] in Fig. 7. It was measured in In+In collisions at top SPS energy. The excess yield up to about $1\,\text{GeV}$ is attributed to emission from the immediate vicinity of T_c. There it is dominated by an in-medium process which is strongly enhanced at the high matter density, prevailing here: two pion annihilation $\pi^+ + \pi^-$ to an intermediate rho vector meson that decays to $\mu^+\mu^-$. The rho turns out to be very strongly broadened in the fireball medium near T_c, thus accounting for the

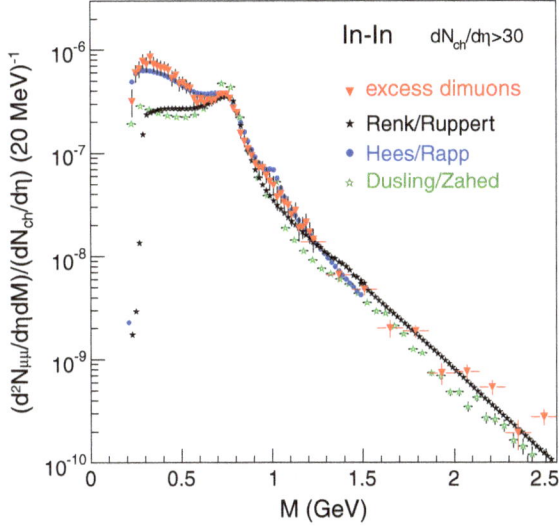

Fig. 7.   Di-muon invariant mass spectrum of the excess yield in In+In, over scaled cocktail.

excess of the data over the cocktail expectation. A first indication of QCD chiral symmetry restoration occurring at the phase boundary to partons. Furthermore, the exponential tail in Fig. 7, upward from about 1 GeV, reveals the contribution of quark–antiquark annihilation to lepton pairs, from the preceding plasma phase. The Planck-like spectrum reflects an average plasma temperature of about 220 MeV (Ref. 30): the first "direct" plasma signal at SPS energy.

4.7.   *The fireball matter exhibits collective hydrodynamic flow*

Consider the very initial stage of a Pb+Pb collision, not head-on but at a finite impact distance between the trajectories of the impinging nuclei (called a semi-peripheral collision). Looking along the incoming projectile we see the overlapping areas of the target and projectile density distributions, as projected onto the plane transverse to the projectile direction. The projectile will carve out an ellipsoidal sector from the target, and a correspondingly shaped fireball will develop next. In it the created energy density falls down faster along the impact vector direction (in the so-called reaction plane) than perpendicular to it. Thus the expansion pressure is higher in the reaction plane, leading to higher expansion momenta of matter emitted in-plane, as compared to the out of plane direction. As a consequence, the initially partonic, and subsequently hadronic expansion pattern will acquire a spatial anisotropy. If a collective, hydrodynamic outward flow of matter sets in at this primordial time it will preserve this particular anisotropy signal. From its origin in an elliptically shaped fireball it is called 'elliptic flow'. If the in-medium viscosity is small the initial anisotropic expansion pattern is carried on to a finally observed hadronic emission anisotropy, as quantified by the second harmonic coefficient ν_2 of a Fourier decomposition of the final, collective hadronic emission pattern. Its measurement by NA49[27] observed both pion and proton ν_2 in semiperipheral Pb+Pb collisions at top SPS energy. Fireball partonic matter flows collectively, in its expansion, much like a liquid, to be described by QCD hydrodynamics. And, moreover, this liquid appears to have rather little dissipative viscosity, or else the primordial pressure anisotropy would not survive in hadrons, emitted much later, after fireball expansion. A topic of very high significance as became obvious in the course of later, more detailed investigations at RHIC and LHC. We shall return to it below.

4.8. *Summary of SPS results and interlude at RHIC*

An appraisal of the SPS programme was made just prior to RHIC turn on, in 2000,[28] based on a 'common assessment' of the results collected and published over the preceding years. It concluded that *'compelling evidence has been found for a new state of matter, featuring many of the characteristics expected for a Quark–Gluon Plasma'*.[29] This conclusion was based primarily on three of the experimental observations mentioned before: the copious production of hadrons containing strange quarks ('strangeness enhancement'), the reduced production of J/Ψ mesons ('anomalous J/Ψ suppression'), and the yields of low mass lepton pairs ('rho melting'). These three signals most closely resembled the predicted hallmarks of the QGP, namely thermalisation, deconfinement, and chiral symmetry restoration.

The experimental results have all stood the test of time and have been amply confirmed and refined in the years thereafter. The essence of the assessment however, seems in hindsight today more compelling than in 2000, given for example much improved low mass lepton pair results from the SPS NA60 experiment,[30] which started taking data only in 2005, and new results and insights gained from the RHIC and LHC programmes which are described below.

In the summer of 2000, the core of heavy ion activity shifted from CERN to BNL, when the dedicated Relativistic Heavy Ion Collider started operation with Au+Au collisions at 130 GeV cm energy. Reaching its design energy of 200 GeV the year after, RHIC stayed at the energy frontier for the next decade until the arrival of LHC. While a detailed description of RHICs scientific legacy is beyond the scope of this article, a short summary of the main highlights is given below to set the stage for the next chapter at CERN with the LHC.

The initial results from RHIC were summarised and assessed in 2005, based on a comprehensive (re)analysis of the first few years of RHIC running.[31] The experiments concluded that at RHIC *'a new state of hot, dense matter'* was created *'out of the quarks and gluons... but it is a state quite different and even more remarkable than had been predicted'*.[32] Unlike the expectation, with hindsight overly naive, that the QGP would resemble an almost ideal gas of weakly coupled quarks and gluons, the hot matter was found to behave like an extremely strongly interacting, almost perfect liquid, sometimes called the sQGP (where the 's' stand for 'strongly interacting'). It is almost opaque and absorbs much of the energy of any fast parton which travels through — a process referred to as 'jet quenching' — and it reacts to pressure gradients by flowing almost unimpeded and with very little internal friction (i.e. has very small shear viscosity).[33] The shear viscosity over entropy ratio, η/s was found to be compatible with a conjectured lower bound of $\eta/s \geq 1/4\pi$ ($\hbar = k_B = 1$), a limiting value reached in a very strongly interacting system when the mean free path approaches the quantum limit, the Compton wavelength.

Also at RHIC, the crucial experimental results as well as the inferred characteristics of the QGP — a 'hot, strongly interacting, nearly perfect liquid' — stood the test of time.[37]

5. Heavy Ion Physics at the LHC

Prior to LHC, 25 years of heavy ion experimentation had already revealed a **'QGP-like'** state at the SPS and not **'the QGP'**, but **'a sQGP'** at RHIC. With the discovery phase considered to be essentially over,[34] a main goal for the heavy ion programme at LHC was increased precision to better characterise this new state of matter, making use of the particular strength of the LHC, i.e. the huge increase in beam energy and a powerful new generation of large acceptance state-of-the-art experiments. These include the dedicated heavy ion detector ALICE,[35] as well as the general purpose pp experiments ATLAS and CMS, which both participate fully in the heavy ion programme, and finally LHCb, which joins in for p–nucleus collisions. The larger cross-section for hard probes and the higher particle density at LHC creates a QGP which should be 'hotter, larger, and longer living'. And indeed, LHC made significant progress towards increasing the precision on shear viscosity (Section 5.2) and plasma opacity (Section 5.3) already during the first two years of ion running.[37, 38] However, when dealing with QCD in the non-pertubative regime, surprises should not come as a surprise, and a number of unexpected findings at LHC have helped shed new light on some old problems or issues, for example on particle production (Section 5.1) and J/Ψ suppression (Section 5.4). And finally the very first discovery made at LHC was the appearance of a mysterious long range 'ridge' correlation in high multiplicity pp reactions (Section 5.5). It reappeared later — and much stronger — in the 2012 p–nucleus run, making it of great interest, and presumably of great relevance, to hot and dense matter physics, even if it is ultimate cause and connection to similar phenomena in nuclear collisions is as of today not finally settled.

5.1. *Hadron formation*

Measuring identified particles at LHC was considered a somewhat boring but necessary exercise, as finding thermal particle ratios essentially identical to the ones measured at SPS and RHIC (save expected differences related to the ratios of particles and antiparticles) was thought to be one of the safest predictions.[36] It therefore came as a surprise when some particle fractions, in particular for the mundane proton, one of the most frequently produced hadrons, were found to differ considerably from expectations (and, to a lesser extent, from the ones measured at RHIC), while others, including those for multi-strange hyperons, were well in line with thermal predictions.

Possible reasons being discussed range from mere adjustments of thermal model parameters, over the consideration of hitherto neglected final state interactions and sequential freeze-out of individual hadron species, to the consideration of different transition temperatures for different quark flavours. The final resolution of the 'proton puzzle' is still outstanding and will probably require a more complete set of particle ratio measurements at LHC as well as revisiting the RHIC results to confirm with better significance if particle ratios in central nuclear collisions indeed evolve with energy. Whichever explanation will finally prevail, the unexpected LHC results are a welcome fresh input likely to advance our understanding of the remarkable success of the statistical model of hadron production.

5.2. *Elliptic flow*

The observation of robust collective flow phenomena in heavy ion reactions at fixed target energies and at RHIC is the most direct evidence for the creation of a strongly interacting, macroscopic (i.e. large compared to the mean free path) and dense matter system in nuclear collisions. Analysed in terms of a Fourier expansion of the azimuthal charged particle density $dN_{ch}/d\phi$ with respect to the reaction plane ($\phi = 0$), the first order component ($v_1 \propto \cos(\phi)$) is called directed and the second order component ($v_2 \propto \cos(2\phi)$) is called elliptic flow (recall Section 4.7). Matter properties like the equation of state, sound velocity or shear viscosity, can be extracted by comparing measurements and hydrodynamic model calculations of elliptic (i.e. azimuth dependent) and radial (azimuthally averaged) flow. Flow however depends not only on the properties of the hydrodynamic evolution but also on initial conditions, in particular the geometrical distribution of energy density within the primordial nuclear overlap zone. The resulting pressure gradients should thus reveal, in particular, possible effects of gluon saturation in the initial stage, as postulated in the Colour Glass Condensate (CGC) model.

When first azimuthal flow data from the LHC became available in early 2011, the evidence from all three experiments, as well as new results shown by the two RHIC collaborations, was overwhelming:[39] The collective flow patterns in heavy ion collisions were much more complex with measurable and significant Fourier coefficients up to at least sixth order ($v_1, v_2, \ldots, v_6$)! Today these patterns are understood to arise from fluctuations, event-by-event, of the initial geometry (i.e. pressure gradients) caused by the stochastic nature of nucleon–nucleon collisions and/or by a CGC initial state.

The complex correlation patterns had actually been strong and clearly visible since many years; however, before 2011, they were in general not recognised as hydrodynamic in origin but discussed in terms of fancy names ('near side ridge, away side cone') and fancy explanations ('gluon Cerenkov radiation, Mach cone, ... ').[40] At LHC, the large acceptance of the experiments, together with the high particle density (as a collective effect, the flow signal increases strongly with multiplicity) made the observation and interpretation straightforward and unambiguous.

The fact that energy density fluctuations on the scale of a fraction of the nuclear radius in the initial state are faithfully converted into measurable velocity fluctuations in the final state was a most amazing, and also most useful, discovery: One could not only identify the on average almond shaped collision zone, but recognise much finer structures of individual nuclear collisions. The analysis of flow has been invigorated and is advancing rapidly ever since,[41] with direct measurements of the fluctuation spectrum,[42] using event-by-event measurement and selection of flow as an analysis tool,[43] and even finding non-linear mode mixing between different harmonics.[44] Like temperature fluctuations in the cosmic microwave background radiation, which can be mapped to initial state density fluctuations in the early Universe, collective flow fluctuations strongly constrain the initial conditions and therefore allows a better measurement of fluid properties. Since 2011, the limit for the shear viscosity has come down by a factor of 2 ($\eta/s < (2-3)x1/4\pi$) and is now precise enough to even see a hint of a temperature dependence, slightly increasing from RHIC to LHC.[45] Future improvements in data accuracy and hydro modelling should either further improve the limit, or give a finite value for η/s. In either case, improved precision is relevant as the shear viscosity is directly related to the in-medium cross-section and therefore contains information about the degrees of freedom relevant in the sQGP via the strength and temperature dependence of their interactions.

5.3. Jet quenching

High energy partons interact with the medium and lose energy, primarily through induced gluon radiation and, to a smaller extent, elastic scattering.[46] The amount of energy lost, ΔE, is expected to depend on medium properties, in particular the opacity and the path length L inside the medium, with different models predicting a linear (elastic ΔE), quadratic (radiative ΔE), and even cubic (AdS/CFT) dependence on L. In addition, ΔE also depends on the parton type via the colour charge (quark versus gluon), the parton mass via formation time and interference effects (light versus heavy quarks), and finally somewhat on the jet energy. The total jet energy is of course conserved and the energy lost by the leading parton appears mostly in radiated gluons, leading in effect to a modified softer fragmentation function. Jet quenching (i.e. measuring the modified fragmentation functions) is

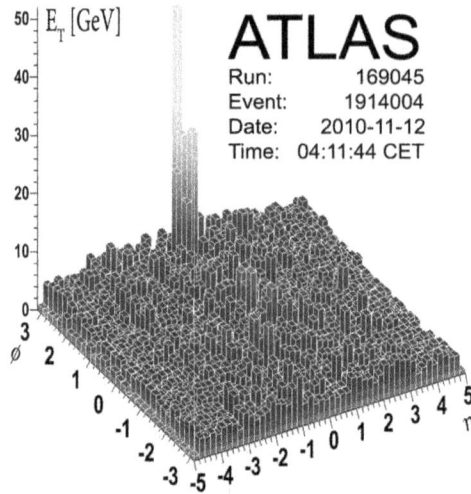

Fig. 8.   Calorimeter display of a very asymmetric (quenched) two-jet event.[47]

therefore a very rich observable which probes not only properties of the medium but also properties of the strong interaction.

Jet quenching was discovered at RHIC not with jets, which are difficult to measure in the high multiplicity heavy ion background environment, but as a suppression of high p_T 'leading' jet-fragments. The effect was experimentally very clean and significant with suppression factors up to five. The high energy of LHC and the correspondingly large cross-sections for hard processes make high energy jets easily stand out from the background even in central nuclear collisions (Fig. 8). Jet quenching is therefore readily recognised and measured, with many unbalanced dijets or even monojets apparent in the data.[47] While the amount of energy lost in the medium can be of the order of tens of GeV and therefore even on average corresponds to a sizeable fraction of the total jet energy, it is nevertheless close to the one expected when extrapolating RHIC results to the higher density matter at LHC. The two jets remain essentially back-to-back (little or no angular broadening relative to pp) and the radiated energy (ΔE) is found in very low p_T particles ($< 2\,\text{GeV}/c$) and at large angles to the jet direction.[48] The latter two findings were initially a surprise, but are now incorporated naturally into models where the energy is lost in multiple, soft scatterings, and the radiated gluons are emitted at large angles. The parton then leaves the matter and undergoes normal vacuum fragmentation, i.e. looking like a normal pp jet but with a reduced energy.

Additional insight into the energy loss process has come from heavy flavours.[49, 50] The suppression of charm mesons is virtually identical to the one of inclusive charged particles; a result which was counterintuitive and initially confusing. The similarity in the energy loss of gluons (the source of the majority of charged particles) and heavy quarks is now understood as an accidental cancellation between the difference in coupling strength (colour charge) of quarks and gluons and their different fragmentation functions. The mass effect however seems to be as predicted: At intermediate p_T, beauty shows less suppression than charm, whereas at very high p_T b-jets and inclusive jets show similar modifications.

5.4. *Quarkonium suppression*

While the 'anomalous' J/Ψ suppression discovered at the SPS was considered one of the strongest indications for the QGP, the RHIC results showed essentially the same suppression at a much higher energy, contrary to most expectations and predictions from both QGP and non-QGP models. These initially very confusing results kept the interpretation of this most direct signal for deconfinement ambiguous for the last ten years.

It had been suggested that J/Ψ suppression actually increases with energy (i.e. from SPS to RHIC), but is more or less balanced by a new production mechanism: upon reaching the parton–hadron phase boundary two independently produced charm quarks from the plasma hadronise along with the lighter quarks, forming J/Ψ.[51] And indeed, LHC data seems to have resolved the J/Ψ puzzle in favour of this coalescence picture:[52] as predicted, the large charm cross-section at LHC leads to *less* J/Ψ suppression at LHC compared to RHIC (Fig. 9). The suppression is also less strong at low p_T, where phase space favours recombination, in clear contrast to the opposite p_T dependence found at SPS and RHIC.

While at first sight, charm quark coalescence may appear as yet another process complicating and masking quarkonium deconfinement, it is actually a respectable

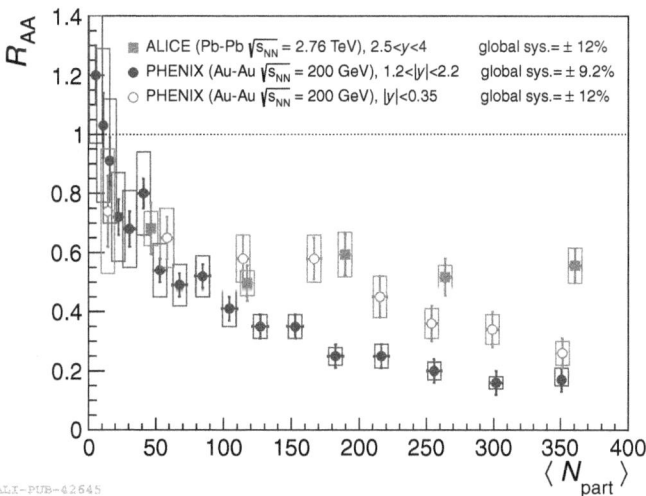

Fig. 9. J/Ψ suppression versus centrality at RHIC and LHC;[52] N_{part} is the number of participating nucleons which increases with increasing centrality (decreasing impact parameter).

and important deconfinement signal in itself: only in a colour conducting, deconfining medium can quarks roam freely over large distances ($\gg 1$ fm), and this is exactly what two charm quarks have to do in order to combine during hadronisation.

The magnitude of the suppression for different quarkonium states should depend on their binding energy, with strongly bound states such as the Υ showing less or no modification. LHC results for the Υ family[53] are fully consistent with the expectation for a deconfining hot medium in which quarkonia survival decreases with binding energy, i.e. in terms of suppression factors: $\Upsilon(3S) > \Upsilon(2S) > \Upsilon(1S)$. The $\Upsilon(1S)$ is suppressed by about a factor of two in central collisions, the $\Upsilon(2S)$ by almost an order of magnitude, and only upper limits have been measured for the $\Upsilon(3S)$. As only about 50% of the observed $\Upsilon(1S)$ are directly produced, these results may be compatible with almost complete melting of all high mass bottonium states and survival of a lone, strongly bound $\Upsilon(1S)$, which according to lattice QCD may melt only at temperatures far above the critical temperature.

5.5. *Discoveries*

The first discovery made at LHC was announced[54] in September 2010 on a subject which was as unlikely as it was unfamiliar to most in the packed audience: The CMS experiment had found a mysterious 'long range rapidity correlation' in a tiny subset of extremely high multiplicity pp collisions at 7 TeV.[56] The correlation in rapidity $\Delta\eta$ and azimuthal angle $\Delta\phi$ between all pairs of particles of intermediate p_T (1–3 GeV/c) in pp collisions is shown in Fig. 10 (left). Besides the so-called 'near side peak' at (0, 0), a feature arising from particle correlations within jets, and the 'away side ridge' at $\Delta\phi = \pi$ in azimuth, where the two particles come — one each — from the members of a pair of back-to-back jets, the correlation structure shows a small but significant second ridge also at $\Delta\phi = 0$. While in the meantime, far eclipsed by the discovery of a Higgs particle, this 'near side ridge' is arguably still

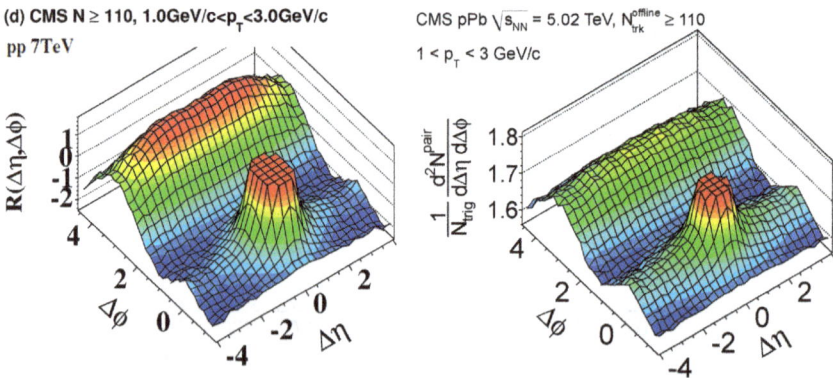

Fig. 10. Two particle correlation function in $\eta - \phi$ for high multiplicity pp collisions[56] (left) and pPb collisions[58] (right).

the most unexpected LHC discovery to date and spawned a large variety of different explanations.[57] The most serious contenders are saturation physics, as formulated in the Colour Glass Condensate model (CGC),[55] and collective hydrodynamic flow. Hydrodynamics is of course a very successful framework to describe long range correlations in the macroscopic hot matter created in heavy ion reactions, but was not supposed to be applicable in small systems like pp collisions, where typically only a few ten particles are produced per unit of rapidity. The CGC is a 'first principles' classical field theory approximation to QCD which is applicable to very dense (high occupation number) parton systems like those found at small-x and small Q^2 in the initial state wave function of hadrons. It has been successfully used to describe some regularities seen, for example, in ep collisions at HERA ('geometric scaling') and to model the initial conditions in heavy ion physics.

Lacking further experimental input, no real progress was made to unravel the origin of these long range pp correlations until the ridge made a robust come-back with the first LHC proton–nucleus run some two years later[58] (Fig. 10 (right), p–Pb at $\sqrt{s_{NN}} = 5\,\mathrm{TeV}$). The correlation strength was actually significantly stronger than in pp at the same multiplicity, and in quick succession it was discovered that:[59] the ridge was actually double-sided, showing correlations between particles both close by in azimuth as well as back-to-back; a Fourier analysis revealed both even (v_2) as well as odd (v_3) components; the dependence of the correlation strength on particle mass was virtually identical to the one expected from hydrodynamic flow; and finally the correlation strength measured with multi-particle methods was almost identical to the one measured with two particles only, convincingly demonstrating that the ridge is a true collective effect which involves *all* low energy particles in *every* event (in contrast to jet correlations, which involve only a *few* particles in *some* events).

All characteristics of the p–Pb ridge are very natural for and in good agreement with a hydrodynamic collective flow origin of the correlation. Even the strength of the signal and its multiplicity dependence are of the correct order of magnitude (within a factor of two) if one uses some reasonable geometrical initial conditions and a standard hydro model and just postulates that the tiny and very short-lived interacting matter system, some 1 fm in size and lifetime, behaves like a macroscopic ideal fluid. Note that the matter created in central Pb–Pb collisions has a size of the order of 5000 fm^3 and therefore is larger by orders of magnitude!

The question how such a tiny (few fm^3) system could thermalise in essentially no time, maybe even become a small droplet of sQGP, has kept the case open, despite what looks like convincing evidence, including very recent and spectacular confirmation of the ridge effect at RHIC using 'elliptical' deuteron projectiles and 'triangular' ^{3}He nuclei.

In any case, the ridge discovery in pp and pA at LHC is definitely more than a curiosity and likely to have profound implications for heavy ion physics. If a sQGP (like) state can be created and studied in much smaller systems than anticipated, we can compare pp, pA, and AA to look for finite size effects, which may reveal information on correlation lengths and relaxation time scales not otherwise easily available. If, on the contrary, initial state effects and saturation physics are the answer, we would have discovered at LHC yet another new state of matter, the Colour Glass Condensate, opening a rich new field of activity for both experiment and theory.

6.   Conclusions

CERN has been an essential player and (mostly) unwavering supporter in the genesis and advance of high energy heavy ion physics. In the incredibly short time span of little over 30 years, the study of the phases of nuclear matter has evolved from light ion reactions at a fixed target energy of some GeV/nucleon to using heavy projectiles at a cm energy of several TeV/nucleon, increasing the available energy by three orders of magnitude.[60] This rapid progress was of course only possible by reusing machines, and initially even detectors, built over a longer time scale for particle physics. Today, with more than 2000 physicists active worldwide in this field, ultra-relativistic heavy ion physics has moved in less than a generation from the periphery into a central activity of contemporary Nuclear Physics. From the early exploratory phase, with sometimes more qualitative than quantitative results and conclusions, the field has grown up and matured, making important and often unexpected discoveries at each new facility. The view of the Quark–Gluon Plasma has dramatically advanced, from a simple weakly interacting parton gas to a strongly interacting ideal fluid that might find a field theoretical description in a so-called "dual", string theoretical framework.[61]

Today, at its height, heavy ion physics has found interest well beyond the circle of its immediate practitioners, with links and cross fertilisation towards neighboring disciplines ranging from plasma physics to string theory. The heavy ion programme is very active and competitive today at both high and low energy, to map the phase diagram, locate the transition between normal matter and the sQGP, and to search for a conjectured 'tri-critical' point somewhere in the region at or below SPS fixed target energy. Two new low energy facilities (FAIR at GSI and NICA at JINR) are being built to study compressed matter, i.e. matter at high baryon density and (comparatively) low temperature where the phase structure may be quite different (1st order phase transition) and the matter is closer related to neutron stars than to the early universe. The LHC however is and will be the energy frontier facility not only of high energy physics but also of nuclear physics for the foreseeable future, with a well-defined and extensive programme and wish list of measurements. And if the first three years can be a guide, strong interaction physics, while firmly rooted in the Standard Model, has shown no end to surprises and discoveries and promises to keep physics with heavy ions interesting (and fun) for quite some time to come.

References

1. S. Weinberg, *The First Three Minutes* (Basic Books, New York, 1977).
2. E. V. Shuryak, *Phys. Rept.* **61**, 71 (1980); *Sov. Phys. JETP* **47**, 212 (1978).
3. G. Baym and S. A. Chin, *Phys. Lett. B* **62**, 241 (1976).
4. J. I. Kapusta, *Nucl. Phys. B* **148**, 461 (1979).
5. J. C. Collins and M. J. Perry, *Phys. Rev. Lett.* **34**, 1353 (1975).
6. D. J. Gross and F. Wilczek, *Phys. Rev. Lett.* **30**, 1343 (1973).
7. S. A. Gottlieb, W. Liu, D. Toussaint, R. L. Renken and R. L. Sugar, *Phys. Rev. D* **35**, 2531 (1987).
8. G. Baym, *Nucl. Phys. A* **418**, 433C (1984).
9. "RHIC and Quark Matter. Proposal for a Relativistic Heavy Ion Collider at Brookhaven National Laboratory," BNL-51801.
10. R. Geller and B. Jacquot, *Nucl. Instrum. Meth.* **184**, 293 (1981).
11. R. Stock *et al.* (GSI-LBL-HEIDELBERG-MARBURG-WARSAW Collaboration), in *Proceedings of Quark Matter Formation and Heavy Ion Collisions*, Bielefeld 1982, pp. 557–582.
12. A. R. Baden *et al.* (GSI-LBL Collaboration), *Nucl. Instrum. Meth.* **203**, 189 (1982).
13. A. Sandoval, R. Bock, R. Brockmann, A. Dacal, J. W. Harris, M. Maier, M. E. Ortiz and H. G. Pugh *et al.*, *Nucl. Phys. A* **400**, 365C (1983).
14. R. Klapisch, *Nucl. Phys. A* **418**, 347C (1984).
15. H. G. Fischer, in *Proceedings of the Future Relativistic Heavy Ion Experiments*, Darmstadt 1980, pp. 528–550; W. J. Willis, *ibid.* p. 499.
16. T. Alber *et al.* (NA49 Collaboration), *Phys. Rev. Lett.* **75**, 3814 (1995).
17. J. D. Bjorken, *Phys. Rev. D* **27**, 140 (1983).
18. S. Hands, *Contemp. Phys.* **42**, 209 (2001), physics/0105022 [physics.ed-ph].
19. M. M. Aggarwal *et al.* (WA98 Collaboration), *Phys. Rev. Lett.* **85**, 3595 (2000), nucl-ex/0006008.
20. D. K. Srivastava and B. Sinha, *Phys. Rev. C* **64**, 034902 (2001), nucl-th/0006018; P. Huovinen, P. V. Ruuskanen and S. S. Rasanen, *Phys. Lett. B* **535**, 109 (2002), nucl-th/0111052.
21. P. Braun-Muninger, K. Redlich and J. Stachel, in *Quark–Gluon Plasma 3*, eds. R.C. Hwa and X.-N. Wang (World Scientific, 2004), pp. 491–599, nucl-th/0304013.
22. F. Becattini, M. Gazdzicki, A. Keranen, J. Manninen and R. Stock, *Phys. Rev. C* **69**, 024905 (2004), hep-ph/0310049.
23. A. Dainese (NA57 Collaboration), *Nucl. Phys. A* **774**, 51 (2006), nucl-ex/0510001.
24. T. Matsui and H. Satz, *Phys. Lett. B* **178**, 416 (1986).
25. M. C. Abreu *et al.* (NA50 Collaboration), *Phys. Lett. B* **477**, 28–36 (2000).
26. S. Damjanovic *et al.* (NA60 Collaboration), *Nucl. Phys. A* **774**, 715 (2006), nucl-ex/0510044.
27. C. Alt *et al.* (NA49 Collaboration), *Phys. Rev. C* **68**, 034903 (2003), nucl-ex/0303001.
28. U. W. Heinz and M. Jacob, Evidence for a new state of matter: An Assessment of the results from the CERN lead beam program, nucl-th/0002042.
29. http://press.web.cern.ch/press-releases/2000/02/new-state-matter-created-cern.
30. H. J. Specht (NA60 Collaboration), *AIP Conf. Proc.* **1322**, 1 (2010), arXiv:1011.0615.

31. The Brahms, Phenix, Phobos, and Star Collaborations, *Nucl. Phys. A* **757**, 1–283 (2005).
32. http://www.bnl.gov/newsroom/news.php?a=1303.
33. B. Muller and J. L. Nagle, *Ann. Rev. Nucl. Part. Sci.* **56**, 93 (2006), nucl-th/0602029.
34. J. Schukraft, *Phil. Trans. Roy. Soc. Lond. A* **370**, 917 (2012), arXiv:1109.4291.
35. C. Fabjan and J. Schukraft, in *The Large Hadron Collider: A marvel technology*, ed L. Evans (EPFL-Press Lausanne, Switzerland, 2009), Chapter 5.4, arXiv:1101.1257.
36. N. Armesto *et al.*, *J. Phys. G* **35**, 054001 (2008), arXiv:0711.0974.
37. B. Muller, J. Schukraft and B. Wyslouch, *Ann. Rev. Nucl. Part. Sci.* **62**, 361 (2012), arXiv:1202.3233.
38. J. Schukraft, *Phys. Scripta T* **158**, 014003 (2013), arXiv:1311.1429.
39. Y. Schutz and U. A. Wiedemann, in *Proceedings of the 22nd International Conference on Ultra-Relativistic Nucleus-Nucleus Collisions (Quark Matter 2011)*, Annecy, France, May 23–28, 2011, *J. Phys. G* **38**, 120301 (2011).
40. J. L. Nagle, *Nucl. Phys. A* **830**, 147C (2009), arXiv:0907.2707.
41. U. Heinz and R. Snellings, *Ann. Rev. Nucl. Part. Sci.* **63**, 123 (2013), arXiv:1301.2826.
42. G. Aad *et al.* (ATLAS Collaboration), *JHEP* **1311**, 183 (2013), arXiv:1305.2942.
43. J. Schukraft, A. Timmins and S. A. Voloshin, *Phys. Lett. B* **719**, 394 (2013), arXiv:1208.4563.
44. Z. Qiu and U. Heinz, *Phys. Lett. B* **717**, 261 (2012), arXiv:1208.1200.
45. B. Muller, *Phys. Scripta T* **158**, 014004 (2013).
46. A. Majumder and M. Van Leeuwen, *Prog. Part. Nucl. Phys. A* **66**, 41 (2011), arXiv:1002.2206.
47. G. Aad *et al.* (ATLAS Collaboration), *Phys. Rev. Lett.* **105**, 252303 (2010), arXiv:1011.6182.
48. S. Chatrchyan *et al.* (CMS Collaboration), *Phys. Rev. C* **84**, 024906 (2011), arXiv:1102.1957.
49. T. Renk, *J. Phys. Conf. Ser.* **509**, 012022 (2014), arXiv:1309.3059.
50. A. Dainese, *J. Phys. Conf. Ser.* **446**, 012034 (2013).
51. P. Braun-Munzinger and J. Stachel, Charmonium from Statistical Hadronization of Heavy Quarks: A Probe for Deconfinement in the Quark-Gluon Plasma, arXiv:0901.2500.
52. B. Abelev *et al.* (ALICE Collaboration), *Phys. Rev. Lett.* **109**, 072301 (2012), arXiv:1202.1383.
53. S. Chatrchyan *et al.* (CMS Collaboration), *Phys. Rev. Lett.* **107**, 052302 (2011), arXiv:1105.4894.
54. http://indico.cern.ch/conferenceDisplay.py?confId=107440.
55. F. Gelis, E. Iancu, J. Jalilian-Marian and R. Venugopalan, *Ann. Rev. Nucl. Part. Sci.* **60**, 463 (2010), arXiv:1002.0333.
56. V. Khachatryan *et al.* (CMS Collaboration), *JHEP* **1009**, 091 (2010), arXiv:1009.4122.
57. W. Li, *Mod. Phys. Lett. A* **27**, 1230018 (2012), arXiv:1206.0148.

58. S. Chatrchyan *et al.* (CMS Collaboration), *Phys. Lett. B* **718**, 795 (2013), arXiv:1210.5482.
59. C. Loizides, *EPJ Web Conf.* **60**, 06004 (2013), arXiv:1308.1377.
60. J. Schukraft, The Future of high energy nuclear physics in Europe, nucl-ex/0602014.
61. J. Casalderrey-Solana, H. Liu, D. Mateos, K. Rajagopal and U. A. Wiedemann, arXiv:1101.0618.

The Measurement of the Number of Light Neutrino Species at LEP

Salvatore Mele

CERN, CH-1211 Geneva 23, Switzerland
salvatore.mele@cern.ch

Within weeks of the start of the data taking at the LEP accelerator, the ALEPH, DELPHI, L3 and OPAL experiments were able to confirm the existence of just three light neutrino species. This measurement relies on the Standard Model relation between the 'invisible' width of the Z-boson and the cross-sections for Z-boson production and subsequent decay into hadrons.

The full data sample collected by the experiments at and around the Z-boson resonance allows a high-precision measurement of the number of light neutrino species as 2.9840 ± 0.0082. The uncertainty is mostly due to the understanding of the low-angle Bhabha scattering process used to determine the experimental luminosity.

This result is independently confirmed by the elegant direct observation of the $e^- e^+ \rightarrow \nu\bar{\nu}\gamma$ process, through the detection of an initial-state-radiation photon in otherwise empty detectors.

This result confirms expectations from the existence of three charged leptons species, and contributes to the fields of astrophysics and cosmology. Alongside other LEP achievements, the precision of this result is a testament to the global cooperation underpinning CERN's fourth decade. LEP saw the onset of large-scale collaboration across experiments totaling over 2000 scientists, together with a strong partnership within the wider high-energy physics community: from accelerator operations to the understanding of theoretical processes.

1. Introduction

The inception, design and approval of the LEP program at CERN, and the subsequent monumental construction of the accelerator and the detectors, represents a watershed in the history of the laboratory. The scientific, organizational and sociological challenges are described first hand in Ref. 1. The largest scientific instrument ever built, LEP was designed to push the frontiers of knowledge and understand the Standard Model of the electroweak interactions, with high-precision measurements of the properties of the recently discovered Z and W bosons.

One in five Z bosons produced at LEP decays into a "light neutrino", that is a neutrino whose mass is less than half the Z-boson mass. The Standard Model relation between this decay width and the cross-section for Z-boson production and subsequent decay into hadrons makes it possible to infer the number of light neutrino species, N_ν.[2]

Table 1   Early measurements of the number of light neutrino species by each of the four LEP experiments, and their average. Uncertainties are dominated by the scale of the observed Z-boson production cross-section.

Experiment	N_ν	Reference
ALEPH	3.27 ± 0.30	4
DELPHI	2.4 ± 0.6	5
L3	3.42 ± 0.48	6
OPAL	3.1 ± 0.4	7
Average	3.10 ± 0.04	8

The first beams circulated in the LEP accelerator on 14th July 1989. The first data-taking campaign was at centre-of-mass energies around 90 GeV, compatible with the mass[a] of the Z boson measured by the UA1 and UA2 experiments.[3] By mid October 1989, the four LEP experiments, ALEPH, DELPHI, L3 and OPAL, had already published their first articles describing the Z boson properties.[4–7] As listed in Table 1, these early observations already allowed to constrain N_ν to be around 3.

This article describes in details the measurement of N_ν in the context of the LEP high-precision physics program. Beyond the physics achievement, this measurement gives an insight on what made the LEP program so successful: a unique combination of exceptional accelerator performance, creative technological achievements in building and operating the detectors, and unprecedented cooperation with the theoretical physics community. These aspects set the scene for turning an important page in the history of CERN, then in its fourth decade. Such an enhanced culture of collaboration would usher the LHC era in CERN's fifth and sixth decades, as a worldwide hub of cooperation and creativity.

The structure of this article is the following. After this introduction, Section 2 recalls the principles of the indirect measurement of N_ν, including some concepts of the Standard Model of the electroweak interactions. Section 3 describes the experimental approach, including a broad-brush description of the LEP detectors, and presents the results of the measurement and a discussion of the uncertainties. Section 4 highlights a complementary direct measurement of N_ν through the detection of spectacular events with a single photon in otherwise empty detectors. Section 5 offers some concluding considerations.

2.   Theoretical Principles

A cornerstone of the LEP physics program is the study of the Z-boson "lineshape". This encompasses the measurement of parameters of the Standard Model of the

---

[a]We assume $h = 2\pi$ and $c = 1$, while using the factor 0.389 to convert GeV^2 into mb^{-1}.

electroweak interactions, and the proof of its internal consistency, through the study of physical observables describing Z-boson production and decay. Among these observables, the 'invisible' width of the Z-boson is related to its decay into neutrinos and gives access to N_ν. This section presents the physical observables leading to the measurement of N_ν and some key theoretical assumptions.

2.1.   The width of the Z boson

The width of the Z boson is defined as

$$\Gamma_Z = \Gamma_{ee} + \Gamma_{\mu\mu} + \Gamma_{\tau\tau} + \Gamma_{had} + N_\nu \Gamma_{\nu\nu}, \tag{1}$$

where the first three terms are the widths of decays into electrons, muons and taus, respectively. Γ_{had} is the sum of the widths of decays into u, d, s, c and b quarks and $\Gamma_{\nu\nu}$ the width of decays into neutrinos. The simultaneous measurement of Γ_Z, and of observables related to the hadronic and leptonic widths of the Z boson, allows one to determine N_ν.[2]

The partial decay widths of the Z boson into each fermion pair are related to the Z-boson couplings and to Standard Model parameters as:[9–11]

$$\Gamma_{f\bar{f}} = N_c^f \frac{G_F m_Z^3}{6\sqrt{2}\pi} \left(|G_{Af}|^2 R_{Af} + |G_{Vf}|^2 R_{Vf} \right) + \Delta_{ew/QCD} \tag{2}$$

where N_c^f is the number of colours (three for quarks and one for leptons), G_F is the Fermi constant determined from muon decay,[12] R_{Af} and R_{Vf} factorise final-state QED and QCD corrections and contributions from non-zero fermion masses to the axial and vector terms, respectively, $\Delta_{ew/QCD}$ accounts for non-factorisable electroweak and QCD corrections, and G_{Af} and G_{Vf} are the axial and vector effective couplings of the Z-boson to fermions, written as[13]

$$G_{Af} = \sqrt{R_f} T_3^f, \tag{3}$$

$$G_{Vf} = \sqrt{R_f} \left(T_3^f - 2Q_f K_f \sin\theta_W \right). \tag{4}$$

The form factors R_f and K_f absorb the overall scale of the coupling and the on-shell corrections to the electroweak mixing angle, θ_W, Q_f and T_3^f are the electric charge and the third component of weak isospin of the fermion, respectively.

2.2.   Experimental observables

Four experimental observables describe the total hadronic and leptonic cross section around the Z-boson resonance, and connect N_ν to the Z-boson lineshape:

(1) the mass of the Z boson, m_z;
(2) the width of the Z boson, Γ_z;

(3) the hadronic pole cross-section

$$\sigma_{\text{had}}^0 = \frac{12\pi}{m_Z^2} \frac{\Gamma_{\text{ee}} \Gamma_{\text{had}}}{\Gamma_Z^2};$$

(5)

(4) the ratio of the Z-boson partial decay widths into hadrons and massless leptons, Γ_{ll}, assuming lepton universality:

$$R_l^0 = \frac{\Gamma_{\text{had}}}{\Gamma_{ll}}.$$

(6)

The non-negligible tau mass is accounted for as $\Gamma_{ll} = \delta_\tau \Gamma_{\tau\tau}$, with $\delta_\tau = -0.23\%$. A fifth experimental observable, less important for the N_ν determination, is the lepton forward–backward asymmetry, again assuming lepton universality:

(5) $A_{\text{FB}}^{0,l}$,

which is defined as the asymmetry at the Z-boson pole of the cross-sections for final state leptons emitted in the forward (i.e. the negative-charged lepton 'continuing' along the direction of the incoming electron) or backward direction, according to the general formula $A_{\text{FB}} = (\sigma_F - \sigma_B)(\sigma_F + \sigma_B)$.

2.3.   Sensitivity to N_ν

A key experimental observable directly related to N_ν is

$$R_{\text{inv}}^0 = \frac{\Gamma_{\text{inv}}}{\Gamma_{ll}} = N_\nu \left(\frac{\Gamma_{\nu\nu}}{\Gamma_{ll}} \right).$$

(7)

The asset of R_{inv}^0 is that uncertainties of experimental and theoretical nature are well-controlled in the ration of the Z-boson widths.

The combination of Eqs. (2) and (5) allows to write R_{inv}^0 as

$$R_{\text{inv}}^0 = \left(\frac{12\pi R_l^0}{\sigma_{\text{had}}^0 m_Z^2} \right)^{\frac{1}{2}} - R_l^0 - (3 + \delta_\tau),$$

(8)

which expresses, together with Eq. (7), the relation between N_ν and the hadronic pole cross-section. This dependency drives the determination of N_ν and is graphically displayed in Fig. 1, which compares the measured hadron production cross-section around the Z-boson resonance with predictions for two, three and four light neutrino species. The curves in Fig. 1 allow to visualise the enormous statistical sensitivity of LEP data to N_ν.

It is important to summarise the assumptions made in describing the dependence of N_ν on the physical observables at LEP: lepton universality holds; Z bosons only decay to known fermions; neutrino masses are negligible; and Z-boson couplings to neutrinos are described by the Standard Model.

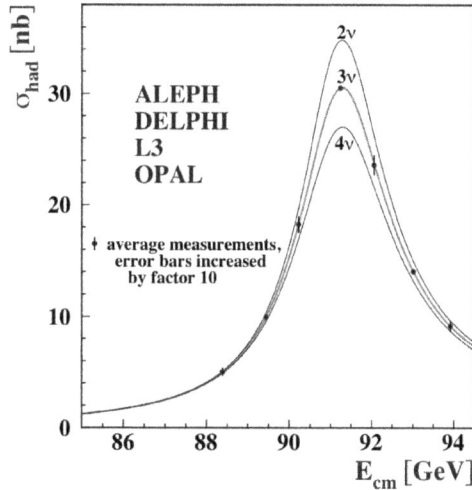

Fig. 1.   Measurement of the hadron production cross-section as a function of the LEP centre-of-mass energy around the Z-boson resonance. Combined results from the four LEP experiments are presented. Curves represent the predictions for two, three and four neutrino species. To further convey the high sensitivity of the measurement, uncertainties are magnified tenfold.[14]

3.   Experimental Measurement

The LEP accelerator and the LEP detectors were unprecedented in their size and complexity. This Section gives a succinct description of how the challenges of high-precision Z-boson detection guided detector design. After recalling the data sample, the measurements of the key observables leading to N_ν are presented, together with the final result, and crucial uncertainties are discussed.

3.1.   *Detection of Z-boson decays*

The design of the four LEP experiments[15–18] was optimised to detect Z-boson decays with high efficiency, within the available budgetary, technological and physical constraints. Teams of several hundred scientists, technicians and engineers designed, prototyped, built and assembled sophisticated apparatuses with dimensions exceeding 10 metres in diameter and length, and weighting several thousand tons. While the basic design principles of the detectors were similar, the choices of particular technologies in some sub-detectors were markedly different and would eventually contribute to reduce combined systematical uncertainties.

Figure 2 presents a cut-away three-dimensional view of the four detectors. All are radially and forward–backward symmetric. The common part of the design is the succession of sub-detectors, moving outwards from the beam axis: tracking chambers, surrounded by calorimeters and bending magnets, with muon spectrometers as the outmost layer. The exception is the L3 detector where the

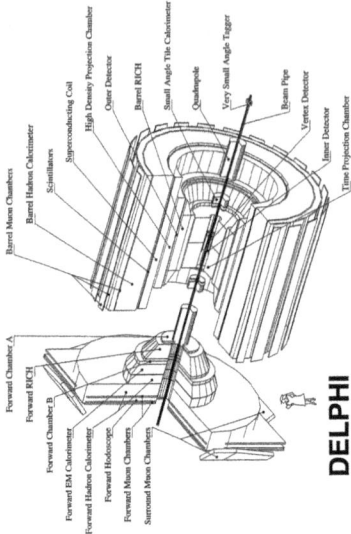

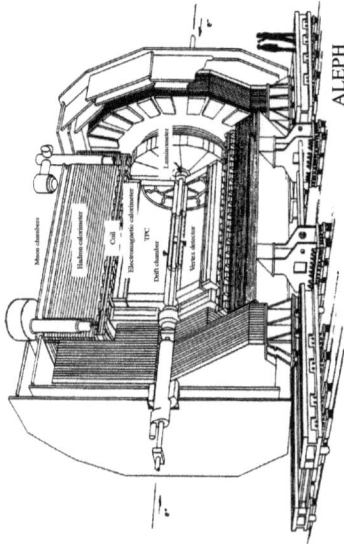

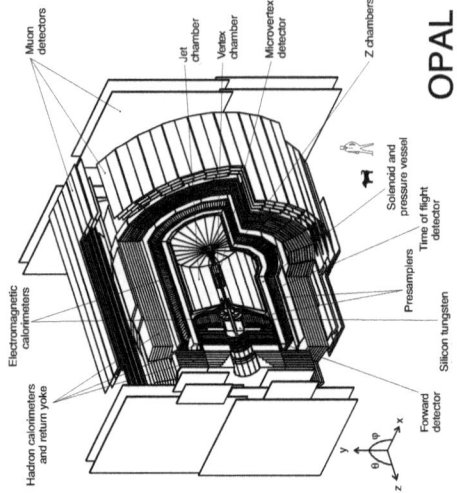

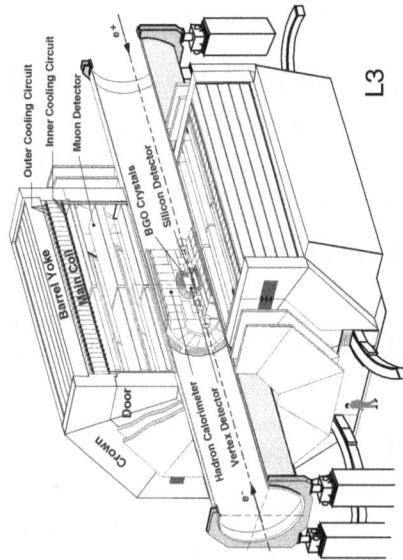

Fig. 2.   Cut-away representation of the four LEP detectors: ALEPH, DELPHI, L3 and OPAL.

entire muon spectrometer is contained in the magnetic field. Some sub-detectors relied on established technology, such as wire chambers for tracking or crystals and scintillator counters for calorimetry, pushing technologies in scale and precision (e.g. the L3 BGO electromagnetic calorimeter, or its high-precision muon spectrometer). Other sub-detector relied on newer technologies, never deployed before on such a large scale (e.g. the ALEPH and DELPHI time-projection chambers, the ALEPH liquid-argon calorimeter and the DELPHI ring imaging Čerenkov detector — RICH).

Some examples of the performance of the LEP detectors are the following:

- the transverse momentum resolution of the ALEPH tracking system, $\sigma(1/p_t) = 0.6 \times 10^{-3}$ GeV^{-1};[19]
- the DELPHI RICH efficiency of 70% to identify K$^\pm$ with a contamination of 30%;[20]
- the energy resolution of the L3 electromagnetic calorimeter $\Delta E/E \approx 1.4\%$ for 45 GeV electrons;[21]
- the momentum resolution of the L3 muon spectrometer $\Delta p/p \approx 2.5\%$ for 45 GeV muons.[21]

Figure 3 illustrates the detection principles for Z-boson decays. Hadronic events are identified from a high multiplicity of tracks in the central trackers and energy deposits in the calorimeters, reconstructed in two back-to-back fully-contained jets. Higher jet multiplicity is possible for rarer higher-order QCD processes. Z-boson decays in electron-positron pairs are characterised by two back-to-back tracks in the central trackers, corresponding to high-energy signals in the electromagnetic calorimeters. Z-boson decays into muons have the unique signature of back-to-back tracks in the central trackers, leaving minimum ionising deposits in the hadronic and electromagnetic calorimeters and tracks in the muon chambers. Z-boson decays in tau pairs are more challenging to detect, requiring a combination of missing energy in the detector, low-multiplicity jets, muons or electrons, according to the tau decay channels.

3.2. *Data sample*

The LEP accelerator operated at and around the Z-boson resonance from its commissioning in 1989 through 1995. In 1990 and 1991, energy scans at a spacing of 1 GeV provided a first mapping of the Z-boson resonance. In the following years, high-luminosity data-taking concentrated on the Z-boson resonance, with two additional "off-peak" energy points in 1993 and 1995, 1.8 GeV above and below the Z-boson resonance, to further constrain the Z-boson lineshape. Further details on the LEP accelerator design and performance are given in Ref. 22.

A total of 17 million Z-boson decays were detected by the four experiments. Table 2 provides a breakdown of the integrated luminosity per each experiment and the total number of events detected in the hadronic and leptonic final states.

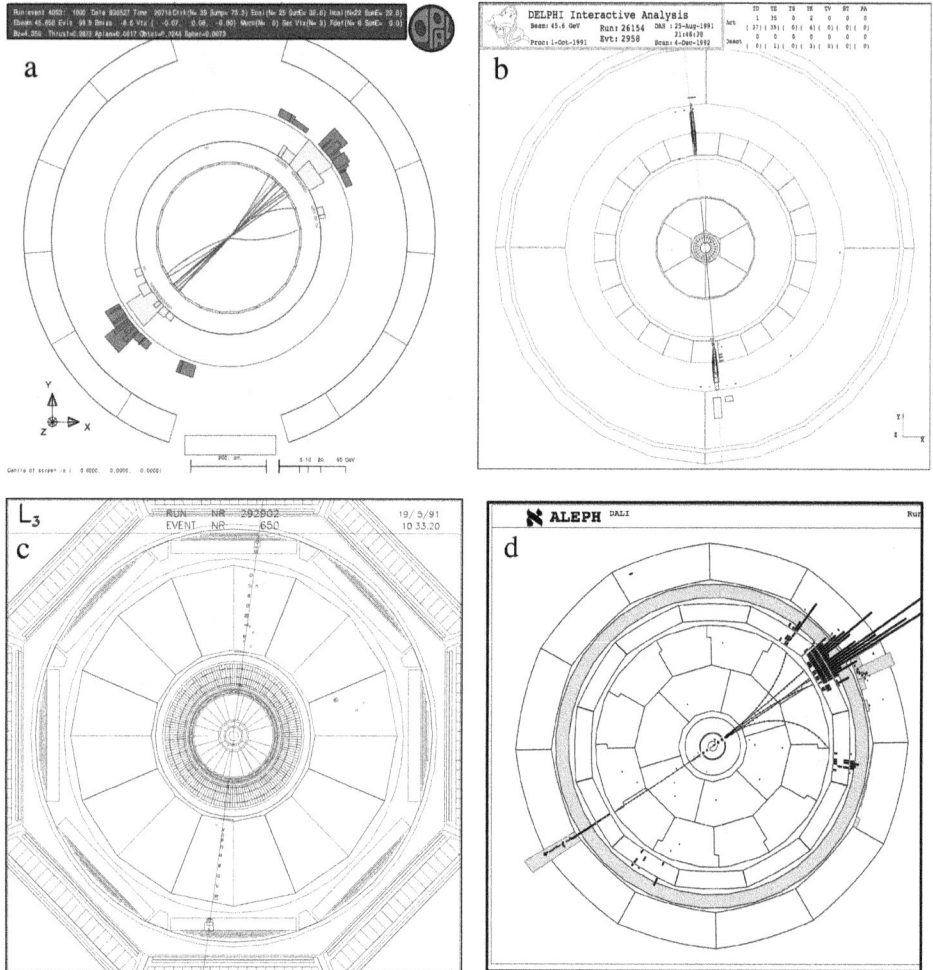

Fig. 3. Event displays of Z-boson decays detected in the four LEP experiments: (a) hadronic decays with the OPAL detector, characterised by two high-multiplicity back-to-back fully-contained jets; (b) electron–positron pairs with DELPHI, with two back-to-back tracks in the central tracker, and two energy deposits in the electromagnetic calorimeter, with energies close to the beam energy; (c) muon pairs with L3, with tracks in the muon chambers (mostly outside the image), minimum ionizing deposits in the hadronic and electromagnetic calorimeters and corresponding tracks in the central chambers, time-of-flight detectors assure such tracks are originating from the collision vertex and not from cosmic rays; (d) tau pairs with ALEPH, in this case with an electron (track and calorimeter deposit) detected in the hemisphere opposite a collimated, low-multiplicity jet, with overall missing energy. In all images the beam axis is perpendicular to the page.

Table 2 Centre-of-mass energy and luminosity delivered to each experiment, and total numbers of events collected by the four experiments in the hadronic and leptonic decay modes. Due to the low integrated luminosity and relative control of the experimental conditions, the 1989 data sample is not used in the study of the Z-boson lineshape.

Year	Centre-of-mass energy [GeV]	Integrated luminosity/experiment [pb^{-1}]	Total detected hadronic events [$\times 10^3$]	Total detected leptonic events [$\times 10^3$]
1990/91	88.2–94.2	27.5	1660	186
1992	91.3	28.6	2741	294
1993	89.4, 91.2, 93.0	40.0	2607	296
1994	91.2	64.5	5910	657
1995	89.4, 91.3, 93.0	39.8	2579	291

3.3. *Measurement of cross-sections and asymmetries*

In each final state of Z-boson decays, cross-sections are measured as $\sigma_{tot} = (N_s - N_b)/\varepsilon\mathcal{L}$, where N_s is the number of selected events, N_b is the number of events expected from background processes, ε is the selection efficiency, which include geometrical acceptance, and $\mathcal{L}$ is the integrated luminosity. The LEP experiments derive N_b and ε from Monte Carlo generators describing the kinematics of both the Z-boson production and decay and of background processes. Events produced with those generators are passed through detailed simulations of the detectors and the same software used to reconstruct collision events. These workflows are cross-checked by using data and refined through the years to give extremely accurate simulation of the detectors.

Asymmetries for each final state are measured as $A_{FB} = (N_F - N_B)/(N_F + N_B)$, where N_F and N_B are event counts for negatively changed leptons 'continuing' along the direction of the incoming electron, or emitted 'backwards', respectively.

The large statistical sample of Z-boson decays collected at LEP results in low statistical uncertainties in the cross-section determinations for each experiment, around 0.5 per mille in the hadronic channel and 2.5 per mille in the leptonic channels. Experiment-dependent systematic uncertainties are mostly due to the calculation of efficiencies and acceptances and the selection procedures, as estimated from data and Monte Carlo simulations. These vary between 0.4 and 0.7 per mille in the hadronic channel and 1 to 7 per mille in the leptonic channel, with the higher value corresponding to tau pairs. For asymmetries, experiment-dependent systematic uncertainties have absolute values between 0.0005 and 0.0030, with the higher value corresponding to tau pairs. Statistical uncertainties are between two and five times larger than the systematic uncertainties.[14, 23–26]

Systematic uncertainties on cross-sections and asymmetries which are common across experiments are irreducible. The main sources are: the LEP energy calibration;[22] the use of the same Monte Carlo generators to simulate signal and background processes; theoretical uncertainties on the parametrisation of Standard Model observables, contributions to the electron–positron final states, and the overall QED final-state corrections. The most important source of common systematical uncertainties affects the determination of luminosity, as discussed in the next section.

3.4. *Measurement of luminosity*

As presented in Eqs. (7) and (8), and in Fig. 1, N_ν depends strongly on the scale of the hadronic cross-section. As detectors are well understood and the large event counts limit statistical uncertainties, the N_ν precision depends on the accuracy of the luminosity measurement. LEP experiments relied on the detection of low-angle Bhabha scattering events for the measurement of instantaneous luminosity.[27] The advantages of this process is a high cross-section and therefore a negligible statistical uncertainty, as well as a low contribution from Z-boson production itself.

Pairs of dedicated calorimeters, completed with tracking devices, were installed close to the LEP beam pipe, in the forward and backward low-angle regions, typically between 30 and 50 mrad from the beam axis. Delicate to operate, these instruments had to be protected from hazardous conditions while beams were manipulated in the machine before stable collisions, and would then count coincidence of energy deposits in the forward and backward regions, originated by charged particles and compatible with the beam energy: the typical signature of Bhabha scattering. Event counts yield a detailed record of the instantaneous luminosity conditions and then allow to extract the total integrated luminosity. Experiment-dependent, systematic uncertainties for the determination of the luminosity are well controlled, in the range 0.03–0.09%.

All experiment relied on the same Monte Carlo generator and state-of-the art theoretical calculations to estimate the accepted low-angle Bhabha scattering cross-section, and derive the luminosity.[28] After intense effort in improving these calculations, a residual theoretical uncertainty of 0.061% remains, mostly originating from vacuum polarisation, higher-order corrections and the production of light fermion pairs.[29] The way the luminosity uncertainty has been reduced over the LEP data-taking campaign tells a success story of highly sophisticated experimental techniques moving in lockstep with dedicated efforts by the theory community to push the understanding of the calculation of low-angle Bhabha scattering.

The LEP-wide combination of cross-sections and asymmetries, in addition to the obvious statistical advantages, allows to reduce several uncorrelated systematic uncertainties of experimental origin. At the same time, the theoretical uncertainty on the determination of the luminosity uncertainty is common to all experiments, and therefore irreducible. It contributes as much as a half of the uncertainty on the hadronic pole cross-section determination and dominates the systematic uncertainty on the determination of the N_ν, as discussed in the following sections.

3.5. *Results*

Each LEP experiment extracted cross-sections and asymmetries in the hadronic and leptonic final states at different energy points, corresponding to about 200 individual measurements. These allowed a precise description of the Z-boson lineshape and the corresponding extraction of parameters of the Standard Model.[23–26]

An additional, through then unprecedented, collaborative efforts across the experiments led to the establishment of the LEP ElectroWeak Working Group.[30] The Group had the mandate to devise and arrange the combination of the Z-boson lineshape measurements across the experiments and thus obtain a considerable reduction of uncertainties, both of a statistical and systematic nature. Each experiments provided results in agreed-upon formats, with full correlation matrices. The LEP ElectroWeak Working Group combined[14] all inputs to both determine the Z-boson lineshape observables with a much higher precision than allowed by each individual experiment statistical sample and check the overall consistency of the results and their implication for the understanding of the Standard Model.[22] Table 3 presents combined results for the observables introduced in Section 2.2, in the hypothesis of lepton universality. The combination shows the compatibility of results across the experiments, with a goodness-of-fit of $\chi^2/\text{d.o.f.} = 36.5/31$.

It is important to remark that the lepton universality hypothesis is tested in the entire LEP data sample by measuring the rations of the Z boson partial decay widths as $\Gamma_{\mu\mu}/\Gamma_{ee} = 1.0009 \pm 0.0028$ and $\Gamma_{\tau\tau}/\Gamma_{ee} = 1.0019 \pm 0.0032$.[14]

Using Eqs. (7) and (8) and the Standard Model value for the ratio of the Z-boson widths to neutrinos and leptons[14]

$$(\Gamma_{\nu\nu}/\Gamma_{ll})_{\text{SM}} = 1.99125 \pm 0.00083, \tag{9}$$

the number of light neutrino species is determined as:

$$N_\nu = 2.9840 \pm 0.0082. \tag{10}$$

It is important to recall the four key assumptions leading to this result:

- lepton universality holds;
- no other Z-boson decays exist beyond those to known fermions;

Table 3   Combined LEP results, and their correlation for key observables (Section 2.1).[14]

| Observable | Combined LEP measurement | Correlations | | | | |
		m_Z	Γ_Z	σ^0_{had}	R^0_l	$A^{0,l}_{\text{FB}}$
m_Z	91.1875 ± 0.021 GeV	1.000				
Γ_Z	2.4952 ± 0.0023 GeV	-0.023	1.000			
σ^0_{had}	41.540 ± 0.037 nb	-0.045	-0.297	1.000		
R^0_l	20.767 ± 0.025	0.033	0.004	0.183	1.000	
$A^{0,l}_{\text{FB}}$	0.0171 ± 0.0010	0.055	0.033	0.006	-0.056	1.000

- neutrino masses are negligible;
- Z-boson couplings to neutrinos are described by the Standard Model.

3.6.   *Uncertainties*

The uncertainty on N_ν is less the three per mille. It is decomposed as the sum in quadrature of three parts:[14]

$$\delta N_\nu \sim 10.5 \frac{\delta n_{\text{had}}}{n_{\text{had}}} \oplus 3.0 \frac{\delta n_{\text{lep}}}{n_{\text{lep}}} \oplus 7.5 \frac{\delta \mathcal{L}}{\mathcal{L}}. \tag{11}$$

The first two are related to uncertainties on the number of events selected for the measurement of cross-section and asymmetries in the hadronic and leptonic channels, respectively. The third term parametrises uncertainties on the scale of the cross-sections deriving from the uncertainties on the luminosity measurement.

The largest contribution to the uncertainty on the luminosity measurement is the theoretical uncertainty (0.061%) discussed in Section 3.4. This uncertainty alone results in an uncertainty on N_ν of 0.0046, accounting for more than half of the total uncertainty on N_ν.

4.   Direct Measurement of N_ν

The LEP experiments pursued an alternative and elegant measurement of N_ν by detecting events with a single visible photon as a signature of the $\text{e}^-\text{e}^+ \to \nu\bar{\nu}\gamma$ process.[2] At the Z-boson resonance, this final state is mostly due to the initial-state radiation of a low-angle photon, with a steeply falling energy spectrum, with a Z boson decaying into neutrinos. Contributions from the t-channel exchange of a virtual W boson are small.

At the Z-boson resonance, the cross-section of the $\text{e}^-\text{e}^+ \to \nu\bar{\nu}\gamma$ process can be written[31] as

$$\sigma^0_{\nu\nu\gamma}(s) = \frac{12\pi}{m_Z^2} \frac{s\Gamma_{\text{ee}}N_\nu\Gamma_{\nu\nu}}{(s - m_Z^2) + s^2\Gamma_Z^2/m_Z^2} + \text{W-boson exchange terms} \tag{12}$$

which is mostly proportional to N_ν. A careful measurement of the cross-section of the process with the control of the residual background sources and the overall acceptance allows to extract N_ν. This cross-section is considerably lower than the Z-boson resonance. The statistical accuracy of the direct measurement of N_ν is therefore over an order of magnitude inferior than the indirect measurement. At the same time, the direct measurement does not rely on the assumption that Z bosons only decay to known fermions. Possible decays into visible 'exotic' particles, conflated within other visible channels and in particular hadronic final states, could in principle alter the Z-boson lineshape and yield an incorrect measurement of N_ν.

The key experimental challenge of the direct measurement is to detect events with a single photon and no other activity in the detector. On the one hand, the cross-section is larger, and therefore the measurement more sensitive, the lower the energy of the photon and the closest the photon is to the beam axis. On the other hand, these exact conditions make both photon detection more complex and experimental backgrounds harder to control. The four LEP experiments devised sophisticated analysis chains and in some cases even dedicated trigger systems to record these "single photon" events (e.g. the one described in Ref. 32). Around 2500 single photon events, with background subtracted, were collectively detected by the four experiments at the Z-boson resonance, with different energy thresholds and fiducial volumes, as summarised in Table 4, which also details data samples and the signal-over-background ratios.

Figure 4 presents an example of the measured cross-section as a function of the centre-of-mass energy and its dependency on N_ν. Fits to the theoretical modeling of the cross-section, with the assumption of Standard Model coupling of the Z-boson to neutrinos, yield the individual direct measurements of N_ν listed in Table 4. These results can be combined as[37]:

$$N_\nu = 3.00 \pm 0.08. \tag{13}$$

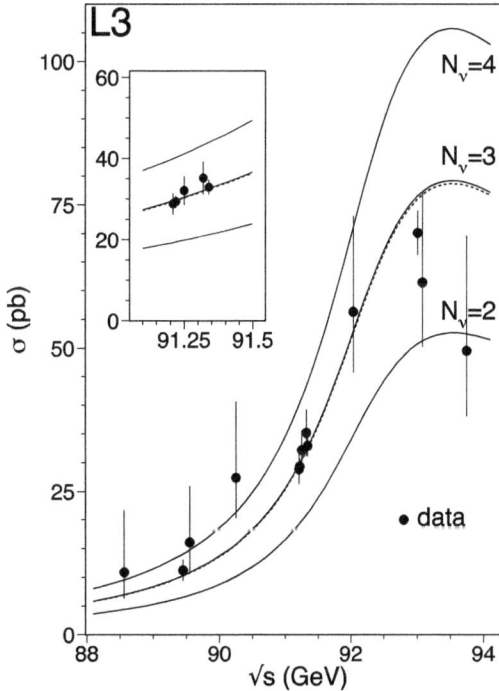

Fig. 4.   Cross-section measured by the L3 experiment for the $e^- e^+ \to \nu\bar\nu\gamma$ process around the Z-boson resonance as a function of the centre-of-mass energy. The lower limit for the photon energy is 1 GeV, and the fiducial volume $|\cos\theta_\gamma| < 0.71$. Theoretical predictions for two, three or four light neutrinos species are also shown. The dashed line represents a fit to the data points.[35]

Table 4   Integrated luminosity, $\mathcal{L}$, photon energy threshold, E_γ, and fiducial volume, $|\cos\theta_\gamma|$, for the four LEP experiments' analyses of single-photon events around the Z-boson resonance. The signal over background ratios, s/b, are also given, together with each experiment direct measurement of N_ν and their average. The first uncertainties are statistical and the second systematic.

Experiment	$\mathcal{L}$ [pb^{-1}]	$E_\gamma >$ [GeV]	$|\cos\theta_\gamma| <$	s/b	N_ν	Reference
ALEPH	15.7	1.5	0.74	1.8	$2.68 \pm 0.20 \pm 0.20$	33
DELPHI	67.6	3.0	0.70	2.7	$2.89 \pm 0.32 \pm 0.19$	34
L3	99.9	1.0	0.71	6.0	$2.98 \pm 0.07 \pm 0.07$	35
OPAL	40.5	1.75	0.70	11.0	$3.23 \pm 0.16 \pm 0.10$	36
				Average	3.00 ± 0.08	37

Table 5   Direct measurement of N_ν at centre-of-mass energies, $\sqrt{s}$, above the Z-boson resonance. Each experiments investigated different observables to extract N_ν.

Experiment	$\sqrt{s}$ [GeV]	Observable(s)	N_ν	Reference
ALEPH	189–207	Missing mass, θ_γ	2.86 ± 0.09 (stat.+syst.)	41
DELPHI	130–209	Cross section	2.84 ± 0.10 (stat.) ± 0.14 (syst.)	42
L3	130–209	Recoil mass, θ_γ	2.95 ± 0.08 (stat.) ± 0.03 (syst.) ± 0.03 (th.)	43
OPAL	130–189	E_γ	3.27 ± 0.30 (stat.+syst.)	44
Average (including lower energies)			2.92 ± 0.05	37

The LEP experiments repeated this measurements at centre-of-mass energies above the Z-boson resonance. At these higher energies, from 130 GeV to 209 GeV, the single-photon energy spectrum exhibits two distinct features. The first feature is a steeply falling behavior similar to that observed at the Z-boson resonance, mostly due to the initial-state radiation of a photon accompanying the t-channel production of a neutrino–antineutrino pair through the exchange of a virtual W boson. The second feature is a peak at the energy corresponding to the difference between the centre-of-mass energy and the Z-boson mass. This structure corresponds to the radiation in the initial state of a photon of the energy needed to lower the centre-of-mass energy back to the Z-boson resonance, with a Z boson decaying into neutrinos. Monte Carlo simulations of these processes[38–40] allow to model the dependence of the photon energy spectrum, and its polar angle, on N_ν.

The four experiments collectively detected about 6200 single photon events above the Z-boson resonance, with relatively low background. The study of various observables allows to extract N_ν, with the results summarised in Table 5. Including lower-energy data, the combined result for the direct determination of the number of light neutrino species across all LEP energies is:[37]

$$N_\nu = 2.92 \pm 0.05. \tag{14}$$

5. Conclusions

In 1989, within the first few weeks of data taking at LEP, the ALEPH, DELPHI, L3 and OPAL collaborations reported the number of light neutrino species to be around three. This is a remarkable achievement which bears witness to the performance of the LEP accelerator, the early understanding of detectors, and the overall planning of the LEP physics program: the most complex CERN had seen in its first four decades. It would take five more years of data-taking, and about a decade more to develop sophisticated analysis techniques to combine results across the LEP experiments, for the final determination of the number of light neutrino species to be published as:[14]

$$N_\nu = 2.9840 \pm 0.0082.$$

The dominating uncertainty is the theoretical control of the low-angle Bhabha scattering process used to determine the experimental luminosity. This result relies on four important assumptions: that lepton universality holds; that Z bosons only decay to known fermions; that neutrino masses are negligible; and finally that Z-boson couplings to neutrinos are as described by the Standard Model. The direct measurement of the $e^-e^+ \to \nu\bar{\nu}\gamma$ process, at the Z-boson resonance and at higher centre-of-mass energies up to 209 GeV, allows an independent verification, obtaining a value $N_\nu = 2.92 \pm 0.05$.

This result stands out as one of the legacies of the LEP physics program. It ruled out for the first time the existence of a fourth generation, and poses stringent limits on theoretical models relevant in astrophysics and cosmology. The high precision of the result further constrains the existence of exotic particles in Z-boson decays. Beyond the tremendous physical importance, the impressive precision of the measurement of the number of light neutrino species at LEP, and the overall determination of the parameters of the Standard Model and the proof of its internal consistency,[22] mark a turning point in the history of CERN as an example of scientific cooperation.

The LEP detectors where the first to be built by truly worldwide collaborations, with large contingents of scientists from the United States and Asia participating to a CERN program. Unprecedented in size, the LEP collaborations were the mold for the true globalisation of particle physics as an enterprise, and of CERN as a laboratory, which ushered the LHC era over the two most recent decades in CERN's history. This example of global scientific collaboration has captured worldwide attention, and imagination, at the time of the first LHC discoveries. It is more than an anecdote, but rather a proof of how scientific cooperation is indispensible to extend human knowledge, that the scientific publication describing the high-precision measurements at LEP[14] was signed by over 2500 authors, the first ever published article to do so.[b]

---

[b]Contrary to what is sometimes heard, the first published article with more than 1000 authors is not on high-energy physics, but about a large-scale Japanese medical study.[45, 46]

The LEP era transformed CERN, with large and crucial contributions from scientists of the then Soviet Union and countries from Eastern Europe, alongside scientists from the United States and Western Europe. This process enshrined the crucial role of CERN as an ambassador of 'Science for Peace', recently recognised by the United Nations in granting CERN observer status at its General Assembly.

On the one hand, the precise determination of the number of light neutrino species is of fundamental importance for our understanding of the Universe. On the other hand, the decade-long global cooperative effort to achieve this result, through the ingenuity and creativity of thousands of dedicated individuals, is part of our collective legacy as the human species.

References

1. H. Schopper, *LEP — The Lord of the Collider Rings at CERN* (Springer, 2009).
2. G. Barbiellini *et al.*, Neutrino counting, in *Proceedings of the Workshop on Z Physics at LEP*, CERN, Switzerland (Sept. 1989), pp. 129–170.
3. C. Rubbia, The discovery of the W and Z particles, in *60 Years of CERN Experiments and Discoveries*. (World Scientific, 2015).
4. D. Decamp *et al.* (ALEPH Collaboration), A precise determination of the number of families with light neutrinos and of the z-boson partial widths, *Phys. Lett. B* **231**, 519–529 (1989).
5. P. Aarnio *et al.* (DELPHI Collaboration), Measurement of the mass and width of the Z^0-particle from multihadronic final states produced in e^+e^- annihilations, *Phys. Lett. B* **231**, 539–547 (1989).
6. B. Adeva *et al.* (L3 Collaboration), A determination of the properties of the neutral intermediate vector boson Z^0, *Phys. Lett. B* **231**, 509–518 (1989).
7. M. Z. Akrawy *et al.* (OPAL Collaboration), A precise determination of the number of families with light neutrinos and of the Z-boson partial widths, *Phys. Lett. B* **231**, 530–538 (1989).
8. J. J. Hernandez *et al.* (Particle Data Group), Review of particle properties, *Phys. Lett. B* **239**, VI.22–VI.23 (1990).
9. K. Chetyrkin *et al.*, QCD corrections to the e^+e^- cross section and the Z-boson decay rate. In *Reports of the working group on precision calculations for the Z resonance*, CERN, Switzerland (Mar. 1993), pp. 175–264.
10. A. Czarnecki and J. Kuhn, Nonfactorizable QCD and electroweak corrections to the hadronic Z boson decay rate, *Phys. Rev. Lett.* **77**, 3955–3958 (1996).
11. R. Harlander *et al.*, Complete corrections of $O(\alpha\alpha_s)$ to the decay of the Z-boson into bottom quarks, *Phys. Lett. B* **426**, 125–132 (1998).
12. K. Olive *et al.* (Particle Data Group), Review of Particle Physics, *Chin. Phys. C* **38**, i–1676 (2014).
13. M. Veltman, Limit on Mass Differences in the Weinberg Model, *Nucl. Phys. B* **123**, 89–99 (1977).
14. S. Schael *et al.* (ALEPH, DELPHI, L3, OPAL and SLD Collaborations and the LEP Electroweak Working Group, SLD Electroweak Group and SLD Heavy Flavour

Group Collaborations), Precision electroweak measurements on the Z Resonance, *Phys. Rept.* **427**, 257–454 (2006).

15. D. Decamp *et al.* (ALEPH Collaboration), ALEPH: A detector for electron-positron annihilations at LEP, *Nucl. Instrum. Meth. A* **294**, 121–178 (1990).

16. P. Aarnio *et al.* (DELPHI Collaboration), The DELPHI detector at LEP, *Nucl. Instrum. Meth. A* **303**, 233–276 (1991).

17. B. Adeva *et al.* (L3 Collaboration), The construction of the L3 experiment, *Nucl. Instrum. Meth. A* **289**, 35–102 (1990).

18. K. Ahmet *et al.* (OPAL Collaboration), The OPAL detector at LEP, *Nucl. Instrum. Meth. A* **303**, 275–319 (1991).

19. D. Busculic *et al.* (ALEPH Collaboration), Performance of the ALEPH detector at LEP, *Nucl. Instrum. Meth. A* **360**, 481–506 (1995).

20. P. Abreu *et al.* (DELPHI Collaboration), Performance of the DELPHI detector, *Nucl. Instrum. Meth. A* **378**, 57–100 (1996).

21. O. Adriani *et al.* (L3 Collaboration), Results from the L3 experiment at LEP, *Phys. Rept.* **236**, 1–146 (1993).

22. W. de Boer, Precision Experiments at LEP, in *60 Years of CERN Experiments and Discovery* (World Scientific, 2015).

23. R. Barate *et al.* (ALEPH Collaboration), Measurement of the Z resonance parameters at LEP, *Eur. Phys. J. C* **14**, 1–50 (2000).

24. P. Abreu *et al.* (DELPHI Collaboration), Cross-sections and leptonic forward backward asymmetries from the Z0 running of LEP, *Eur. Phys. J. C* **16**, 371–405 (2000).

25. M. Acciarri *et al.* (L3 Collaboration), Measurements of cross-sections and forward backward asymmetries at the Z resonance and determination of electroweak parameters, *Eur. Phys. J. C* **16**, 1–40 (2000).

26. G. Abbiendi *et al.* (OPAL Collaboration), Precise determination of the Z resonance parameters at LEP: 'Zedometry', *Eur. Phys. J. C* **19**, 587–651 (2001).

27. G. M. Dallavalle, Review of precision determinations of the accelerator luminosity in LEP experiments, *Acta Phys. Pol. B* **28**, 901–923 (1997).

28. S. Jadach *et al.*, Upgrade of the Monte Carlo program BHLUMI for Bhabha scattering at low angles to version 4.04, *Comput. Phys. Commun.* **102**, 229–251 (1997).

29. B. Ward *et al.*, New results on the theoretical precision of the LEP/SLC luminosity, *Phys. Lett. B* **450**, 262–266 (1999).

30. LEP Electroweak Working Group, http://lepewwg.web.cern.ch/LEPEWWG/, last accessed February 6th, 2015.

31. O. Nicrosini and L. Trentadue, Structure Function Approach to the Neutrino Counting Problem, *Nucl. Phys. B* **318**, 1–21 (1989).

32. R. Bizarri *et al.*, The First level energy trigger of the L3 experiment: Description of the hardware, *Nucl. Instrum. Meth. A* **317**, 463–473 (1992).

33. D. Buskulic *et al.* (ALEPH Collaboration), A Direct measurement of the invisible width of the Z from single photon counting, *Phys. Lett. B* **314**, 520–534 (1993).

34. P. Abreu *et al.* (DELPHI Collaboration), Search for new phenomena using single photon events in the DELPHI detector at LEP, *Z. Phys. C* **74**, 577–586 (1997).

35. M. Acciarri *et al.* (L3 Collaboration), Determination of the number of light neutrino species from single photon production at LEP, *Phys. Lett. B* **431**, 199–208 (1998).

36. R. Akers *et al.* (OPAL Collaboration), Measurement of single photon production in $e^+ e^-$ collisions near the Z0 resonance, *Z. Phys. C* **65**, 47–66 (1995).

37. C. Amsler *et al.* (Particle Data Group), Review of Particle Physics, *Phys. Lett. B* **667**, 1–1340 (2008).

38. S. Jadach *et al.*, The Precision Monte Carlo event generator K K for two fermion final states in e^+e^- collisions, *Comput. Phys. Commun.* **130**, 260–325 (200).

39. S. Jadach *et al.*, The Monte Carlo program KORALZ, version 4.0, for the lepton or quark pair production at LEP / SLC energies, *Comput. Phys. Commun.* **79**, 503–522 (1994).

40. G. Montagna *et al.*, Single photon and multiphoton final states with missing energy at e^+e^- colliders, *Nucl. Phys. B* **541**, 31–49 (1999).

41. A. Heister *et al.* (ALEPH Collaboration), Single photon and multiphoton production in e^+e^- collisions at $\sqrt{s}$ up to 209 GeV, *Eur. Phys. J. C* **28**, 1–13 (2003).

42. J. Abdallah *et al.* (DELPHI Collaboration), Photon events with missing energy in e^+e^- collisions at $\sqrt{s} = 130$ GeV to 209 GeV, *Eur. Phys. J. C* **38**, 395–411 (2005).

43. P. Achard *et al.* (L3 Collaboration), Single photon and multiphoton events with missing energy in e^+e^- collisions at LEP, *Phys. Lett. B* **587**, 16–32 (2004).

44. G. Abbiendi *et al.* (OPAL Collaboration), Photonic events with missing energy in e^+e^- collisions at $\sqrt{s} = 189$ GeV, *Eur. Phys. J. C* **18**, 253–272 (2000).

45. C. King, Multiauthor Papers: Onward and Upward, in *ScienceWatch Newsletter*, July 2012, http://archive.sciencewatch.com/newsletter/2012/201207/multiauthor_papers, last accessed February 6th, 2015.

46. H. Nakamura, *et al.* (MEGA Study Group), Design and baseline characteristics of a study of primary prevention of coronary events with pravastatin among Japanese with mildly elevated cholesterol levels, *Circulation J.* **68**, 860–7 (2004).

Precision Experiments at LEP

W. de Boer

Karlsruhe Institute of Technology,
Institut für Experimentell Kernphysik, Gaedestr. 1,
76131 Karlsruhe, Germany
wim.de.boer@kit.edu

The Large Electron–Positron Collider (LEP) established the Standard Model (SM) of particle physics with unprecedented precision, including all its radiative corrections. These led to predictions for the masses of the top quark and Higgs boson, which were beautifully confirmed later on. After these precision measurements the Nobel Prize in Physics was awarded in 1999 jointly to 't Hooft and Veltman "for elucidating the quantum structure of electroweak interactions in physics".

Another hallmark of the LEP results were the precise measurements of the gauge coupling constants, which excluded unification of the forces within the SM, but allowed unification within the supersymmetric extension of the SM. This increased the interest in Supersymmetry (SUSY) and Grand Unified Theories, especially since the SM has no candidate for the elusive dark matter, while SUSY provides an excellent candidate for dark matter. In addition, SUSY removes the quadratic divergencies of the SM and *predicts* the Higgs mechanism from radiative electroweak symmetry breaking with a SM-like Higgs boson having a mass below 130 GeV in agreement with the Higgs boson discovery at the LHC. However, the predicted SUSY particles have not been found either because they are too heavy for the present LHC energy and luminosity or Nature has found alternative ways to circumvent the shortcomings of the SM.

1. Introduction

The Standard Model is a relativistic quantum field theory describing the strong and electroweak interactions of quarks and leptons, which up to now are considered to be elementary particles. The complexity and non-triviality of the Standard Model (SM) of particle physics is lucidly described in the 36 Nobel Lectures unraveling the stepwise discovery of the SM.[1] The first example of a relativistic quantum field theory was quantum electrodynamics, which describes the electromagnetic interactions by the exchange of a massless photon. The short range of the weak interactions implies that they are mediated by massive gauge bosons, the W and Z bosons, which were discovered at the SPS, as described elsewhere in this volume.

Relativistic quantum field theories based on local gauge symmetries had two basic problems: (i) explicit gauge boson mass terms are not allowed in the SM, since they break the symmetry and (ii) the high energy behaviour leads to infinities in the cross-sections, masses and couplings. The first problem was solved in 1964 by Higgs and others,[2-5] who proposed that gauge boson masses are generated by interactions with an omnipresent scalar (Higgs) field in the vacuum, so no explicit mass terms are needed in the Lagrangian for these dynamically generated masses. The quantum of the Higgs field, the Higgs boson, was discovered at the LHC in 2012, as described elsewhere in this volume. After this discovery Englert and Higgs were awarded the Nobel Prize in 2013. The second problem was solved by "renormalising" the divergent masses and couplings to observable quantities. In this way the electroweak theory becomes a "renormalisable" theory, as proven by 't Hooft and Veltman in the years 1971–1974.[6] This worked well, as demonstrated by the excellent agreement between the calculated and observed radiative corrections, leading to correct predictions for the masses of the top quark and Higgs boson from the electroweak precision experiments at the LEP collider at CERN. 't Hooft and Veltman were awarded the Nobel Prize in 1999 after the confirmation of their calculations at LEP.

How does this contribution fit into this picture? First I will discuss the electroweak precision experiments at LEP, which tested the quantum structure of the SM in great detail. A second topic has to do with physics beyond the SM. The SM is based on the product of the $SU(3) \otimes SU(2) \otimes U(1)$ symmetry groups, so a natural question is: why three groups? And why can we not unify these groups into a larger group, like $SU(5)$, having the SM groups as subgroups[7-9]? The consequences are dramatic: since each $SU(n)$ group is predicted to have $n^2 - 1$ gauge bosons, it doubles the number of gauge bosons (12 in the SM; 24 in $SU(5)$). In $SU(5)$ the leptons from $SU(2)$ and quarks from $SU(3)$ are contained in the same multiplet, which leads automatically to new lepton- and baryon-number violating interactions between leptons and quarks. This inevitably leads to the proton decaying into leptons and quarks via the interactions with the new gauge bosons. In the standard $SU(5)$ the proton lifetime was estimated to be of the order of 10^{31} years.[8] The experimental limits[a] are two orders of magnitude above this prediction,[10, 11] thus excluding grand unification in the SM, but not in the supersymmetric extension of the SM, which predicts a longer proton lifetime.[12]

To explain the long proton lifetime in a unified theory, the new gauge bosons must be heavy. How heavy? Presumably these gauge bosons get a mass by the breaking of the $SU(5)$ symmetry into the $SU(3) \otimes SU(2) \otimes U(1)$ symmetry, just like the W and Z bosons get a mass by breaking of the $SU(2) \otimes U(1)$ symmetry into the $U(1)$ symmetry. Above the $SU(5)$ breaking scale one has a Grand Unified Theory (GUT) with a single gauge coupling constant. Extrapolating the precisely measured gauge couplings at LEP to high energies showed that unification is excluded in the SM, but in

---

[a]Since the background for proton decay experiments is provided by neutrinos, the discovery of different backgrounds for up-going and down-going neutrinos led to the discovery of neutrino oscillations, which implies neutrino masses. This led to the Nobel Prize for Koshiba in 2002.

the supersymmetric extension of the SM the gauge couplings unify and interestingly, at a scale consistent with the long proton lifetime. This result, obtained by simultaneously estimating the GUT scale and the scale of Supersymmetry (SUSY) from a fit to the unification of the gauge couplings,[13] became quickly on the top-ten citation list and was discussed in widely read scientific journals[14-16] and the daily press.

SUSY was developed in the early 70s as a unique extension of the rotational and translational symmetries of the Poincaré group by a symmetry based on an internal quantum number, namely spin, see Ref. 17 for a historical review and original references. SUSY requires an equal number of bosons and fermions, which can be realised only, if every fermion (boson) in the SM gets a supersymmetric bosonic (fermionic) partner. This doubles the particle spectrum, but the supersymmetric partners have not been observed so far, so if they exist, they must be heavier than the SM particles. Not only gauge coupling unification made SUSY popular, since it removes several shortcomings of the SM as well. Especially it provides a dark matter candidate with the correct relic density,[18, 19] see e.g. Refs. 20–23 for reviews. On the other hand, the main shortcoming of SUSY is the fact that none of the predicted supersymmetric partners of the SM particles have been observed, which could be a lack of luminosity or energy at the LHC, as will be discussed in the last section. And of course, other DM candidates exist as well.[24]

2. The Electron–Positron Colliders

After the discovery of neutral currents in elastic neutrino–electron scattering in the Gargamelle Bubble Chamber, as discussed elsewhere in this volume, it was clear that a heavy neutral gauge boson must exist, as predicted by Weinberg.[25] The weak gauge bosons were indeed observed at CERN's proton–antiproton collider SPS, as discussed elsewhere in this volume. But it was clear, that precision experiments would need the clean environment of an e^+e^- collider. The CERN director, John Adams, who had just finished building the SPS, established in 1976 a study group to look into a Large Electron–Positron Collider (LEP) for the production and study of the W and Z bosons, predicted to have masses around 65 and 80 GeV. The group was led by Pierre Darriulat[26] and the famous Yellow Report was delivered half a year later.[27] It contained already many ideas on the physics potential and first design ideas for LEP, which was finally approved in 1982 and started taking data in 1989. The difficulties in realizing such a large project has been described in the book entitled "LEP: The Lord of the collider rings at CERN 1980–2000: The making, operation and legacy of the world's largest scientific instrument" by Herwig Schopper, who was director-general at CERN during the construction of LEP. The book not only covers the technical, scientific, managerial and political aspects, but also discusses the sociological enterprises of building the large experimental collaborations of the LEP experiments with about 500 physicists per collaboration. It also mentions the World-Wide-Web, which was invented during the LEP operation by Berners-Lee and Cailliau in the IT department of CERN to improve the communication and data handling in the large LEP collaborations.

During the same period SLAC set out to build a linear collider by equipping the existing linear accelerator with damping rings and bending sections at the end to bring the sequentially accelerated bunches of electrons and positrons into collision. Although on paper SLAC was expected to be ready before LEP, the pioneering task of colliding bunches of electrons and positrons in a linear collider took longer than anticipated, so finally, in the summer of 1989, the MARK-II collaboration observed its first few hundred Z events[28] just before LEP came into operation.

With a 45 kHz bunch crossing rate at LEP versus a 120 Hz repetition rate at the SLC the data sample at LEP quickly outgrew the one at the SLC, since at its peak luminosity of 10^{32} cm^{-2}s^{-1} each LEP experiment collected about 1000 Z bosons per hour. A brief review of all the ups and downs on the way to reach a luminosity at LEP above its design value was given at the Topical Seminar on "The legacy of LEP and SLC" in Sienna in 2001.[29] This review on the LEP accelerator describes also the precise beam energy determination via spin depolarisation techniques, which can determine the beam energy to 0.2 MeV or a relative accuracy of $5 \cdot 10^{-6}$. In addition, the many surprises, like the correlation of the tides from the gravitational interaction between the moon and the earth or the amount of water in Lake Geneva with the beam energy, are described. These effects of a few MeV in the beam energy correspond to a change in the orbit length of a few mm, caused by the elasticity of the earth's crust. Also the short term energy fluctuations from the fast TGV train between Geneva and Paris, for which the LEP magnets turned out be a good current return path, were finally understood after these fluctuations were absent during a railway strike in France. The final uncertainty of about 2 MeV in the Z mass from the beam energy is considerably larger, mainly because the field of the dipole magnets varies with time. A schematic picture of the 27 km long LEP tunnel and its experiments is shown in Fig. 1, together with the joyful faces after the start of the operation in July 1989.

After LEP started running the SLC made an amazing improvement by providing highly polarised beams, which are a sensitive probe of the weak interactions, in which left- and right-handed particles have different couplings. These data were largely collected by the SLD detector, which could determine the electroweak mixing angle with comparable precision in spite of the much smaller data sample of about half a million Z bosons (in comparison with 17 million events for the combined LEP experiments). At LEP a polarisation scheme had been studied in great detail as well,[30] but finally it was discarded in favour of going to higher energies as quickly as possible.

In 1995, LEP was upgraded to reach the WW and ZZ pair production threshold and later on up to 208 GeV (by adding more accelerating cavities) in the hunt for the Higgs. One could set a 95% C.L. lower limit of 114.4 GeV on the Higgs mass,[31] just 11 GeV short of the Higgs mass found at the LHC in 2012. This higher

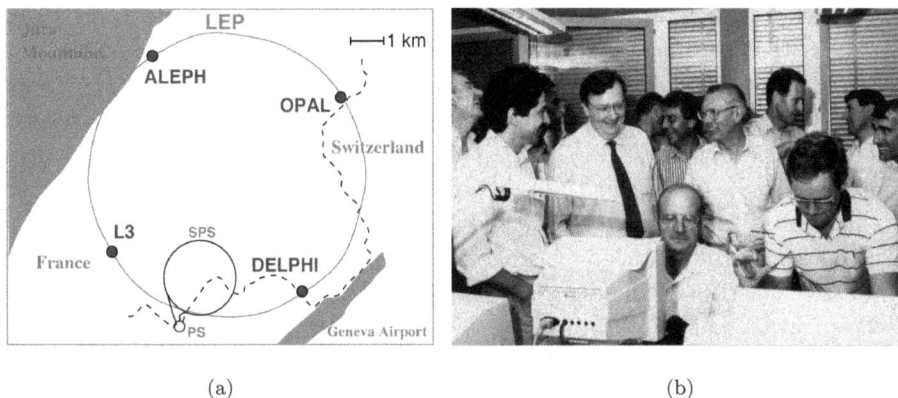

(a)   (b)

Fig. 1. (a): The LEP storage ring with the four experiments and its pre-accelerators (PS and SPS). (b): Happy faces during the start of LEP in July 1989.

energy could have been reached, if all available space at LEP would have been filled with superconducting cavities, in which case Higgs masses up to the SUSY upper limit of 130 GeV[32] could have been reached, see e.g. the review on LEP and SLC results.[33] However, the time and financial pressure from the LHC in competition with a Tevatron upgrade (the SCC had been abandoned two years before in 1993 due to budget problems) led to the decision to stop LEP operation in 2000. Of course, in retrospect, the Higgs boson could have been discovered 10 years earlier at LEP and studied in the clean environment of an e^+e^- collider.

3. The Four LEP Detectors

In total four LEP detectors were approved: ALEPH (Appartus for LEP Physics),[34] DELPHI (Detector with Lepton and Hadron Identification),[35] L3 (Letter of Intent 3)[36] and OPAL (Omnipurpose Apparatus for LEP).[37] All detectors are large 4π detectors with sizes of typically 10 m in each direction and a weight of up to thousand medium-sized cars. They are designed to study the hadronic, electromagnetic and leptonic components of the final states of the Z boson, but they differ in experimental techniques, like resolution of the magnetic spectrometers, the electromagnetic- and hadronic calorimeters and the extent of particle identification. In addition, all detectors were upgraded to have silicon based vertex detectors just outside the beam pipe (see Ref. 38 for a review), which allowed to locate the primary collision vertex typically with a precision of a few μm. This allowed to tag jets from b- and c-quarks by their secondary vertex, since the long-lived B- and D-mesons travel on average several mm before decaying and producing a secondary vertex.

The resources and manpower needed for large detectors require large collaborations, typically 250 at the start of LEP and climbing to 500 physicists at the end. Around 20–50 institutions were involved, most of them from the European member states, but also from Asia, Isreal, Russia and the US. The ALEPH and DELPHI detectors were considered "risky" by the LEP Experiments Committee, since they used superconducting magnets[b] and time-projection-chambers as 3D tracking devices. In addition, ALEPH used large liquid argon electromagnetic calorimeters, while DELPHI applied the 3D time-projection idea also to the electromagnetic calorimeter and installed in addition Ring Imaging Cherenkov (RICH) detectors for hadron identification. The L3 and OPAL detectors used more conventional techniques, like wire chambers for tracking, a warm magnet and scintillating crystals as electromagnetic calorimeters.

One may wonder why one needed as many as four experiments at LEP. Would two not have been enough? The four detectors do not only provide redundancy, but have different systematic uncertainties. The redundancy turned out to be of utmost importance to investigate fluctuations, like the many standard deviations excess in 4-jets[39] and the Higgs-like signal with a mass around 115 GeV.[40] If the Higgs-like signal, mainly based on three ALEPH events, was combined with all other experiments, the significance was less than 2σ. We now know from the observed Higgs mass that it was indeed a statistical fluctuation. Also the 4-jet excess turned out to be a fluctuation, as was clear from the combined data of all experiments.[41]

And last, but not least, in spite of the impressive data sample, in ratios involving leptonic decay modes, the statistical errors still dominate, so they profit from a factor two lower error after combining the data from the four experiments. The combination holds also the risk of dominating common systematic theory errors, which, if not correctly estimated, may change the results. We will see examples in the discussion of the coupling constants.

In spite of being competitors, the four experiments collaborated in working groups to combine all experimental data in order to get the most precise answers to the questions asked. Prominent working groups were the Electroweak Working Group (EWWG), the Heavy Flavour Working Group, the Higgs Working Group and the Working Group on searches. This working in large collaborations and even combining data from different collaborations was a turning point in the history of high energy physics, not only important for LEP, but also a sociological exercise for LHC, where the largest collaborations grew to about 3000 collaborators.

4. Quantum Corrections to the W and Z Boson Masses

The interaction between two matter particles can be mediated by a gauge boson, which leads for massless gauge bosons to a propagator factor $g^{\mu\nu}/q^2$ in the Feynman diagram, where q is the momentum flowing through the propagator and $g^{\mu\nu}$ is the Minkowski metric with Lorentz indices μ and ν. For a massive gauge boson with

---

[b]The DELPHI solenoid was with 6.2 m in diameter, 7.2 m in length and a field of 1.2 T the world's largest superconducting magnet.

mass m the propagator gets an additional factor $k_\mu k_\nu / m^2$. This factor, originating from the longitudinal spin degree of freedom of the gauge boson, becomes infinite, if the momenta k of the incoming and outgoing particles become infinite. This infinity can only be compensated by adding a counterpiece, so in general the propagator of a massive particle is:

$$\frac{g_{\mu\nu} - \frac{k_\mu k_\nu}{m^2}}{q^2 - m^2 + i\epsilon} + \frac{\frac{k_\mu k_\nu}{m^2}}{q^2 - \frac{m^2}{\lambda} + i\epsilon}, \tag{1}$$

where the gauge parameter λ can be chosen as 0, 1 or infinity, which corresponds to the unitarity gauge, Feynman or 't Hooft gauge and Landau or Lorentz gauge, respectively. The last term in Eq. (1) represents the propagator of a scalar particle for $\lambda = 1$, i.e. in the Feynman or 't Hooft gauge. In this case the physics behind the compensation of the $k_\mu k_\nu / m^2$ term is simple: the infinity in the amplitude of longitudinal W boson exchange is compensated by the exchange of a Higgs boson, so the calculated cross-section will not pass the unitarity limit.[c] As 't Hooft noted in his Nobel Lecture:[1] people knew that gauge boson masses can be generated by the Higgs mechanism, but they did not know that this was a *unique* solution, since at the same time it removes the infinities, thus making the theory renormalisable.[d] An important aspect of proving the renormalisability of the SM is a recipe on how to technically handle the divergences. This was done most conveniently by dimensional regularisation, as discussed by 't Hooft and Veltman.[6] But what was of utmost importance for LEP: with such a renormalisation scheme Veltman could calculate the radiative corrections from Higgs boson and fermion loops to the weak gauge bosons, depicted in Fig 2(a), and found surprisingly that the corrections depend quadratically on the top mass.[42] For the Higgs mass the quadratic term happens to have zero amplitude,[e] so only a logarithmic dependence is left. After electroweak symmetry breaking via the Higgs mechanism the mass eigenstates become linear combinations of the gauge bosons of the original (symmetric) Lagrangian ($W^i, i = 1, 2, 3$ for SU(2) and B for U(1)): $W^\pm = (W^1 \mp W^2)/\sqrt{2}$, $Z = -B \cos\theta_W + W^3 \sin\theta_W$, $\gamma = B \sin\theta_W + W^3 \cos\theta_W$, where the electroweak mixing angle θ_W is determined by the ratio of the coupling constants of the U(1) and SU(2) groups: $\tan\theta_W = g'/g$ and its relation to the electric charge is depicted in Fig. 2(b), implying $e = g \sin\theta_W$.

---

[c] Weinberg noted in his Nobel Prize lecture,[1] that he did not succeed in proving the renormalisability, since he was using the unitarity gauge, which has the advantage of exhibiting the true particle spectrum, but the disadvantage of obscuring the renormalisability, as is obvious from Eq. (1).

[d] 't Hooft noted also that the unitarity problem did not bother him, since he had discovered already that the SU(3) group had a negative β-function, thus decreasing the cross-section at high energy. However, he did not realise "what treasure he had here", so he did not connect it to asymptotic freedom a discovery for which Gross, Wilczek and Politzer got the Nobel Prize in 2004. He expected anyway that all experts would know about the different signs of the β-function in QED and QCD.

[e] Veltman called this the "screening theorem", since the Higgs boson "screens" itself against detection via observable radiative corrections.

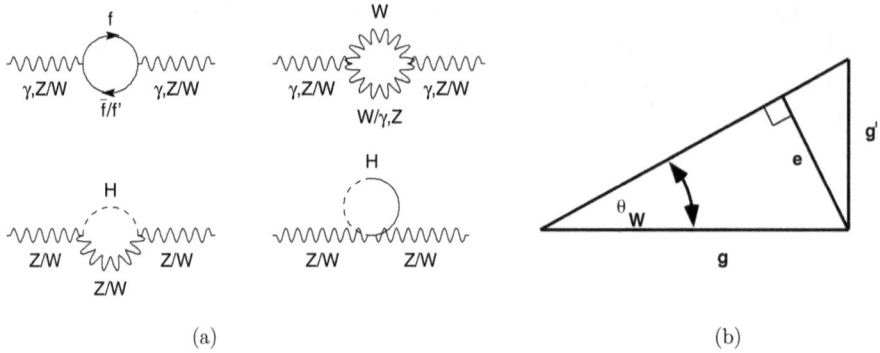

Fig. 2.   (a): Loop corrections to the SM propagators. (b): Relations between gauge couplings.

Since the Higgs mechanism predicts the gauge boson masses to be proportional to the gauge couplings one finds:

$$\cos\theta_W = \frac{g}{\sqrt{g'^2 + g^2}} = \frac{M_W}{M_Z} \quad \text{or} \quad \rho_0 = \frac{M_W^2}{M_Z^2 \cos^2\theta_W}. \tag{2}$$

In the SM $\rho_0 = 1$, but it can deviate from 1 for a more complicated Higgs structure. The muon decay proceeds via W exchange, so the W mass is related to the muon decay constant: $G_F = \pi\alpha/(\sqrt{2}\sin^2\theta_W M_W^2)$, which leads to $M_W^2 = A^2/\sin^2\theta_W$, $M_Z^2 = A^2/(\sin^2\theta_W \cos^2\theta_W)$ with $A = \sqrt{\pi\alpha/\sqrt{2}G_F} = 37.2805$ GeV. This value of A leads with $\sin^2\theta_W = 0.2314$ to $M_Z = 88$ GeV. However, these relations hold only at tree level and are modified by loop corrections (see Fig. 2(a)):

$$\sin^2\theta_W = \left(1 - \frac{M_W^2}{M_Z^2}\right) = \frac{A^2}{1 - \Delta r}, \tag{3}$$

where the radiative corrections have been lumped into Δr, which depends quadratically on the top mass and logarithmically on the Higgs mass. These definitions are valid in the so-called on-shell renormalisation scheme,[43–46] in which case the electroweak mixing angle is defined by the on-shell masses of the gauge bosons: $\sin^2\theta_W \equiv 1 - M_W^2/M_Z^2$. In this scheme $\Delta r \approx \Delta r_0 - \rho_t/\tan^2\theta_W$, where $\Delta r_0 = 1 - \alpha/\alpha(M_Z) = 0.06637(11)$ and $\rho_t = 3G_F M_t^2/8\sqrt{2}\pi2 = 0.00940(M_t/173.24\,\text{GeV})^2$. The latter term shows the quadratic top quark dependence, which is enhanced by $1/\tan^2\theta_W = 3.32$, so the negative M_t corrections are almost 50% of the dominant Δr_0 correction.

The on-shell renormalisation scheme has been used by the EWWG for the analysis of the LEP electroweak precision data. An alternative scheme, the modified minimal subtraction $\overline{MS}$ scheme,[47] is extensively used in QCD. In this scheme the electroweak mixing angle is not defined by the masses $(\sin^2 \theta_W \equiv 1 - M_W^2/M_Z^2)$, but defined by the tree level values of the couplings: $\sin \theta_{\overline{MS}} \equiv g'/\sqrt{g'^2 + g^2}$ (see Fig. 2(b)) with all couplings defined at the Z mass.[f] The total cross-section must be independent of such a choice, so the masses in the $\overline{MS}$ scheme must be redefined to: $M_W^2 = A^2/(\sin^2 \theta_{\overline{MS}}(1 - \Delta r_{\overline{MS}}))$ and $M_Z^2 = M_W^2/(\rho_{\overline{MS}} \cos^2 \theta_{\overline{MS}})$, where $\Delta r_{\overline{MS}} \approx \Delta r_0$ and $\rho_{\overline{MS}} \approx 1 + \rho_t$. With these definitions M_W becomes practically independent of the top mass. This is reasonable, since its value is determined by G_F, which has the radiative corrections absorbed in the measurement. All top mass dependent corrections are now included in M_Z and the couplings between the Z boson and the fermions.

The W bosons couple only to left-handed particles and right-handed antiparticles with a strength given by the weak charge I_3, which is $+1/2$ for the neutrinos and up-type quarks, $-1/2$ for the charged leptons and down-type quarks. The right-handed particles have vanishing weak charge, i.e. $I_3 \approx 0$.[g] The photon couples equally to left- and right-handed particles, so after mixing of W^3 and B the Z couplings obtain an electromagnetic component $-Q_f \sin^2 \theta_W$: $g_L^f = \sqrt{\rho_f}(I_3^f - Q_f \sin^2 \theta_W)$ and $g_R^f = -Q_f \sin^2 \theta_W$. The vector and axial vector couplings are defined as:

$$g_V^f = g_L^f + g_R^f = \sqrt{\rho_f}(I_3^f - 2Q_f \sin^2 \theta_{\text{eff}}) \qquad g_A^f = g_L^f - g_R^f = \sqrt{\rho_f}I_3^f, \qquad (4)$$

where $\sin^2 \theta_{\text{eff}}^f = \kappa^f \sin^2 \theta_W$ is the effective mixing angle, i.e. the one including radiative corrections. At tree level $\rho_f = \rho_0 = 1$, except for the b quark, since the vertex correction from a triangle loop with top quarks and a W boson changes slightly the b quark production cross-section. In this case[48]

$$\rho_b \approx 1 + \frac{4}{3}\rho_t \quad \text{and} \quad \kappa_b \approx 1 + \frac{2}{3}\rho_t. \qquad (5)$$

The difference between the effective mixing angle and the $\overline{MS}$ mixing angle for $f \neq b$ is small and almost independent of the Higgs and top mass: $\sin^2 \theta_{\text{eff}}^f - \sin^2 \theta_{\overline{MS}}^f = 0.00029$,[48] an important relation, since the LEP electroweak working group always determines $\sin^2 \theta_{\text{eff}}^l$, but for gauge coupling unification one needs the value in the $\overline{MS}$ scheme.

---

[f]The values of the electroweak mixing angles are related in both schemes by $\sin^2 \theta_{\overline{MS}} = c(M_t, M_H) \sin^2 \theta_W = 1.0344 \pm 0.0004) \sin^2 \theta_W$, where $c(M_t, M_H) = 1 + \rho_t$, so in this case the couplings become dependent on the top mass.

[g]The difference in the weak charge between left and right is the basis for the famous parity violation, observed in 1954 by C. S. Wu and explained by Yang and Lee, who received for this fundamental discovery the Nobel Prize in 1957.[1]

5.   SM Cross-Sections, Asymmetries and Branching Ratios

The differential cross-section for e^+e^- annihilation into fermion pairs can be written as:[48]

$$\frac{2s}{\pi}\frac{1}{N_c^f}\frac{d\sigma_{ew}}{d\cos\theta}(e^+e^- \to f\bar{f}) = \alpha^2(s)\left[F_1(1+\cos^2\theta)+2F_2\cos\theta\right]+B, \qquad (6)$$

where $F_1 = Q_e^2Q_f^2\chi Q_eQ_fg_V^eg_V^f\cos\delta_R + \chi^2(g_V^{e2}+g_A^{e2})(g_V^{f2}+g_A^{f2})$, $F_2 = -2\chi Q_eQ_f$ $g_A^eg_A^f\cos\delta_R + 4\chi^2g_V^eg_A^eg_V^fg_A^f$, $\tan\delta_R = M_Z\Gamma_Z/(M_Z^2-s)$, $\chi(s) = (G_FsM_Z^2)/$ $\left(2\sqrt{2}\pi\alpha(s)\left[(s-M_Z^2)^2+\Gamma_Z^2M_Z^2\right]^{1/2}\right)$, $\alpha(s)$ is the energy dependent electromagnetic coupling and θ is the scattering angle of the out-going fermion with respect to the direction of the e^- beam. The colour factor N_c^f is either one (for leptons) or three (for quarks), and $\chi(s)$ is the propagator term; B represents small contributions from the electroweak box graphs. The cross-section is asymmetric around the peak, as illustrated in Fig. 3(a): at energies above the peak the cross-section is higher, because of QED corrections, mainly from single photon radiation off the incoming beams. After radiating a photon the effective CM energy is reduced, thus increasing the cross-section at the effective CM energy. The asymmetry in the cross-section can be described by a radiator function,[49] which is usually taken into account in the fitting function.

Since an axial vector changes its sign in a mirror, the axial vector coupling is responsible for the cosine term in Eq. (6), which leads to asymmetries in the angular dependence of the cross-section or in the polarisation asymmetry in case of polarised beams. Defining for a fermion f:

$$A_f = \frac{2g_V^fg_A^f}{g_V^{f2}+g_A^{f2}} = \frac{2g_A^f/g_V^f}{1+(g_A^f/g_V^f)^2}, \qquad (7)$$

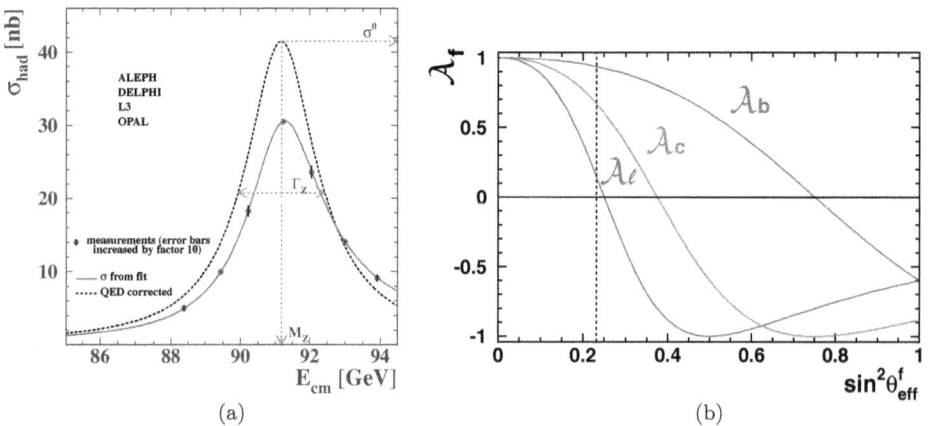

(a)   (b)

Fig. 3.   (a): Hadronic cross-section with and without radiation. (b): Sensitivity of the asymmetry to $\sin^2\theta_W$ for various fermionic final states.

Table 1   Z branching ratios for $x = \sin^2\theta_W = 0.2315$.

Particles Symbol	Couplings (Eq. (4))			Branching ratios	
	g_V	g_A	$\sum(g_V^2 + g_A^2)$	calc.	obs.
ν_e, ν_μ, ν_τ	$\frac{1}{2}$	$\frac{1}{2}$	$3(\frac{1}{2})^2 + 3(\frac{1}{2})^2$	20.5%	$20.00 \pm 0.06\%$
e, μ, τ	$-\frac{1}{2} + 2x$	$-\frac{1}{2}$	$3(-\frac{1}{2} + 2x)^2 + 3(\frac{1}{2})^2$	10.3%	$10.097 \pm 0.0069\%$
u,c	$\frac{1}{2} - \frac{4}{3}x$	$\frac{1}{2}$	$6(\frac{1}{2} - \frac{4}{3}x)^2 + 6(\frac{1}{2})^2$	23.6%	$23.2 \pm 1.2\%$
d,s	$-\frac{1}{2} + \frac{2}{3}x$	$-\frac{1}{2}$	$6(-\frac{1}{2} + \frac{2}{3}x)^2 + 6(\frac{1}{2})^2$	30.3%	$31.68 \pm 0.8\%$
b	$-\frac{1}{2} + \frac{4}{3}x$	$-\frac{1}{2}$	$3(-\frac{1}{2} + \frac{2}{3}x)^2 + 3(\frac{1}{2})^2$	15.3%	$15.12 \pm 0.05\%$

one finds for the forward–backward asymmetries A_{FB} from the cross-sections integrated over the forward (σ_F) and the backward (σ_B) hemisphere, $A_{FB} = (\sigma_F - \sigma_B)/(\sigma_F + \sigma_B) = 3A_e/4A_f$ and the left–right asymmetry from the cross-sections $\sigma_{L,R}$ for left- and right-handed polarised electrons, $A_{LR} = (\sigma_L - \sigma_R)/(\sigma_L + \sigma_R) = A_e$, all of them being determined by the ratio g_A/g_V, so they are sensitive to the electroweak mixing angle $\sin^2\theta_W$ (see Eq. (4)), especially for the leptons, since g_V changes sign for $\sin^2\theta_W = 1/4$, while for quarks the zero-crossing happens at much larger values, as shown in Fig. 3(b). However, for quarks the asymmetries are larger for $\sin^2\theta_W = 1/4$, thus reducing the relative systematic errors.

The weak mixing angle completely determines the branching fractions $\sum(g_V^2 + g_A^2)/\sum_{tot}$, where the numerator is summed over the fermions considered and $\sum_{tot}$ is the sum over all possible fermions. The branching fractions, calculated for $x = \sin^2\theta_W = 0.2315$, agree reasonably well with observations, as demonstrated in Table 1. The small discrepancies with the observed values originate from neglected fermion masses and missing higher order radiative corrections, since only the dominant radiative correction at the b-vertex from the top loop (Eq. (5)) has been taken into account.

6.   LEP I Electroweak Results

The final legacy papers describing and interpreting the results in the framework of the SM were published in *Physics Reports* in 2006 for the Z production at LEP I[50] and in 2013 for the W pair production at LEP II.[51] The four LEP experiments, shortly described in Section 3, collected between 1990 and 1995 a total of 17 million Z events distributed over seven CM energies with most of the luminosity taken at the peak. The total cross-section is given by $\sigma_{tot} = (N_{sel} - N_{bg})/(\epsilon_{sel}\mathcal{L})$, where N_{sel} is number of selected events in a final state, N_{bg} the number of background events, ϵ_{sel} the selection efficiency including acceptance, and $\mathcal{L}$ the integrated luminosity. We shortly discuss the uncertainties in these variables. The combination of magnetic spectrometers with good tracking, electromagnetic and hadronic calorimeters and muon tracking allows a good discrimination of q$\bar{\text{q}}$ from $\ell^+\ell^-$ final states and a strong reduction of background, which was typically below 1% for all final states (except for hadronic tau final states, where the background went up to 3%). Since the background is largely independent of the LEP energy it provides a constant background, so it can be determined experimentally from off-peak measurements and is small, as discussed above.

The luminosity is determined from small angle Bhabha scattering using the acceptance calculations and cross-section from the the the program BHLUMI, which was used by all experiments leading to a correlated common error from the higher order uncertainties in the Bhabha scattering cross-section of 0.061%.[52] From calorimeters with high angular resolution silicon detectors the experiments obtained a luminosity error of about 0.1%, which led to an experimental error in the cross-sections from the global fit comparable to the theoretical uncertainty from the higher order corrections.

The acceptance is limited largely by the geometrical acceptance. The electromagnetic calorimeters have typically a geometrical acceptance of $|\cos\theta| \leq 0.7$, the muon trackers typically $|\cos\theta| \leq 0.9$. For the hadronic final states the jets do not have a sharp angular edge for the acceptance, so the acceptance is limited by requiring a fraction of the total CM energy to be visible in the detector (typically 10%). Since the simulation programs of the Z decays and the detector simulation[h] are realistic inside the acceptance, the total efficiency can be extrapolated reliably to the full acceptance. Inside the acceptance the trigger efficiency is usually high, since events can be triggered by a multitude of signals, like track triggers, calorimetric triggers and combinations thereof. The selection efficiencies inside the acceptances are high, above 95% for electrons and muon pairs and 70–90% for tau pair final states. The symmetric Breit–Wigner function can be described by the mass, the width and the peak height. The leptonic cross-sections can be parametrised by the ratio of hadronic and leptonic widths: $R_\ell^0 = \Gamma_{\text{q}\bar{\text{q}}}/\Gamma_{\ell\ell}$. Since lepton universality was compatible with all observations, we quote only results including lepton universality. The fitted values for these parameters from the various experiments and their combination are shown in Fig. 4. One observes that for the combined values of the experiments the common systematic errors are large in case of the hadronic final states, but for the leptonic final state the statistical error is still significant. The systematic errors on mass and width are dominated by the uncertainty of the LEP energy (around 2 MeV, as discussed in Section 2) and for the cross-section by the luminosity error discussed above.

The combined fit to all data requires a knowledge of the correlated errors between the observables and the experiments. These correlations can be taken into account by minimizing $\chi^2 = \Delta^T V^{-1}\Delta$, where Δ is the vector containing the N residuals between the N measured and fitted values, V is the $N \times N$ error matrix where the

---

[h]Details about the simulation software can be found in Ref. 50.

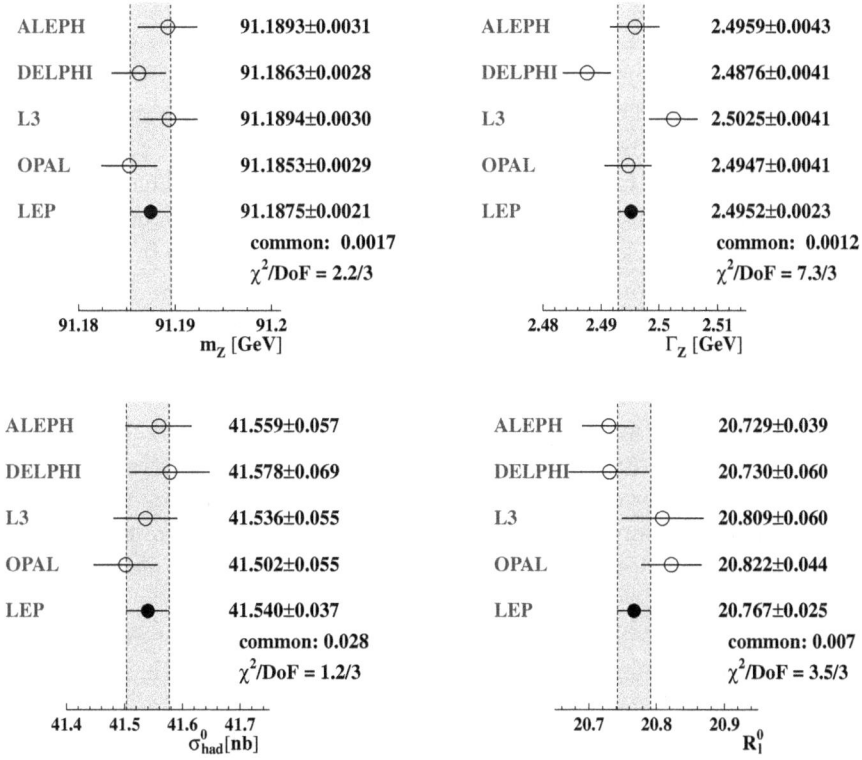

Fig. 4. The fitted values of the mass, width (top row), peak cross-section and ratio of hadronic to leptonic width (bottom row) of the Z boson. From Ref. 50.

diagonal elements σ_{ii}^2/O_i^2 represent the relative total error squared for observable O_i and the off-diagonal elements $\sigma_{ij}^2/(O_iQ_j)$ the relative correlated error. For example, the correlated error of the Bhabha luminosity of 0.061% is added in quadrature to all off-diagonal elements of observables depending on the luminosity. This method was pioneered for e^+e^- annihilation data from the DORIS and PETRA colliders at DESY and the TRISTAN collider at KEK, where the tail of the Z resonance increases the hadronic cross-section already by 50% at the highest energy of 57 GeV.[53] The complete correlation matrices for all LEP data can be found in the final report from the EWWG.[50]

7. Constraints on the SM

The measurements of the cross-sections and asymmetries discussed above can all be predicted in the SM, if one knows the three gauge couplings, the gauge boson masses and the masses of the top quark and Higgs boson. Since the electromagnetic and weak couplings are related via the gauge boson masses, only two coupling constants are needed: $\alpha(M_Z)$ and $\alpha_s(M_Z)$. Furthermore, M_W can be traded for G_F, which was recently measured from the muon lifetime to 0.5 ppm: $G_F = 1.1663787(6) \cdot 10^{-5}\,\text{GeV}^{-2}$.[54] This value is precise enough to be considered a constant in the fit. The masses of the light fermions have only a small effect on the cross-section and their effect can be calculated with sufficient precision. $\alpha(M_Z)$ is in principle known from the running from its low energy value, but the loop corrections including quarks have a significant uncertainty. Therefore, the hadronic contribution for 5 quarks to $\Delta\alpha_{\text{had}}^{(5)}(M_Z^2)$ is taken as a parameter in the fit (instead of $\alpha(M_Z)$) with the constraint from the experimental knowledge on $\Delta\alpha_{\text{had}}^{(5)}(M_Z^2)$. The SM parameters to be fitted to the measured observables are then: M_Z, M_t, M_H, α_s, $\Delta\alpha_{\text{had}}^{(5)}(M_Z^2)$. Given these parameters all observables can be calculated, e.g. with the programs TOPAZ0,[55] ZFITTER[56] or GAPP.[57]

The quadratic top quark dependence of the loop corrections to the gauge boson masses led quickly to first estimates of the top mass from the precise Z boson mass measurements, as shown in Fig. 5(a). These top mass estimates were confirmed latter by direct measurements, as shown by the data points from the Tevatron experiments in Fig. 5(a), which in turn agree with the LHC measurements, as shown in Fig. 5(b).[58]

From a fit to the Z-pole data and preliminary data for M_t and M_W the EWWG finds for these parameters:[50] $M_Z = 91.1874 \pm 0.0021$, $M_t = 178.5 \pm 3.9$ GeV, $M_H = 129^{+74}_{-49}$ GeV, $\alpha_s = 0.1188 \pm 0.0027$ and $\Delta\alpha_{\text{had}}^{(5)}(M_Z^2) = 0.02767 \pm 0.00034$.[i] These five parameters describe the data quite well, as shown in Fig. 6(a), which displays

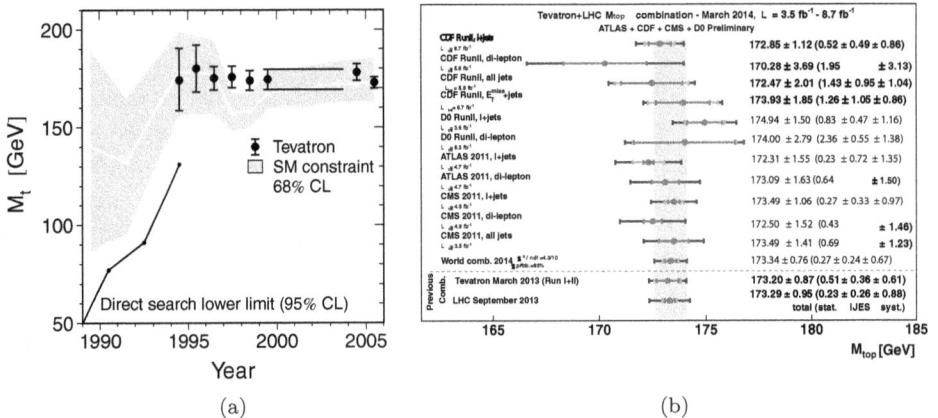

Fig. 5. (a): The measured top quark mass as function of time.[50] The indirect determinations from the electroweak fits (shaded area) to LEP data predicted a heavy top quark mass before it was discovered at the Tevatron (data points). (b): A summary of direct top quark measurements.[58]

---

[i]With newer data the value quoted in the Particle Data Book[48] has a considerably smaller error: 0.02771 ± 0.00011.

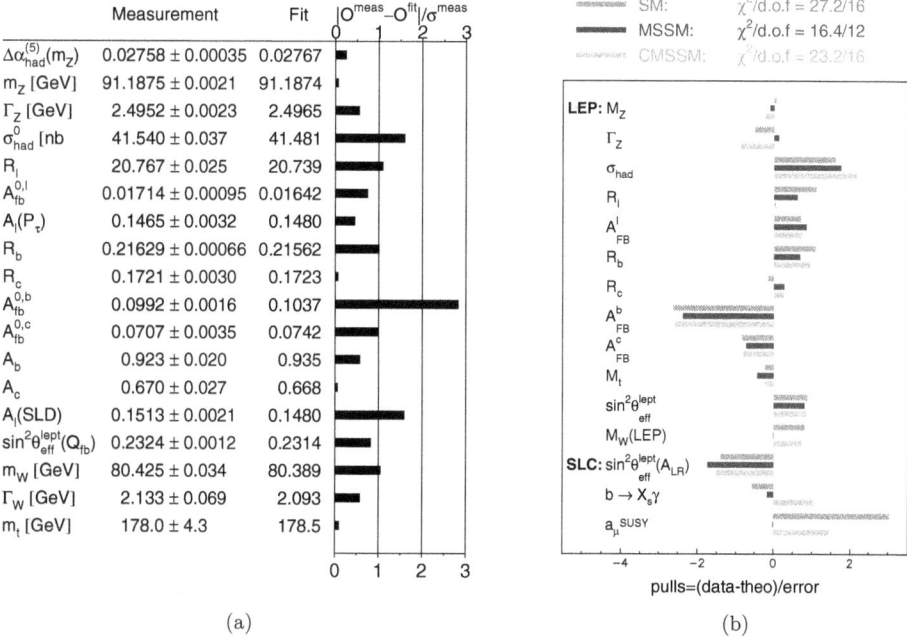

Fig. 6. (a): A comparison of the measured and calculated values of the precision electroweak observables and a graphical presentation of the difference, expressed in standard deviations ("pulls"). The fit has a χ^2/d.o.f of 18.3/13, corresponding to a probability of 15%. (b): A comparison of the pulls in the SM, the minimal supersymmetric SM (MSSM) and constrained MSSM (CMSSM).

the difference between the calculated and observed values of the observables. The largest pull of 2.8σ is caused by the forward–backward asymmetry of the b quarks, followed by 1.6σ for the peak cross-section and the left–right polarisation asymmetry from SLC. The correlation between the Higgs mass and $\sin^2\theta_W$ is demonstrated in Fig. 7(a), where the diagonal shows the SM prediction. The two horizontal bands show the $\sin^2\theta_W$ values from A_{LR} and $A_{FB}^{0,b}$, which lead to quite different Higgs mass values, as is apparent from the crossing with the SM prediction. The vertical (yellow) band shows the expected Higgs mass in the supersymmetric extension of the SM. The Higgs boson mass observed at the LHC falls inside this band, which crosses the SM prediction at a $\sin^2\theta_W$ value close to the value from the averaged asymmetries.

In addition to the discrepancy in the asymmetries, the anomalous magnetic moment of the muon a_μ shows a 3σ deviation from the SM.[59] Supersymmetric loop corrections to a_μ reduce the observed difference between theory and experiment, so many groups have tried to improve the fit in supersymmetric extensions of the SM, both in the Minimal Supersymmetric SM (MSSM) and in the constrained version (CMSSM), see reviews in Refs. 60–61 for details. Here minimal means the minimum extension of the SM, i.e. one superpartner for each SM particle and a minimal Higgs sector of two Higgs doublets. In the CMSSM one assumes in addition unification

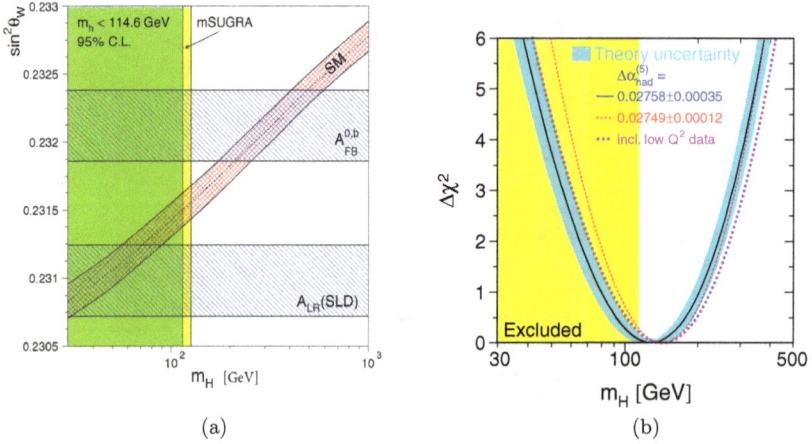

Fig. 7. (a): The values of $\sin^2 \theta_W$ versus the Higgs mass. The two horizontal bands correspond to the $\sin^2 \theta_W$ values from A_{LR} and $A_{FB}^{0,b}$. The diagonal band corresponds to the SM prediction for the parameters from the global fit. The shaded (green) area for $M_H < 114.3$ GeV is excluded by LEP data. (b): The $\Delta\chi^2$ distribution as function of the Higgs mass from the LEP I and SLC data before the Higgs boson discovery, but including the constraints from M_W and M_t.[50] The minimum corresponds to $M_h = 129^{+74}_{-49}$ GeV.

of gauge couplings and SUSY masses at the GUT scale.[j] All other observables are fitted approximately as good in the SM and (C)MSSM, especially the value of $A_{FB}^{0,b}$ does not improve with SUSY, as shown by the "pulls" in Fig. 6(b).[63] Although the χ^2 is smaller in the (C)MSSM, the probability stays similar, because of the larger number of parameters.

7.1. *Constraints on the SM after the Higgs discovery*

The global fits have been repeated after the Higgs discovery and the results have been described by Erler and Freitas in the electroweak review of the Particle Data Group.[48] Also newer values from M_W, G_F and M_t have been included. The anomalous muon magnetic moment has been fitted as well. The global fit including the measured top quark and Higgs boson masses yields a good $\chi^2/\text{d.o.f}$ of 48.3/44. The probability to obtain a larger χ^2 is 30%.

To check the consistency between direct mass measurements of M_W, M_t and M_H and the SM predictions via indirect measurements we show two examples from the Particle Data Group.[48] Figure 8(a) shows the SM prediction for M_W versus the top quark mass as the light (green) diagonal contour, which shows the quadratic dependence of the gauge boson mass on the top quark for a Higgs boson of 125 GeV. This contour almost collapses into a line, because the precisely measured

---

[j]With the present lower limits on SUSY masses the deviations of a_μ cannot be explained in the CMSSM, for details, see e.g. Ref. 62 and references therein.

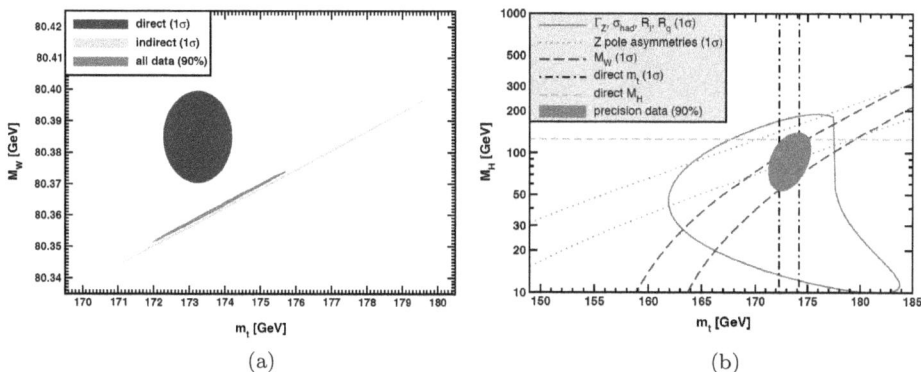

Fig. 8. (a): Allowed 1σ contours with a probability of 39.35% in the M_W versus M_t plane for the direct (dark (blue) ellipse) and indirect measurements (light (green) "line"). The dark (red) "line" is the 90% C.L. contour ($\Delta\chi^2 = 4.605$) allowed by all data. From Ref. 48. (b): Allowed 1σ contours with a probability of 39.35% in the M_H versus M_t plane for various observables. The dark (red) ellipse corresponds to the 90% C.L. contour ($\Delta\chi^2 = 4.605$) from a global fit to all data. From Ref. 48.

Higgs mass was included in the fit. Otherwise the line would have been a band in this plane, since higher Higgs boson masses would shift this line parallel to lower W masses. The direct measurements of M_W and M_t are bounded by the dark (blue) ellipse. These contours of the direct and indirect measurements (green and blue) correspond to 1σ with a probability of 39%. Combining the direct and indirect measurements leads to the dark (red) "line", for which $\Delta\chi^2 = 4.61$ or a probability of 90% was chosen. The value of the directly measured M_W mass is 1.5σ above the SM prediction,[48] which implies some tension between M_W and M_H, since lower values of M_H would shift the SM prediction upwards. This tension is also visible in Fig. 8(b), which shows the allowed 1σ contours in the M_H versus M_t plane from various indirect measurements. The direct measurements are indicated by the horizontal and vertical lines. The error on the Higgs mass is not visible on this scale. The dark (red) ellipse corresponds to the 90% C.L. contour ($\Delta\chi^2 = 4.605$) from a global fit to all data.[48] The central value of the ellipse (indirect measurements) is slightly below the direct measurement of the Higgs boson mass, since the slightly high value of M_W pulls the Higgs mass to lower values. Although the indirectly measured Higgs mass is not precise, it indicated for the first time that a Higgs boson is needed with a mass around the electroweak scale, a value predicted by SUSY.[32] In the SM, the Higgs boson mass is not predicted.[64]

8. LEP II Electroweak Results

The LEP II data allowed to investigate the selfcoupling of the gauge bosons by studying W pairs, which can be produced in e^+e^- annihilation via t-channel neutrino-exchange and s-channel photon, Z and Higgs exchange. As mentioned in

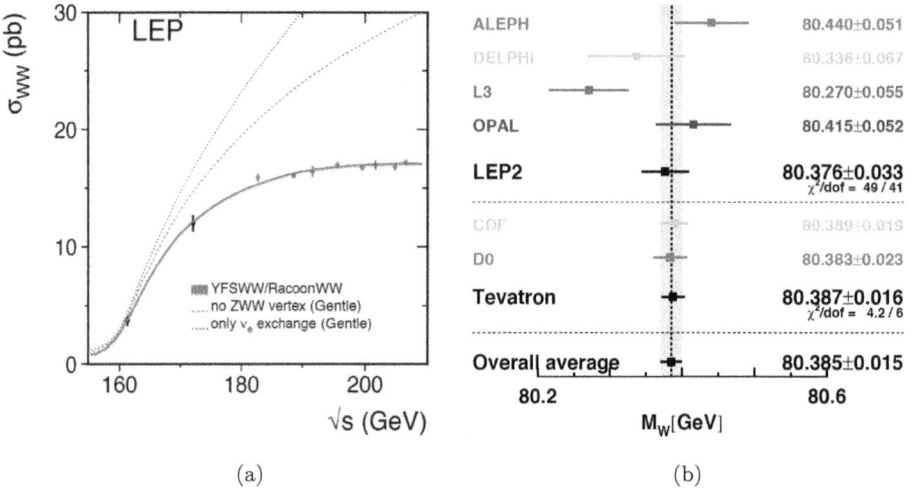

Fig. 9. (a): The W pair production cross-section at LEP II as function of the centre-of-mass energy. Without ZWW vertex the cross-section would diverge as function of energy, as shown by the dotted lines for the cases that only the t-channel neutrino exchange or neutrino and photon exchange ("no ZWW") would be present. (b) A comparison of the directly measured W boson masses. From Ref. 51.

Section 4 the Higgs exchange is needed to compensate the divergences from the longitudinal components of the gauge bosons. One can indeed verify by explicit calculations that the amplitudes cancel at high energies, i.e. $A_\nu + A_\gamma + A_Z = -A_H$. However, the Higgs exchange is proportional to $m_e \sqrt{s}/M_W^2$, so this term becomes only important for $\sqrt{s} \approx M_W^2/m_e \approx 10^7$ GeV. At LEP II energies the longitudinal cross-section can be neglected and only A_γ, A_ν and A_Z are important. Each of them increases with the energy squared, but A_Z interferes negatively with the other amplitudes. The energy dependence of A_ν, $A_\nu + A_\gamma$ and the total cross-section are displayed in Fig. 9(a). The negative interference leads to a rather slow energy dependence of the total W pair production cross-section by virtue of the fact that the triple gauge boson vertex in A_Z has the same gauge coupling as the coupling to fermions, a feature imposed by the gauge invariance of the SM. One observes excellent agreement between the SM prediction and data. The shape of the cross-section in Fig. 9(a) is sensitive to M_W. Combining this shape with invariant mass distributions of W final states leads to: $M_W = 80.376 \pm 0.033$ GeV and $\Gamma_W = 2.195 \pm 0.083$ GeV,[51] which agrees with mass measurements at the Tevatron, as shown in Fig. 9(b). The world average of the directly measured W masses ($M_W = 80.385 \pm 0.015$ GeV) is slightly higher than the indirectly measured W masses from the global electroweak fit ($M_W = 80.363 \pm 0.006$ GeV), as shown before in Fig. 8(a), but the discrepancy is only at the 1.5σ level, as discussed before.

9. QCD Results

The LEP I data were an eldorado for studying QCD given the high Z boson cross-section and large branching ratio into hadrons ($\approx 70\%$, see Table 1). Among the milestones: (i) a direct demonstration of the self interaction of gluons, thus confirming experimentally the basis for asymptotic freedom; (ii) The precise experimental measurement of the strong coupling constant; (iii) From a comparison with lower energy data evidence for the running of the bottom quark mass and the running of the strong coupling constant. We shortly describe these impressive results.

9.1. *The gluon self-interaction*

Four jet events in e^+e^- annihilation originate either from the radiation of two gluons or radiation of a single gluon with subsequent spitting either into two quarks or two gluons. All three contributions have a different angular distribution and different cross-section, so with the clean and high statistics of 4-jet events at LEP one can disentangle the various contributions. The contribution from the triple gluon vertex is clearly established[67-70] and agrees with the SU(3) prediction, as shown in Fig. 10(a) by the filled circles. In addition, the gluon self-coupling increases the gluon jet multiplicity and changes the averaged thrust with increasing energy, as determined by the beta function of the RGE. Combining all these

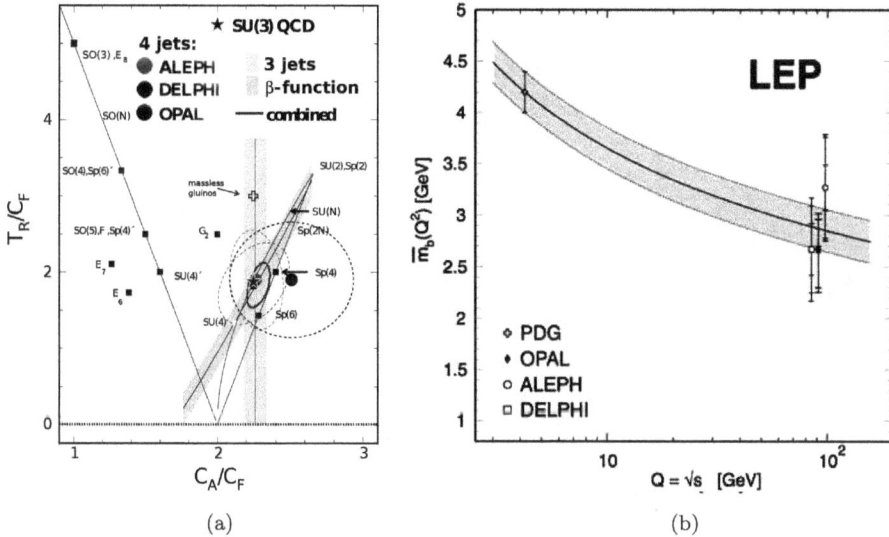

(a)   (b)

Fig. 10. (a): T_R/C_F versus C_A/C_F, where the colour factors T_R, C_F and C_A are associated with $g \to q\bar{q}$, $q \to qg$ and $g \to gg$, respectively. The combined fit to all data (dark (red) ellipse) agrees with the SU(3) group from QCD, but excludes many other groups, see Ref. 65 for details and further references. (b): The running of the b quark mass. From Ref. 66.

measurements[65] constrains the gauge group of the strong interactions to SU(3), as shown in Fig. 10(a).

9.2.   Running of the b quark mass

A bare quark is surrounded by a cloud of gluons, which increases its mass in an energy dependent way. The energy dependence can be calculated by taking into account the running of the coupling constant and the scale, at which the quark mass is probed. For the b quark mass one expects a change from 4.2 GeV at the b mass to 3 GeV at the Z mass. The b quark mass can be measured by a comparison of the 3-jet rate for b quarks and light quarks,[72–74] since the b quark mass effect reduces the cross-section by about 5%.[41] Comparing the LEP value with the measurements at low energy clearly shows the running,[41, 75, 76] see Fig. 10(b).

9.3.   Determination of the strong coupling constant

Gluon radiation from quarks increases the hadronic Z cross-section by a factor $1 + \alpha_s/\pi + \cdots \approx 1.04$, where the dots indicate the higher order corrections, known up to α_s^4.[77] A precise determination of the cross-section allows one to extract the strong coupling constant at the Z scale. The hadronic peak cross-section σ_{had}^0 can be determined either by normalising to the luminosity or to the leptonic cross-section. In the latter case one determines R_ℓ^0, the ratio of the hadronic and leptonic decay width of the Z boson. The different normalisations yield different values of the strong coupling constant: $\alpha_s = 0.1154 \pm 0.0040$ and $\alpha_s = 0.1225 \pm 0.0037$, if one uses σ_{had}^0 or R_ℓ^0, respectively. Here only M_Z, Γ_{tot} and σ_{had}^0 from all LEP experiments are used in the fit.[63] The low value obtained from the cross-section normalised to the luminosity is correlated with the low value of the number of neutrino generations, determined as $N_\nu = 2.982(8)$, which is 2.3σ below the expected value of three neutrino generations. The error is dominated by the common theoretical error on the luminosity, as discussed before. In contrast, the ratio R_ℓ^0 does not depend on the luminosity. If we require the number of neutrino generations to be three, this is most easily obtained by changing the common Bhabha cross-section for all LEP experiments by 0.15% (3σ), which leads to $\alpha_s = 0.1196 \pm 0.0040$, a value close to $\alpha_s = 0.1225 \pm 0.0037$ from R_ℓ^0 and also close to the value from the ratio of the hadronic and leptonic widths of the τ lepton, R_τ, which yields $\alpha_s = 0.1197 \pm 0.0016$.[78] These α_s values are slightly above the world average of $\alpha_s = 0.1185 \pm 0.0006$, quoted in the Partice Data Book. However, this value is dominated by the lattice calculations, for which the correlations between the different groups were not taken into account. Instead, only the weighted average was taken, implying that the groups estimating the systematic error from the "window" problem conservatively,[78] have a small weight. The window problem is, stated simply, the problem of transferring the strong coupling from the non-perturbative regime of fitted quark masses, as used in lattice calculations, to the $\overline{MS}$ scheme, which relies on a perturbative expansion. If one would take the spread in the values from the different lattice calculations as a window for the correct values, as is done in the α_s determination from the τ-data, the error would be a factor three larger, implying consistency between all measurements.

10. Gauge Coupling Unification

Shortly after the first high statistics data from LEP became available, the gauge couplings were determined with unprecedented precision and by using renormalisation group equations (RGEs)[79] the couplings can be extrapolated up to high energies. If second order effects are included, one has to consider the interactions between Yukawa and gauge couplings as well as the running of the SUSY- and Higgs-masses, which leads to a set of coupled differential equations. They can be solved numerically, see Ref. 21 for a compilation of the many RGEs and references therein. However, the second order effects are small and in first order the running of the coupling constants as function of the energy scale Q is proportional to $1/\beta \log(Q^2)$, so the inverse of the coupling constant versus $\log(Q^2)$ is a straight line with a slope given by the β coefficient of the RGE. The fine structure constant is calculated from the RGE to change from $1/137.035999074$ at low energy to $1/(127.940 \pm 0.014)$ at LEP I energies, which agrees with data, as shown in Fig. 11(a).[71] Also the running of the strong coupling constant agrees with data, as shown in Fig. 11(b).[48] One can obtain the gauge couplings at the Z scale from $\alpha_1 = (5/3)g'^2/(4\pi) = 5\alpha/(3\cos^2\theta_W)$, $\alpha_2 = g^2/(4\pi) = \alpha/\sin^2\theta_W$,

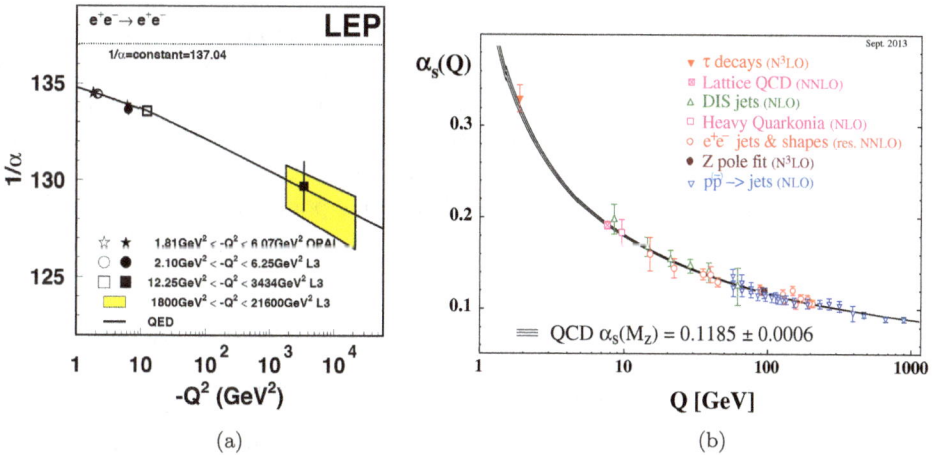

Fig. 11. Running of the electroweak[71] (a) and strong coupling[48] (b) in comparison with the expected running from the RGEs.

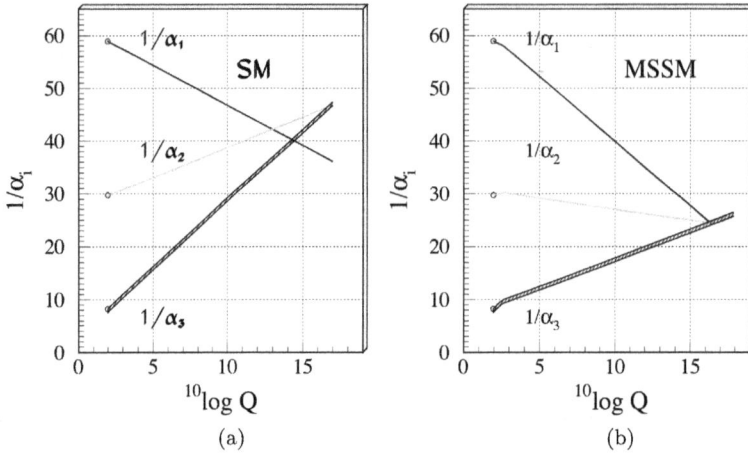

Fig. 12.   The running of the couplings in the SM (a) and MSSM (b).[13] Note that the running in the MSSM is slower, which leads to an order of magnitude larger GUT scale. Since the proton lifetime is for the Born diagram proportional to M_{GUT}^4, the predicted proton lifetime in the MSSM is four orders of magnitude above the SM value, see e.g. Ref. 21. The width of the lines corresponds to the experimental error.

$\alpha_3 = g_s^2/(4\pi)$, where g', g and g_s are the $U(1)$, $SU(2)$ and $SU(3)$ coupling constants.[k] The connection between the first two couplings and the electroweak mixing angle can be obtained from Fig. 2(b). The factor $5/3$ in the definition of α_1 is needed for the proper normalisation of the gauge groups, whose operators are required to be represented by traceless matrices, see e.g. Ref. 21. Figure 12(a) demonstrates that the gauge coupling constants do not meet in a single point, at least of the RGEs from the SM are used.[l] However, the running of the couplings changes, if one includes SUSY particles in the loops. Allowing the SUSY mass scale and GUT scale to be free parameters in a fit requiring unification allows to derive these scales and their uncertainties.[13] Perfect unification is possible at a scale above 10^{16} GeV, which is consistent with the lower limits on the proton lifetime, as shown in Fig. 12(b), in agreement with unification results from other groups.[63, 82–87] Such a unification is by no means trivial, even from the naive argument, that two lines always meet, so three lines can always be brought to a single meeting point with one additional free parameter, like the SUSY mass scale. However, since new mass scales effect all three

---

[k]The couplings are usually given in the $\overline{MS}$ scheme. However, for SUSY the dimensional reduction $\overline{DR}$ scheme is more appropriate.[80] It has the advantage that the three gauge couplings meet exactly at one point. The $\overline{MS}$ and $\overline{DR}$ couplings differ by a small offset $1/\alpha_i^{\overline{DR}} = 1/\alpha_i^{\overline{MS}} - C_i/12\pi$, where $C_i = N$ for $SU(N)$ and 0 for $U(1)$, so α_1 stays the same.

[l]The tests for unification in the SM were done before LEP in 1987 by Amaldi *et al.*,[81] but the precision of the couplings was not high enough to exclude unification in the SM. Amaldi suggested to repeat the analysis with the new LEP data, which showed that within the SM unification is excluded. However, we found that it is perfectly possible in SUSY.

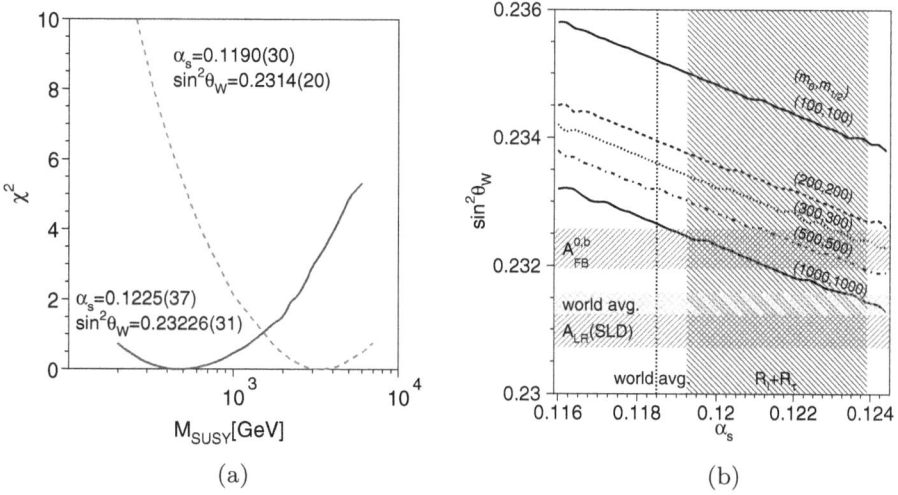

(a)   (b)

Fig. 13. (a): The χ^2 distribution of M_{SUSY}.[63] The two different sets of $\alpha_s(M_Z)$ and $\sin^2\theta_W$ yield quite different SUSY masses needed for unification, as indicated by the minima. (b): The inclined lines, with the SUSY masses of the CMSSM indicated in brackets in GeV, yield perfect gauge unification. The horizontal shaded bands indicate the $\sin^2\theta_W$ measurements from LEP and SLC, respectively, while the vertically shaded band indicates the value of the strong coupling constant from R_ℓ^0 and R_τ. These values are above the world average, but well motivated (see text) and they lead more easily to unification. From Ref. 63.

couplings simultaneously, unification is only reached in rare cases.[88] For example, a fourth family with an arbitrary mass scale changes all slopes by the same amount, thus never leading to unification.

The SUSY mass scale depends on the values of the couplings at the Z scale, as can be seen from the minima of the χ^2 distributions in Fig. 13(a) for slightly different couplings leading to variations in the SUSY scale from 0.5 to 3.5 TeV. Hence, the values of α_s, $\sin^2\theta_W$ and M_{SUSY} are correlated. The combination of these three parameters yielding perfect unification are indicated by the diagonal lines for given values of M_{SUSY} in the α_s, $\sin^2\theta_W$ plane in Fig. 13(b).[63] Here the full second order RGEs were used with step functions in the beta coefficient at the threshold for each SUSY particle using the particle spectrum from the constrained minimal supersymmetric model (CMSSM), which assumes equal masses m_0 ($m_{1/2}$) for the spin 0 (1/2) particles at the GUT scale. Low energy mass differences originate from the running of the masses from the GUT scale to the low energy scale, taken to be the mass of the SUSY particle. The horizontal bands indicate the value of $\sin^2\theta_W$ from $A_{FB}^{0,b}$ and A_{LR}. For $\sin^2\theta_W$ from A_{LR} no unification is possible with the central value of α_s. However, this of $\sin^2\theta_W$ value is inconsistent with the value of $\sin^2\theta_W$ from $A_{FB}^{0,b}$ at the 3σ level (Section 7). With $\sin^2\theta_W$ from $A_{FB}^{0,b}$ unification is possible for $\alpha_s \approx 0.12$ and $M_{\text{SUSY}} > 1$ TeV. These values are consistent with the α_s value from observables not depending on the luminosity, (R_ℓ^0 and R_τ indicated by the shaded vertical band in Fig. 13(b)) and present limits on M_{SUSY} from LHC.

Clearly, new data from a future Z-factory would be highly welcome to settle the minor, but important discrepancies in α_s and $\sin^2 \theta_W$ displayed in Fig. 13(b).

11.   Summary

The electroweak precision data from LEP and SLC have provided a remarkable verification of the quantum structure of the SM. Not only the masses of the top quark and Higgs boson mass could be inferred from the quantum corrections, but also a possible hint for the SM being part of a Grand Unified Theory was obtained from the running of the gauge couplings in case the symmetries of the SM are extended by another symmetry, namely SUSY. SUSY solves several shortcomings of the SM (see e.g. Refs 20–23 for reviews): (i) Electroweak symmetry breaking (EWSB) does not need to be introduced ad hoc, but is induced via radiative corrections; (ii) EWSB predicts a SM-like Higgs boson mass below 130 GeV; (iii) EWSB explains the large difference between the GUT and electroweak scale, because of the slow running of the Higgs mass terms from positive to negative values; (iv) EWSB requires the top quark mass to be between 140 and 190 GeV for a correct running of these Higgs mass terms; (v) The quadratic divergences in the loop corrections of the SM disappear in SUSY, because of the cancellations between an equal number of fermions and bosons in the loops; (vi) The mass ratio of bottom quark over tau lepton is predicted in SUSY, if one presumes Yukawa coupling unification at the GUT scale; (vii) The lightest SUSY particle is a perfect DM candidate, since its self-annihilation cross-section is of the right order of magnitude to provide the correct relic density.

The only troublesome question: where are all the predicted SUSY particles? LHC has excluded squarks and gluinos below the TeV scale. However, as shown in Fig. 13(a), gauge unification for SUSY masses up to several TeV is perfectly possible. Also the argument that for heavier SUSY masses the cancellation of the quadratic divergences is impacted, is only qualitative. Anyway, the squarks and gluinos are expected to be the heaviest particles because of the gluon clouds surrounding them, so the gauginos and additional Higgs particles may be considerably lighter. These have only weak production cross-sections at the LHC, so we do not have the sensitivity, even if the energy might be sufficient. For example, for the associated WZ production in the 3-lepton channel the LHC has typically produced 2500 events per experiment for the present luminosity of about 20/fb at 8 TeV. Assuming the SUSY partners to be a factor 4 heavier reduces the cross-section roughly by a factor $1/M^4$ or more than two orders of magnitude, bringing them to the edge of discovery. Even at the full LHC energy and an integrated luminosity of 3000/fb the discovery reach for charginos will only be 800 GeV.[89] This integrated luminosity can be reached around 2030, but of course, nothing may be found, either because the SUSY particles are still heavier or Nature may have found ways different from SUSY to circumvent the shortcomings of the SM.

Acknowledgments

I thank Ugo Amaldi, Jens Erler, Klaus Hamacher, Hans Kühn, Herwig Schopper, Greg Snow, Dmitri Kazakov and Wilbur Venus for useful discussions and comments.

References

1. *Nobel Lectures*, http://www.nobelprize.org/nobel_prizes/physics/laureates/year/ name-lecture.pdf (or html), where 'name' is one of the following: yang, lee, schwinger, feynman, tomonaga, gellmann, glashow, salam, weinberg, 'thooft, veltman, gross, politzer, wilczek, kobayashi, maskawa, nambu, englert, higgs, ting, richter, fitch, cronin, meer, rubbia, lederman, schwartz, steinberger, friedman, kendall, taylor, perl, koshiba, alvarez, davis, charpak, (1957–2013).
2. P. W. Higgs, Broken symmetries, Massless particles and gauge fields, *Phys. Lett.* **12**, 132–133 (1964). doi: 10.1016/0031-9163(64)91136-9.f
3. P. W. Higgs, Broken Symmetries and the Masses of Gauge Bosons, *Phys. Rev. Lett.* **13**, 508–509 (1964). doi: 10.1103/PhysRevLett.13.508.
4. F. Englert and R. Brout, Broken Symmetry and the Mass of Gauge Vector Mesons, *Phys. Rev. Lett.* **13**, 321–323 (1964). doi: 10.1103/PhysRevLett.13.321.
5. G. Guralnik, C. Hagen, and T. Kibble, Global Conservation Laws and Massless Particles, *Phys. Rev. Lett.* **13**, 585–587 (1964). doi: 10.1103/PhysRevLett.13.585.
6. G. 't Hooft and M. Veltman, Regularization and Renormalization of Gauge Fields, *Nucl. Phys. B* **44**, 189–213 (1972). doi: 10.1016/0550-3213(72)90279-9.
7. H. Georgi and S. Glashow, Unity of All Elementary Particle Forces, *Phys. Rev. Lett.* **32**, 438–441 (1974). doi: 10.1103/PhysRevLett.32.438.
8. H. Georgi and S. Glashow, Unified Theory of Elementary Particle Forces, *Phys. Today.* **33N9**, 30–39 (1980). doi: 10.1063/1.2914275.
9. H. Georgi, H. R. Quinn, and S. Weinberg, Hierarchy of Interactions in Unified Gauge Theories, *Phys. Rev. Lett.* **33**, 451–454 (1974). doi: 10.1103/PhysRevLett.33.451.
10. C. McGrew *et al.*, Search for nucleon decay using the IMB-3 detector, *Phys. Rev. D* **59**, 052004 (1999). doi: 10.1103/PhysRevD.59.052004.
11. K. Abe *et al.* (Super-Kamiokande Collaboration), Search for proton decay via $p \to \nu K^+$ using 260kilotonyear data of Super-Kamiokande, *Phys. Rev. D* **90** (7), 072005 (2014). doi: 10.1103/PhysRevD.90.072005.
12. W. J. Marciano and G. Senjanovic, Predictions of Supersymmetric Grand Unified Theories, *Phys. Rev. D* **25**, 3092 (1982). doi: 10.1103/PhysRevD.25.3092.
13. U. Amaldi, W. de Boer, and H. Fürstenau, Comparison of grand unified theories with electroweak and strong coupling constants measured at LEP, *Phys. Lett. B* **260**, 447–455 (1991). doi: 10.1016/0370-2693(91)91641-8.
14. G. G. Ross, Evidence of supersymmetry, *Nature* **352**, 21–22 (1991). doi: 10.1038/352021a0.
15. D. Hamilton, A Tentative vote for supersymmetry: Do new measurements offer indirect support for an elegant attempt to unify fundamental forces of nature?, *Science.* **253**, 272 (1991). doi: 10.1126/science.253.5017.272.

16. S. Dimopoulos, S. Raby, and F. Wilczek, Unification of couplings, *Phys. Today.* **44**(10), 25–33 (1991). doi: 10.1063/1.881292.

17. P. Ramond, SUSY: The Early Years (1966–1976), *Eur. Phys. J. C* **74**, 2698 (2014). doi: 10.1140/epjc/s10052-013-2698-x.

18. G. Jungman, M. Kamionkowski, and K. Griest, Supersymmetric dark matter, *Phys. Rept.* **267**, 195–373 (1996). doi: 10.1016/0370-1573(95)00058-5.

19. P. Ade *et al.* (Planck Collaboration), Planck 2013 results. XVI. Cosmological Parameters, *Astron. Astrophys.* **571**, A16 (2014). doi: 10.1051/0004-6361/201321591.

20. H. E. Haber and G. L. Kane, The Search for Supersymmetry: Probing Physics Beyond the Standard Model, *Phys. Rept.* **117**, 75–263 (1985). doi: 10.1016/0370-1573(85)90051-1.

21. W. de Boer, Grand Unified theories and Supersymmetry in Particle Physics and Cosmology, *Prog. Part. Nucl. Phys.* **33**, 201–302 (1994). doi: 10.1016/0146-6410(94)90045-0.

22. S. P. Martin, A Supersymmetry Primer, in *Perspectives on Supersymmetry II, ed. G. Kane* (World Scientific, 2010).

23. D. Kazakov, Supersymmetry on the Run: LHC and Dark Matter, *Nucl. Phys. B (Proc. Suppl.)* **203–204**, 118 (2010). doi: 10.1016/j.nuclphysbps.2010.08.007.

24. G. Bertone (ed.), *Particle Dark Matter: Observations, Models and Searches* (Cambridge University Press, 2010).

25. S. Weinberg, Effects of a Neutral Intermediate Boson in Semileptonic Processes, *Phys. Rev. D* **5**, 1412–1417 (1972). doi: 10.1103/PhysRevD.5.1412.

26. P. Darriulat, The Discovery of the W and Z, a Personal Recollection, *Eur. Phys. J. C* **34**, 33–40 (2004). doi: 10.1140/epjc/s2004-01764-x.

27. L. Camilleri, D. Cundy, P. Darriulat, J. R. Ellis, J. Field, *et al.*, Physics with Very High-Energy e+e− Colliding Beams, CERN-YELLOW-REPORT-76-18. (1976).

28. G. Abrams *et al.* (Mark-II Collaboration), First Measurements of Hadronic Decays of the Z Boson, *Phys. Rev. Lett.* **63**, 1558 (1989). doi: 10.1103/PhysRevLett.63.1558.

29. R. Assmann, M. Lamont, and S. Meyers, A Brief History of the LEP Collider, *Nucl. Phys. B (Proc. Suppl.)* **109**, 17 (2002). doi: 10.1016/S0920-5632(02)90004-6.

30. A. Blondel, A Scheme to Measure the Polarization Asymmetry at the Z Pole in LEP, *Phys. Lett. B* **202**, 145 (1988). doi: 10.1016/0370-2693(88)90869-6.

31. R. Barate *et al.* (LEP Working Group for Higgs Boson Searches, ALEPH-, DELPHI-, L3- and OPAL Collaborations), Search for the Standard Model Higgs boson at LEP, *Phys. Lett. B* **565**, 61–75 (2003). doi: 10.1016/S0370-2693(03)00614-2.

32. A. Djouadi, The Anatomy of Electroweak Symmetry Breaking. II. The Higgs Bosons in the Minimal Supersymmetric Model, *Phys. Rept.* **459**, 1–241 (2008). doi: 10.1016/j.physrep.2007.10.005.

33. D. Treille, LEP/SLC: What did we expect? What did we achieve? A very quick historical review, *Nucl. Phys. B (Proc. Suppl.)* **109**, 1 (2002). doi: 10.1016/S0920-5632(02)90004-6.

34. D. Decamp *et al.* (ALEPH Collaboration), ALEPH: A Detector for Electron–Positron Annihilations at LEP, *Nucl. Instrum. Meth. A* **294**, 121–178 (1990). doi: 10.1016/0168-9002(90)91831-U.

35. P. Aarnio *et al.* (DELPHI Collaboration), The DELPHI detector at LEP, *Nucl. Instrum. Meth. A* **303**, 233–276 (1991). doi: 10.1016/0168-9002(91)90793-P.

36. O. Adriani *et al.* (L3 Collaboration), Results from the L3 experiment at LEP, *Phys. Rept.* **236**, 1–146 (1993). doi: 10.1016/0370-1573(93)90027-B.

37. K. Ahmet *et al.* (OPAL Collaboration), The OPAL detector at LEP, *Nucl. Instrum. Meth. A* **305**, 275–319 (1991). doi: 10.1016/0168-9002(91)90547-4.

38. F. Hartmann, Evolution of Silicon Sensor Technology in Particle Physics, *Springer Tracts Mod. Phys.* **231**, 1–204 (2009).

39. D. Buskulic *et al.* (ALEPH Collaboration), Four jet final state production in e+ e− collisions at center-of-mass energies of 130-GeV and 136-GeV, *Z. Phys. C* **71**, 179–198 (1996). doi: 10.1007/s002880050163.

40. A. Heister *et al.* (ALEPH Collaboration), Final Results of the Searches for Neutral Higgs Bosons in e+ e- Collisions at s**(1/2) up to 209-GeV, *Phys. Lett. B* **526**, 191–205 (2002). doi: 10.1016/S0370-2693(01)01487-3.

41. P. Abreu *et al.* (DELPHI Collaboration), Study of the four-jet Anomaly observed at LEP center-of-mass Energies of 130-GeV and 136-GeV, *Phys. Lett. B* **448**, 311–319 (1999). doi: 10.1016/S0370-2693(99)00066-0.

42. M. Veltman, Radiative Corrections to Vector Boson Masses, *Phys. Lett. B* **91**, 95 (1980). doi: 10.1016/0370-2693(80)90669-3.

43. A. Sirlin, Radiative Corrections in the SU(2)-L x U(1) Theory: A Simple Renormalization Framework, *Phys. Rev. D* **22**, 971–981 (1980). doi: 10.1103/PhysRevD.22.971.

44. D. Kennedy and B. Lynn, Electroweak Radiative Corrections with an Effective Lagrangian: Four Fermion Processes, *Nucl. Phys. B* **322**, 1 (1989). doi: 10.1016/0550-3213(89)90483-5.

45. D. Y. Bardin *et al.*, A Realistic Approach to the Standard Z Peak, *Z. Phys. C* **44**, 493 (1989). doi: 10.1007/BF01415565.

46. W. Hollik, Radiative Corrections in the Standard Model and their Role for Precision Tests of the Electroweak Theory, *Fortsch. Phys.* **38**, 165–260 (1990). doi: 10.1002/prop.2190380302.

47. S. Fanchiotti, B. A. Kniehl, and A. Sirlin, Incorporation of QCD Effects in Basic Corrections of the Electroweak Theory, *Phys. Rev. D* **48**, 307–331 (1993). doi: 10.1103/PhysRevD.48.307.

48. K. Olive *et al.* (Review of Particle Physics), *Chin. Phys. C* **38**, 090001 (2014). doi: 10.1088/1674-1137/38/9/090001.

49. D. Y. Bardin and G. Passarino, The Standard Model in the Making: Precision Study of the Electroweak Interactions. (Clarendon, Oxford, UK, 1999).

50. S. Schael *et al.* (LEP Electroweak Working Group, ALEPH-, DELPHI-, L3- and OPAL-Collaborations), Precision Electroweak Measurements on the Z resonance, *Phys. Rept.* **427**, 257–454 (2006). doi: 10.1016/j.physrep.2005.12.006.

51. S. Schael *et al.* (LEP Electroweak Working Group, ALEPH-, DELPHI-, L3- and OPAL-Collaborations), Electroweak Measurements in Electron–Positron Collisions at W-Boson-Pair Energies at LEP, *Phys. Rept.* **532**, 119–244 (2013). doi: 10.1016/j.physrep.2013.07.004.

52. B. Ward, S. Jadach, M. Melles, and S. Yost, New results on the Theoretical Precision of the LEP/SLC Luminosity, *Phys. Lett. B* **450**, 262–266 (1999). doi: 10.1016/S0370-2693(99)00104-5.

53. G. D'Agostini, W. de Boer, and G. Grindhammer, Determination of $\alpha^- s$ and the Z^0 Mass From Measurements of the Total Hadronic Cross-section in e^+e^- Annihilation, *Phys. Lett. B* **229**, 160 (1989). doi: 10.1016/0370-2693(89)90176-7.

54. V. Tishchenko *et al.* (MuLan Collaboration), Detailed Report of the MuLan Measurement of the Positive Muon Lifetime and Determination of the Fermi Constant, *Phys. Rev. D* **87** (5), 052003 (2013). doi: 10.1103/PhysRevD.87.052003.

55. G. Montagna, O. Nicrosini, F. Piccinini, and G. Passarino, TOPAZ0 4.0: A New Version of a Computer Program for Evaluation of Deconvoluted and Realistic Observables at LEP-1 and LEP-2, *Comput. Phys. Commun.* **117**, 278–289 (1999). doi: 10.1016/S0010-4655(98)00080-0.

56. A. Arbuzov *et al.*, ZFITTER: A Semi-analytical Program for Fermion Pair Production in e+ e− Annihilation, from Version 6.21 to Version 6.42, *Comput. Phys. Commun.* **174**, 728–758 (2006). doi: 10.1016/j.cpc.2005.12.009.

57. J. Erler, GAPP: Global Analysis of Particle Properties. http://www.fisica. unam.mx/erler/GAPPP.html.

58. ATLAS, CDF, CMS, and D0 Collaborations, First Combination of Tevatron and LHC Measurements of the Top Quark Mass, arXiv:1403.4427. (2014).

59. F. Jegerlehner and A. Nyffeler, The Muon $g - 2$, *Phys. Rept.* **477**, 1–110 (2009). doi: 10.1016/j.physrep.2009.04.003.

60. S. Heinemeyer, W. Hollik, and G. Weiglein, Electroweak precision observables in the minimal supersmmetric standard model, *Phys. Rept.* **425**, 265–368 (2006).

61. W. Hollik, Electroweak theory, *J. Phys. Conf. Ser.* **53**, 7–43 (2006).

62. C. Beskidt, W. de Boer, D. Kazakov, and F. Ratnikov, Constraints on Supersymmetry from LHC data on SUSY Searches and Higgs Bosons combined with Cosmology and Direct Dark Matter Searches, *Eur. Phys. J. C* **72**, 2166 (2012). doi: 10.1140/epjc/s10052-012-2166-z.

63. W. de Boer and C. Sander, Global Electroweak Fits and Gauge Coupling Unification, *Phys. Lett. B* **585**, 276–286 (2004). doi: 10.1016/j.physletb.2004.01.083.

64. A. Djouadi, The Anatomy of Electroweak Symmetry Breaking. I: The Higgs Boson in the Standard Model, *Phys. Rept.* **457**, 1–216 (2008). doi: 10.1016/j.physrep. 2007.10.004.

65. J. Abdallah *et al.* (DELPHI Collaboration), Charged particle multiplicity in three-jet events and two-gluon systems, *Eur. Phys. J. C* **44**, 311–331 (2005). doi: 10.1140/epjc/s2005-02390-x.

66. P. Zerwas, W & Z physics at LEP, *Eur. Phys. J. C* **34**, 41–49 (2004). doi: 10.1140/epjc/s2004-01765-9.

67. B. Adeva *et al.* (L3 Collaboration), A Test of QCD Based on Four Jet Events from Z0 Decays, *Phys. Lett. B* **248**, 227–234 (1990). doi: 10.1016/0370-2693(90) 90043-6.

68. P. Abreu *et al.* (DELPHI Collaboration), Experimental Study of the Triple Gluon Vertex, *Phys. Lett. B* **255**, 466–476 (1991). doi: 10.1016/0370-2693(91)90796-S.

69. P. Abreu *et al.* (DELPHI Collaboration), Measurement of the Triple Gluon Vertex from Four-jet Events at LEP, *Z. Phys. C* **59**, 357–368 (1993). doi: 10.1007/BF01498617.

70. D. Decamp *et al.* (ALEPH Collaboration), Evidence for the Triple Gluon Vertex from Measurements of the QCD Color Factors in Z Decay into Four Jets, *Phys. Lett. B* **284**, 151–162 (1992). doi: 10.1016/0370-2693(92)91941-2.

71. S. Mele, Measurements of the Running of the Electromagnetic Coupling at LEP, *XXVI. Phys. in Collision, Rio de Janeiro,* hep-ex/0610037. (2006). http://www.slac.stanford.edu/econf/C060706/pdf/0610037.pdf.

72. W. Bernreuther, A. Brandenburg, and P. Uwer, Next-to-leading Order QCD Corrections to Three Jet Cross-Sections with Massive Quarks, *Phys. Rev. Lett.* **79**, 189–192 (1997). doi: 10.1103/PhysRevLett.79.189.

73. G. Rodrigo, A. Santamaria, and M. S. Bilenky, Do the Quark Masses Run? Extracting m-bar(b) (m(z)) from LEP Data, *Phys. Rev. Lett.* **79**, 193–196 (1997).

74. M. S. Bilenky *et al.*, $m_b(m_z)$ from Jet Production at the Z Peak in the Cambridge Algorithm, *Phys. Rev. D* **60**, 114006 (1999). doi: 10.1103/PhysRevD.60.114006.

75. R. Barate *et al.* (ALEPH Collaboration), A Measurement of the b Quark Mass from Hadronic Z Decays, *Eur. Phys. J. C* **18**, 1–13 (2000). doi: 10.1007/s100520000533.

76. G. Abbiendi *et al.* (OPAL Collaboration), Determination of the b Quark Mass at the Z Mass Scale, *Eur. Phys. J. C* **21**, 411–422 (2001). doi: 10.1007/s100520100746.

77. P. Baikov, K. Chetyrkin, J. Kühn, and J. Rittinger, Complete $\mathcal{O}(\alpha_s^4)$ QCD Corrections to Hadronic Z-Decays, *Phys. Rev. Lett.* **108**, 222003 (2012).

78. S. Aoki *et al.*, Precise Determination of the Strong Coupling Constant in $N(f) = 2 + 1$ Lattice QCD with the Schrödinger Functional Scheme, *JHEP* **0910**, 053 (2009).

79. K. Wilson and J. B. Kogut, The Renormalization Group and the Epsilon Expansion, *Phys. Rept.* **12**, 75–200 (1974). doi: 10.1016/0370-1573(74)90023-4.

80. I. Antoniadis, C. Kounnas, and K. Tamvakis, Simple Treatment of Threshold Effects, *Phys. Lett. B* **119**, 377–380 (1982). doi: 10.1016/0370-2693(82)90693-1.

81. U. Amaldi *et al.*, A Comprehensive Analysis of Data Pertaining to the Weak Neutral Current and the Intermediate Vector Boson Masses, *Phys. Rev. D* **36**, 1385 (1987). doi: 10.1103/PhysRevD.36.1385.

82. J. R. Ellis, S. Kelley, and D. V. Nanopoulos, Probing the Desert using Gauge Coupling Unification, *Phys. Lett. B* **260**, 131–137 (1991). doi: 10.1016/0370-2693(91)90980-5.

83. C. Giunti, C. Kim, and U. Lee, Running Coupling Constants and Grand Unification Models, *Mod. Phys. Lett. A* **6**, 1745–1755 (1991). doi: 10.1142/S0217732391001883.

84. P. Langacker and M.-X. Luo, Implications of Precision Electroweak Experiments for M_t, ρ_0, $\sin^2 \theta_W$ and Grand Unification, *Phys. Rev. D* **44**, 817–822 (1991). doi: 10.1103/PhysRevD.44.817.

85. J. R. Ellis, S. Kelley, and D. V. Nanopoulos, A Detailed Comparison of LEP Data with the Predictions of the Minimal Supersymmetric SU(5) GUT, *Nucl. Phys. B* **373**, 55–72 (1992). doi: 10.1016/0550-3213(92)90449-L.

86. M. S. Carena, S. Pokorski, and C. Wagner, On the Unification of Couplings in the Minimal Supersymmetric Standard Model, *Nucl. Phys. B* **406**, 59–89 (1993). doi: 10.1016/0550-3213(93)90161-H.

87. J. Bagger, K. T. Matchev, and D. Pierce, Precision Corrections to Supersymmetric Unification, *Phys. Lett. B* **348**, 443–450 (1995). doi: 10.1016/0370-2693(95)00207-2.

88. U. Amaldi, W. de Boer, P. H. Frampton, H. Fürstenau, and J. T. Liu, Consistency Checks of Grand Unified Theories, *Phys. Lett. B* **281**, 374–383 (1992). doi: 10.1016/0370-2693(92)91158-6.

89. CMS-Collaboration, SUSY future analyses for Technical Proposal, CMS–PAS–SUS–14–012 (2014).

The Discovery of the W and Z Particles

Luigi Di Lella[1] and Carlo Rubbia[2]

[1] *Physics Department, University of Pisa, 56127 Pisa, Italy*
luigi.di.lella@cern.ch
[2] *GSSI (Gran Sasso Science Institute), 67100 L'Aquila, Italy*
carlo.rubbia@cern.ch

This article describes the scientific achievements that led to the discovery of the weak intermediate vector bosons, $W^\pm$ and Z, from the original proposal to modify an existing high-energy proton accelerator into a proton–antiproton collider and its implementation at CERN, to the design, construction and operation of the detectors which provided the first evidence for the production and decay of these two fundamental particles.

1. Introduction

The first experimental evidence in favour of a unified description of the weak and electromagnetic interactions was obtained in 1973, with the observation of neutrino interactions resulting in final states which could only be explained by assuming that the interaction was mediated by the exchange of a massive, electrically neutral virtual particle.[1] Within the framework of the Standard Model, these observations provided a determination of the weak mixing angle, θ_w, which, despite its large experimental uncertainty, allowed the first quantitative prediction for the mass values of the weak bosons, $W^\pm$ and Z. The numerical values so obtained ranged from 60 to 80 GeV for the W mass, and from 75 to 92 GeV for the Z mass, too large to be accessible by any accelerator in operation at that time.

The ideal machine to produce the weak bosons and to measure their properties in the most convenient experimental conditions is an e^+e^- collider, as beautifully demonstrated by the success of the LEP program at CERN. However, while LEP was still far in the future, in 1976 Rubbia, Cline and McIntyre[2] proposed the transformation of an existing high-energy proton accelerator into a proton–antiproton collider as a quick and relatively cheap way to achieve collisions above threshold for W and Z production. In such a scheme a proton (p) and an anti-proton ($\bar{p}$) beam, each of energy E, circulate along the same magnetic path in opposite directions, providing head-on $\bar{p}p$ collisions at a total centre-of-mass energy $\sqrt{s} = 2E$.

Such a scheme was suggested both at Fermilab and CERN. It was adopted at CERN in 1978 for the 450 GeV proton synchrotron (SPS), and the first $\bar{p}p$ collisions at $\sqrt{s} = 540$ GeV were observed in July 1981. By the end of 1982, the $\bar{p}p$ collision rate was high enough to permit the observation of W → $e\nu$ decays. In a subsequent run during the spring of 1983, the decays Z → e^+e^- and Z → $\mu^+\mu^-$ were also observed.

After a short description of the collider itself and of the two detectors, UA1 and UA2, which took data at this new facility, this article describes the experimental results which led to the first observation of the W and Z bosons. This major discovery was awarded the 1984 Nobel Prize for Physics.

2.   The CERN Proton–Antiproton Collider

The conception, construction and operation of the CERN proton–antiproton collider was a great achievement in itself. It is useful, therefore, to give a short description of this facility.

The production of W and Z bosons at a $\bar{p}p$ collider is expected to occur mainly as the results of quark–antiquark annihilation $\bar{d}u$ → W^+, $d\bar{u}$ → W^-, $u\bar{u}$ → Z, $d\bar{d}$ → Z. In the parton model ∼ 50% of the momentum of a high-energy proton is carried, on average, by three valence quarks, and the remainder by gluons. Hence a valence quark carries about 1/6 of the proton momentum. As a consequence, W and Z production should require a $\bar{p}p$ collider with a total centre-of-mass energy equal to about six times the boson masses, or 500–600 GeV. The need to detect Z → e^+e^- decays determines the minimal collider luminosity: the cross-section for inclusive Z production at ∼600 GeV is ∼1.6 nb, and the fraction of Z → e^+e^- decays is ∼3%, hence a luminosity $L = 2.5 \times 10^{29}$ cm^{-2}s^{-1} would give an event rate of ∼1 per day. To achieve such luminosities one would need an antiproton source capable of delivering daily ∼3 × 10^{10} $\bar{p}$ distributed in few (3–6) tightly collimated bunches within the angular and momentum acceptance of the CERN SPS.

The CERN 26 GeV proton synchrotron (PS) is capable of producing antiprotons at the desired rate. The PS accelerates ∼10^{13} protons per pulse which are transported every 2.4 s to the $\bar{p}$ production target. Approximately 7 × 10^6 $\bar{p}$ with a momentum of 3.5 GeV/c are then produced at 0° in a solid angle of 8 × 10^{-3} sr in a momentum interval $\Delta p/p = 1.5\%$. These antiprotons are sufficient in number, but they occupy a phase space volume which is too large by a factor $\geq 10^8$ to fit into the SPS acceptance, even after acceleration to the SPS injection energy of 26 GeV. It is necessary, therefore, to increase the $\bar{p}$ phase space density at least 10^8 times before sending the $\bar{p}$ beam to the SPS. This process is called "cooling" because a bunch of particles occupying a large phase space volume appears as a hot gas, with large velocities in all directions, when viewed by an observer at rest in the centre-of-mass frame of the bunch itself.

The CERN collider project used the technique of stochastic cooling, invented by S. van der Meer in 1972.[3] A central notion in accelerator physics is phase space, well-known from other areas of physics. An accelerator or storage ring has an acceptance that is defined in terms of phase space volume. Traditional particle

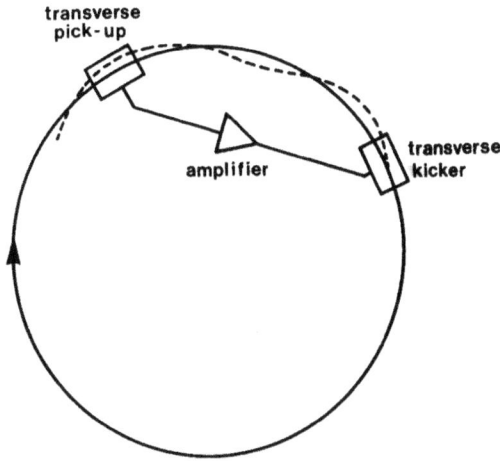

Fig. 1.   Cooling of a single particle horizontal oscillation.

acceleration is generally dominated by the so-called Liouville theorem that forbids any compression of phase volume by conservative forces such as the electromagnetic fields that are used by accelerator builders. In fact, all that can be done in treating particle beams is to distort the phase space volume without changing the density anywhere. Already at MURA in the 1950s[4] it was quickly realised that some beam phase-space compression was required from the source to the collisions (O'Neill, Piccioni, Symon).

The principle of stochastic cooling is illustrated in Fig. 1 for the cooling of horizontal "betatron" oscillations. Particles which do not follow exactly the central orbit in a magnetic ring undergo oscillations around the central orbit under the influence of focusing magnetic fields. A pick-up electrode mounted in a location where the oscillation amplitude is maximum provides a signal proportional to the particle distance from the central orbit. This signal is amplified and applied to a "kicker" mounted in a location where the particle crosses the central orbit. The signal must arrive at the kicker at the same time as the particle, hence the cable connecting pick-up electrode and kicker must follow as much as possible a straight path. In practice, the pick-up electrode measures the average distance of a group of particles from the central orbit, instead of a single particle. The size of this group depends on the sensitivity of the pick-up system, and especially on its frequency response.

The specific feature of the stochastic cooling is based on the fact that particles are points in phase space with empty spaces in between. We may push each particle towards the centre of the distribution, squeezing the empty space outwards. The small-scale density is strictly conserved, but in a macroscopic sense the particle density appears as increased. In this way, and maintaining the Liouville theorem, the information about the individual particle's position can be used, pushing together the individual particles against empty space. As a result the density in 6-dimensional phase space has been boosted by a factor as large as 10^9 using simple methods in which sensors acquiring electric signal from the particles are excited in order to influence the amplified pick-up signals.

In practice, the sensor will not see one particle, but a very large number (e.g. 10^6 to 10^{12}). Each particle's individual signal will be overlapped to the perturbing signal of the other particles. Fortunately, this effect is proportional to the square of the gain, whereas the cooling effect (each particle acting on itself) varies linearly with gain and one can choose it so that the cooling effect predominates.

Following the success of the so-called Initial Cooling Experiment (ICE)[5] providing the experimental demonstration that stochastic cooling could indeed achieve the increase of $\bar{p}$ phase space density required to detect the W and Z bosons, the CERN proton–antiproton collider project was approved on May 28, 1978.

For the CERN collider stochastic cooling is achieved in a purpose-built machine called Antiproton Accumulator (AA), which includes several independent cooling systems to cool both horizontal and vertical oscillations, and also to decrease the beam momentum spread (cooling of longitudinal motion), by using pick-up electrodes which provide signals proportional to Δp. The AA is a large aperture magnetic ring. A picture of the AA during construction is shown in Fig. 2. Figure 3 illustrates cooling and accumulation of a $\bar{p}$ stack in the AA.

Fig. 2.   View of the Antiproton Accumulator during construction.

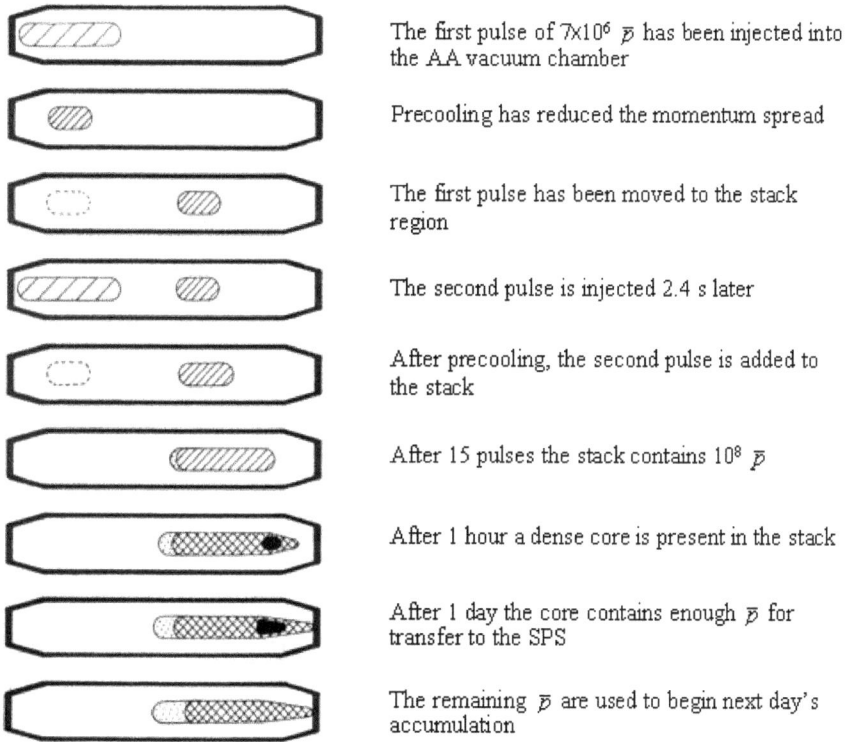

The first pulse of 7×10⁶ p̄ has been injected into the AA vacuum chamber

Precooling has reduced the momentum spread

The first pulse has been moved to the stack region

The second pulse is injected 2.4 s later

After precooling, the second pulse is added to the stack

After 15 pulses the stack contains 10^8 p̄

After 1 hour a dense core is present in the stack

After 1 day the core contains enough p̄ for transfer to the SPS

The remaining p̄ are used to begin next day's accumulation

Fig. 3.   Schematic sequence illustrating antiproton cooling and accumulation in the AA.

When a sufficiently dense p̄ stack has been accumulated in the AA, beam injection into the SPS is achieved using consecutive PS cycles. Firstly, three proton bunches (six after 1986), each containing $\sim 10^{11}$ protons, are accelerated to 26 GeV in the PS and injected into the SPS (see Fig. 4). Then three p̄ bunches (six after 1986), of typically $\sim 10^{10}$ p̄ each, are extracted from the AA and injected into the PS. Here they are accelerated to 26 GeV in a direction opposite to that of the protons, and then injected into the SPS. The relative injection timing of the bunches is controlled with a precision of ~ 1 ns to ensure that bunch crossing in the SPS occurs in the centre of the detectors.

The CERN experiment with proton–antiproton collisions has been the first storage ring in which bunched protons and antiprotons collided head on. Although the CERN proton–antiproton collider uses *bunched* beams, as do the e⁺e⁻ colliders, a continuous phase-space damping due to synchrotron radiation is now absent. Furthermore, since antiprotons are scarce, one has to operate the collider in conditions of relatively large beam–beam interaction, which was not the case for the continuous proton beams of the previously operated Intersecting Storage Rings (ISR) at CERN.

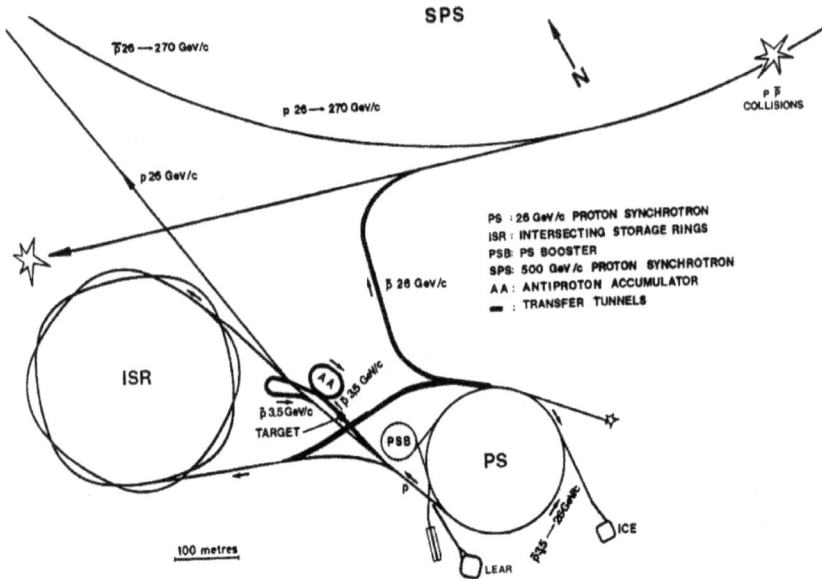

Fig. 4.   Layout of the three machines initially involved in the operation of the CERN proton–antiproton collider: the PS, the AA, the SPS, and the interconnecting transfer lines.

In the early days of construction, very serious concern had been voiced regarding the potential instability of the beams due to beam–beam interaction. It had been observed, for instance at SLAC, that in an e^+e^- collider, the maximum allowed tune-shift, and hence the luminosity, dropped dramatically when decreasing the energy and consequently the simultaneous synchrotron damping rate. Under the assumption that successive proton–antiproton kicks were randomised like it is the case for e^+e^-, the extrapolation to the CERN collider conditions would have implied a catastrophically large $1/e$ growth and a maximum viable tune-shift of $\Delta Q \approx 10^{-6}$.

This bleak prediction was not confirmed since the optimum tune-shift per crossing of $\Delta Q \approx 3 \times 10^{-3}$ and six crossings have been routinely achieved with a beam luminosity lifetime approaching one day.

What, then, is the reason for such a striking contradiction between electrons and proton colliders? In an e^+e^- collider the emission of photons is at the same time a major source of quick randomisation between crossings and of constant damping due to synchrotron radiation. Fortunately, the proton–antiproton collider remains stable because *both the randomising and the damping mechanisms are absent*. The beam has a very long and persistent 'memory' which allows these strong kicks to be added coherently rather than at random. This unusually favourable combination of effects has ensured that proton–antiproton colliders have become viable devices.

In 1987, the CERN $\bar{p}$ source was improved by the addition of a second ring, called Antiproton Collector (AC), built around the AA (see Fig. 5). The AC had a much larger acceptance than the AA to single $\bar{p}$ pulses. It could accept and cool single

Fig. 5. View of the Antiproton Collector surrounding the Antiproton Accumulator.

Table 1 CERN proton–antiproton collider operation, 1981–1990.

Year	Collision energy (GeV)	Peak luminosity $(cm^{-2}\ s^{-1})$	Integrated luminosity (cm^{-2})
1981	546	$\sim 10^{27}$	2×10^{32}
1982	546	5×10^{28}	2.8×10^{34}
1983	546	1.7×10^{29}	1.5×10^{35}
1984–85	630	3.9×10^{29}	1.0×10^{36}
1987–90	630	3×10^{30}	1.6×10^{37}

pulses of $\sim 7 \times 10^7$ $\bar{p}$, thus increasing the $\bar{p}$ stacking rate by a factor of ~ 10. Table 1 summarises the evolution of the main collider parameters between 1981 (the first physics run) and 1990 (the last physics run). The collider was shut down at the end of 1990 because it was no longer competitive with the 1.8 TeV proton–antiproton collider at Fermilab which had started operation in 1987.

3. The Experiments

Since the SPS is built in an underground tunnel at an average depth of ~ 100 m, the project also required the excavation of underground experimental areas to house the detectors. The first experiment, named UA1 for "Underground Area 1" was soon

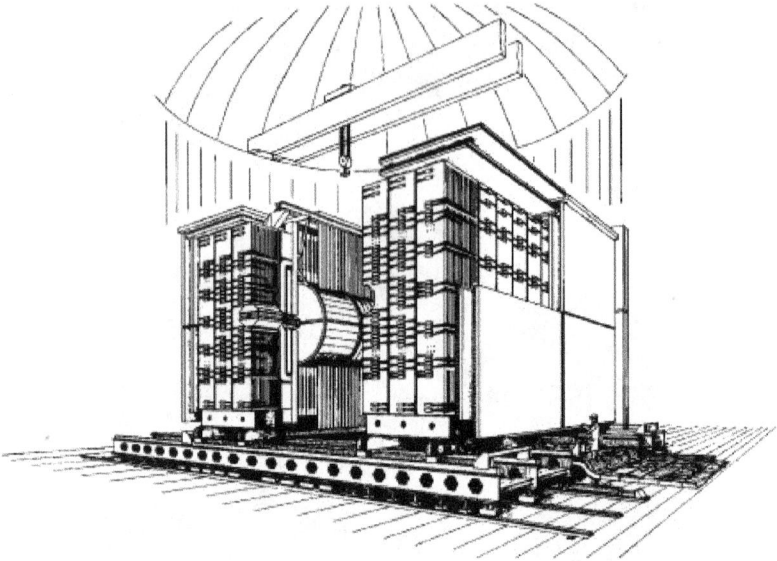

Fig. 6.   View of the UA1 detector with the two magnet halves opened up.

approved on June 29, 1978. It was followed by a second experiment, named UA2, which was approved at the end of the same year.

3.1.   *The UA1 experiment*

The UA1 experiment was designed as a general-purpose magnetic detector[6] with an almost complete solid-angle coverage. A view of the detector with the two halves of the magnet opened up is shown in Fig. 6. The magnet is a dipole with a horizontal field of 0.7 T perpendicular to the beam axis over a volume of $7 \times 3.5 \times 3.5\,\mathrm{m}^3$, produced by a warm aluminium coil to minimise absorption.

The magnet contains the central track detector, which is a system of drift chambers filling a cylindrical volume 5.8 m long with a 2.5 m diameter reconstructing charged particle trajectories down to polar angles of $\sim 6°$ with respect to the beams. Tracks were sampled approximately every centimetre and could have up to 180 hits. This detector, at the cutting edge of technology in those days, was surrounded by electromagnetic and hadronic calorimeters down to $0.2°$ to the beam line. This "hermeticity", as it was called later, turned out to be very effective to reconstruct undetected neutrinos from $W \to e\nu$ decay, and also to search for possible new, as yet undiscovered neutral particles escaping direct detection. It became one of the basic features of all general-purpose detectors at the next-generation e^+e^- and hadron colliders (LEP, the Fermilab $\bar{p}p$ collider and the LHC).

Electromagnetic calorimeters, consisting of Pb–scintillator multi-layer sandwich are also mounted inside the magnet. In the central region, they consist of two cylindrical half-shells surrounding the tracker, each subdivided into 24 elements ("gondolas") covering $180°$ in ϕ and $24\,\mathrm{cm}$ along the beam line, with a total thickness of 26.4 radiation lengths (X_0). Two similar structures are present at smaller angles to the beam line, each consisting of 32 radial sectors. The energy resolution for electrons was $\sigma(E)/E = 0.15/\sqrt{E}$ (E in GeV).

The magnet return yoke, and two iron walls located symmetrically at the two ends of the magnet, are laminated and scintillator is inserted between the iron plates to form a hadronic calorimeter, which is subdivided into 450 independent cells. Muon detectors, consisting of systems of drift tubes, surround the magnet yoke. The momentum resolution for a $40\,\mathrm{GeV}/c$ muon track is typically $\pm20\%$ (for comparison, the energy resolution for a $40\,\mathrm{GeV}$ electron, as measured by the electromagnetic calorimeter, is $\pm2.5\%$).

In the days of initial construction, the UA1 collaboration consisted of about 130 physicists from Aachen, Annecy, Birmingham, CERN, Collège de France, Helsinki, London (QMC), UCLA-Riverside, Rome, RAL, Saclay and Vienna. In the history of particle physics, it was the first time that so many physicists were seen to work together on a common project, thus paving the way to the much larger collaborations of the LEP and LHC experiments in the following years.

A picture of UA1 during assembly is shown in Fig. 7.

Despite the general scepticism in the particle physics community that such a complex detector could be built and operated in time, it was essentially functional by the summer of 1981, in time for the first physics run.

Fig. 7. The UA1 detector during assembly.

3.2. *The UA2 detector*

UA2 was not designed as a general-purpose detector, but rather optimised for the detection of electrons from W and Z decays. The emphasis was on highly granular calorimetry with spherical projective geometry, which was well adapted also to the detection of hadronic jets.

Figure 8 shows the layout of the UA2 detector for the collider runs between 1981 and 1985.[7] The central region contains a "vertex detector", which consists of various types of cylindrical tracking chambers. A "preshower" counter, located just behind the last chamber, and consisting of a tungsten cylinder followed by a multi-wire proportional chamber, is crucial for electron identification. The vertex detector is surrounded by the central calorimeter, which covers the full azimuth and is subdivided into 240 independent cells, each subtending the angular interval $\Delta\theta \times \Delta\phi = 10° \times 15°$ and consisting of an electromagnetic (Pb–scintillator) and a hadronic (Fe–scintillator) section. The calorimeter energy resolution for electrons was $\sigma(E)/E = 0.14/\sqrt{E}$ (E in GeV), and was $\sim$10% for an 80 GeV hadron in the central calorimeters. There is no magnetic field in this region.

The two forward detectors, covering the polar angle interval between 20° and 37.5° with respect to the beams, consist of twelve azimuthal sectors in which a toroidal magnetic field is generated by twelve coils equally spaced in azimuth. Each sector includes tracking chambers, a "preshower" detector and an electromagnetic

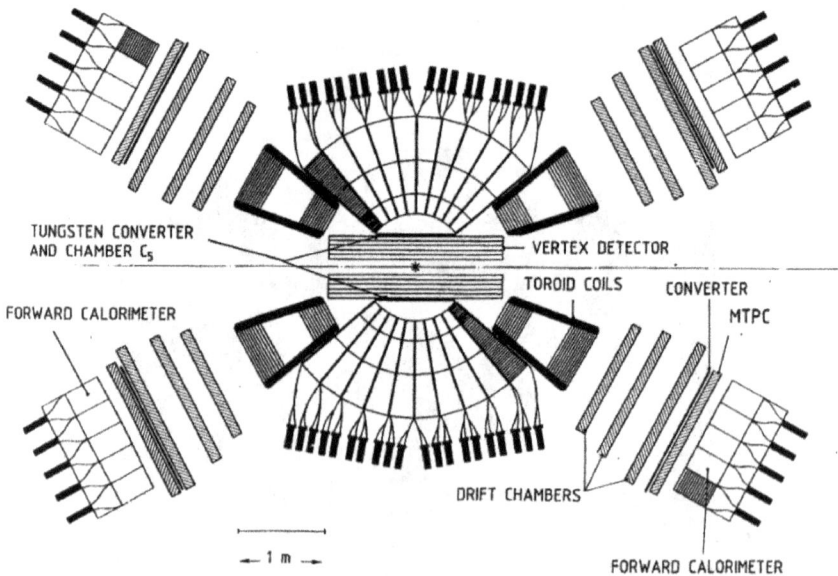

TUNGSTEN CONVERTER
AND CHAMBER C_5

VERTEX DETECTOR

TOROID COILS   CONVERTER

FORWARD CALORIMETER

MTPC

DRIFT CHAMBERS

1 m

FORWARD CALORIMETER

Fig. 8.   Sketch of the UA2 detector in the 1981–85 configuration.

calorimeter. There is no muon detector in UA2. The initial UA2 collaboration consisted af about 60 physicists from Bern, CERN, Copenhagen, Orsay, Pavia and Saclay.

For the initial running (1981–83) the azimuthal coverage of the central calorimeter was 300°, with an interval of ±30° around the horizontal plane being covered by a single arm magnetic spectrometer at 90° to the beams.

At the end of 1985, the two forward magnetic detectors were replaced by calorimeters with full angular coverage down to 5° to the beams, thus greatly improving the detector hermeticity. These calorimeters were subdivided into cells with segmentation similar to that of the central calorimeter, and contained both an electromagnetic and a hadronic section. The central tracker was also upgraded, with silicon pad detectors, trackers and preshower counter made of scintillating fibres, and drift chambers detecting X-rays from the transition radiation produced by electrons in traversing many thin Lithium layers. This detector took data between 1987 and 1990. Figure 9 displays a picture of UA2 in the 1987–90 configuration. At that time groups from Cambridge, Heidelberg, Milano, Perugia and Pisa had joined the collaboration which had grown to about 100 physicists.

Fig. 9.   The UA2 detector in the 1987–90 configuration.

4.   The Discovery of the W and Z Bosons

The first physics run of the CERN collider took place at the end of 1981. The total integrated luminosity recorded by the two experiments during that run was not yet sufficient to detect the W and Z bosons, but that run demonstrated that there were no conceptual obstacles to further increase the luminosity to the required values by a careful tuning of all the machines involved in the collider operation (PS, AA, SPS) and of the interconnecting beam transfer lines. In those days the CERN collider caught the attention not only of the scientific community, but also of a part of the public opinion. The physics results from the 1982 and 1983 collider runs were eagerly awaited, as demonstrated by the many articles on this subject appearing in the world press. There was even the British Prime Minister, Margaret Thatcher, who asked the CERN Director-General to be personally informed of the W and Z discovery before the public announcements.

4.1.   *Discovery of the W boson*

The W boson decays predominantly ($\sim$70%) to quark–antiquark pairs ($q\bar{q}'$), which appear as two hadronic jets. Such configurations are overwhelmed by two-jet production from hard parton scattering,[8] hence both experiments have chosen to detect the W by identifying its leptonic decays: $W^{\pm} \rightarrow e^{\pm}\nu_e(\bar{\nu}_e)$ in both UA1 and UA2, and $W^{\pm} \rightarrow \mu^{\pm}\nu_\mu(\bar{\nu}_\mu)$ in UA1 only.

The signal from $W \rightarrow e\nu_e$ decay is expected to have the following features:

- the presence of a high transverse momentum (p_T) isolated electron;
- a peak in the electron p_T distribution at $m_W/2$ (the "Jacobian" peak);
- the presence of high missing transverse momentum from the undetected neutrino.

These features are the consequence of the main mechanism of W production (quark-antiquark annihilation), which results mainly in W bosons almost collinear with the beam axis, hence the decay electron and neutrino emitted at large angles to the beam axis have large p_T. We note that the missing longitudinal momentum cannot be measured at hadron colliders because of the large number of high-energy secondary particles emitted at very small angles to the beam which cannot be detected because their trajectories are inside the machine vacuum pipe.

The missing transverse momentum vector ($\vec{p}_T^{\text{miss}}$) is defined as

$$\vec{p}_T^{\text{miss}} = -\sum_{\text{cells}} \vec{p}_T$$

where $\vec{p}_T$ is the transverse component of a vector associated with each calorimeter cell, with direction from the event vertex to the cell centre and length equal to the energy deposition in that cell, and the sum is extended to all cells with an energy deposition larger than zero. In an ideal detector with no measurement errors, for events with an undetected neutrino in the final-state it follows

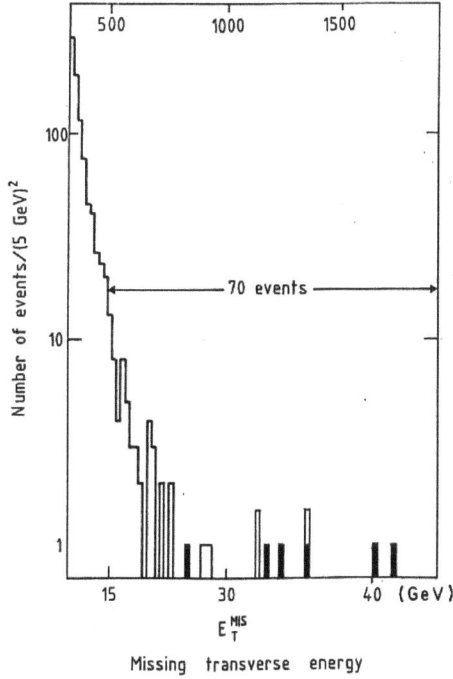

Fig. 10. UA1 distribution of the missing transverse momentum (called E_T^{MIS} in this plot) for equal bins of $(E_T^{MIS})^2$. The events shown as dark areas in this plot contain a high p_T electron.

from momentum conservation that $\vec{p}_T^{\,miss}$ is equal to the neutrino transverse momentum.

Figure 10 shows the $|\vec{p}_T^{\,miss}|$ distribution, as measured by UA1 from the 1982 data.[9] There is a component decreasing approximately as $|\vec{p}_T^{\,miss}|^2$ due to the effect of calorimeter resolution in events without significant $|\vec{p}_T^{\,miss}|$, followed by a flat component due to events with genuine $|\vec{p}_T^{\,miss}|$. Six events with high $|\vec{p}_T^{\,miss}|$ in the distribution of Fig. 10 contain a high-p_T electron. The $\vec{p}_T^{\,miss}$ vector in these events is almost back-to-back with the electron transverse momentum vector, as shown in Fig. 11. These events are interpreted as due to W → $e\nu_e$ decay. This result was first announced at a CERN seminar on January 20, 1983. Figure 12 shows the graphics display of one of these events.

The results from the UA2 search for W → $e\nu$ events[10] was presented at a CERN seminar on the following day (January 21, 1983). Six events containing an electron with $p_T > 15\,\text{GeV}/c$ were identified among the 1982 data. Figure 13 shows the distribution of the ratio between $|\vec{p}_T^{\,miss}|$ and the electron p_T for these events. Also shown in Fig. 13 is the electron p_T distribution for the events with a $|\vec{p}_T^{\,miss}|$ value comparable to the electron p_T (four events). These events have the properties expected from W → $e\nu$ decay.

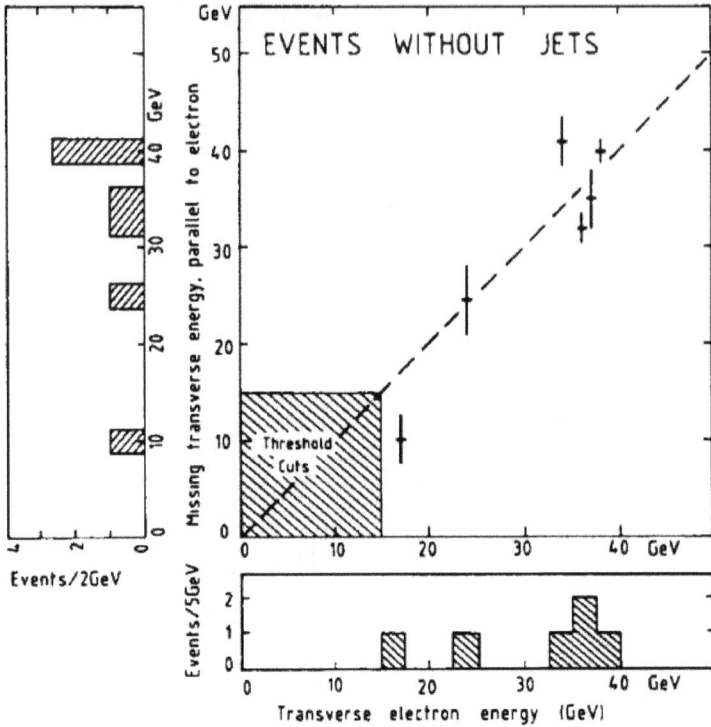

Fig. 11.　UA1 scatter plot of all the events from the 1982 data which contain a high-pT electron and large $|\vec{p}_T^{\,\text{miss}}|$. The abscissa is the electron $|\vec{p}_T|$ and the ordinate is the $\vec{p}_T^{\,\text{miss}}$ component antiparallel to the electron $\vec{p}_T$.

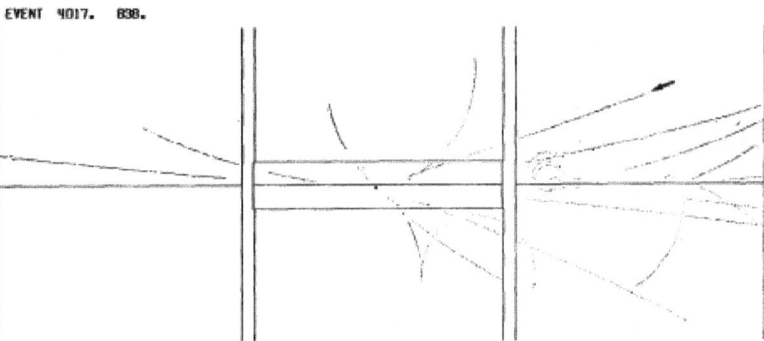

Fig. 12.　Display of a UA1 W $\rightarrow$ eν event. The arrow points to the electron track.

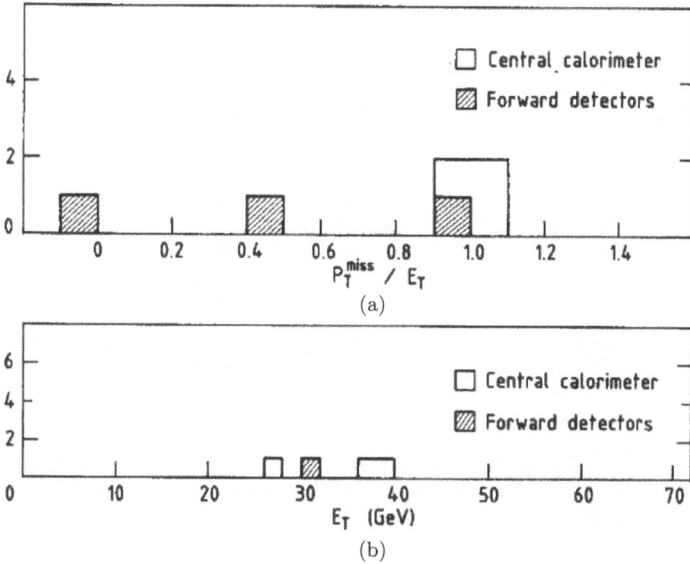

Fig. 13.   (a) Display of the ratio between $|\vec{p}_T^{\text{miss}}|$ and the electron transverse momentum (called E_T in this plot) for six UA2 events containing an electron with $E_T > 15\,\text{GeV}$; (b) Electron distribution for the four events with the highest $|\vec{p}_T^{\text{miss}}|/E_T$ ratio.

4.2.   *Discovery of the Z boson*

Figure 14 illustrates the search for the decay $Z \rightarrow e^+e^-$ in UA1.[11] The first step of the analysis requires the presence of two calorimeter clusters consistent with electrons and having a transverse energy $E_T > 25\,\text{GeV}$. Among the data recorded during the 1982–83 collider run, 152 events are found to satisfy these conditions. The next step requires the presence of an isolated track with $p_T > 7\,\text{GeV}/c$ pointing to at least one of the two clusters. Six events satisfy this requirement, showing already a clustering at high invariant mass values, as expected from $Z \rightarrow e^+e^-$ decay. Of these events, four are found to have an isolated tracks with $p_T > 7\,\text{GeV}/c$ pointing to both clusters. They are consistent with a unique value of the e^+e^- invariant mass within the calorimeter resolution. One of these events is displayed in Fig. 15.

An event consistent with the decay $Z \rightarrow \mu^+\mu^-$ was also found by UA1 among the data collected in 1983 (see Fig. 16). Figure 17 shows the mass distribution of all lepton pairs found by UA1 from the analysis of the 1982–83 data. The mean of these values is

$$m_Z = 95.2 \pm 2.5 \pm 3.0 \text{ GeV}$$

where the first error is statistical and the second one originates from the systematic uncertainty on the calorimeter energy scale.

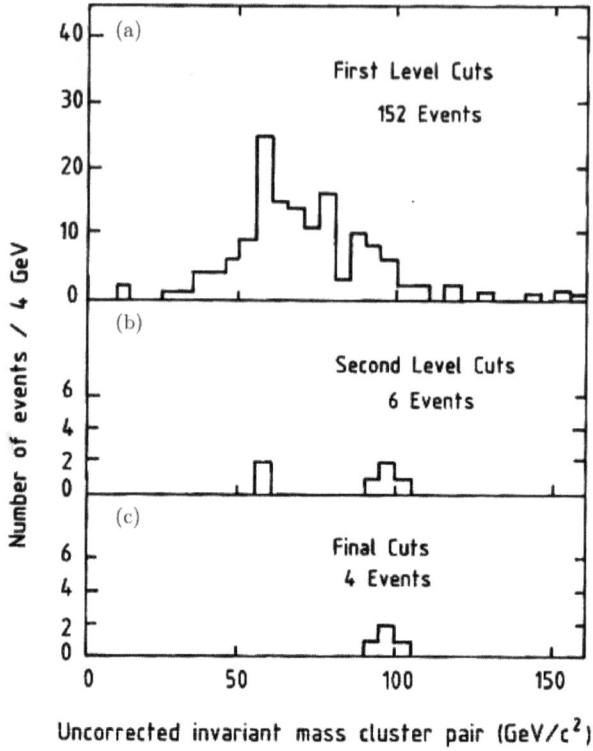

Fig. 14.   Search for the decay Z → e⁺e⁻ in UA1 (see text).

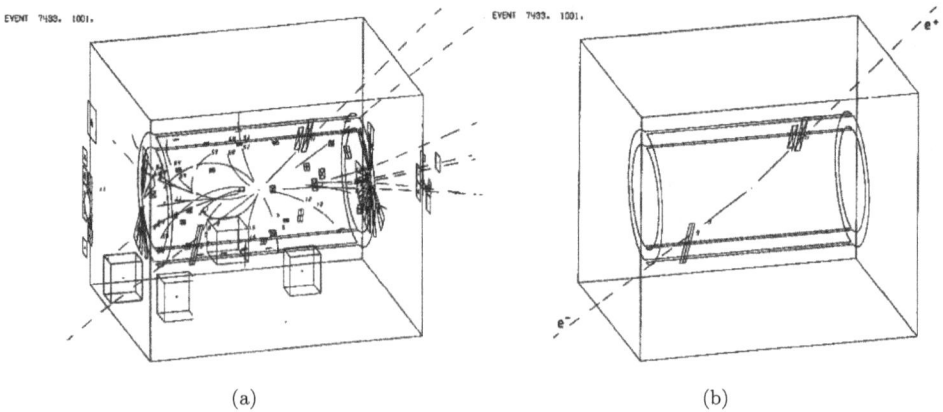

Fig. 15.   One of the Z → e⁺e⁻ events in UA1: (a) display of all reconstructed tracks and calorimeter hit cells; (b) only tracks with $p_T > 2\,\text{GeV}/c$ and calorimeter cells with $E_T > 2\,\text{GeV}$ are shown.

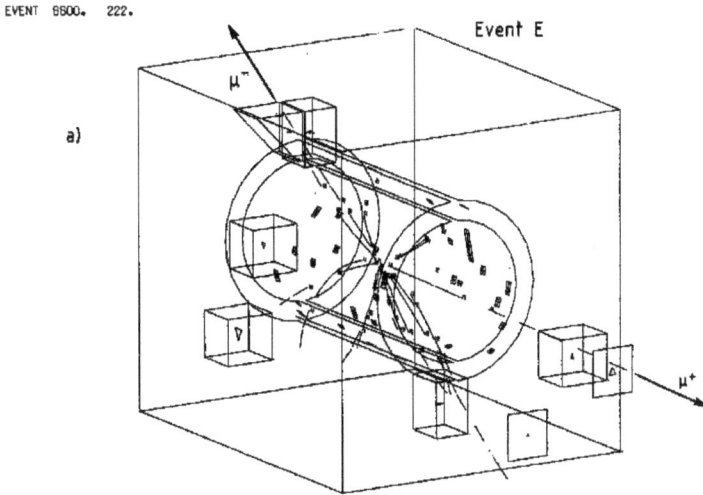

Fig. 16. $Z \rightarrow \mu^+\mu^-$ event in UA1.

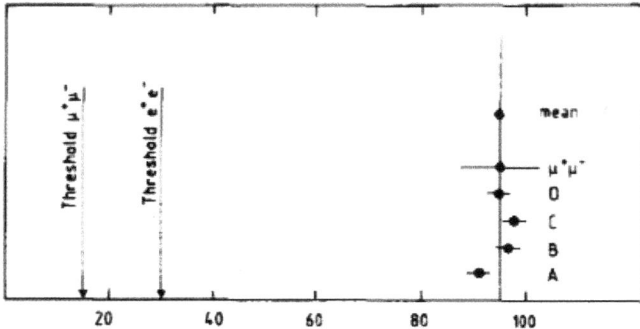

Fig. 17.   Invariant mass distribution of all lepton pairs found by UA1 in the 1982–83 data.

The UA2 search for the decay $Z \rightarrow e^+e^-$ among the 1982–83 data[12] is illustrated in Fig. 18. First, pairs of energy depositions in the calorimeter consistent with two isolated electrons and with $E_T > 25$ GeV are selected. Then, an isolated track consistent with an electron (from preshower information) is required to point to at least one of the clusters. Eight events satisfy these requirements: of these, three events have isolated tracks consistent with electrons pointing to both clusters. The weighted average of the invariant mass values for the eight events is

$$m_Z = 91.9 \pm 1.3 \pm 1.4 \text{ GeV}$$

where the first error is statistical and the second one originates from the systematic uncertainty on the calorimeter energy scale. The latter is smaller than the corresponding UA1 value because the smaller size of the UA2 calorimeter, and

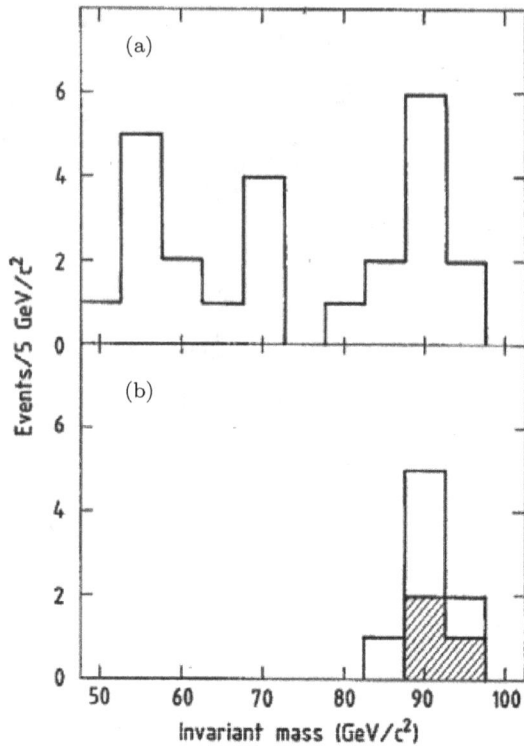

Fig. 18.   Search for the decay Z → e⁺e⁻ in UA2 (see text). The shaded area shows the three events with isolated electron tracks pointing to both calorimeter energy clusters.

its modularity, allow frequent recalibrations on electron beams of known energies from the CERN SPS.

Figure 19 shows the energy deposited in the UA2 calorimeter by a W → eν and by a Z → e⁺e⁻ event. Such distributions, usually called "Lego plots", illustrate the remarkable topologies of such events, with large amounts of energy deposited in a very small number of calorimeter cells, and little or no energy in the remaining cells.

5.   Physics Results from Subsequent Collider Runs

Following the historical runs in 1982–83 which led to the discovery of the W and Z bosons, additional runs took place in the following years.

In a first phase, up to the end of 1985, with the two detectors basically unchanged, the collider energy was raised from $\sqrt{s} = 540\,\text{GeV}$ to $630\,\text{GeV}$ and the peak luminosity doubled (see Section 2, Table 1). The new data allowed more detailed studies of the W and Z production and decay properties, beautifully confirming the Standard Model expectations.

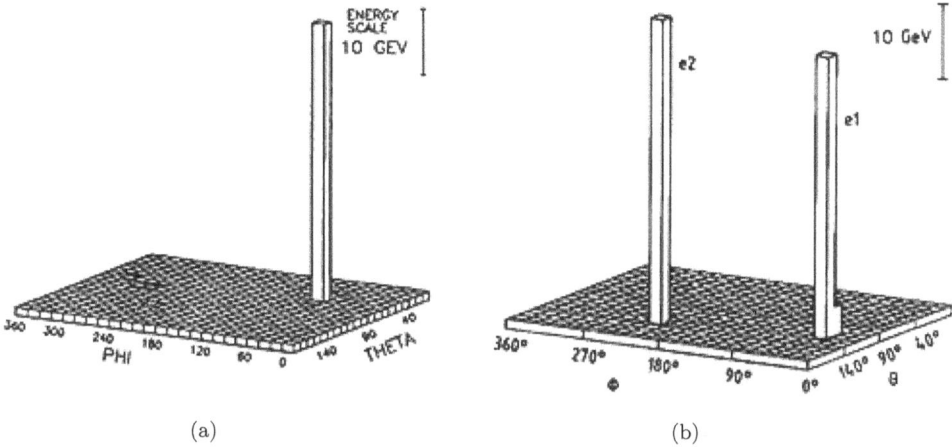

Fig. 19. The energy deposited in the UA2 calorimeter for a W $\to$ eν (a) and a Z $\to$ e$^+$e$^-$ event (b).

As mentioned in Section 2, more physics runs took place at the CERN Collider between 1987 and its shut-down at the end of 1990. Other important physics results obtained by the two experiments between 1982 and 1990 are described in the next sub-sections.

5.1. *W and Z masses and production cross-sections*

At the end of 1985 UA1 had recorded 290 W $\to$ eν, 33 Z $\to$ e$^+$e$^-$, 57 W $\to$ $\mu\nu$ and 21 Z $\to$ $\mu^+\mu^-$ events.[13] As an example, Fig. 20 shows the W $\to$ eν transverse mass (M_T) distribution, where $M_T = [2p_T^e p_T^\nu (1 - \cos\phi_{e\nu})]^{1/2}$, and $\phi_{e\nu}$ is the azimuthal separation between electron and neutrino (the transverse mass is used instead of the electron transverse momentum because its distribution is less sensitive to the W transverse momentum).

Figure 21 shows the invariant mass distribution of all e$^+$e$^-$ pairs recorded by UA1 during the same period. The W and Z mass values obtained from fits to the distributions of Figs. 20 and 21 were

$$m_W = 82.7 \pm 1.0 \pm 2.7 \text{ GeV},$$

$$m_Z = 93.1 \pm 1.0 \pm 3.1 \text{ GeV},$$

where the first error is statistical and the second one reflects the uncertainty on the calorimeter energy scale.

The W and Z production cross-sections, multiplied with the corresponding decay branching ratios (BR), as measured by UA1, were

$$\sigma_W \text{BR}(W \to e\nu) = 630 \pm 50 \pm 100 \text{ pb};$$

$$\sigma_Z \text{BR}(Z \to e^+e^-) = 74 \pm 14 \pm 11 \text{ pb}.$$

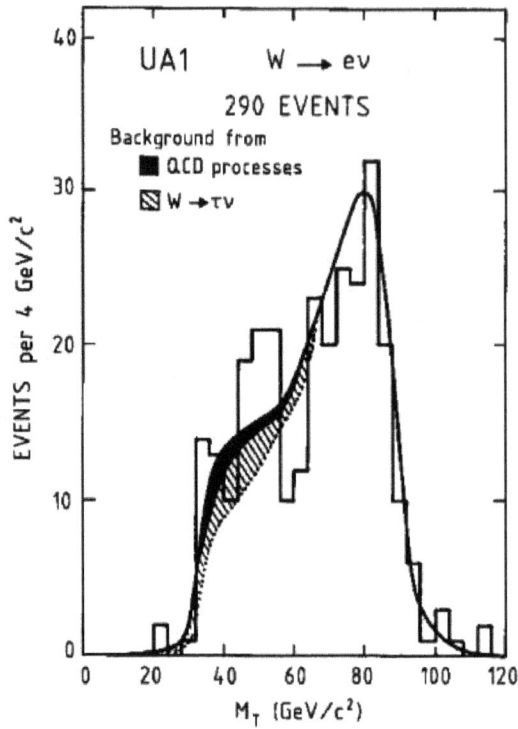

Fig. 20.   Transverse mass distribution for all W → eν events recorded by UA1 between 1982 and 1985.

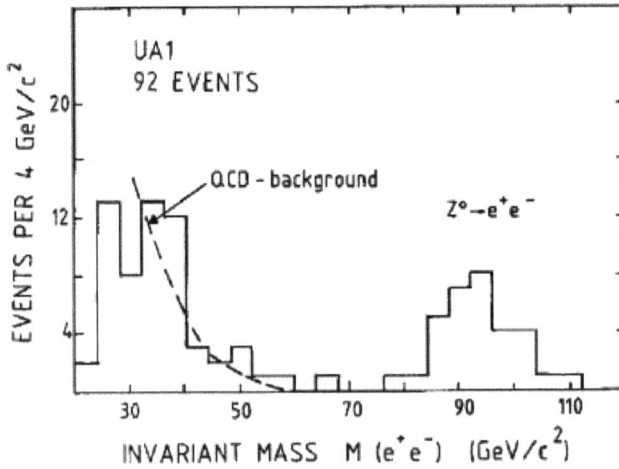

Fig. 21.   Invariant mass distribution of all e⁺e⁻ pairs recorded by UA1 between 1982 and 1985.

UA1 has also observed 32 W $\rightarrow \tau\nu$ decays followed by τ hadronic decay.[14] These events appear in the detector has a highly collimated, low multiplicity hadronic jet approximately back-to-back in azimuth to a significant missing p_T.

In the same physics runs, from 1982 to 1985, the UA2 experiment had recorded samples of 251 W $\rightarrow e\nu$ and 39 Z $\rightarrow e^+e^-$ events.[15] The measured properties of these events were in good agreement with the UA1 results. The W and Z mass values, as measured by UA2, were

$$m_W = 80.2 \pm 0.8 \pm 1.3 \text{ GeV},$$

$$m_Z = 91.5 \pm 1.2 \pm 1.7 \text{ GeV},$$

where, as usual, the first error is statistical and the second one reflects the uncertainty on the calorimeter energy scale.

5.2. *Charge asymmetry in the decay* W $\rightarrow e\nu$

At the energies of the CERN $\bar{p}p$ collider, W production is dominated by $q\bar{q}$ annihilation involving at least one valence quark or antiquark. As a consequence of the $V - A$ coupling, which violates parity conservation, the helicity of the quarks (antiquarks) is -1 ($+1$) and the W is almost fully polarised along the $\bar{p}$ beam. Similar helicity arguments applied to W $\rightarrow e\nu$ decay predict that the leptons (e^-, μ^-, ν) should be preferentially emitted opposite to the direction of the W polarisation, and antileptons (e^+, μ^+, $\bar{\nu}$) along it.

The angular distribution of the charged lepton in the W rest frame can be written as

$$\frac{dn}{d\cos\theta^*} \propto (1 + q\cos\theta^*)^2$$

where θ^* is the angle of the charged lepton measured with respect to the W polarisation, and $q = -1$ ($+1$) for electrons (positrons). This axis is practically collinear with the incident $\bar{p}$ direction if the W transverse momentum is small.

A complication arises from the fact that the neutrino longitudinal momentum is not measured, and the requirement that the invariant mass of the $e\nu$ pair be equal to the W mass gives two solutions for θ^*. The UA1 analysis[13] retains only those events for which one solution is unphysical (W longitudinal momentum inconsistent with kinematics), and the lepton charge sign is unambiguously determined. Figure 22 shows the distribution of the variable $q\cos\theta^*$ for 149 unambiguous events. The distribution agrees with the expected $(1 + q\cos\theta^*)^2$ form. It must be noted that this result cannot distinguish between $V - A$ and $V + A$ because in the latter case all helicities change sign and the angular distribution remains the same.

5.3. *A test of QCD: The W boson transverse momentum*

To lowest order the W and Z bosons produced by $q\bar{q}$ annihilation are emitted with very low transverse momentum. However, gluon radiation from the initial

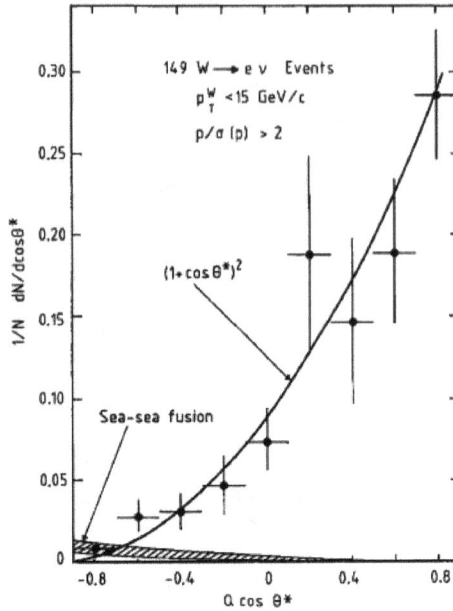

Fig. 22. Decay angular distribution for the final UA1 $W \to e\nu$ event sample (see text). The shaded band shows the expected contribution of wrong polarisation from the annihilation of a sea quark with a sea antiquark.

quarks (or antiquarks) may result in W and Z production with a sizeable transverse momentum, which is equal and opposite to the total transverse momentum of all hadrons produced in association with the intermediate bosons.

Figure 23 shows distributions of the W transverse momentum, p_T^W, as measured by UA1[13] using the $W \to e\nu$ event sample. A QCD prediction,[16] also shown in Fig. 23, agrees with the data over the full p_T^W range. The W bosons produced with high p_T^W are expected to recoil against one or more jets, and such jets are indeed observed experimentally.

5.4. Hadronic decays of the W and Z bosons

As mentioned earlier (see Section 4), the W and Z bosons decay predominantly (∼70%) to quark–antiquark pairs which appear as two hadronic jets. Such configurations are overwhelmed by the QCD background of two-jet production from hard parton scattering. However, despite the unfavourable signal-to-noise ratio, the detection of $W \to q\bar{q}'$ and $Z \to q\bar{q}$ decays at the collider was considered as interesting not only as an experimental challenge, but also as a demonstration that the reconstruction of the two-jet invariant mass at next-generation hadron colliders could be useful to detect new particles decaying to two hadronic jets.

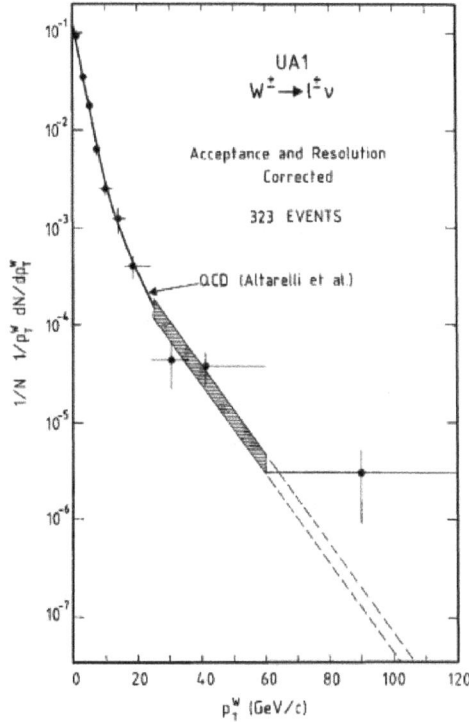

Fig. 23. Distribution of p_T^W, as measured by UA1.[13] The curve is a QCD prediction,[16] and the shaded band shows the theoretical uncertainty in the region of high p_T^W.

In the UA2 calorimeter jet energies are measured with a resolution $\sigma E/E \approx 0.76/\sqrt{E}$ (E in GeV). The two-jet invariant mass distribution measured by UA2[17] from the data collected between 1983 and 1985 is shown in Fig. 24. This distribution has a clear bump structure in the mass region where two-jet final states from W and Z decays are expected to fall (the bump contains 632 ± 190 events over the continuous background from parton–parton scattering). We note that the ordinates are multiplied by the fifth power of the two-jet mass value in order to remove most of the fast decrease and to use a linear scale. The W and Z peaks are not resolved.

5.5.   *Precision measurement of the W to Z mass ratio*

During the last three years of collider operation (1988–90) UA2 collected large samples of W $\rightarrow$ eν and Z $\rightarrow$ e$^+$e$^-$ decay events. As shown in Subsection 5.1, the systematic error from the uncertainty on the calorimeter energy scale affecting the W and Z masses, as measured by UA1 and UA2 using the 1982–85 data samples, was already comparable to, or even larger than the statistical error. However, the error on the measurement of the ratio m_W/m_Z is mainly statistical, because the systematic uncertainty on the calorimeter energy scale largely cancels in this ratio.

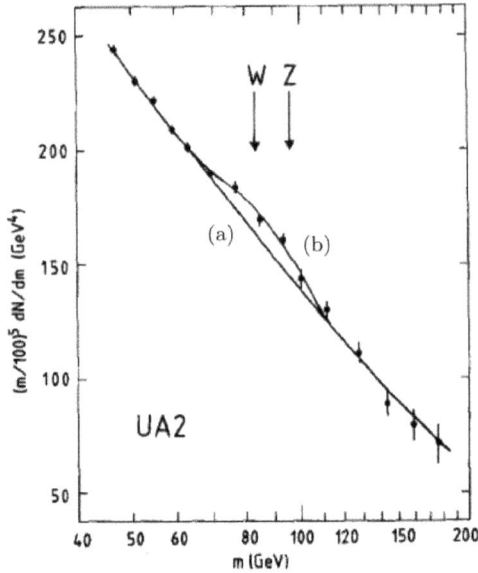

Fig. 24. Two-jet invariant mass distribution, as measured in the UA2 central calorimeter. Curve (a) is a best fit to the data excluding the mass interval $65 < m < 105\,\mathrm{GeV}$. Curve (b) is a fit to all data points with the addition of two Gaussians centred at the nominal W and Z mass values.

An additional reason for measuring precisely the ratio $m_\mathrm{W}/m_\mathrm{Z}$ was the start of LEP operation in July 1989, with the expectation that a precise measurement of m_Z would soon become available. Then the two measurements could be combined to obtain a precise determination of m_W.

Figure 25 shows the transverse mass distribution for 2065 W $\rightarrow$ eν decays with the electron measured in the UA2 central calorimeter.[18] A best fit to this distribution using m_W as a free parameter gives $m_\mathrm{W} = 80.84 \pm 0.22\,\mathrm{GeV}$ (statistical error only).

The measured e$^+$e$^-$ invariant mass distribution is shown in Fig. 26, which displays two spectra: one containing 95 events in which both electrons fall in a fiducial region of the central calorimeter and their energies are accurately measured; and another spectrum containing 156 events in which one of the two electrons falls outside the fiducial region of the central calorimeter, resulting in a broader mass resolution. Best fits to the two spectra give $m_\mathrm{Z} = 91.65 \pm 0.34\,\mathrm{GeV}$ and $m_\mathrm{Z} = 92.10 \pm 0.48\,\mathrm{GeV}$, respectively. The weighted mean of these two values is $m_\mathrm{Z} = 91.74 \pm 0.28\,\mathrm{GeV}$ (statistical error only).

The two independent measurements of m_W and m_Z give

$$\frac{m_\mathrm{W}}{m_\mathrm{Z}} = 0.8813 \pm 0.0036 \pm 0.0019$$

where the first error is statistical and the second one is a small systematic uncertainty which takes into account a possible calorimeter non-linearity.

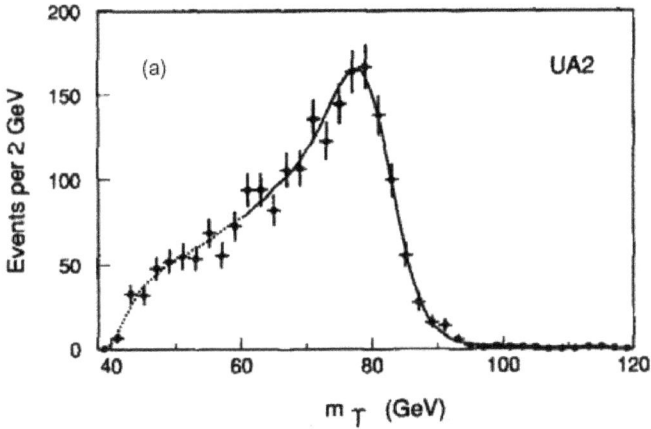

Fig. 25.   Transverse mass distribution for 2065 W → eν decays (see text). The curve is the best fit to the experimental distribution using m_W as a free parameter.

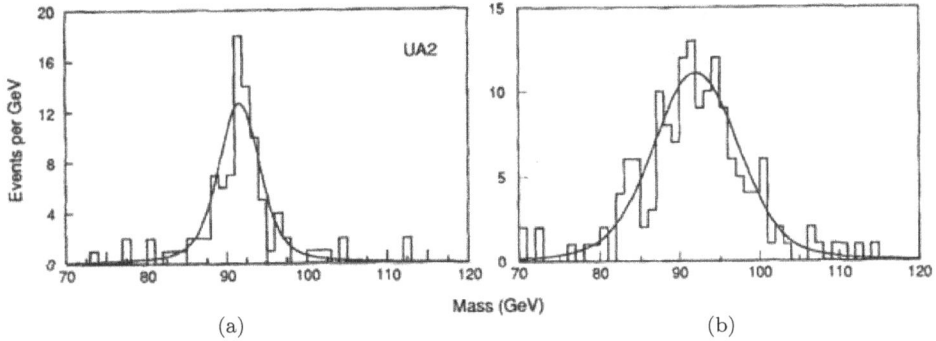

Fig. 26.   Invariant mass spectra for two Z → e⁺e⁻ event samples, as measured by UA2 (see text). The curves are best fits to the data using m_Z as a free parameter.

By 1991 a precise measurement of m_Z from LEP experiments had become available, $m_Z = 91.175 \pm 0.021$ GeV.[19] Multiplying this value with the ratio m_W/m_Z measured by UA2 provided a determination of m_W with a precision of 0.46%:

$$m_W = 80.35 \pm 0.33 \pm 0.17 \text{ GeV},$$

in agreement with a direct measurement, $m_W = 79.91 \pm 0.39$ GeV, by the CDF experiment at the Fermilab $\bar{p}p$ collider.[20]

The precise determination of m_W was used to obtain bounds on the top quark mass, for which early direct searches at the CERN and Fermilab $\bar{p}p$ colliders had only provided the lower bound $m_{top} > 89$ GeV.[21] As shown by Veltman,[22] within the frame of the Standard Model the value of m_W for fixed m_Z depends quadratically on the mass of the top quark through electroweak radiative corrections from virtual fermion loops (and also, to a much smaller extent, on the mass of the Higgs boson).

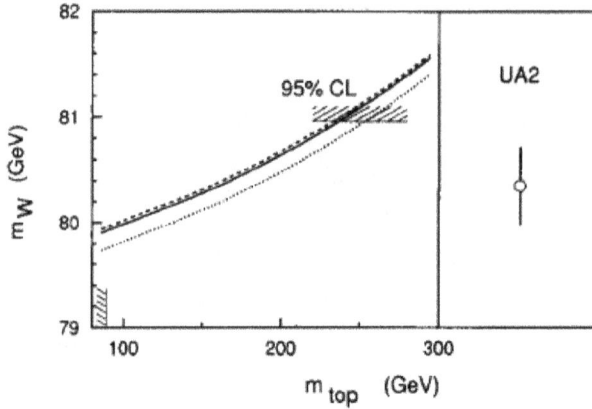

Fig. 27. m_W versus m_{top} and the determination of m_W obtained by combining the precise UA2 measurement of m_W/m_Z with an early precise measurement of m_Z at LEP (see text). The curves are Standard Model predictions for fixed m_Z (as measured at LEP), and for different values of the Higgs boson mass: 50 GeV (dashed curve); 100 GeV (solid curve); and 1000 GeV (dotted curve). Also shown are the 95% confidence level (CL) upper bound $m_{top} < 250$ GeV obtained from the error on m_W, and the lower bound $m_{top} > 89$ GeV from early direct searches at the CERN and Fermilab $\bar{p}p$ colliders.

As illustrated in Fig. 27, the UA2 result gave

$$m_{top} = 160^{+50}_{-60} \text{ GeV},$$

suggesting a heavy top quark well before its discovery at the Fermilab 1.8 TeV collider with a measured mass $m_{top} = 174 \pm 10 \pm 13$ GeV[23] (the present world average of measurements from the experiments at the Fermilab $\bar{p}p$ collider and, more recently, at the LHC, is $m_{top} = 173.21 \pm 0.51 \pm 0.71$ GeV[24]).

6.   Conclusions

The CERN proton–antiproton collider was initially conceived as an experiment to detect the W and Z bosons. Not only it beautifully fulfilled this task, but it also tested the electroweak theory to a level of few percent and provided important verifications of QCD predictions. In the end, it turned out to be a first-class general-purpose accelerator facility with a very rich physics programme. It cannot be excluded that the construction of the LHC would not have been approved if the CERN proton–antiproton collider had not been so successful.

References

1. F. J. Hasert *et al. Phys. Lett. B* **46**, 121 (1973); *Phys. Lett. B* **46**, 128 (1973); D. Haidt, *The Discovery of Weak Neutral Currents*, in this book, pp. 165–183.

2. C. Rubbia, P. McIntyre and D. Cline, in *Proc. International Neutrino Conference*, ed. H. Faissner, H. Reither and P. Zerwas, Vieweg, Braunschweig (1977), 683.

3. S. van der Meer, *CERN-ISR-PO* 72-31(1972); S. Van der Meer, *Rev. Mod. Phys.* **57**, 689 (1985).

4. For a review see: L. Jones, F. Mills, A. Sessler, K. Symon and D. Young, *Innovation Is Not Enough: A History of the Midwestern Universities Reasearch Association (MURA)*, (World Scientific, Singapore, 2010).

5. G. Carron *et al.*, *Phys. Lett. B* **77**, 353 (1978).

6. A. Astbury *et al.* (UA1 Collaboration), *Phys. Scripta* **23**, 397 (1981).

7. B. Mansoulié (UA2 Collaboration), in *Proc. Moriond Workshop on Antiproton–Proton Physics and the W Discovery*, La Plagne, Savoie, France, 1983 (Ed. Frontières, 1983), p. 609.

8. For a review see: P. Darriulat and L. Di Lella, *Revealing Partons in Hadrons: From the ISR to the SPS Collider*, in this book, pp. 313–341.

9. G. Arnison *et al.* (UA1 Collaboration), *Phys. Lett. B* **122**, 103 (1983).

10. M. Banner *et al.* (UA2 Collaboration), *Phys. Lett. B* **122**, 476 (1983).

11. G. Arnison *et al.* (UA1 Collaboration), *Phys. Lett. B* **126**, 398 (1983).

12. P. Bagnaia *et al.* (UA2 Collaboration), *Phys. Lett. B* **129**, 130 (1983).

13. C. Albajar *et al.* (UA1 Collaboration), *Z. Phys. C* **44**, 15 (1989).

14. C. Albajar *et al.* (UA1 Collaboration), *Phys. Lett. B* **185**, 233 (1987); and Addendum, *Phys. Lett. B* **191**, 462 (1987).

15. R. Ansari *et al.* (UA2 Collaboration), *Phys. Lett. B* **186**, 440 (1987).

16. G. Altarelli, R. K. Ellis, M. Greco and G. Martinelli, *Nucl. Phys. B* **246**, 12 (1984).

17. R. Ansari *et al.* (UA2 Collaboration), *Phys. Lett. B* **186**, 452 (1987).

18. J. Alitti *et al.* (UA2 Collaboration), *Phys. Lett. B* **276**, 354 (1992).

19. J. Carter, in *Proc. Joint Lepton–Photon Symp. & Europhys. Conf. on High-Energy Physics,* Geneva (Switzerland) 25 July–1 August 1991, (World Scientific, 1992), Vol. 2, p. 3.

20. F. Abe *et al.* (CDF Collaboration), *Phys. Rev D* **43**, 2070 (1991).

21. T. Åkesson *et al.* (UA2 Collaboration), *Z. Phys. C* **46**, 179 (1990); C. Albajar *et al.* (UA1 Collaboration), *Z. Phys. C* **48**, 1 (1990); F. Abe *et al.* (CDF Collaboration), *Phys. Rev. Lett.* **64**, 142 (1990).

22. M. Veltman, *Nucl. Phys. B* **123**, 89 (1977).

23. F. Abe *et al.* (CDF Collaboration), *Phys. Rev. D* **50**, 2966 (1994).

24. K. A. Olive *et al.* (Particle Data Group), *Chin. Phys. C* **38**, 090001 (2014), p. 739.

The Discovery of Weak Neutral Currents

Dieter Haidt

Deutsches Elektronen-Synchrotron (DESY)
Notkestrass 85, D-22603 Hamburg, Germany
dieter.haidt@desy.de

The beginning of high energy neutrino physics at CERN is outlined followed by the presentation of the discovery of weak neutral currents in the bubble chamber Gargamelle.

1. Preface

Neutrino physics has played a prominent role in the history of CERN. The very first large project at the Proton Synchrotron starting in 1960 was a neutrino experiment aimed at solving one of the urgent questions in the understanding of weak interactions. It was the beginning of a long range program. Its highlight was the discovery of weak neutral currents in the bubble chamber Gargamelle. Four decades passed since then and the huge impact of the discovery both for CERN and worldwide stands out clearly.

This article begins with a glance at the first neutrino experiment at CERN and focusses then on the discovery of weak neutral currents in the Gargamelle experiment. For personal testimonies of the beginning neutrino physics at CERN see Refs. 1 and 2. The discovery of weak neutral currents has been the subject of dedicated conferences[3–5] and several reviews, e.g. Refs. 6–9.

2. The Beginning of High Energy Neutrino Physics at CERN

2.1. *Status of weak interactions at the end of the 1950s*

It was beyond imagination, when Pauli invented in 1930 the neutrino in a stroke of genius, that once it would become the tool par excellence to investigate the leptonic sector of weak interactions. Right after sending the famous letter to his *radioactive friends in Tübingen*, Pauli told his astronomer friend Walter Baade:[10] *"I have today done something terrible which no theoretician ever should do and proposed something which never will be possible to be verified experimentally."* Shortly afterwards, Fermi formulated his theory of β-decay[11] on the basis of Pauli's neutrino hypothesis and the recently discovered neutron. Bethe and Peierls[12] calculated in the following year the cross-section for a neutrino induced process and found it hopelessly small. Only much later, in 1946, it occurred to Pontecorvo,[14] that with the advent of powerful nuclear plants and their high antineutrino fluxes there may be a chance. Indeed, Cowan and Reines succeeded in detecting the first neutrino induced reactions at the Savannah River reactor. They observed the inverse β-decay: $\bar{\nu}_e + p \rightarrow e^+ + n$. So, 26 years after the formulation of the neutrino hypothesis, June 14, 1956, Cowan and Reines could send a telegram to Pauli saying: "*We are delighted to tell you that we have definitely found neutrinos through observing inverse β-decay.*" Pauli prompted: "*Everything comes to him who knows how to wait.*"

The Dirac equation for fermions written in terms of 4-spinors can conveniently be written as a set of coupled equations with Weyl 2-spinors. These equations have the interesting property to decouple for massless fermions, such as it was assumed for the neutrino. The Lorentz structure in the original Fermi theory of beta decay was not specified, it could involve scalar, pseudoscalar, vector, axialvector or tensor contributions. The experimental investigation of nuclear and particle decays have shown that the interaction is of the type V, A. The demonstration in 1957 that parity is maximally violated in weak interactions, has prompted the 2-component theory of the neutrino and the formulation of the $V - A$ theory of weak interactions.

This inspired immediately the idea of a weak intermediate vector boson as the analog to the photon in electromagnetic interactions. The processes at that time, mainly decays, involved only small momentum transfers and thus appeared as effective 4-fermion interactions. This raised interest in experiments at much larger momentum transfer soon accessible at the planned accelerators of CERN, Dubna and BNL for the investigation of the existence of an intermediate vector boson and the properties of weak interactions in general.

Another fundamental question arose in the study of muon decays: $\mu^+ \rightarrow e^+ + \nu + \tilde{\nu}$ and $\mu \rightarrow e + \gamma$. It was known that the leptonic muon decay is a 3-body decay consisting of an electron and two light nonidentical neutrals. They could be the known neutrino and its antiparticle. However, there was no compelling reason for being particle and antiparticle, there could also exist two distinct neutrino species. The same question appeared also in the attempt to understand the absence of the decay $\mu \rightarrow e + \gamma$. If the decay is assumed to involve an intermediate vector boson, then it should not be suppressed. Feinberg[16] argued that the decay could nevertheless be suppressed, if the neutrinos associated with the two vertices are different. Pontecorvo devoted in Refs. 17 and 18 a thorough discussion of the 2-neutrino question and proposed ways to tackle it experimentally.

The idea of high energy neutrino beams derived from pion decays for answering these outstanding questions were considered by Pontecorvo,[19, 20] Markov[21] with his young collaborators Zheleznykh and Fakirov and by Schwartz[22] and T. D. Lee.[23] Pontecorvo recalls in Ref. 24 how he came to propose a neutrino beam at high energy from meson factories and from very high energy accelerators.

2.2. The first neutrino experiment at CERN

In 1960 the time has come to realise the idea of high energy neutrino beams at the new accelerators of CERN and BNL. The first operation at the CERN proton synchrotron was at the end of 1959, the Brookhaven AGS started a year later in autumn 1960. Bernardini[25] was at that decisive time director of research at CERN and recognised the potential of neutrino experiments to open a new and promising field of research for exploring the properties of weak interactions in a hitherto unprecedented energy regime with particular emphasis on solving the two burning questions, namely whether there exist two neutrinos and whether there exists an intermediate vector boson. Bernardini[26] reported the program of neutrino experiments and their feasibility at CERN to the 1960 Rochester conference.[a] Two weeks after his return to CERN appeared the proposal by Steinberger, Krienen and Salmeron[27] for an *experiment at CERN to detect neutrino induced reactions.* In a recent letter Steinberger[28] recalls: *"I am personally indebted to Pontecorvo for proposing, in 1959, to check experimentally if the neutrinos associated with muons in pion and kaon decay are the same, or not, as those in β decay, and that the higher energy accelerators, then under construction at Brookhaven and CERN, would permit neutrino beams of energy high enough to allow such an experiment (Pontecorvo 1959) — the experiment for which M Schwartz, L Lederman and I later shared the Nobel prize (Danby et al. 1962). Independently, Schwartz had proposed that neutrino beams would permit the study of weak interactions at higher energy, but he did not consider the particular question of the possible inequality of the two neutrinos, proposed by Pontecorvo (Schwartz 1960)."* The three authors studied the feasibility of a neutrino experiment at the CERN PS using a heavy liquid bubble chamber as detector. Basic questions addressed were:

- *Neutrino source*
  The protons from the PS strike a thin target in one of the straight sections. The pions produced at an angle of 6 degrees generate by their decay in flight the neutrino beam. The alternative would have been to postpone the experiment by about one year, until an external proton beam would be available. It has been argued that there is no compelling reason against a setup with an internal target.
- *Neutrino flux*
  The evaluation of the neutrino flux involves the initial pion flux and the pion decay kinematics. For the estimate of the actual number of events in the bubble chamber various efficiency factors had to be taken into account and, of course, the theoretical cross section of the process to be measured. A detailed consideration was devoted to the determination of the pion trajectories in the presence of the fringing field of the next magnet.

---

[a]He acknowledged Pontecorvo and Schwartz for the idea of this kind of experiment and added in the list of references and notes that also Markov and Fakirov had such ideas.

- *Size of the shielding*

  All hadrons and charged leptons travelling in the direction of the neutrinos have to be strongly absorbed, otherwise the scanning of the pictures and the interpretation of the events will be difficult. Furthermore, one has to worry about background from cosmic rays and neutrons. The requirements for the size of the shielding were considerable: 650 tons of iron and 4000 tons of heavy concrete have been estimated adequate.
- *Event rate and aim*

  The estimated rate was 1 event per day per ton of sensitive detecting material. A run of 2 to 3 weeks would be sufficient to settle the question of whether or not there are two types of neutrinos.

The authors concluded their proposal with the recommendation that CERN should make the effort to realise the experiment.

Bernardini presented the status of the neutrino program to the Scientific Policy Committee[29] in November 1960. The setting up of an experiment of this size was a real challenge for the young laboratory, since it required the coordination of several teams. The original layout was later on modified and finally three detectors came into operation, the Ecole Polytechnique bubble chamber, a cloud chamber complemented with electronic devices and the newly built NPA bubble chamber (the Ramm 1.2 m chamber). The next status report on the neutrino experiment to the 19th SPC in April 1961 from an engineering run was quite encouraging. However, three months later Bernardini had to announce at the 20th SPC meeting:[30] *"It is probably well known that the initial programme of experiments, with which at CERN we intended to open the field of the high-energy neutrino physics, is going through a crisis."* In fact, Guy von Dardel demonstrated that the flux was overestimated by an order of magnitude and thus no neutrino candidate could be expected. The failure was attributed to limitations in the beam optics at the internal target and to the simplified decay kinematics of the pions. Immediate remedies to increase the flux were discussed. Although solutions to increase the flux by one or two orders of magnitude were at hand, their realization on short terms was impossible. So, the race with the BNL group was lost. They[31] made the discovery of two neutrinos in 1962.

Even though this first experiment did not bring the expected success, it was nevertheless the beginning of high energy neutrino physics at CERN. In a second attempt the weaknesses have been overcome. An important achievement on the machine side was the fast extraction of the proton beam. The external proton beam was now hitting a thin and long target. The produced secondary pions and kaons were focussed efficiently by Van der Meer's newly conceived magnetic horn. The neutrino flux increased by more than two orders of magnitude. The shielding was improved and the CERN NPA heavy liquid bubble chamber (the Ramm chamber) was operated together with the spark chamber array.[32] Results were ready for the

Siena 1963 Conference.[33] Further runs with the Ramm chamber followed in 1964 with freon filling and 1967 with propane filling.

The initiative of Bernardini was the beginning of a long term program, which eventually brought about a fundamental result with the discovery of weak neutral currents.

2.3. *Early searches for weak neutral currents*

At the end of the 1950s weak processes were described as the interaction of two weak *charged* currents. This stimulated theoreticians quickly to think about a possible neutral current and a neutral intermediate field. Feynman and Gell-Mann merely noted in their famous publication:[34] "*We deliberately ignore the possibility of a neutral current, containing terms like $\bar{e}e$, $\bar{\mu}e$, etc. and possibly coupled to a neutral intermediate field. No weak coupling is known that requires the existence of such an interaction.*" Others speculated about implications of weak neutral currents, see for instance Refs. 35–37.

The successful description of all known low energy weak processes within the $V - A$ theory called the attention to the behaviour at higher energies. Lee and Yang[23] published in 1961 a catalog of fundamental questions to be addressed in the upcoming neutrino experiments. Among them was also the search for weak neutral currents. The experimental situation was however rather discouraging. The presence of weak neutral currents was first checked by examining the decay rates of elementary particles. Decays without change of the electric charge Q and strangeness S, i.e. $\Delta Q = 0$ and $\Delta S = 0$, were not useful, because they were dominated by electromagnetic interactions, therefore decays obeying $\Delta Q = 0$ and $\Delta S \neq 0$ were considered. However, both leptonic and hadronic kaon decays turned out to be dismayingly small. A new way of searching for weak neutral currents became possible in the CERN neutrino experiment 1963. The bubble chamber group has searched for the elastic process $\nu p \rightarrow \nu p$, i.e. a process with $\Delta Q = 0$ and $\Delta S = 0$. It turned out that neutron interactions represented a dangerous background, thus only an upper limit was obtained. Figure 1 shows Bernardini in the CERN auditorium with the upper limit of 5% (point 3 on the right black board) relative to the quasielastic process $\nu + n \rightarrow \mu + p$. A later revision[30] yielded 12±6%. The spark chamber group could not look for weak neutral currents, because they were running without the appropriate trigger. However, both groups searched for the existence of the intermediate vector boson. There was no sign of a resonance nor of an effect in the energy dependence of the total neutrino nucleon cross section. It had to be concluded, that the W, if it exists, must be heavier than a few GeV. A dedicated search[38] for weak neutral currents with the data of the NPA 1.2 m bubble chamber remained inconclusive because of the neutron background.

Fig. 1.   Bernardini reporting in the CERN auditorium results from the Siena Conference 1963. © 1964 CERN.

3.   The Discovery of Weak Neutral Currents

3.1.   *The bubble chamber Gargamelle*

The results presented at the Siena conference demonstrated a great potential for future investigations of weak interactions. With the experience gained in the first neutrino experiment, Lagarrigue — like others — noted that a next generation experiment should be based on much larger statistics. His dream was to build a bubble chamber satisfying the requirements:

- *An order of magnitude more events*:
  need large target mass and intense flux (booster, focussing).
- *Good identification of muons and electrons*:
  must distinguish muons from charged pions requiring long path lengths in the chamber.
- *Detailed knowledge about final state*:
  must identify hadrons, neutral pions through their decay in two gammas (short conversion length), kaons through their decay, neutrons through interactions inside the chamber, charged hadrons through a visible interaction.

The result was a cylindrical bubble chamber 5 m in length and 1 m in diameter filled with a heavy liquid. When Leprince-Ringuet saw the giant chamber he called it after

Fig. 2.   André Lagarrigue, the father of Gargamelle.

Rabelais *Gargamelle*. Figure 2 shows André Lagarrigue, the father of Gargamelle; he became professor at the university of Orsay in 1964 and director of LAL Orsay in 1969. He formed a European collaboration consisting of seven laboratories: III.Phys.Institut RWTH Aachen, ULB Bruxelles, CERN, Ecole Polytechnique Paris, Istituto di Fisica dell'Università di Milano, LAL Orsay and University College London. They met in 1968 for a two-day meeting at Milan to discuss the physics program. Although the search for the W, the carrier of weak interactions, remained at high priority, the discovery of the substructure of the proton by SLAC attracted the attention. Would the weak current in the neutrino experiment reveal the partonic structure of the proton as does the electromagnetic current in the ep experiment? New and additional information should then come from the fact that in a neutrino and antineutrino exposure probes with different charges are involved. Today the Gargamelle experiment is famous for the discovery of weak neutral currents, but while preparing the physics program this topic was not even discussed and ranged in the proposal[40] submitted in 1970 at low priority.

Figure 3 shows the chamber body inserted in the coils. One notices already here the huge amount of heavy material around the chamber body. The exposures to the improved neutrino and antineutrino wide band beams started in 1971. The films were shared among the seven laboratories. Strict scanning and measuring rules ensured the same standards in all laboratories. Based on the experience of the previous neutrino experiments with the Ramm chamber the events were classified in four categories:

(A) Events with a muon candidate
(B) Multi-prong events without muon candidate
(C) Proton stars
(D) Single electron or positron or gamma

Fig. 3.   The body of Gargamelle installed inside the magnetic coils.

At that time, a neutrino nucleon interaction was supposed to proceed as $\nu_\mu N \to \mu^- + X$ with X being a hadron system and was registered as event of type A. Neutrino induced events are characterised as multi-prong events with muon *candidate* defined phenomenologically as negatively charged noninteracting particle. Since muons are not explicitly identified, any charged particle with the appropriate charge will simulate a muon as long as it does not show a visible interaction. Therefore, the event sample A is unavoidably contaminated and must be corrected. The dominant background source are neutron induced events in the chamber. These neutrons are generated by neutrino interactions in the upstream heavy material. They produce interactions in the chamber, called *neutron stars*, and contribute to the class B, if *all* final state charged particles are identified as hadrons, whereas they contribute to class A, if one of the charged pions with the right charge does not interact in the visible part of the chamber. This contamination can be readily evaluated from the observed neutron stars in class B.

3.2.   *The challenge*

The data analysis for investigating the parton structure of the nucleon was well in progress, when the theory friends of Gargamelle, in particular Jacques Prentki and Mary-Kay Gaillard, pointed out to the collaboration that a breakthrough in the theory of weak interactions had been achieved: the Glashow–Salam–Weinberg model encompassing both electromagnetic and weak phenomena in a local gauge theory. The immediate excitement arose from the fact that the model is renormalisable and that it predicted weak neutral currents, i.e. the process $\nu_\mu N \to \nu_\mu + X$, in addition to the well known charged current process $\nu_\mu N \to \mu^- + X$. If so, one should observe in Gargamelle neutrino induced events without a charged lepton in the final state. Although the collaboration was not prepared for such a search, it took up the challenge without losing time in view of the highly relevant topic. This was possible, because neutral current induced events, if they really existed, should already be present among the events in class B and just waiting to be identified. It was, however, clear from the outset, that the neutron background would be *the* problem.

A dedicated search for neutral current candidates was started. In order to reduce the background from neutrons a strong energy cut of 1 GeV was imposed on the hadronic final state. For future comparison a reference sample was formed from charged current events, where the hadron system respects the same criteria as for neutral current candidates. While the work was going on an exciting event in the antineutrino film was found at Aachen in December 1972. It consisted of a single completely isolated electron and was interpreted as a *leptonic neutral current* candidate $\bar{\nu}_\mu e \to \bar{\nu}_\mu e$, since all conventional interpretations could be safely excluded (see Ref. 41). This extremely clean event became later on famous and served as textbook example. No such event was found in the neutrino film. The interpretion within the Glashow–Salam–Weinberg model provided the very first constraint on the weak mixing angle.

3.3. *Status in March 1973*

Within less than one year a sizeable sample of hadronic neutral current candidates has been obtained. Lagarrigue was chairing the collaboration meeting in March 1973 at CERN. The status of the analysis is summarised in Table 1 and Fig. 4.

Figure 5 shows a neutral current candidate. There is evidently no lepton in the final state. Following track by track one notices a strong interaction and thus verifies its nature as hadron.

There were good reasons to be euphoric. In fact, three arguments seemed to hint at a new effect:

- *The distributions of the neutral current candidates look neutrino-like.*
  Their shapes are compared to the reference sample of neutrino induced CC events with the same properties as the NC candidates ignoring the muon.
- *The ratio of neutral current candidates over charged current events.*
  It is not small and it is flat both along the beam direction (X) and radially (R).

Table 1   The neutral current (NC) and charged current (CC) event samples in the ν and $\bar{\nu}$ films.

Event Type	ν-exposure	$\bar{\nu}$-exposure
# NC	102	64
# CC	428	148

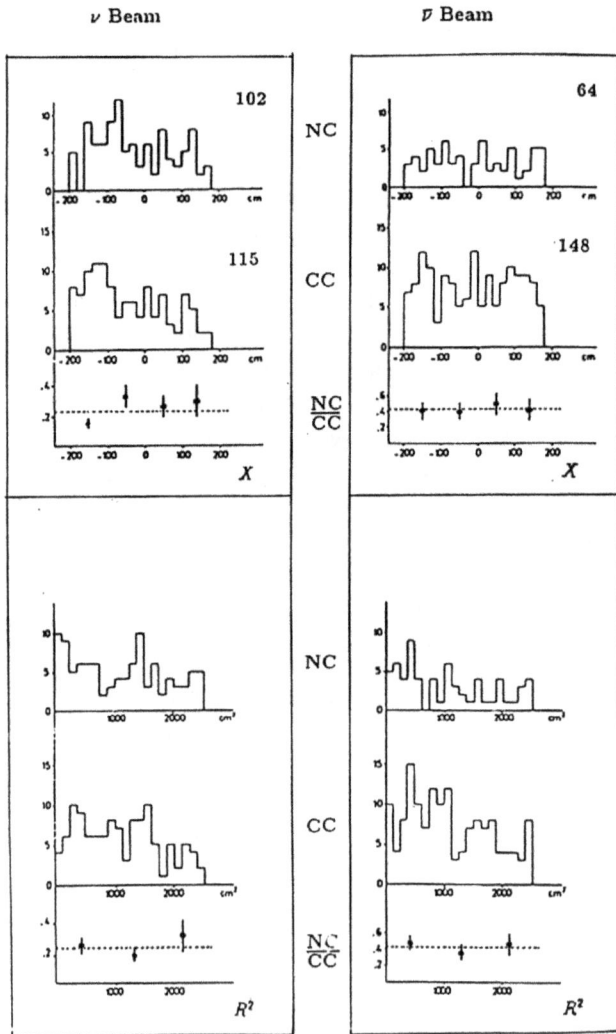

Fig. 4.   Various distributions[43] of the neutral current (NC) and charged current (CC) samples; R denotes the radial and X the longitudinal position.

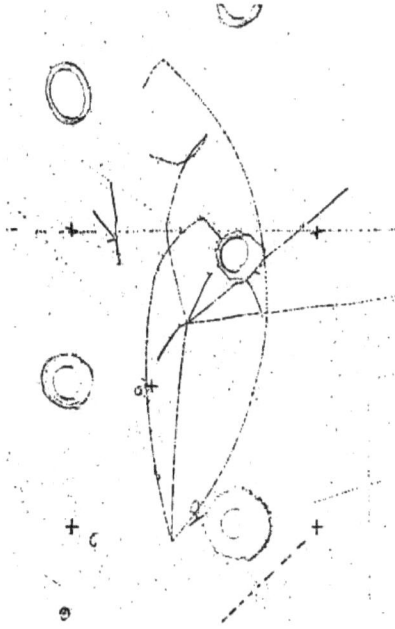

Fig. 5.   Neutral current candidate.

● *The neutral current candidates do not look neutron-like.*

Otherwise the entering neutrons would produce a fall off in the first half of the chamber due to their interaction length being small compared to the chamber dimensions. This was corroborated by a Monte Carlo calculation of the Orsay group assuming simply a source of neutrons at the entrance window of the chamber.

The euphory was damped by two counter-arguments:

● *The neutrino flux has a broad radial distribution.*

The neutrons originating from upstream central neutrino interactions generate indeed a fall-off in the fiducial volume of the chamber, but a substantial fraction of the neutrino flux extends radially way beyond the fiducial volume and produces neutron sources distributed all along the nonvisible part of the chamber and further out to the coils. The net effect is that neutrons enter also laterally and thus generate a flat distribution along the chamber just as genuine neutrino-induced events do. The potential danger is obvious, since the outside material acting as source is a multiple of that contributing at the front (see Fig. 6).

● *Energetic neutrons in the iron shielding propagate in cascades.*

Neutrons entering the chamber and depositing there more than 1 GeV may be the result of a hadron cascade induced by the original neutrino interaction in the shielding. This means that the neutron background is not proportional

Side view

1m

Top view

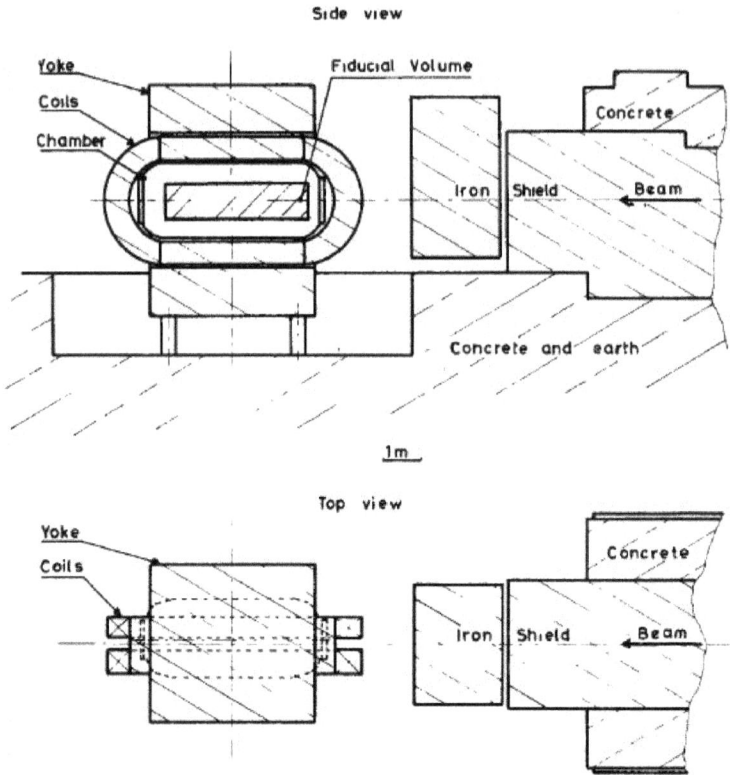

Fig. 6. The experimental setup is sketched in side and top views. The neutrino beam enters from the right through the shielding into the bubble chamber Gargamelle which is located inside the magnetic coils and the yoke. The fiducial volume inside the chamber body is also indicated.

to the interaction length, but rather to the cascade length which is bigger and energy dependent.

At the end of the hot meeting, it was clear that a quantitative estimate of the neutron background[b] was indispensable. A new effect can only be claimed, once it is unambiguously demonstrated that the contributing neutron background is small compared to the number of observed neutral current candidates.

3.4. *The neutron background*

Figure 6 displays the side and top views of the experimental setup. The neutrino beam passing the iron shielding from right to left enters the chamber, which is located inside huge copper coils. The chamber is filled with heavy freon of 1.5 g/cm^3. The cylindrical fiducial volume 0.5 m in radius and 4 m long is indicated in the

---

[b]Other background sources were studied, but found to be of no relevance.

upper view of Fig. 6. The target mass with about 5 tons is very small compared to the surrounding heavy material. The neutrons originate from upstream neutrino interactions. Their sources are therefore located according to the neutrino flux distribution. The neutrino flux has been determined experimentally by measuring the muon flux in the shielding and exploiting the constraint of the known meson flux and decay kinematics. The energy and angular distributions of the produced neutrons has been obtained from the observed neutrino events themselves.

Thus, the spatial and kinematic properties of the neutron source distribution could be safely established. The crucial aspect of calculating the neutron interactions simulating neutral current candidates in the chamber volume consisted in the treatment of hadron propagation in matter. The final state hadrons of an upstream neutrino interaction usually generate a shower in the shielding implying an increase in multiplicity. It has to be decided which of the particles leaving the shielding and entering the chamber would be able to simulate a neutral current candidate. It looked almost hopeless to come up in a short time with a reliable prediction, until it was realised[42] that only the nucleon component of the shower is relevant, since the mesons are unable to generate fast neutrons. Furthermore, it was recognised that the nucleon cascade is linear. So, the task was reduced to determine the elasticity distribution of fast nucleons in matter. This could be achieved from published nucleon–nucleon interactions. In conclusion, the prediction of the neutron background was free of unknown parameters.

A neutron interaction in the chamber can occur in two topologies, as illustrated in Fig. 7, depending on whether the neutron's origin is visible or not. The two event topologies are called *associated event* and denoted as AS, resp. *nonassociated, i.e. background event* and denoted as B. The interaction length of a neutron in the chamber liquid is about 80 cm, therefore a sizeable sample of AS events could be collected thanks to the large longitudinal extension of Gargamelle, namely 15 events in the neutrino and 12 in the antineutrino film. The observed numbers of AS and B events imply a constraint about the properties of the nucleon cascade, since the B-events represent the *end* of a nucleon cascade, while the AS represent the *beginning* of a cascade. At the beginning of July 1973 the background program was ready. First, the hypothesis *all neutral current candidates are background* was examined. This is the worst possible assumption. Then the ratio B/AS is for the neutrino film 102/15. The background program predicted for the ratio 1 ± 0.3 in manifest disagreement with the measured ratio. The data of the antineutrino film yielded the same conclusion. The hypothesis had to be rejected and the observed neutral current candidates are definitely *not all neutron background*. On the contrary, the neutron background accounts only for a small part. The next step was to evaluate the background using the angular and energy distributions appropriate for neutrons emitted in neutrino interactions. The result was $B/AS = 0.7 \pm 0.3$. The absolute number of neutron background events among the 102 neutral current candidates could then be predicted using the calculated ratio

AS - Event

B - Event

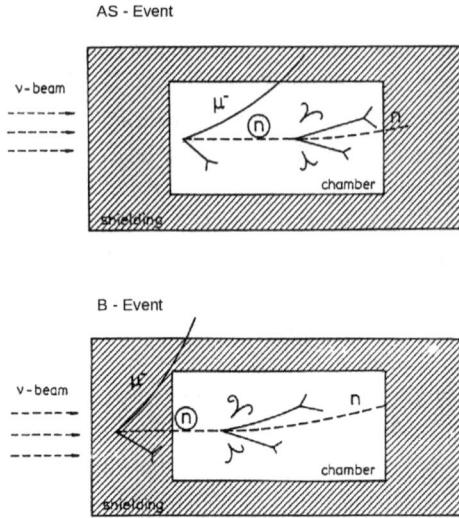

Fig. 7.   Sketch of the tow topologies of a neutron interaction in the chamber.

B/AS and the observed number of AS events and yielded 10 for the neutrino data and similarly for the antineutrino data, thus a genuine new effect could be claimed.

This conclusion was intensively discussed within the collaboration. All ingredients of the background calculation were critically scrutinised. The modular structure of the program allowed for an immediate answer to the consequences of the proposed *ad hoc* modifications, particularly regarding the treatment of the cascade. At the end of July 1973 the collaboration was convinced that the observed *events without final state charged lepton* constitute a genuine new effect and sent the paper for publication to *Physics Letters*.[43] The single electron event[41] had already been sent off a few weeks earlier.

3.5.   *The hot autumn*

A month later the discovery has been reported to the Electron–Photon Conference at Bonn. As a last-minute contribution also the Harvard–Pennsylvania–Wisconsin (HPW) collaboration contributed their observation. In a parallel session the new results were intensely debated. In his final talk C. N. Yang announced the discovery of weak neutral currents as the highlight of the conference.

Nevertheless some prominent physicists questioned the validity of the background calculation arguing that its underestimation, particular with regard to an optimistic treatment of the nucleon cascade, reduces the claim to nothing. Although Gargamelle's replies were safe and sound, the disbelief was strong and further increased, when the rumour got spread around that the HPW collaboration did not reproduce the effect in their modified setup. Given the implications of a failure the CERN management decided to perform an *experimentum crucis* to prove or disprove the validity of the neutron background calculation.

3.6.   *The proton experiment*

Single proton pulses of fixed momentum (4, 7, 12 and 19 GeV) were extracted from the Proton Synchrotron and were sent to Gargamelle. Two runs were allocated, one at the end of November and another one mid-December 1973. The incoming protons initiate cascades just as do neutrons. The properties of these cascades could now be observed and investigated. An example of a cascade induced by a 7 GeV proton is shown in Fig. 8. For the application of the neutron cascade program only the initial condition had to be set to a proton with given momentum. Thus it was ensured that the crucial aspects of the program are really tested. Several critical questions to be answered were set up beforehand and the background program had to anticipate the expectations.

The answer to the two most important questions, namely the measurement of the apparent interaction length and of the cascade length, is shown in Fig. 9. The prediction of the apparent interactions depends upon the use of the relevant cross section, which is not just the total cross-section. A neutron is identified by a visible

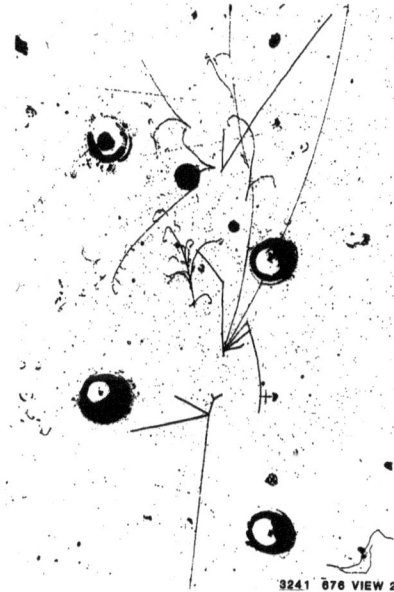

Fig. 8.   Example of a multistep cascade initiated by a 7 GeV proton entering from below in the Gargamelle chamber. After the first interaction a charge exchange occurs and the cascade is continued by a fast secondary neutron, which in turn interacts, emits a fast proton interacting again and generating a π^0 and a neutron which interacts further downstream near the end of the visible volume.

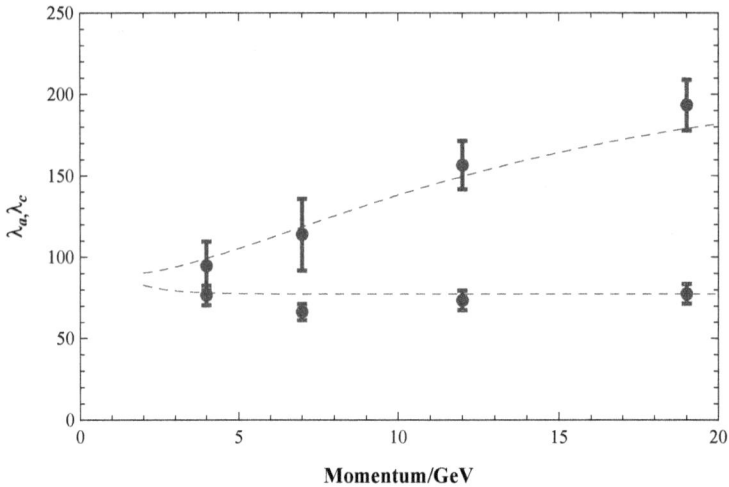

Fig. 9.   Comparison of measured (points) and predicted (dotted lines) apparent interaction length (below) resp. cascade length (above) as a function of the proton momentum.

interaction with an energy deposition of at least 150 MeV. The apparent interaction length was measured as the distance to the *first* visible interaction with an energy deposition of at least 150 MeV, whereas the cascade length as the distance to the *last* interaction with an energy deposition of at least 1 GeV, otherwise it does not qualify for a neutral current candidate.

The good agreement between these and other measurements (see Refs. 6 and 42) and their predictions by the neutron background program confirmed the validity of the background evaluation in the discovery paper and dissipated all criticisms as unfounded.

The analysis of the two runs was final by the end of March 1974 and was reported to the APS Meeting[44] at Washington in April 1974.

3.7.   *Confirmations*

By Spring 1974 there was ample additional evidence for the existence of weak neutral currents. First of all the Gargamelle collaboration has increased their event samples[45] corroborating the original findings, moreover it confirmed the neutron background calculation by the proton experiment and presented a further independent background determination based on the event position and the different interaction lengths of neutrino and neutron induced events in the chamber.[6, 45] Figure 10 shows a likelihood analysis of the apparent interaction lengths of charged current (CC), neutral current (NC) and associated (AS) events. The CC events are genuine neutrino-induced events and their interaction length is indeed consistent with infinity, whereas the NC events have a shorter apparent interaction length

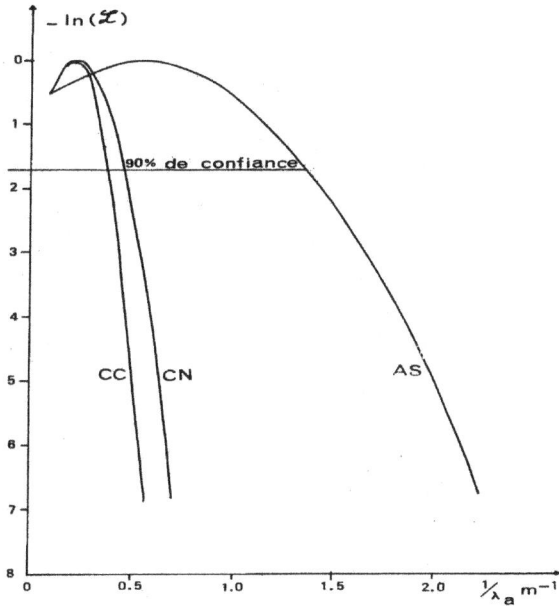

Fig. 10.   Log likelihood distributions of charged current (CC), neutral current (CN) and associated (AS) events. The horizontal line indicates the 90% confidence level.

by an amount determined by the contributing neutron component. The estimated amount agrees with the previous direct determination of the neutron background.

The CalTech–Fermilab experiment[46] running in a dichromatic neutrino beam peaking at 45 and 125 GeV has observed a clear signal of muonless events. Charged and neutral current events were distinguished by their event length in the calorimeter. This new method enjoyed many later applications.

A significant number of events ascribed to $\nu n \to \nu p \pi^-$ and $\nu p \to \nu n \pi^+$ has been observed in the 12 ft ANL bubble chamber.[47] This was the first observation of an exclusive neutral current channel.

Finally, the HPW collaboration has understood the reason, why they lost their initial neutral current signal, and came up also with a clear signal.[48]

3.8.   *Conclusion*

The Gargamelle collaboration published their discovery in 1973 and stood firm against all criticisms. One year later also the last skeptic was convinced.

The discovery of weak neutral currents initiated a long-lasting boost to high energy physics. The experimental and theoretical investigation of weak neutral currents has led to unprecedented progress on the fundamental scientific frontier as well as in technology and the energy frontier. All this is evident in the retrospect of

40 years. The outstanding achievement is that weak and electromagnetic phenomena are now commonly described by an electroweak gauge theory.

Acknowledgments

It is a pleasure to thank Luisa Cifarelli for granting to me as gift the book on Pontecorvo with his profound insight in weak interactions. Email exchanges on historical aspects with S. Bilenky, A. Bettini, M. Block, G. Fidecaro and E. Fiorini are gratefully acknowledged. I would like to thank P. Zerwas for valuable comments.

References

1. M. Veltman, *Facts and Mysteries in Elementary Particle Physics* (World Scientific Publishing, 2003) Chapter 7.
2. D. H. Perkins, An early neutrino experiment, *EPJH* **38**, 713 (2013).
3. U. Nguyen-Khac and A. M. Lutz, (eds.) *Neutral Currents Twenty Years Later*, Proceedings of the International Conference at Paris 1993, Paris, France July 6–9, 1993 (World Scientific, 1994).
4. A. K. Mann and D. B. Cline, (eds.) *Discovery of Weak Neutral Currents: The Weak Interaction Before and After*, AIP Conference Proceedings 300, Santa Monica CA February, 1993.
5. R. Cashmore, L. Maiani and J. P. Revol, Prestigious Discoveries at CERN, *EPJC* **34**(1) (2004).
6. D. Haidt and A. Pullia, The Weak Neutral Current — Discovery and Impact, *Rivista del Nuovo Cimento*, **36**(8) (2013).
7. D. H. Perkins, Gargamelle and the Discovery of Neutral Currents, in *Proc. of the Third International Symposium on the History of Particle Physics*, SLAC 24–27 June, 1992.
8. D. B. Cline, *Weak Neutral Currents: The Discovery of the Weak Force* (Westview Press, 1997).
9. D. Haidt and H. Pietschmann, *Electroweak Interactions*, Landolt-Börnstein New Series I/10 (Springer, 1988).
10. H. Pietschmann, *Geschichten zur Teilchenphysik* (Ibera Verlag, 2007), pp. 30, 41, 42.
11. E. Fermi, *Z. Phys.* **88** 161 (1934).
12. H. Bethe and R. Peierls, *Nature* **133** 689 (1934).
13. S. M. Bilenky, T. D. Blokhintseva, L. Cifarelli, V. A. Matveev, I. G. Pokrovskaya and M. G. Sapozhnikov (eds.), *Bruno Pontecorvo Selected Scientific Works and Recollections on Bruno Pontecorvo*, 2nd edn. (Società Italiana di Fisica, 2013).
14. B. M. Pontecorvo, see Ref. 13, p. 402.
15. J. Steinberger, *A personal debt to Bruno Pontecorvo*, in Ref. 13 p. 455.
16. G. Feinberg, *Phys. Rev.* **110** 1482 (1958).
17. B. M. Pontecorvo, Electron and Muon Neutrinos *Zh. Exp. Teor. Fiz* **37** 1751 (1959); *JETP* **10**, 1236 (1960); Ref. 13, p. 167.
18. B. M. Pontecorvo, Experiments with neutrinos emitted by mesons, *Zh. Exp. Teor. Fiz* **39** 1166 (1060); see also Ref. 13, p. 181.

19. B. M. Pontecorvo, Una nota autobiographica in Ref. 13, p. 424.

20. G. Fidecaro, Bruno Pontecorvo: From Rome to Dubna, see Ref. 13, p. 480.

21. M. A. Markov, in *Proc.Int.Conference on Neutrino Physics and Neutrino Astrophysics* (neutrino'77), Baksan Valley 18–24 June, 1977; *On high energy neutrino physics*, 10th Annual Int. Conf. on High Energy Physics at Rochester, August 25–September 1, 1960.

22. M. Schwartz, Feasibility of using High-energy neutrinos to study the weak interactions, *Phys. Rev. Lett.* **4**, 306 (1960).

23. T. D. Lee and C. N. Yang, Theoretical discussions on possible high-energy neutrino experiments, *Phys. Rev. Lett.* **4**, 307 (1960).

24. B. M. Pontecorvo, see Ref. 13, p. 405.

25. The 40th anniversary of EPS — Gilberto Bernardini's contributions to the physics of the XX century, prometeo.sif.it:8080/papers/online/sag/024/05-06/pdf/08.pdf

26. G. Bernardini, The program of Neutrino Eperiments at CERN, 10th Annual Int. Conf. on High Energy Physics at Rochester, August 25–September 1, 1960, p. 581.

27. F. Krienen, R. Salmeron and J. Steinberger, Proposal for an experiment to detect neutrino induced reactions, PS/Int. EA 60-10 (September 12, 1960).

28. J. Steinberger, Pontecorvo and neutrino physics, CERN Courier Letters February 24, 2014, cerncourier.com/cws/article/cern/56229.

29. G. Bernardini, Report to the Scientific Policy Committee, CERN/SPC/121, cds.cern.ch/record/39801/files/CM-P00094942-e.pdf

30. G. Bernardini, The Neutrino Experiment, CERN/SPC/138 (July 21, 1961).

31. G. Danby *et al.*, *Phys. Rev. Lett.* **9**, 36 (1962).

32. C. Franzinetti (eds.), The 1963 NPA Seminars: The Neutrino Experiment, CERN 63-37, Februry 1963.

33. J. S. Bell, J. Løvseth and M. Veltman, Conclusions at the Siena 1963 conference: The CERN experiment, dspace.library.uu.nl/bitstream/handle/1874/4793/13782.pdf.

34. R. Feynman and M. Gell-Mann, *Phys. Rev.* **109**, 193 (1958).

35. B. M. Pontecorvo, see Ref. 13, p. 196.

36. S. B. Treiman, Weak Global Symmetry, *Il Nuovo Cimento* **15**, 916 (1960).

37. T. D. Lee, Intermediate Boson Hypothesis of Weak Interactions, 10th Annual Int.Conf. on High Energy Physics at Rochester, August 25–September 1, 1960, p. 567.

38. E. Young, PhD thesis, CERN Yellow Report 67–12 (1967).

39. D. C. Cundy *et al.*, *Phys. Lett.* **77B**, 478 (1070).

40. Gargamelle Collaboration, ν-proposal, CERN TCC/70–12 (1970).

41. F. J. Hasert *et al.*, *Phys. Lett.* **46B**, 121 (1973).

42. W. F. Fry and D. Haidt, CERN Yellow Report 75–1 19751; see also Ref 3.

43. F. J. Hasert *et al.*, *Phys. Lett.* **46B**, 138 (1973);

44. D. Haidt, Contribution to the *APS* Meeting at Washington, April 1974.

45. F. J. Hasert *et al.*, *Nucl. Phys. B* **73**, 1 (1974).

46. B. Barish, Contribution to the London Conference, June 1974; *Phys. Rev. Lett.* **33**, 538 (1975).

47. S. J. Barish *et al.*, Contribution to the APS Meeting at Washington, April 1974; *Phys. Rev. Lett* **33**, 1454 (1974).

48. A. Benvenuti *et al.*, *Phys. Rev. Lett.* **32**, 800 (1974).

Highlights from High Energy Neutrino Experiments at CERN

W.-D. Schlatter

CERN, CH-1211 Geneva, Switzerland
dieter.schlatter@cern.ch

Experiments with high energy neutrino beams at CERN provided early quantitative tests of the Standard Model. This article describes results from studies of the nucleon quark structure and of the weak current, together with the precise measurement of the weak mixing angle. These results have established a new quality for tests of the electroweak model. In addition, the measurements of the nucleon structure functions in deep inelastic neutrino scattering allowed first quantitative tests of QCD.

1. Introduction

High energy neutrino beams were used successfully in the 1970s and 1980s to study the weak interaction as well as probing the nucleon with deep inelastic scattering experiments without the interference of strong interaction. At CERN's PS accelerator this was highlighted with the discovery of neutral currents with the Gargamelle heavy liquid bubble chamber experiment[1] in 1973. In the late 1970s, with even higher energy neutrino beams up to $200\,\text{GeV}$, new opportunities opened up for experiments using deep inelastic neutrino–nucleon scattering to test the foundations of the Standard Model which had been formulated in the decade before. This article recalls the highlights from the CERN neutrino experiments at that time.[a] Historic reviews of early neutrino experiments by D. H. Perkins and J. Steinberger can be found in Refs. 3 and 4 respectively.

Deep inelastic neutrino–nucleon scattering is usually described by four kinematic variables, Q^2, ν, x and y; for convenience their definitions are repeated in Fig. 1.

The neutrino and antineutrino cross-sections are described by three nucleon structure functions, $2xF_1(x,Q^2)$, $F_2(x,Q^2)$, and $xF_3(x,Q^2)$. In the Parton Model spin $1/2$ partons imply $2xF_1(x) = F_2(x)$ and with $q(x)$ and $\bar{q}(x)$ the sum of all quark and antiquark structure functions, the cross-sections depend on only two structure functions: $F_2(x) = q(x) + \bar{q}(x)$ and $xF_3(x) = q(x) - \bar{q}(x)$.

---

[a]The story of the discovery of neutral currents by Gargamelle is recalled in a separate article in this book.[2]

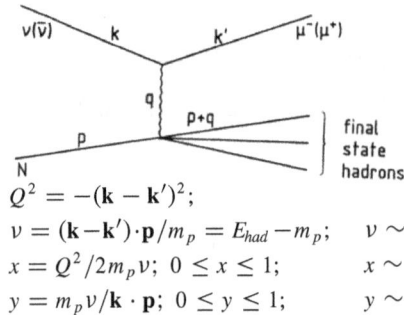

$$Q^2 = -(\mathbf{k} - \mathbf{k}')^2;$$
$$\nu = (\mathbf{k} - \mathbf{k}') \cdot \mathbf{p}/m_p = E_{had} - m_p; \quad \nu \sim E_{had};$$
$$x = Q^2/2m_p\nu; \ 0 \le x \le 1; \quad x \sim Q^2/(2m_p E_{had});$$
$$y = m_p\nu/\mathbf{k} \cdot \mathbf{p}; \ 0 \le y \le 1; \quad y \sim E_{had}/E_\nu$$

Fig. 1.   Definition of kinematic variables in deep inelastic neutrino–nucleon scattering.

2.   Early Gargamelle Results on the Quark Parton Model

In the late 1960s the Parton Model[5] was formulated by R. Feynman, prompted by the new SLAC electron–nucleon scattering experiments.[b] The observed scaling behaviour was best explained by point-like constituents of the nucleon, called partons. One of the important elements of the Parton Model is the idea that "scaling" of deep inelastic lepton–nucleon scattering is understood as the sum of elastic scatterings of the lepton on free partons within the nucleon. As a consequence the structure functions of the nucleon scale, i.e., $F_i(x, Q^2) \rightarrow F_i(x)$ in the limit of very large Q^2 and ν with x fixed.

In the early 1970s deep inelastic neutrino–nucleon scattering experiments made with Gargamelle, the heavy liquid bubble chamber at the CERN PS, could clarify some of the open questions. Two important observations were made. Firstly, a linearly rising cross-section with energy in deep inelastic neutrino and antineutrino interactions[7] confirming the evidence for point-like constituents of the nucleon (see Fig. 2).[c]

Secondly, the structure function F_2 from the Gargamelle neutrino data[8] agrees with F_2 from the ep scattering experiment at SLAC,[9] when divided by a charge factor 5/18, $F_2^{\nu N} = F_2^{eN}[\frac{1}{2}((2/3)^2 + (1/3)^2)]^{-1}$, the mean square charge of the u and d quarks in the nucleon, as predicted by the Quark Model for fractionally charged quarks. The point-like partons are really quarks. This is illustrated in Fig. 3 showing F_2 from Gargamelle neutrino data as a function of the scaling variable x compared to parametrisations of the SLAC/MIT electron–proton data.

Furthermore, two important sum rules of the Quark Parton Model were also evaluated by the Gargamelle experiment. The momentum sum rule,

---

[b]A personal recollection by J. I. Friedman of those experiments can be found in Ref. 6.

[c]Actually, a linearly rising cross-section in neutrino scattering had been observed before in a heavy liquid bubble chamber experiment at CERN[10] "but the significance had not been appreciated at the time".[11]

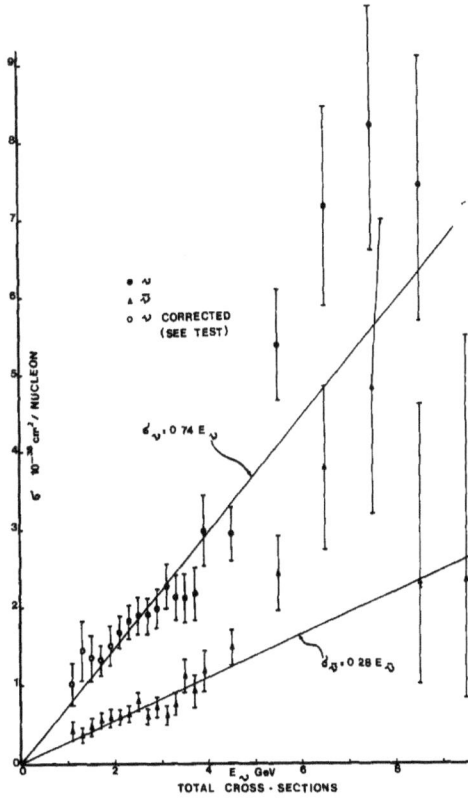

Fig. 2. Total cross-section for neutrino and antineutrino scattering as a function of energy.[7] The linear rise is a consequence of the point-like interaction of the constituents.

$\frac{1}{2} \int (F_2^{\nu p}(x) + F_2^{\nu n}(x))dx = 0.49 \pm 0.07$, in good agreement with the earlier results from the electron scattering, implied that the momentum fraction carried by quarks in the nucleon is about $1/2$ of the nucleon momentum, indicating the existence of a new partonic constituent, the gluon. In addition, the number of valence quarks in the nucleon $= \frac{1}{2} \int (F_3^{\nu p}(x) + F_3^{\nu n}(x))dx$ was measured to be 3.2 ± 0.6, consistent with the Quark Model expectation of three.

3. Neutrino Beams and Experiments

Progress in deep inelastic neutrino scattering experiments came with higher energy neutrino beams and larger, more powerful detectors. At CERN, high energy neutrino beams became available with the construction of the SPS, which was completed by 1976 and the first neutrino beam was commissioned in December of that year.

There were two types of neutrino beams at the SPS, a narrow band beam (NBB), using momentum selected charged hadrons (pions and kaons) and a

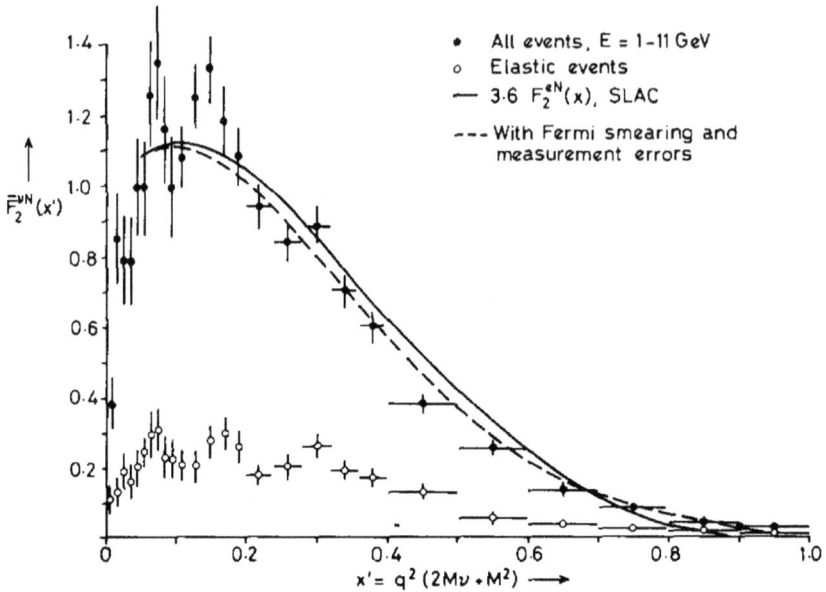

Fig. 3.   The structure function $F_2(x)$ for neutrino scattering from Gargamelle.[7] The curves show empirical fits of quark momentum distributions from electron scattering,[9] multiplied by 18/5.

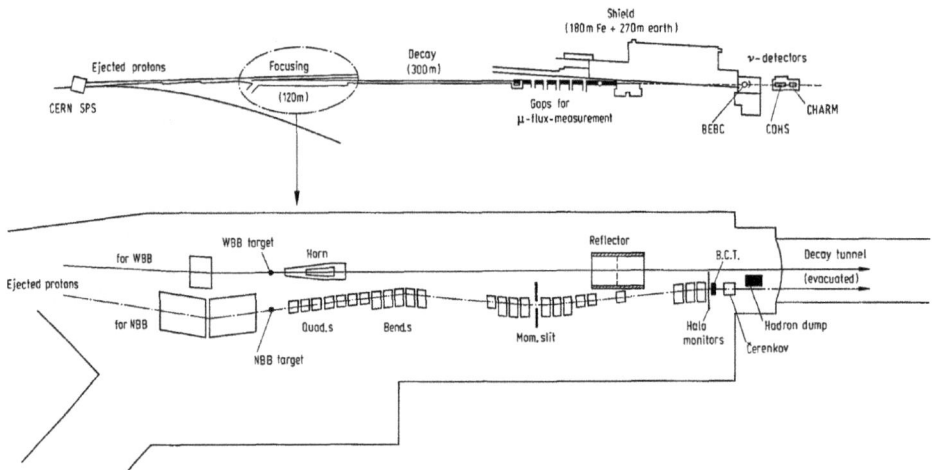

Fig. 4.   Layout of the SPS neutrino beams. The lower half shows the focussing part enlarged.[12]

more intense wide-band beam (WBB), using van der Meer focussing horns. A layout of the neutrino area is shown in Fig. 4 and the spectra of these beams for neutrinos and antineutrinos are shown in Fig. 5. Positive hadrons produce neutrinos, negative ones antineutrinos. Narrow band beams permit the determination of the energy of the events, also of neutral currents, using the radial

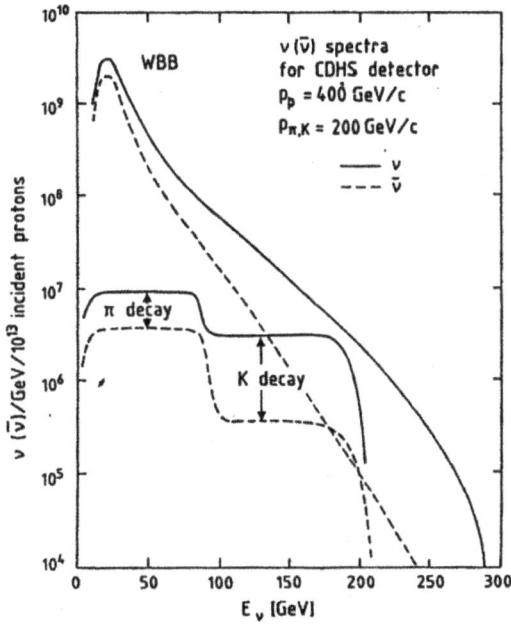

Fig. 5. Fig. 5. Neutrino and antineutrino energy spectra of the wide band beam (falling spectra) and the narrow band beam (flat spectra).

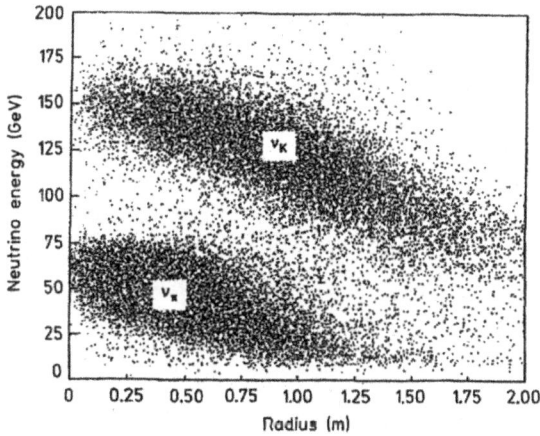

Fig. 6. Narrow band beam energy vs. radius of events in the detector, top group from kaon decay, bottom from pion decay.

position of the event in the detector. This is shown for charged current events in Fig. 6.

After the 300 m long decay tunnel and the 400 m long iron shield, followed four detectors: the Big European Bubble Chamber, BEBC, which could be filled with hydrogen, deuterium or neon, and the two new electronic detectors, CDHS, and CHARM. The fourth detector was Gargamelle which was moved from the PS neutrino beam to the SPS beams in 1977. However, only one year later the experiment had to be terminated due to a crack in the chamber body.

The CDHS detector[13] combined the function of target, hadron calorimeter, and muon spectrometer integrally in 19 similar modules, forming a scintillator calorimeter with toroidally magnetised iron plates as absorber. Between the modules were drift chambers for track reconstruction. The total weight was 1200 t; the detector started data taking in Spring 1977. The layout is shown in Fig. 7. The second electronic detector, CHARM,[14] consisted of a fine grained calorimeter surrounded by a magnetised iron frame followed by a muon spectrometer. The calorimeter was composed of scintillators, drift and streamer tubes in between marble absorber plates. The total weight was 100 t and the detector layout is shown in Fig. 8.

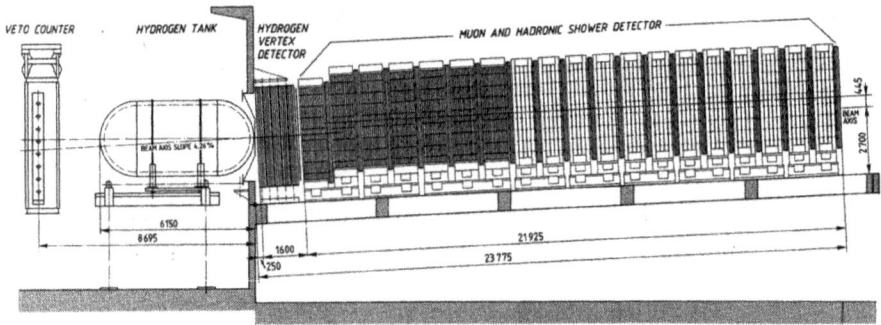

Fig. 7.   Layout of the CDHS detector.[13]

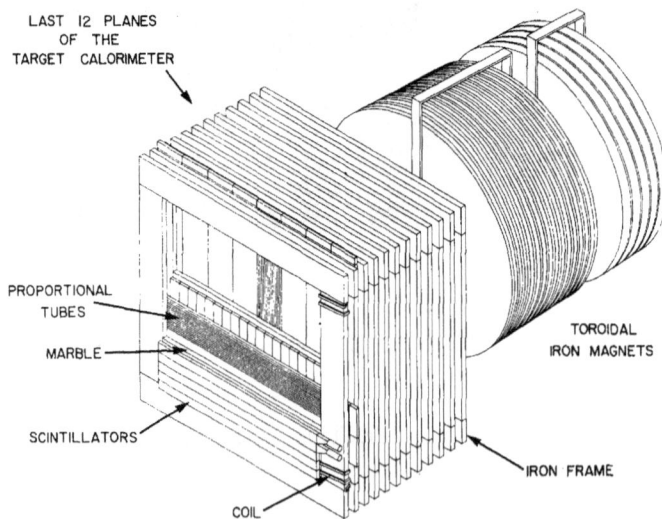

Fig. 8.   Layout of the CHARM detector.[14]

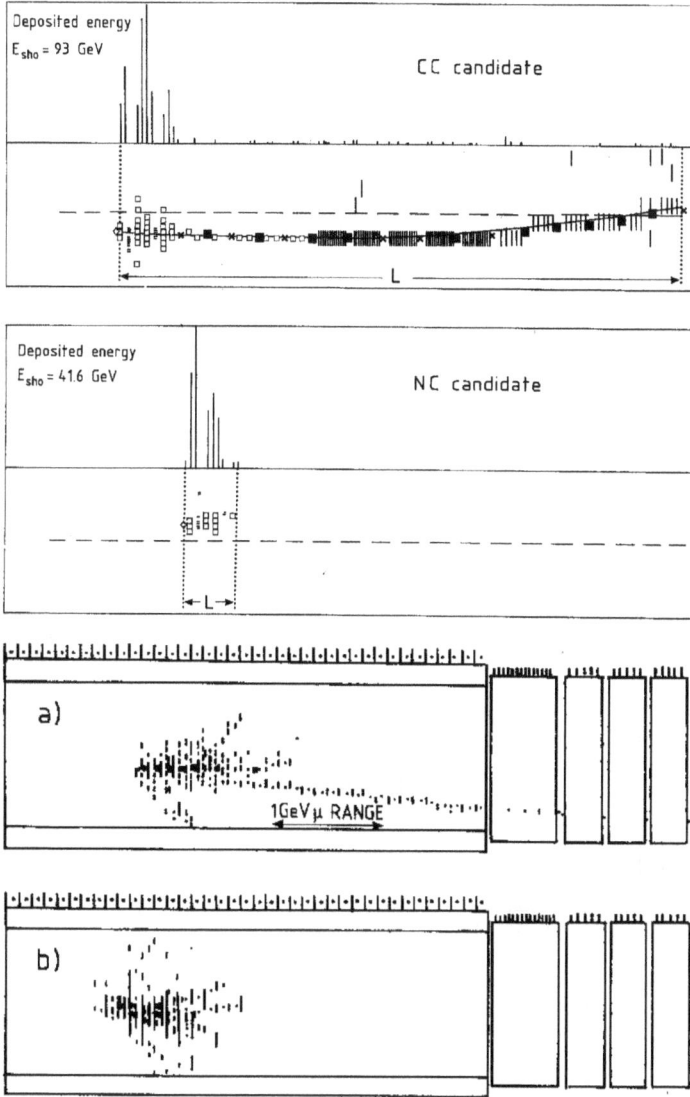

Fig. 9. Displays of CC and NC events in the CDHS (top) and CHARM (bottom) detectors.

Typical charged-current (CC) and neutral-current (NC) events in these detectors are shown in Fig. 9.

4. Nuclear Structure and Quark Parton Model

In the late 1970s, high energy neutrino data were collected at the CERN SPS with narrow band beams for which neutrino fluxes could be measured much

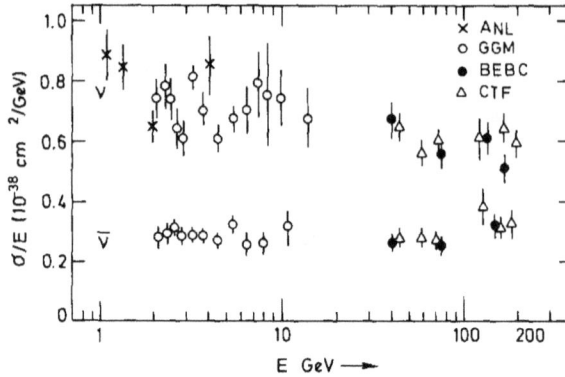

Fig. 10.   Results on total cross-sections, (σ/E), for neutrinos and antineutrinos from BEBC[17] and Gargamelle.[15, 16] Data from the Caltech-Fermilab and ANL 12-foot BC experiments are also shown.

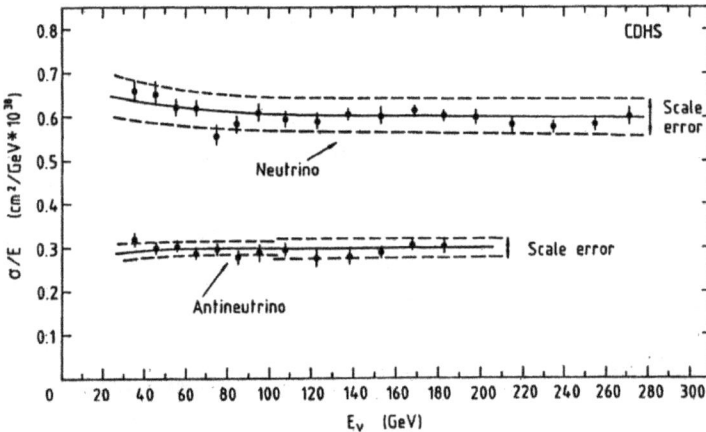

Fig. 11.   Total cross-sections, divided by the neutrino energy, for neutrinos and antineutrinos,[18] illustrating scaling behaviour in the Parton Model.

more reliably. Cross-section measurements at the SPS have been presented from BEBC[17] and CDHS,[18] the most precise data came from the 1200 t calorimeter of CDHS. Results are shown in Fig. 10 for BEBC and Fig. 11 for CDHS. The high energy behaviour of the ratio σ/E illustrates the Parton Model prediction of scaling. The expected scaling violation from QCD is too small in this energy range ($<5\%$) to be seen. The y distribution, y being approximately equal to the relative hadron energy, is another convenient way to compare to the predictions of the Parton Model. Figure 12 illustrates a remarkable agreement with the Quark Parton Model assumption of point-like structure of the nucleon.

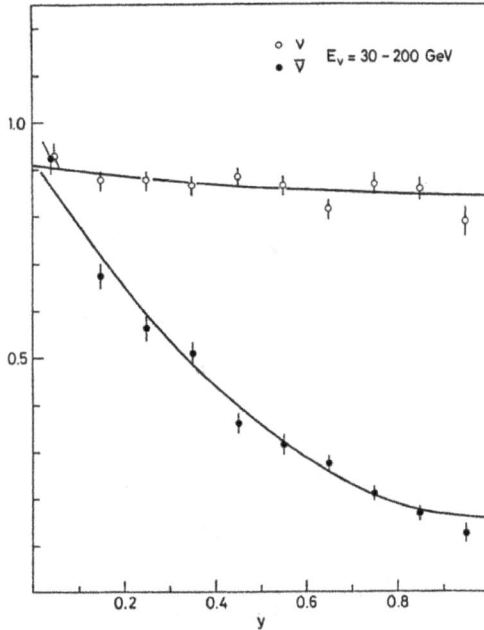

Fig. 12. CDHS results for the y distributions for neutrino and antineutrino deep inelastic scattering.[21] The lines are the predictions of the Quark Parton Model.

The structure function $F_2(x)$ measured with high statistics data has been compared for neutrinos from CDHS[18] with results from eN scattering from SLAC-MIT[20] with a charge factor of 18/5 and with μN scattering from EMC[19] with a charge factor of 9/5. Figure 13 shows that the notion of quarks being the point-like partons could be further strengthened.

5. Electroweak Measurements

5.1. *Weak mixing angle*

After the discovery of the neutral current interaction by the Gargamelle experiment[1] interest was concentrated on measurements of its strength and structure.

From the ratios of neutral to charged current inclusive cross-sections for neutrinos (R_ν) and antineutrinos ($R_{\bar\nu}$) the electroweak mixing angle, called the "Weinberg angle", can be extracted within the electroweak theory. During the years 1974 to 1976 Gargamelle presented first estimates of the Weinberg angle, the result was $\sin^2 \theta_W = 0.3$–0.4. During 1977 CDHS has measured R_ν and $R_{\bar\nu}$ with high statistics data collected with the NBB at the SPS. CDHS was able to extract a first precise measurement of the Weinberg angle with the value of $\sin^2 \theta_W = 0.24 \pm 0.02$ (Fig. 14).

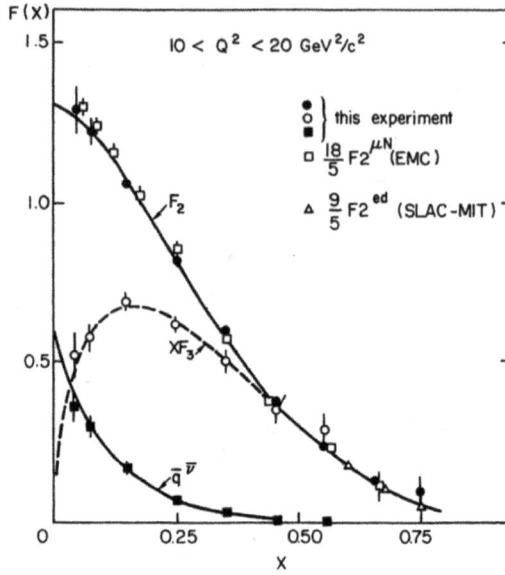

Fig. 13.   CDHS results for the structure functions $F_2(x)$, $xF_3(x)$ and $\bar{q}(x)$, the sum of all antiquark structure functions.[18] Superimposed for F_2 are the results for μN and ed scattering, multiplied with the corresponding charge factors. The lines are the predictions of the Quark Parton Model.

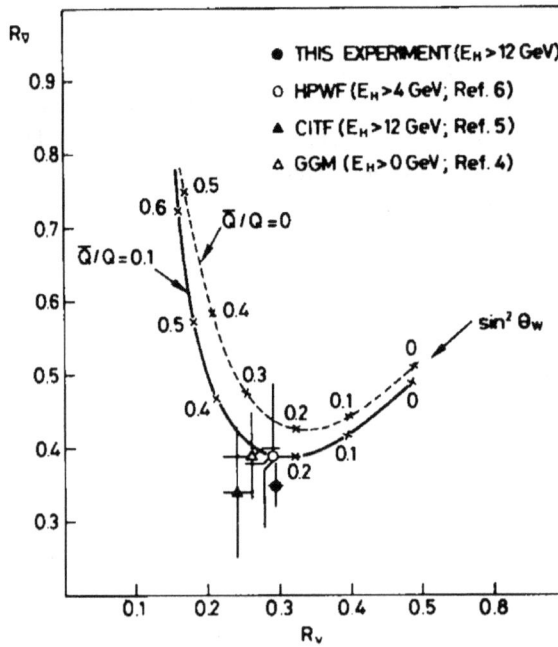

Fig. 14.   CDHS cross-section ratios[22] R_ν and $R_{\bar{\nu}}$ compared to the Weinberg–Salam model. An antiquark contribution of $\bar{q}/q = 0.1$ (solid line) was assumed and, for comparison, $\bar{q}/q = 0$ (dashed line).

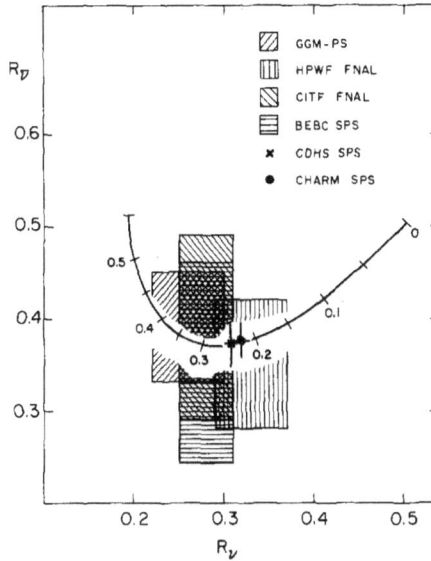

Fig. 15. CHARM cross-section ratios[23] R_ν and $R_{\bar{\nu}}$ compared to the Weinberg–Salam model.

Table 1 Various measurements of the weak mixing angle from the CDHS and CHARM experiments.

	$\sin^2\theta_W$
CDHS 1977[22]	0.24 ± 0.02
CHARM 1981[23]	0.230 ± 0.023
CDHS 1986[24]	$0.225 \pm 0.006 + 0.013(m_c - 1.5 \text{ GeV})$
CHARM 1986[25]	$0.236 \pm 0.006 + 0.012(m_c - 1.5 \text{ GeV})$

A few years later, the CDHS measurement was nicely confirmed by the CHARM experiment (Fig. 15). The results from both experiments are listed in Table 1. The comparison with the earlier Gargamelle results is visualised in Fig. 16, it illustrates the progress made by the large electronic detectors since the earlier bubble chamber results.

The precision of the analysis was improved with more statistic and the introduction of QED radiative corrections[27] in the analysis. The dominant uncertainty in the measurement of the Weinberg angle in neutrino scattering became the poor knowledge of the value of the charm quark mass, m_c. Therefore, the results were presented as function of m_c (see Table 1).

5.2. *Charm production and GIM mechanism*

Oppositely charged dimuon events provide access to open charm production, as predicted by the GIM mechanism[28] for production and semi-leptonic decay

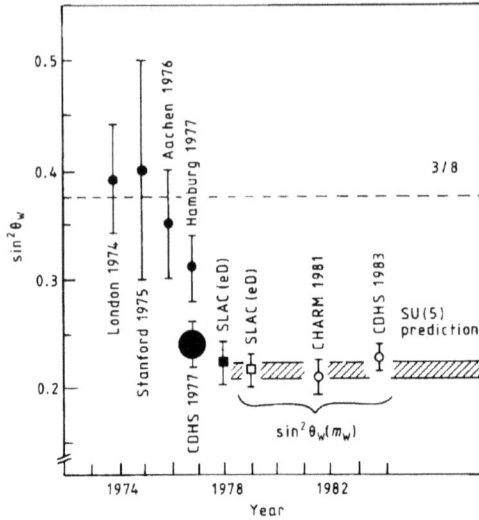

Fig. 16.   Different results of early measurements of the Weinberg mixing angle as function of time.[26] Before 1977, the Gargamelle results are shown.

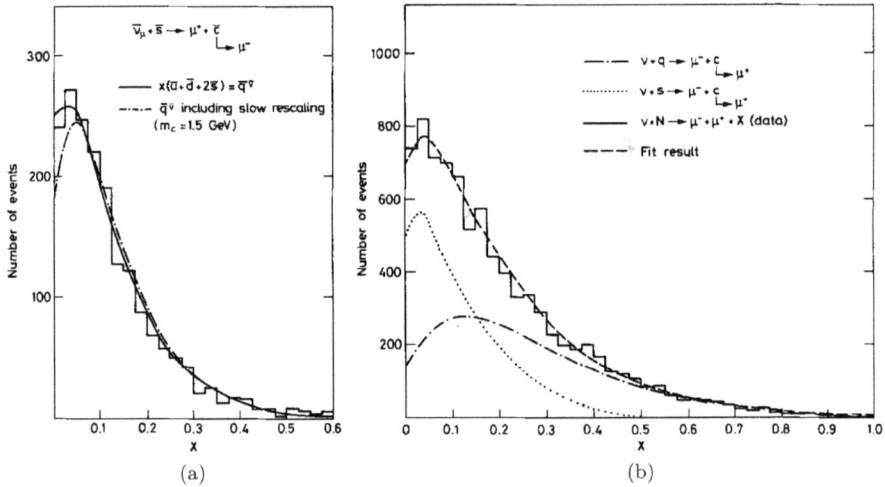

Fig. 17.   Distribution of the x variable for dimuon events.[29] (a) Antineutrino, the solid curve is the "sea" distribution, $\bar{q}(x)$, from single muon events. (b) Neutrino, the curves show the decomposition into 48% strange-sea from the data of (a) (dotted curve) and 52% quark contribution (dashed dotted curve). The dashed curve is the sum.

of charmed particles. The x-distribution of dimuons are different from ordinary CC events and agree well with the specific mixture of quark and antiquark distributions (see Fig. 17). As expected for heavy quarks[30] charm fragmentation turned out to be hard with an average relative momentum $\langle z \rangle \approx 0.7$.

6. QCD and Structure Functions

The theory of strong interactions of quarks and gluons, QCD, and the notion of "asymptotic freedom" was formulated in 1972/73. Neutrino deep inelastic scattering on the nucleon provided an excellent opportunity for quantitative tests. The analysis of the nucleon structure functions was used to test QCD in detail, to determine the scale parameter λ_{QCD} and the gluon momentum distribution in the nucleon, $g(x)$.

During 1977/78 BEBC and the CDHS experiments showed that the scaling of the naive Quark Parton Model is violated for higher Q^2. The effect from radiation of hard gluons from the quarks in QCD leads to logarithmic scaling violations and the shape of the nucleon structure function F_2 is dependent on the neutrino energy (see Fig. 18) with a rise at small x for higher energies and a drop for low energies. Similarly, the Q^2 dependence at small and large x could clearly be seen in the Gargamelle/BEBC data[17] (Fig. 19) and the early CDHS data[31] (Fig. 20).

One of the few cases in which QCD makes an absolute prediction which could be experimentally tested early on is for the structure function F_3. The moments of the x-distribution of xF_3 (defined as $M_n(Q^2) = \int dx \, x^{n-2} xF_3(x, Q^2)$) have a simpler Q^2 dependence in QCD than the distributions themselves, they are predicted to vary as $\log Q^2$ to a certain power, called anomalous dimension. In a 2-dimensional log–log representation different pairs of moments plotted for different values of Q^2 should

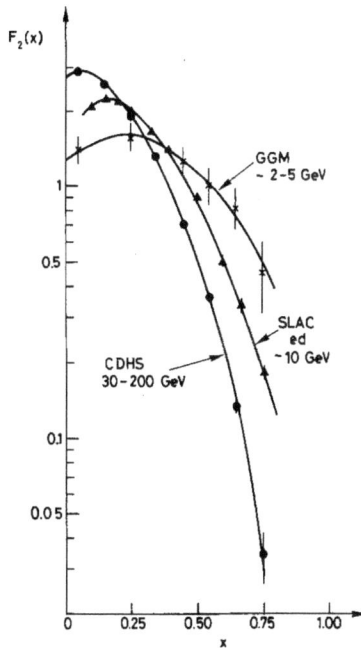

Fig. 18. Comparison of F_2 structure function seen in different lepton energy domains.[31]

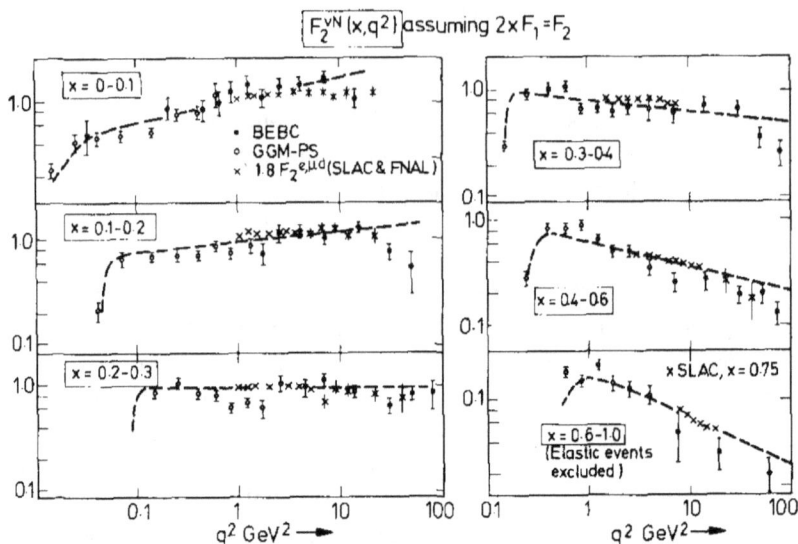

Fig. 19. Gargamelle and BEBC results F_2^ν for various x ranges versus Q^2.[17] Results from the SLAC electron and muon scattering experiments, multiplied by the quark charge factor 9/5 are shown by crosses.

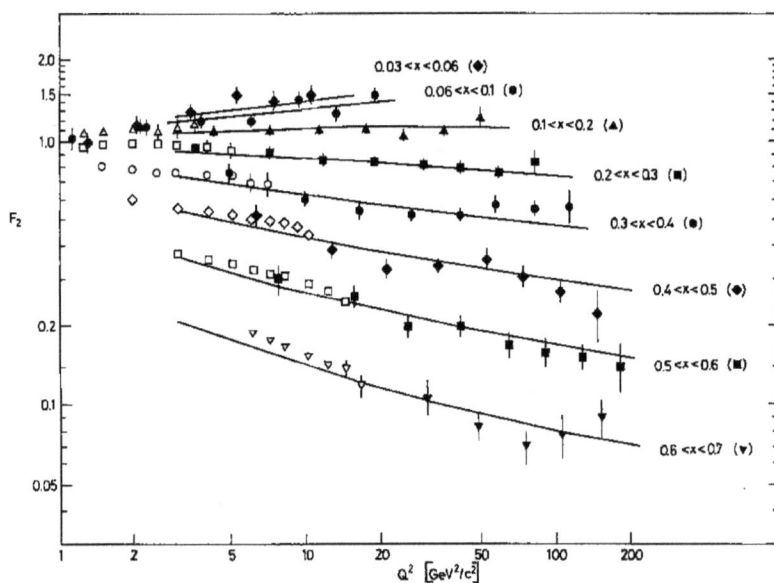

Fig. 20. First CDHS results on $F_2(x, Q^2)$[31] (solid symbols), with the fits of the DGLAP evolution equations superimposed. Open symbols are for ed scattering.

P.C. Bosetti et al. / Nucleon structure functions

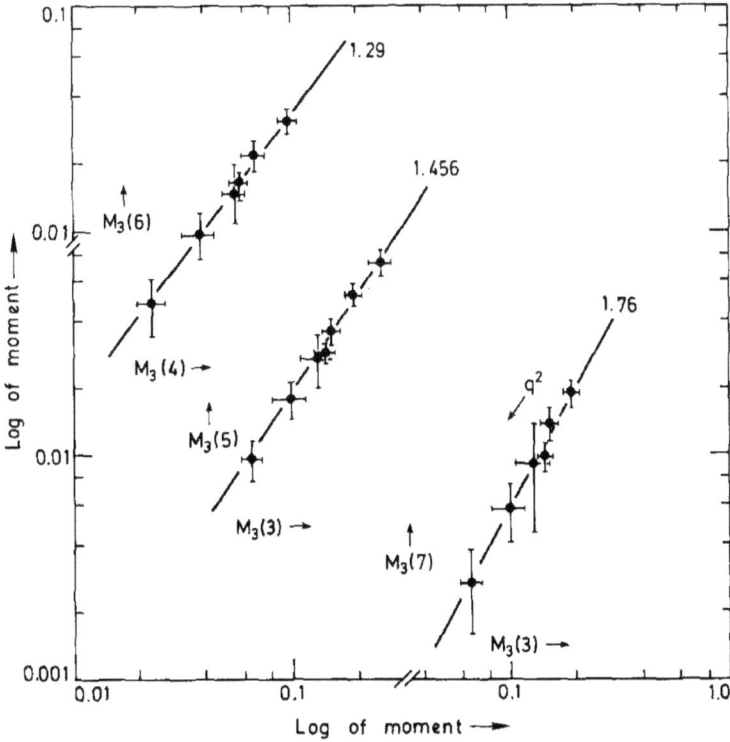

Fig. 21. Log–log plots of various moments of xF_3.[17] QCD predicts a linear relation with the slopes indicated. The logarithm of one moment is plotted against another as Q^2 varies over the range 5–50 GeV^2/c^2. Note, the errors on any pair of moments are strongly correlated.

lie on a straight line with a slope given by the ratio of the anomalous dimensions. This is shown in Fig. 21. The agreement of the observed slopes with the QCD expectation is remarkable, in spite of the relatively low Q^2.

More stringent quantitative tests of perturbative QCD became possible with more precise high energy neutrino and antineutrino data of F_2 and xF_3. In Fig. 22 $F_2(x, Q^2)$ and $xF_3(x, Q^2)$ are shown for the CDHS data. Fits of the DGLAP evolution equations[33] for a scale parameter $\lambda_{QCD} = 250$ MeV describe well the observed Q^2 evolution. The logarithm of the scale parameter is related to the running coupling constant of QCD.[d] For the typical Q^2 range of the data from $3\,\text{GeV}^2/c^2$ to $200\,\text{GeV}^2/c^2$ and $\lambda_{QCD} = 250\,\text{MeV}$ the corresponding strong coupling constant drops from 0.30 to 0.20.

Gluons do not take part directly in the deep inelastic neutrino–nucleon scattering process. QCD predicts that their interactions with quarks inside the nucleon leads

---

[d] $\alpha_s(Q^2) = \frac{12\pi}{33-2N_f} / \ln(Q^2/\lambda_{QCD}^2)$, N_f is the number of quark flavors.

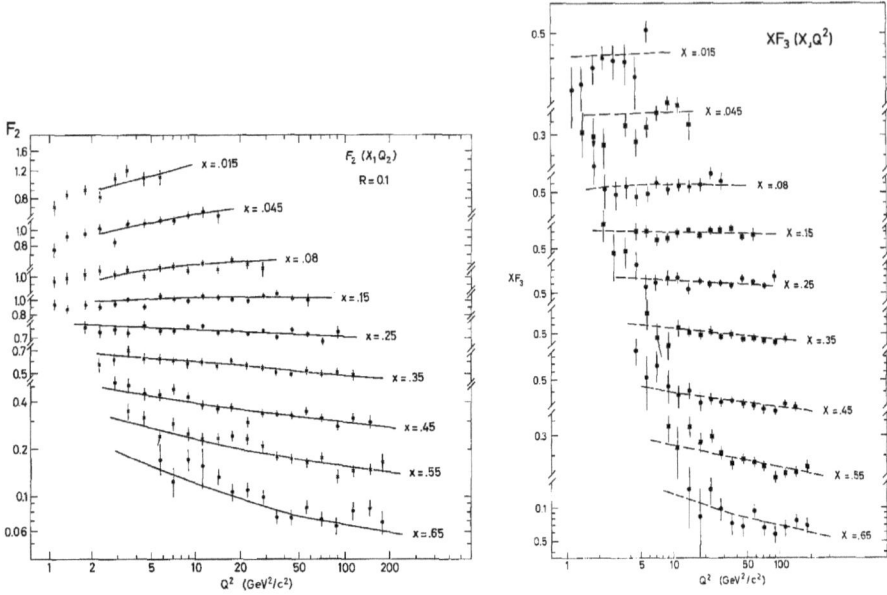

Fig. 22.   CDHS results on the structure functions $F_2(x, Q^2)$ and $xF_3(x, Q^2)$.[32] Lines are fits of the DGLAP evolution equations for $\lambda_{QCD} = 250$ MeV.

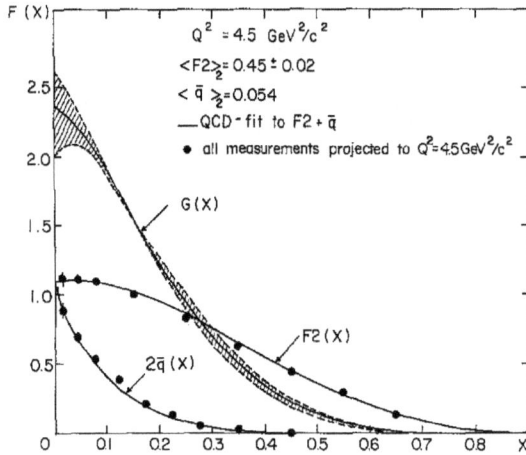

Fig. 23.   CDHS results on the gluon function, $g(x)$,[34] extracted from QCD fits to F_2 and $\bar{q}$.

to scaling violations of the structure functions. A combined analysis of $F_2(x)$ and the antiquark distribution $\bar{q}(x)$ extracted from antineutrino data at large y did allow a simultaneous extraction of the x-distribution of the gluon function and λ_{QCD} by fits of the QCD evolution equations. The results are shown in Fig. 23 for the CDHS

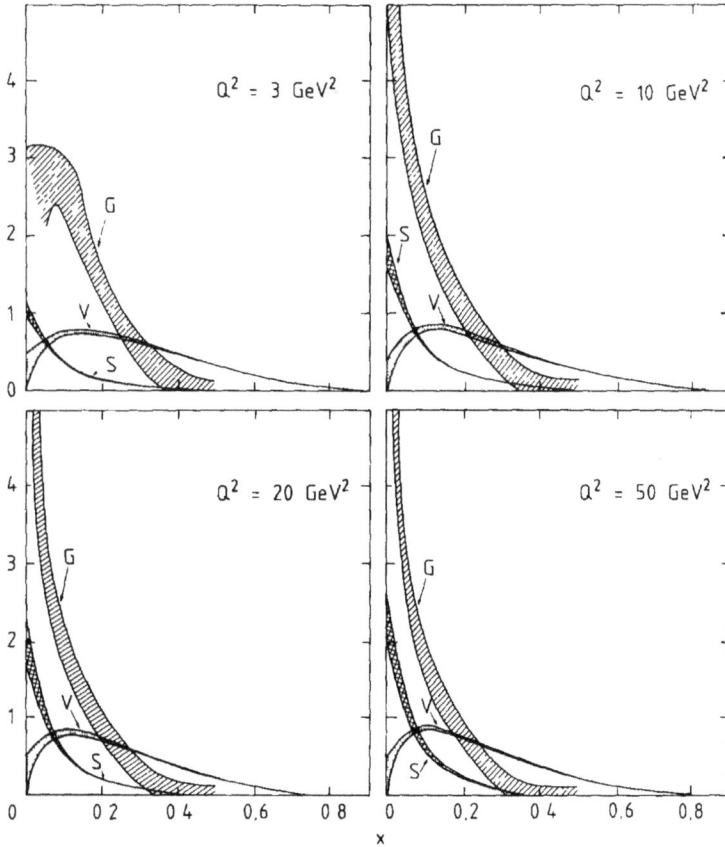

Fig. 24. CHARM results on $g(x)$,[35] by fitting F_2, xF_3 and $\bar{q}$ at different Q^2 values.

analysis at $Q^2 = 4.5$ GeV$^2/c^2$,[34] and in Fig. 24 for the CHARM analysis at several Q^2 values.[35] These were the first determinations of the x dependence of the gluon function.

7. Epilogue

The first phase of neutrino scattering experiments at the SPS lasted about a decade, culminating in quantitative tests of QCD by means of precise measurements the nucleon structure functions. The CCFR neutrino experiment at Fermilab[36] with a neutrino beam of 600 GeV energy continued these measurements. The detector used the wide band mixed ν_μ and $\bar{\nu}_\mu$ beam. Due to the rising cross-section the measurements of the structure functions F_2 and xF_3 were statistically much more powerful. Similarly, the muon nucleon scattering experiment at the SPS, BCDMS, became leading in the F_2 measurements.

After these successful tests of the Standard Model, the interest in neutrino physics moved to the search for neutrino oscillations. Two new experiments were built at CERN, the CERN Hybrid Oscillation Research apparatUS, CHORUS (1993/97) and the Neutrino Oscillation MAgnetic Detector, NOMAD (1995/98). Unfortunately, their sensitivity was not high enough to observe neutrino oscillations. Finally, the first strong experimental evidence for atmospheric neutrino oscillations was announced in 1998 by the Super-Kamiokande experiment in Japan.[37] This was the first experimental observation demonstrating that the neutrino has non-zero mass. Now, at the forefront of neutrino oscillation research is the Daya Bay Nuclear Reactor experiment in China with their latest 5σ measurement of the mixing angle between the first and third generation of neutrinos, $\sin^2(2\theta_{13})$.[38]

The legacy of the high energy neutrino experiments at CERN remains the precise confirmation of the Quark Parton Model, the precise measurement of the Weinberg angle and the first quantitative tests of QCD with the observation of "scaling violations" in the Q^2 evolution of the nucleon structure functions and the determination of the QCD interaction strength.

Since then, the ultimate tests of the Standard Model were performed at the e^+e^- collider LEP at CERN and the ep collider HERA at DESY and more recently culminating in the discovery of the Higgs boson at the LHC in 2012.

References

1. F. J. Hasert *et al.* (Gargamelle), *Phys. Lett. B* **46**, 121 (1973); **46**, 138 (1973).
2. D. Haidt, *The Discovery of Weak Neutral Currents*, in this book, pp. 165–183.
3. D. H. Perkins, *PoS* **HEP2001**, 305 (2001).
4. J. Steinberger, *Annals of Physics* **327**, 3182 (2012).
5. J. D. Bjorken and E. A. Paschos, *Phys. Rev.* **185**, 1975 (1969); R. P. Feynman, *Photon Hadron Interactions* (Benjamin, NewYork, 1972).
6. J. I. Friedman, *Eur. Phys. J. H* **36**, 469 (2011).
7. T. Eichten *et al.* (Gargamelle), *Phys. Lett. B* **46**, 274 (1973).
8. H. Deden *et al.* (Gargamelle), *Nucl. Phys. B* **85**, 269 (1975).
9. G. Miller *et al.* (SLAC-MIT), *Phys. Rev. D* **5**, 528 (1972).
10. I. Budagov *et al. Phys. Lett. B* **30**, 364 (1969).
11. J. H. Mulvey, *Nucl. Phys. B* (*Proc. Suppl*). **36**, 427 (1994).
12. D. Haidt, and H. Pietschmann, *Electroweak Interactions. Experimental Facts and Theoretical Foundation*, Landolt-Börnstein Elementary Particles, Nuclei and Atoms, Vol. 10, 146 (Springer, Berlin, 1988).
13. M. Holder *et al.* (CDHS), *Nucl. Instrum. Meth.* **148**, 235 (1978).
14. A. N. Diddens *et al.* (CHARM), *Nucl. Instrum. Meth.* **178**, 27 (1980); C. Bosio *et al.*, *Nucl. Instrum. Meth.* **157**, 35 (1978).
15. T. Eichten *et al. Phys. Lett. B* **46**, 281 (1973).
16. P. C. Bosetti *et al. Phys. Lett. B* **70**, 273 (1977).
17. P. C. Bosetti *et al.* (BEBC), *Nucl. Phys. B* **142**, 1 (1978).

18. H. Abramowicz *et al. Z. Phys. C* **17**, 283 (1983).
19. J. J. Aubert *et al.* (EMC), *Phys. Lett. B* **105**, 322 (1981).
20. A. Bodek *et al. Phys. Rev. D* **20**, 1471 (1979).
21. J. G. H. de Groot *et al.* (CDHS), *Z. Phys. C* **1**, 143 (1979).
22. M. Holder *et al.* (CDHS), *Phys. Lett. B* **71**, 222 (1977).
23. M. Jonker *et al.* (CHARM), *Phys. Lett. B* **99**, 265 (1981).
24. H. Abramowicz *et al.* (CDHS), *Phys. Rev. Lett.* **57**, 298 (1986).
25. J. V. Allaby (CHARM), *Phys. Lett. B* **177**, 446 (1986).
26. F. Eisele, *Rep. Prog. Phys.* **49**, 233 (1986).
27. J. F. Wheater and C. H. Llewellyn Smith, *Nucl. Phys. B* **208**, 27 (1982); D. Yu. Bardin and O. M. Fedorenko, *Sov. J. Nucl. Phys.* **30**, 418 (1979); D. Yu. Bardin, P. Ch. Christova and O. M. Fedorenko, *Nucl. Phys. B* **197**, 1 (1982); A. Sirlin and W. J. Marchiano, *Nucl. Phys. B* **189**, 442 (1981).
28. S. L. Glashow, J. Iliopoulos, L. Maiani, *Phys. Rev. D* **2**, 1285 (1970).
29. H. Abramowicz *et al.* (CDHS), *Z. Phys. C* **15**, 19 (1982).
30. J. D. Bjorken, *Phys. Rev. D* **17**, 171 (1978); M. Suzuki, *Phys. Lett. B* **71**, 189 (1977).
31. J. G. H. de Groot *et al.* (CDHS), *Z. Phys. C* **1**, 143 (1979).
32. H. Abramowicz *et al.* (CDHS), *Z. Phys. C* **17**, 283 (1983).
33. Y. L. Dokshitzer, *Sov. Phys. JETP* **46**, 641 (1977); V. N. Gribov, L. N. Lipatov, *Sov. J. Nucl. Phys.* **15**, 675 (1972); V. N. Gribov, L .N. Lipatov, *Sov. J. Nucl. Phys.* **15**, 438 (1972); G. Altarelli, G. Parisi, *Nucl. Phys. B* **126**, 298 (1979).
34. H. Abramowicz *et al.* (CDHS), *Z. Phys. C* **12**, 289 (1982).
35. F. Bergsma *et al.* (CHARM), *Phys. Lett. B* **123**, 269 (1983).
36. E. Oltman *et al.* (CCFR), *Z. Phys. C* **53**, 51 (1992).
37. Y. Fukuda *et al.* (Super-Kamiokande), *Phys. Rev. Lett.* **81**, 1562 (1998).
38. F. P. An *et al.* (Daya Bay), *Phys. Rev. Lett.* **112**, 061801 (2014).

The Discovery of Direct CP Violation

L. Iconomidou-Fayard[1] and D. Fournier[2]

LAL, Univ Paris-Sud, CNRS/IN2P3, Orsay 91400, France
[1] *lyfayard@in2p3.fr*
[2] *daniel.fournier@cern.ch*

Soon after the discovery in 1964 of the non-conservation of CP symmetry in the neutral kaon system, the hunt was launched for a component arising from direct violation on top of the dominant effect due to mixing. It took almost 20 years until the first evidence of a signal was reported by NA31 and another 10 years to establish the effect with a significance of more than 5 standard deviations. This article describes the beams, detectors and analysis methods used by the two CERN experiments, NA31 and NA48, which made key contributions to these results and established new standards for precision measurements.

1. Introduction

1.1. *The early days of CP violation*

In a world where CP is conserved, a K^0 produced by the interaction of a proton beam on a target evolves as the coherent superposition, in equal proportions at the time of production, of a K_1^0 and a K_2^0, where $K_1^0 = (K^0 + \bar{K}^0)/\sqrt{2}$ is the positive CP eigenstate, and $K_2^0 = (K^0 - \bar{K}^0)/\sqrt{2}$ is the negative one. Being positive under CP, the K_1^0 decays rapidly to the $\pi\pi$ final state. On the contrary K_2^0 has a much longer lifetime being forced to decay mainly to 3π or semi-leptonic $\pi l \nu$ final states, disfavoured by phase space. The two lifetimes are very different, $c\tau_S = 2.68\,\text{cm}$ for the former and $c\tau_L = 15.34\,\text{m}$ for the latter, which allows a very clean separation of the two eigenstates: far enough from the target, for exemple after 20 or more τ_S, one does not expect to find any two-pion decay.

In 1964, at BNL, Christenson, Cronin, Fitch and Turlay[1] observed for the first time a significant number of two pion decays after enough $c\tau_S$ to demonstrate that CP is not conserved. The simplest explanation, confirmed in 1967 by the observation[2] of an asymmetry between $\pi^+ l^- \bar{\nu}$ and $\pi^- l^+ \nu$ rates in neutral kaon decays, is to consider that the long and short lifetimes mass eigenstates, K_L^0 and K_S^0, are not pure CP eigenstates, but contain a small admixture of the opposite CP species, described by a complex parameter ϵ:

$$K_L^0 = \frac{(K_2^0 + \epsilon K_1^0)}{\sqrt{(1+\epsilon^2)}}, \tag{1}$$

$$K_S^0 = \frac{(K_1^0 + \epsilon K_2^0)}{\sqrt{(1+\epsilon^2)}}. \tag{2}$$

This is called "CP violation in the mixing" or "indirect CP violation". The CP violating parameter ϵ was soon after measured with good precision. Today $|\epsilon| = (2.228 \pm 0.011) \times 10^{-3}$ and $\phi_\epsilon = (43.52 \pm 0.02)°$.

1.2. Basic phenomenology

L. Wolfenstein[3] postulated in 1964 that there exists a "superweak" interaction which manifests itself only in the $\Delta S = 2$ mixing between K^0 and $\bar{K}^0$, and does not appear in the $\Delta S = 1$ transitions of the neutral kaon decays.

A direct consequence of this assumption is that $\eta^{00} = \eta^{+-}$ with:

$$\eta^{00} = \frac{A(K_L \to \pi^0 \pi^0)}{A(K_S \to \pi^0 \pi^0)}, \tag{3}$$

$$\eta^{+-} = \frac{A(K_L \to \pi^+ \pi^-)}{A(K_S \to \pi^+ \pi^-)}. \tag{4}$$

Contrary to the superweak hypothesis, one can expect on general grounds that CP violation may be present in any weak decay, leading to a difference between the amplitude of a particle X decaying to a final state f, $A(X \to f)$ and the amplitude of the CP-conjugate particle $\bar{X}$ to $\bar{f}$, $A(\bar{X} \to \bar{f})$. In the case of the neutral kaon system this means that the decay amplitude of the CP-odd combination K_2^0 to a $\pi\pi$ final state might differ from zero, corresponding to a "direct CP violation".

Given the conservation of CPT, direct CP violation may not lead to visible effects. In a general manner, the direct CP-violating transition to a given final state can be non-zero provided there are at least two amplitudes leading to this same final state, each with different phases. This is the case for a kaon decaying into two-pions through two decay amplitudes, one in the $I = 0$ and the second in the $I = 2$ isospin states. As the Clebsch–Gordan coefficients projecting the $I = 0$ and $I = 2$ states onto the $\pi^+\pi^-$ and $2\pi^0$ final states are different, in the presence of direct CP violation η^{00} and η^{+-} are no longer equal.

More precisely:

$$\eta^{+-} = \epsilon + \epsilon', \qquad \eta^{00} = \epsilon - 2\epsilon' \tag{5}$$

with

$$\epsilon' = \frac{i}{\sqrt{2}} \, \mathrm{Im}\left(\frac{A_2}{A_0}\right) e^{i(\delta_2 - \delta_0)} \tag{6}$$

where $A_{0,2}$ and $\delta_{0,2}$ are the amplitudes and the strong phases of the two-pion final isospin states $I = 0$ and $I = 2$. Conversely:

$$\mathrm{Re}\left(\frac{\epsilon'}{\epsilon}\right) = \frac{\left(1 - |\frac{\eta^{00}}{\eta^{+-}}|^2\right)}{6} = \frac{(1 - \mathrm{RR})}{6} \tag{7}$$

where RR is the so-called "double ratio", defined as:

$$\text{RR} = \frac{\Gamma(K_L^0 \to \pi^0\pi^0)}{\Gamma(K_S^0 \to \pi^0\pi^0)} \bigg/ \frac{\Gamma(K_L^0 \to \pi^+\pi^-)}{\Gamma(K_S^0 \to \pi^+\pi^-)}. \tag{8}$$

Measurements of $\pi\pi$ phase shifts[4] show that the phase of ϵ' is approximately equal to that of ϵ.

As demonstrated by Kobayashi and Maskawa, CP violation is possible in the Standard Model when three generations of weakly interacting quark doublets are present.[5] The quark mass eigenstates and the quark flavour eigenstates are related by the 3×3 complex and unitary CKM matrix V_{ij}, parametrised by three rotation angles and one phase δ. The amplitude of CP violation is determined by the Jarlskog invariant,[6]

$$J_{CP} = \sin\phi_{12} \times \sin\phi_{13} \times \sin\phi_{23} \times \cos\phi_{12} \times \cos\phi_{13}^2 \times \cos\phi_{23} \times \sin\delta \tag{9}$$

representing the common area of triangles resulting from the unitarity conditions $V_{ij}^* V^{ik} = 0$ (with $j \neq k$). The index i (j,k) runs over the three up (down) quarks. The key parameter is the phase δ, origin of the CP violation.

In the late 70s to early 80s, several calculations of ϵ and ϵ' were available, which however were not very precise for the latter, mainly due to "long range" effects. The most representative calculations of ϵ'/ϵ at this time[7-9] were ranging from 0.002 to 0.02.

Later on in the early 90s, new theoretical developments brought the various teams working in the field[10, 11] to converge towards a quite small central value, typically below 5×10^{-4}. This was due to large cancellations between the amplitudes of two $\Delta S = 1$ diagrams, namely the so-called "penguins" electromagnetic and QCD diagrams,[12] and to the increasing lower bound on the top quark mass.[13] The uncertainties on the predictions were, however, quite important.

1.3. *Experimental situation on ϵ'/ϵ in the 80s–90s*

The experimental precision is, in general, statistically limited by the measurement of η^{00} which requires the detection of the four photons from the two π^0 decays. The most precise results available in the early 1980s were: $|\eta^{00}/\eta^{+-}| = 1.00 \pm 0.06$ obtained at the CERN PS[14] with about 45 reconstructed $K_L^0 \to \pi^0\pi^0$ decays, $|\eta^{00}/\eta^{+-}| = 1.03 \pm 0.07$[15] obtained at the BNL AGS with about 120 reconstructed $K_L^0 \to \pi^0\pi^0$ decays, and $|\eta^{00}/\eta^{+-}| = 1.00 \pm 0.09$ obtained at the BNL AGS[16] in the region of K_S^0–K_L^0 interference which allows one to measure both the magnitude and phase of η^{00} (see Section 3.3).

Based on the rather large theoretical predictions that were valid in early 80s, several experiments were proposed in the US, at BNL[17] and at Fermilab.[18, 19] At CERN, an initiative was undertaken in the early 80s which gave birth to the NA31 experiment, which was formally approved in 1982. NA31 was aiming at a total accuracy on ϵ'/ϵ of about one permil, using more intense beams than in the past and methods that partially cancel the systematic uncertainties of the measurement.

After this round, in the early 90s, the need for a new ϵ'/ϵ measurement with precision as high as 10^{-4} became clear from both the theoretical and the experimental status and this resulted in a new generation of experiments, KTeV at Fermilab and NA48 at CERN.

The NA31 and NA48 experiments were both installed on beams from the 450 GeV Super Proton Synchrotron (SPS), in the EHN1 and EHN2 areas respectively.

1.4. The main challenges in the measurement of ϵ'/ϵ

A precise measurement requires both high accumulated statistics for all the four channels, especially for the most suppressed $K_L^0 \to 2\pi^0$ mode, and small and well-controlled systematic uncertainties.

The statistical accuracy can be improved by means of intense proton beams hitting fixed targets and producing K_S^0 and K_L^0 beams. At CERN, this was possible due to the availability of the 450 GeV SPS. Neutral beams produced by proton interactions on a target contain practically equal amount of short-lived and long-lived kaons. Because of the very different lifetimes, the K_S^0 component will rapidly decay, unlike the K_L^0 one. This requires the production of the K_S^0 and K_L^0 beams at two different distances, close and far from the detectors respectively. To create the K_S^0 beam, both CERN experiments prefered the solution of protons hitting a second target, instead of regenerating[a] the K_S^0 component from the K_L^0 beam as done traditionally by the pioneering groups[1, 14, 15] and also in the Fermilab experiments.[19, 20] The CERN choice was made in order to avoid the drawbacks of the regeneration technique: the inelastic K_L^0 interactions and the interference occuring between the original CP violating K_L^0 component and the regenerated K_S^0 one, that requires several auxiliary measurements of the regeneration parameters.

A second challenge concerns the collection of the events of interest. The measurement of ϵ'/ϵ consists in counting events from the four modes, K_S^0 and K_L^0 decaying into $2\pi^0$ and $\pi^+\pi^-$, in order to build the double ratio RR defined in Eq. (8). While in K_S^0 the two-pion channels saturate the decays, in K_L^0 these final states count for less than 0.3% of the total decay rate. The CP-violating channels in K_L^0 decay must therefore be efficiently identified in the presence of the large amount of three-body final states which have to be suppressed by means of a powerful trigger and accurate measurements.

---

[a]When a pure K_L^0 beam traverses a thick enough block of matter, a K_S^0 component appears in the emerging beam. This happens because the cross-section of $\bar{K}^0$ (negative strangeness) is larger than for K^0. The "regenerated" K_S^0 component is dominated by coherent kaon scattering from the target.

Third, despite the severe collimation, intense beams are often accompanied by particles that can affect a good event resulting in its loss. While small, these effects may impact the measurement of the double ratio if they alter differently the four decay modes. The measurement of the accidental activity was one of the challenges for precision experiments which developed a series of subtle analyses, in particular the so-called overlay method, to investigate the effect. This method consists in superimposing by software specific so-called "random" triggers onto each event in the data, where the "random" triggers are recorded in proportion to the beam intensities to give an accurate picture of the ambient activity.

Finally, particular care is needed to reduce the systematic effects, for exemple by concurrently recording — whenever possible — the four modes of interest thus leading to cancellations, and by designing the experiments so as to minimise the need for corrections.

2. First Generation: The NA31 Beams and Detectors

The project proposed by four institutes aggregated from 1982 to 1993 up to 60 physicists from seven European institutes. The first important data taking period took place in 1986. Additional data were recorded in 1988 and 1989, after beam and detector upgrades, in order to improve both the statistical and the systematic uncertainties.

2.1. *The K_L^0 and K_S^0 beams*

The choice of the NA31 collaboration[21] was to alternate the data taking between coaxial K_S^0 and K_L^0 beams, with a typical cycle of 30 hours. As the apparatus registers concurrently charged and neutral decays, a part of systematic uncertainties cancels out.

The K_L^0 beam is produced from an extracted proton beam from the SPS hitting a far target placed ~120 m upstream the decay volume. The construction of a high intensity K_L^0 beam requires the sweeping of the charged secondaries and careful multistage collimation of the neutral component to precisely define the beam aperture and to remove scattered halo particles. The K_S^0 beam is created by transporting an attenuated proton beam to a second target, located very close to the decay region. The beam parameters are listed in Table 1 and a sketch of the experimental layout is shown in Fig 1. In the K_S^0 beam, it is essential to define precisely the beginning of the decay volume against the earlier decays inside the collimators. This is done by positioning a scintillator, used as veto, right after the K_S^0 defining collimator. In order for this counter to veto also neutral decays with high efficiency, a 7 mm lead sheet is positionned in front of it. This "antiKS counter" (AKS) also plays an essential role in the energy scale determination and control.

Table 1   Parameters of K_L^0 and K_S^0 beams. In parentheses the modified values used after the 1986 run are shown.

Beam type	K_L^0	K_S^0
Beam energy (GeV)	450	450 (360)
Length of beam from target to		
-defining collimator (m)	48.0	
-exit of final collimator(m)	120.0	7.1
Dist final coll-LAr Calo(m)	123.8	76.7 to 124.7
Be target diam/length(mm)	2/400	2/400
Production angle (mrad)	3.6 (2.5)	3.6 (4.5)
Beam acceptance (mrad)	±0.2	± 0.5
Protons on target per pulse	1×10^{11}	3×10^7
K^0 per pulse, at prod., in beam accept.	1.8×10^6	3.3×10^3
Neutron per pulse in beam accept.	1.5×10^7	3×10^4

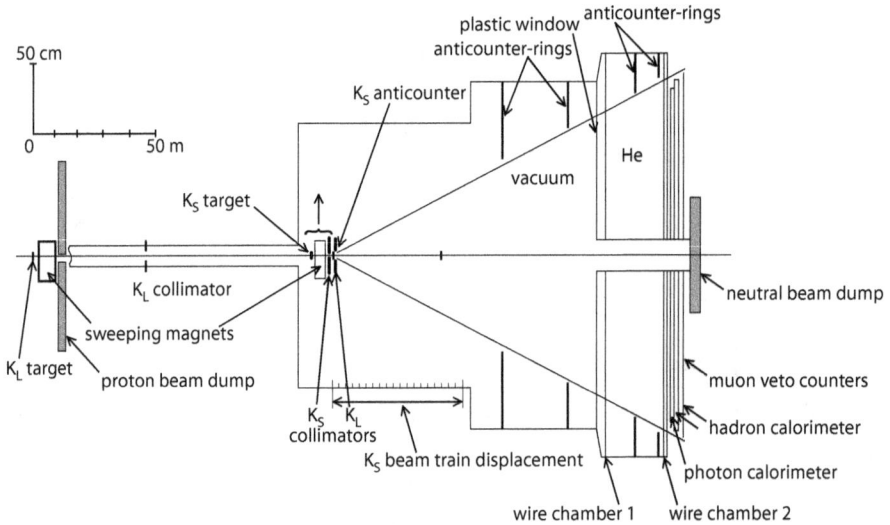

Fig. 1.   Layout of the NA31 experiment with, from left to right: the K_L^0 beam, the K_S^0 beam, the evacuated decay volume with anticounters, the thin window, the two wire chambers, the liquid argon calorimeter, the hadronic calorimeter and the muon veto.

In the K_S^0 mode the distribution of the decay vertices along the beam axis is an exponential with a slope of about 5 m at 100 GeV. This is by far different from the flat K_L^0 distribution. To mitigate this effect, the K_S^0 beam elements were installed on supports movable on rails, the "XTGV", positioned for the necessary running time at fixed stations every 1.2 m along the 50 m decay region. Thus, after summing the data from all stations the overall vertex distribution of the K_S^0 events became to a large extent similar to the K_L^0 one.

Despite this trick, the energy spectra still differ between the two beams, because the K_S^0 collimation length selects preferentially the more energetic kaons. This difference amplifies the sensitivity of the result by about a factor of 3 to a possible difference of the energy scale between charged and neutral events. To reduce this effect, some beam parameters were modified in 1988: the energy of the proton beam striking the K_S^0 target and the incident angle of the protons on both K_S^0 and K_L^0 targets were tuned so that the two energy spectra measured in the decay volume became more similar, reducing in a significant way the sensitivity of the result to their difference.

2.2. *The NA31 experimental layout*

The K_S^0 XTGV is enclosed in a 130 m long cylinder of 2.4 m diameter that contains the decay volume.[21] The first 100 m are evacuated down to about 3×10^{-3} Torr in order to prevent K_S^0 regeneration from K_L^0 and multiple scattering of charged decay pions. A thin kevlar window separates the evacuated part from an enclosure filled with Helium where the two wire chambers are installed. The Helium enclosure ends with an Aluminium cover, shaped to withstand the forces generated by a breakdown of the kevlar window in case of an accident. Downstream of it are the scintillator hodoscopes, the electromagnetic liquid argon calorimeter (LAC) and hadronic (HAC) calorimeters, and the muon veto. All detectors are traversed by a central evacuated pipe for the beam, ending at the beam dump.

Four rings of anticounters are used to veto three-body decays. A hodoscope of scintillation counters is placed in front of the LAC to start the trigger for charged decays. A valid charged trigger requires two hits in opposite quadrants. A second hodoscope, inserted at mid depth in the LAC is used to trigger the acquisition of neutral candidates. A valid neutral trigger requires a left–right coincidence.

Downstream of the first (second) 80 cm iron slab of the muon filter which follows the calorimeters is a hodoscope of horizontal (vertical) scintillator slabs each covering the full width of 2.7 m and read out at both ends. This detector is used as muon veto.

The measurement of both decay modes is based on calorimetry (electromagnetic and hadronic) and uses the wire chambers for reconstructing the decay vertex and trajectories of charged particles.

2.3. *Measuring the neutral decays: Liquid argon calorimeter*

To achieve the needed discrimination between the $2\pi^0$ signal and the many times more abundant $3\pi^0$ background requires a calorimeter with a high level of energy and space resolution. A liquid argon ionisation calorimeter (LAC) has been used for this purpose. Its body consists of a stack of lead conversion plates, aluminium clad, alternating with printed-circuit boards for signal readout. Together with the liquid argon, its total thickness is about 24 radiation lengths (X_0). The calorimeter is physically divided into a left and a right part, each of them split in a front and a back part, with the neutral hodoscope fitted in between. The important parameters

Table 2   Main parameters of the liquid argon calorimeter.

Size of the Pb/Al sandwich plates (mm)	$1204 \times 2408 \times 2.3$
Size of the readout boards (mm)	$1200 \times 2400 \times 0.8$
Length of one cell (mm/rad.length)	7.3/0.3
Number of cells in depth (front+back)	$40 + 40$
Strip pitch (mm)	12.5
Number of cells per quadrant (X–Y)	96–96
Number of electronics channels	1536

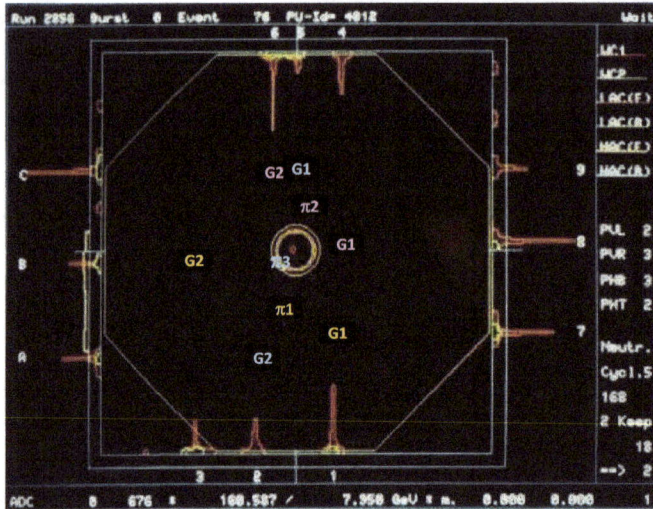

Fig. 2.   Display of a $3\pi^0$ candidate event. On the borders of the squared area representing the LAC surface are drawn peaks corresponding to the energy deposition in the horizontal and vertical calorimeter strips. The reconstructed photon positions are denoted by the letters G_i. The positions of the three π^0 are indicated by π_i. Different colours are used for each π^0 and its decay products.

are given in Table 2. The readout boards are etched to form 96 parallel strips per quadrant, alternately vertical and horizontal. Corresponding strips are ganged by groups of 20 in depth, to form readout cells. With this scheme showers are measured by their horizontal and vertical projections, separately for each quadrant.

In order to optimise the signal to noise ratio, each readout cell is coupled to its preamplifier through a low impedance cable and a transformer. The electronic circuit features a 1.6 μs differentiation and two 50 ns integration steps. As a result, the charge signal fed to a 12-bit ADC has a parabolic rise, lasting about 500 ns, followed by a slow decay. A second output is fed to the "peak-finder" system used to count the number of photons in the calorimeter, at the trigger level. Figure 2 shows the event display of a $3\pi^0$ candidate.

A zero-suppression system, with a threshold of 150 MeV (about 9 times the noise of typically 16 MeV per channel) combined to a neighbourhood logic, is used to limit the readout bandwidth.

The whole calorimeter front face was exposed several times to an electron beam to measure *in situ* the response on a grid of points. From these measurements, the sampling term of the energy resolution was found to be equal to $7.5\%/\sqrt{(E(\text{GeV})}$, the uniformity within $\pm 0.5\%$, and the linearity within $\pm 0.3\%$ between 12 and 120 GeV. The electronic noise contributes about 100 MeV in the resolution.

Electronic calibrations were taken regularly and revealed small drifts of the pedestal and channel gains, the latter being associated to temperature variations in the hall (typically 0.15% per degree). An hourly correction based on the measured temperature was applied to compensate this effect.

For the 1989 run, it was decided to complement the charge readout by a precise time measurement based on "zero-crossing" TDCs (ZTDC). Any calorimeter activity in the time interval from 0.9 μs before to 3.0 μs after the passage of particles is recorded, with an average time resolution for photons of 10 ns. This implementation allowed to cross check the effect of accidental activity in the measurement of neutral decays, otherwise estimated by the overlay method.

2.4. Measuring the charged mode

The direction of charged particles is measured by the two wire chambers. Each track energy is computed combining the energies measured in the LAC and HAC. Each chamber consists of four planes with wires (at a 6 mm pitch) along the vertical, horizontal, U (53 degrees from horizontal) and V (perpendicular to U) directions, positioned so as to minimise ambiguities in space-point reconstruction. Three hits among four planes are enough to reconstruct a space point. The average efficiency per plane is 99.3% with a precision on the space points of 750 μm. The decay vertex is reconstructed with a longitudinal precision of 80 cm and an rms of 5 mm on the closest distance of approach.

The HAC is a sampling device, with 25 mm iron sheets alternating with planes made out of $1.3\,\text{m} \times 0.12\,\text{m}$ scintillator slabs of 4.5 mm thickness forming a quadrant structure. There are 24 (25) planes, alternately with horizontal and vertical scintillator slabs in the front (back) module with a total of 176 channels.

After the analysis of the first data taking period, a transition radiation detector was built and installed in 1988 between the second wire chamber and the Aluminium cover of the large tube, to independently cross check the estimation of the Ke3 ($K_L^0 \rightarrow \pi e \nu$) background. The detector consists of four identical ensembles, each made of a radiator of polypropylene foils followed by a chamber operated with a Xe–He–CH_4 mixture. The detector offered an additional electron–pion separation resulting in a rejection power of 10 against electrons for a 98% pion efficiency.[22]

2.5. Trigger, online background rejection and data acquisition

The trigger and data acquisition were designed to cope with the memory time of the detector (2.1 μs, determined by the LAC), the single rate in the detectors, about

100 kHz, to allow a throughput of about 1 Mbyte/s corresponding to about 2000 accepted events per burst in the K_L^0 mode, mostly of three-body decay background despite a high rejection by the three levels of the trigger.

The first level performs a fast selection of events consistent with either two charged particles, or at least two electromagnetic showers using signals from the hodoscopes, LAC and HAC. Furthermore, for the neutral trigger, less than 5 peaks in both the horizontal and the vertical strips of the front part of the LAC are required, using the "peakfinder" system.

At the second level, a custom made hard-wired processor (AFBI) with 150 ns cycle time treats the neutral and the charged conditions in two parallel streams. The neutral stream uses all readout strips to calculate the total energy, the barycenter position in the transverse plane and the longitudinal decay vertex position Z_V. This latter quantity is calculated from the second moments, assuming that the clusters correspond to the decay of an object with the kaon mass. The $3\pi^0$ decays with missing clusters are reconstructed with a vertex closer to the calorimeter than the real decay position, and are thus preferentially rejected by the condition $Z_V < 50$ m counted from the K_S^0 collimator. The charged stream requires conditions on the energy deposited in the LAC and HAC compartments. About 50% (30%) of neutral (charged) triggers are rejected by the AFBI, while the loss of good 2π decays is less than 0.1%. Events which have both a neutral and a charged trigger are all accepted.

Finally, at the third level, data from events accepted by the AFBI are loaded into the input memory of a dual 168E processor where some of the offline tracking calculations are made for charged triggers, which allows to cut on Z_V, on the acoplanarity, and on the energy assuming a kaon mass. In the K_L^0 mode, the 168E accepts about 50% of the charged triggers, including 15% which are too complicated to be treated in the allowed time of 1ms, and 10% of downscaled events used for efficiency calculations. Accepted events are written to tape for offline analysis.

3. The NA31 Analysis and Result

In the following, the descriptions and numerical results are given mainly for the first data taking period (1986). Tables allow the comparison between the running periods.

3.1. *Analysis*

The $K^0 \rightarrow 2\pi^0 \rightarrow 4\gamma$ decays are reconstructed from the measured energies and positions of photons in the calorimeter. The energy of a photon results from the sum of the cluster energies in the two projections within a fixed size window, and is corrected for leakage outside. Positions result from a barycenter using a maximum of 15 strips in each projection. Events with a fifth photon of more than 2.5 GeV are rejected. Valid photons have to be above 5 GeV, and to be more than 5 cm

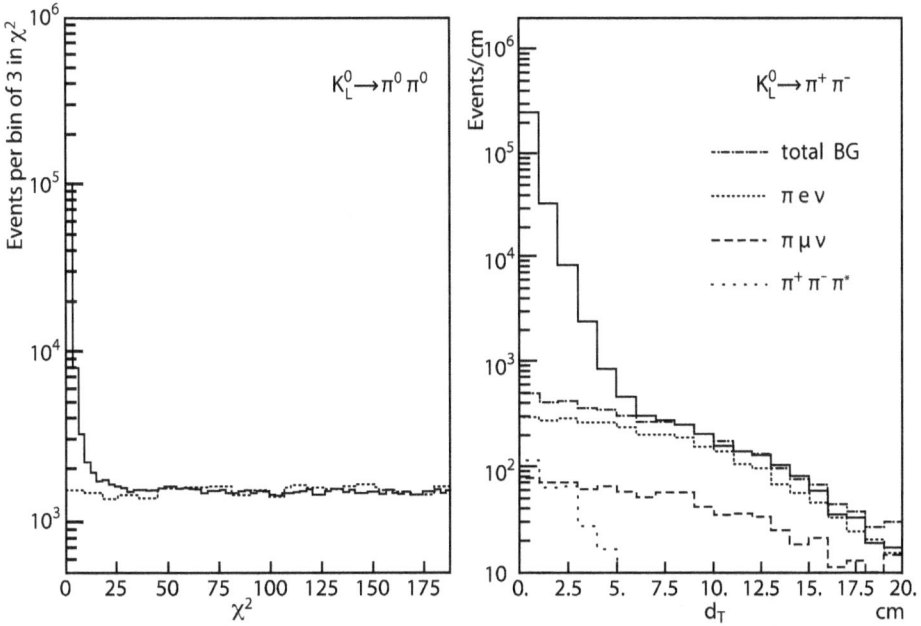

Fig. 3. Left: For the 4γ events in the K_L^0 sample, distribution of the smallest χ^2 against a $2\pi^0$ hypothesis. Results from a Monte Carlo study of $3\pi^0$ background are also shown as a dotted histogram. Right: For the $\pi^+\pi^-$ sample in the K_L^0 beam, distribution of the transverse distance d_T to the K_L^0 target. Distributions from Monte Carlo simulations of Ke3 and Kμ3 backgrounds are also shown, while the $\pi^+\pi^-\pi^0$ background is estimated from a subsample of data with an extra photon.

apart from each other. The K^0 energy is measured as the sum of energies of the four photons with about 1% precision. Assuming a kaon mass, the distance from the decay vertex to the calorimeter is computed with a similar relative precision. To reject the remaining background mostly due to $K^0 \to 3\pi^0 \to 6\gamma$ in the K_L^0 beam, a π^0 mass constraint[b] on the three two-photon combinations is applied. The best two-photon pairing is given by the smallest χ^2, shown in Fig. 3. The signal region is taken as $\chi^2 < 9$. The residual background is estimated by linear extrapolation from the large χ^2 region to the signal region, to be $(4.0 \pm 0.2)\%$, where the error includes statistics and systematics. The numbers of neutral events available in the K_S^0 and K_L^0 beams are given in Table 3.

The $K^0 \to \pi^+\pi^-$ decays are reconstructed from space points in the chambers. Events with more than two space points in the first chamber are rejected.[c] The $\pi^+\pi^-$ invariant mass, calculated from the angle between the two tracks and the calorimeteric energy associated with each of them, is required to be within 2.1σ

---

[b]The resolution is about 2 MeV.

[c]For consistency, neutral events with one or more space points in the first chamber are also rejected.

Table 3   Event statistics accumulated in the NA31 data taking periods for the four modes, together with the estimated background fractions.

Decay mode	1986 data		1988 + 1989 data	
	Events($\times 1000$)	Bakground(%)	Events($\times 1000$)	Background(%)
$K_L^0 \to \pi^0\pi^0$	109	4.0	319	2.67
$K_L^0 \to \pi^+\pi^-$	295	0.6	847	0.63
$K_S^0 \to \pi^0\pi^0$	932	<0.1	1322	0.07
$K_S^0 \to \pi^+\pi^-$	2300	<0.1	3241	0.03

($\sigma \sim 20\,\text{MeV}$) from the neutral kaon mass. The K^0 energy is calculated from the opening angle of the two tracks, and the ratio R of their calorimetric energies, assuming a kaon mass

$$E = \frac{1}{\theta}\sqrt{\left(2 + R + \frac{1}{R}\right)\left(m_K^2 - m_\pi^2\left(2 + R + \frac{1}{R}\right)\right)}. \tag{10}$$

The ratio R is limited to 2.5 in order to achieve a 1% resolution and to reduce Λ decays to a negligible level. Events with additional photons are rejected. Ke3 decays are suppressed by requiring that none of the two tracks pass "electron-like" shower-shape criteria. About half of $\pi^+\pi^-$ decays are rejected by the cuts. The response of the HAC, monitored by sending laser pulses to all photomultipliers at a regular pace, was estimated to stay constant within $\pm 0.5\%$, which ensures a stability of the acceptance ratio $K_L \to \pi^+\pi^-$ to $K_S \to \pi^+\pi^-$ better than 0.1%. The remaining background from three-body K_L^0 decays is estimated using the distribution of the transverse distance d_T from the target to the decay plane (Fig. 3). Extrapolating from the control region $7 < d_T < 12$ cm under the signal region defined as $d_T < 5$ cm, a background fraction of $(0.65 \pm 0.2)\%$ is estimated and subtracted. The numbers of charged events available in the K_S^0 and K_L^0 samples are given in Table 3.

Fitting the decay vertex spectrum of K_S^0 decays in charged or neutral mode to the simulated distribution shaped by the AKS located at different longitudinal positions allows to fix the relative energy scale between charged and neutrals with an uncertainty of $\pm 0.1\%$. An illustration of this fit, for energies between 70 and 170 GeV, is given in Fig. 4(left).

3.2.   The NA31 results

Once the energy scale is fixed, events are counted in 10 bins of energy between 70 and 170 GeV, and 32 bins in vertex position between 10.5 and 48.9 m. The chosen decay range is defined to minimise the pollution from kaon scattering and regeneration on the final collimator and to limit the $3\pi^0$ background which increases with Z_V. The weighted average of the double ratio, background subtracted and corrected for acceptance and resolution, is RR $= 0.977$. The effect of accidental

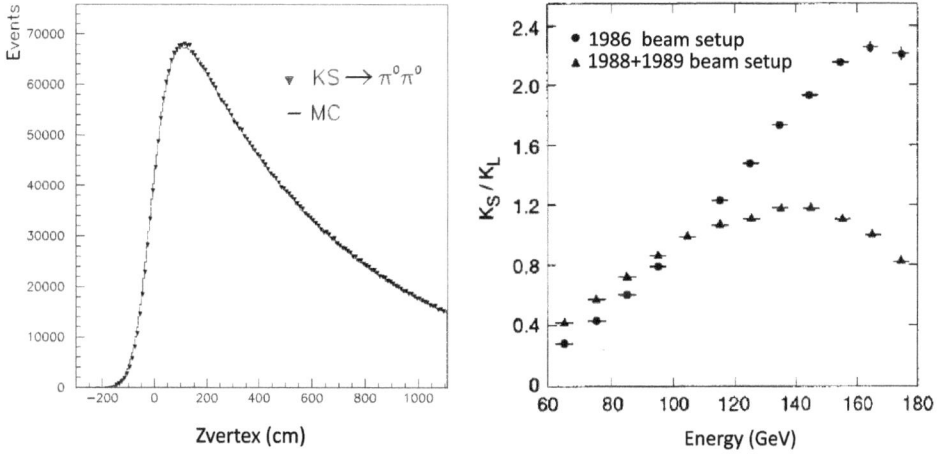

Fig. 4. Left: Reconstructed position of decay vertices in the K_S^0 beam, for the neutral mode. The fitted curve correspond to the sum of exponential decays modified by resolution. Right: Ratio of the K_S^0/K_L^0 energy spectra with the 1986 and the 1988 improved set-up.

coincidences between a kaon decay and some activity in the beam is estimated by overlaying data with random events. The net effect (losses-gains) on the double ratio is $-0.34 \pm 0.10\%$. Other small corrections are applied for acceptance and scattering differences between the four modes, trigger and AKS anticounter inefficiencies, leading to an overall correction of 0.003. The final result is RR $=$ 0.980 ± 0.004 (stat) ± 0.005 (syst), giving:

$$\mathrm{Re}(\epsilon'/\epsilon) = (3.3 \pm 1.1) \times 10^{-3}. \tag{11}$$

For the first time a 3-standard deviation effect had been observed. This result was published in 1988[23] with the conservative title "First evidence for direct CP violation".

The main sources of systematic uncertainties are related to the accidental losses, to the energy scale difference between charged and neutral modes and to the background estimation.

After this first succesful run, the collaboration decided to improve the beam and the detector in order to reduce both the main systematic and the statistical uncertainties, in two new data-taking periods which took place in 1988 and 1989.

In practice, the use of a lower proton energy to produce the K_S^0 beam and the tuning of the production angle of both beams (Table 1), resulted in a much flatter ratio of K_S^0/K_L^0 energy spectra, as shown in Fig. 4(right). This change reduced the systematic uncertainty from the energy scale from 0.3% to 0.12%.

On the charged decay mode, the information provided by the new transition radiation detector allowed an independent estimation of the main component of the background, that is Ke3 decays. With this additional handle, the systematic uncertainty on the subtracted background in the signal region was reduced to 0.1% (see Table 4).

Table 4   Systematic uncertainties on the double ratio (in %) for the two NA31 data taking periods.

Source of uncertainty	1986 data	1988 + 1989 data
Background in $K^0_L \to \pi^0\pi^0$	0.2	0.13
Background in $K^0_L \to \pi^+\pi^-$	0.2	0.10
Energy scale	0.3	0.13
Accidental losses	0.2	0.14
Monte Carlo acceptance	0.1	0.10
Trigger and AKS efficiency	0.2	0.09
Total systematic uncertainty	0.5	0.3

The installation in 1989 of the ZTDC system provided the time measurement of the calorimetric clusters. Using this information the effect of accidentals was found to be in nice agreement with the overlay method. Overall, the systematic uncertainty associated with accidental activity was reduced to 0.14%.

The evolution of the systematic uncertainties between the run periods is shown in Table 4.

The final double ratio obtained with the upgraded detectors and beams using the 1988+1989 data taking is RR = 0.9878±0.0026 (stat)±0.0030 (syst). This value of RR translates into $\mathrm{Re}(\epsilon'/\epsilon) = (2.0 \pm 0.7) \times 10^{-3}$. Taking into account that some systematic uncertainties are common to the two data taking periods, the average is

$$\mathrm{Re}(\epsilon'/\epsilon) = (2.30 \pm 0.65) \times 10^{-3}. \tag{12}$$

This final result, now 3.5-standard deviation away from zero, was published[24] in November 1993.

A few months earlier, the E731 collaboration at Fermilab came out with a result of a similar precision,[20] but much lower central value, namely $\epsilon'/\epsilon = (0.74 \pm 0.61) \times 10^{-3}$. This E731 result was both compatible with zero and with the NA31 result and thus rather inconclusive. Given the smallness of the observed effect, another round of experiments was justified.

3.3. *Phase measurement*

In 1987, the NA31 experiment devoted a special run to the measurement of the phases, ϕ^{00} and ϕ^{+-}.

As early as 1965, it was shown by Bell and Steinberger[25] using unitarity and CPT conservation, that the phase of ϵ, ϕ_ϵ, should have the "natural value" given by

$$\phi_\epsilon = \tan^{-1}(2\Delta M/\Gamma_S). \tag{13}$$

When CPT is conserved, and neglecting the small contribution of $\mathrm{Im}(\epsilon'/\epsilon)$, the phases of η^{00} and of η^{+-} should be both equal to ϕ_ϵ.[d] In 1987, ϕ^{+-} was

---

[d]With $\Delta M = M_L - M_S = 3.48 \times 10^{-12}$ MeV $= 0.53 \times 10^{10}$ s^{-1} and $\Gamma_S = 1.11 \times 10^{10}$ s^{-1} one has $\Delta M/\Gamma_S = 0.477$ resulting to $\phi_\epsilon = 43.5°$.

already rather well measured, unlike ϕ^{00} which was not precisely known, leading to a difference $\phi^{00} - \phi^{+-}$ of $12 \pm 6°$.

In a K^0 beam, after about $12\tau_S$, the $\pi\pi$ rate from the K_S^0 component and the CP violating component of K_L^0 decays become comparable and can interfere. In the interference region, the two-pion decay rate as a function of time in the kaon rest frame can be written as

$$I(t) = S(p)[e^{-t/\tau_S} + |\eta|^2 e^{-t/\tau_L} + 2D(p)|\eta|e^{-t/2(1/\tau_S+1/\tau_L)} \cos(\Delta M t - \phi)] \quad (14)$$

where $S(p)$ is the momentum spectrum of $(K^0 + \bar{K}^0)$ and $D(p) = (K^0 - \bar{K}^0)/(K^0 + \bar{K}^0)$ is the dilution factor.

In order to have an optimised acceptance in the region between $10\tau_S$ and $15\tau_S$, most sensitive to the interference term, NA31 used an improved beam scheme with two target stations, "far" and "near", which were installed at 48 m and 33.6 m respectively, upstream the K_L^0 cleaning collimator. This beam set-up, with two-stage collimation and improved shielding, allowed intensities up to 2×10^{10} protons per pulse. The same target dimensions, proton energy, and incidence angle as for the 1986 period shown in Table 1 were used.

Data were taken with the same apparatus and trigger as for ϵ'/ϵ, but downscaling all events with early decay vertices ($<7\tau_S$), in order to maximise the number of events in the interference region. The data sample consists of about 140 million triggers written to tape. Some additional data with the XTGV placed in its most upstream position were also taken at regular intervals to determine the energy scale of the neutral sample. An error of 0.1% on the energy scale induces a phase difference of about $1°$. The measured kaon decay rates in the charged and neutral modes are shown in Fig. 5.

The value of the phase difference is obtained by a combined fit (charged and neutral modes) to the ratio of event rates between the two targets, as a function of lifetime counted from the mid-point between the two targets.

The result[26] was published in 1990: $\phi^{00} - \phi^{+-} = (0.20 \pm 2.6 \text{ (stat)} \pm 1.3 \text{ (syst)})° = (0.20 \pm 2.9)°$, where the dominant systematic uncertainty comes from the energy scale.[e] Combined with the exceedingly small value of ΔM compared to M_K, the phase difference measurements provide the strongest test of CPT conservation, at the level of $\pm 4 \times 10^{-19}$.[26]

CP violation associated to CPT conservation implies that T is violated. Such a violation has been observed with a significance of 6 standard deviations in the CPLear[28] experiment at CERN, by comparing the rates of tagged $\bar{K}^0 \to \pi^- e^+ \nu$ and $K^0 \to \pi^+ e^- \bar{\nu}$ in $\bar{p}p$ annihilations. This observation implies that the transition amplitudes $K^0 \to \bar{K}^0$ and $\bar{K}^0 \to K^0$ are not identical, demonstrating the violation of T symmetry.

---

[e]This result is now superseded by KTeV[27] with $\phi^{00} - \phi^{+-} = (0.29 \pm 0.31)°$.

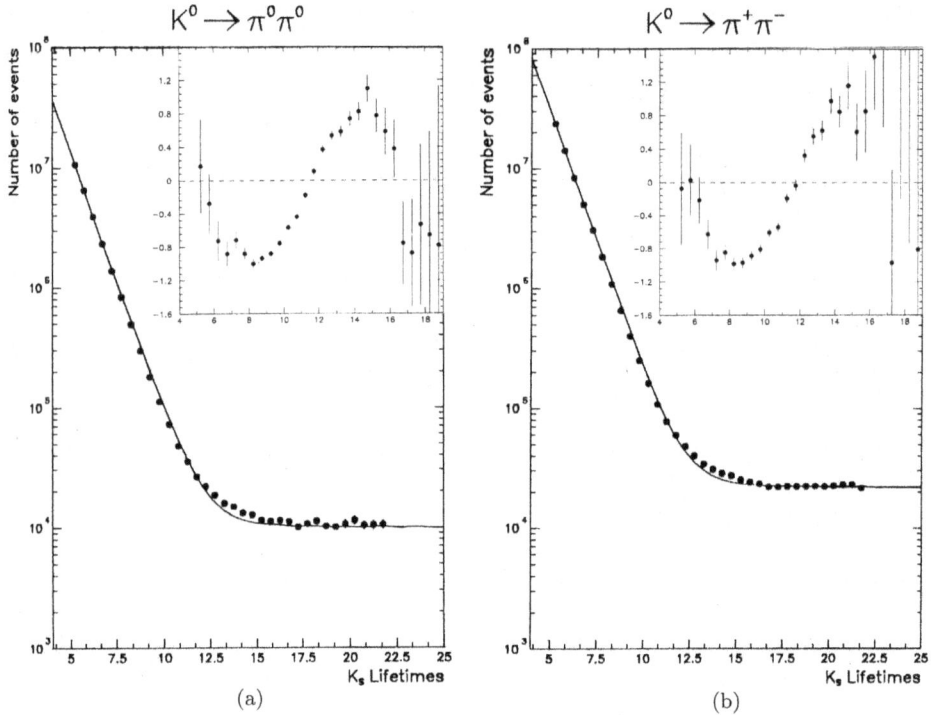

Fig. 5. Decay rate in the neutral (a) and charged (b) modes as a function of lifetime. Superimposed are the fitted lifetime distributions without the interference term. The inserts show the interference terms extracted from data.

4. The Second Generation: The NA48 Beams and Detectors

The NA48 experiment aimed at achieving a precision of a few 10^{-4} on ϵ'/ϵ. Ten new European institutes on top of the six from the NA31 community contributed to this new collaboration.[29] Data for the ϵ'/ϵ measurement have been recorded in three periods in 1997, 1998–1999 and in 2001. A special run took place in 2000, allowing auxiliary studies to improve the control of specific systematic effects.

The collaboration focussed on a method allowing an even better cancellation of systematic uncertainties than in NA31, by collecting at the same time and in the same decay volume all the four decay modes. To allow this, the decay volume had to be shortened to $3.5\tau_S$ counted from the end of the K^0_S defining collimator.

Acceptance differences between K^0_L and K^0_S decays are minimised by weighting each K^0_L event by a function of its proper time, so that the K^0_S and K^0_L decay spectra become almost identical. Unlike NA31, a magnetic spectrometer is used for the measurement of charged decays with a better precision. A quasi-homogeneous liquid krypton calorimeter with high granularity provides excellent position, energy and time information for the neutral decays. A schematic drawing of the NA48 beam line and detectors[30] is given in Fig. 6.

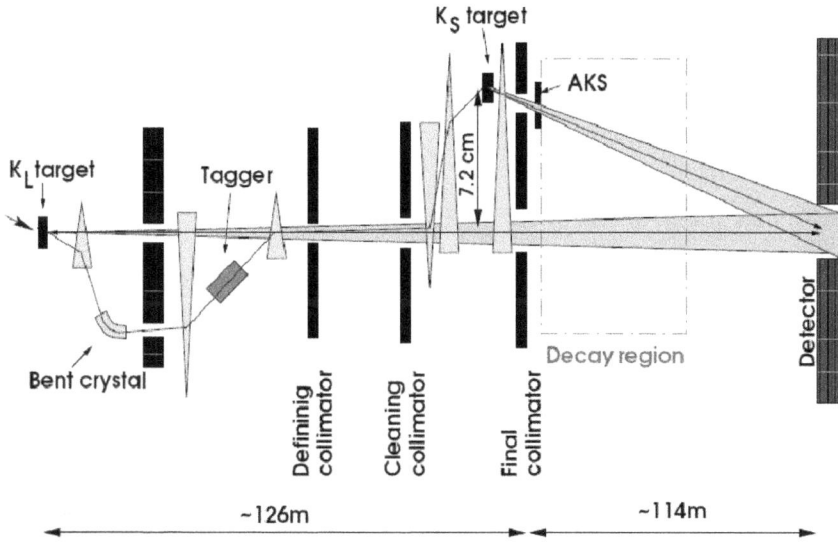

Fig. 6. Schematic view of the NA48 beams (top) and detectors (bottom).

4.1. The NA48 beams

The K_L^0 beam is produced by a ten times more intense 450 GeV proton beam than for NA31, impinging on a beryllium target under an angle of 2.4 mrad. The protons

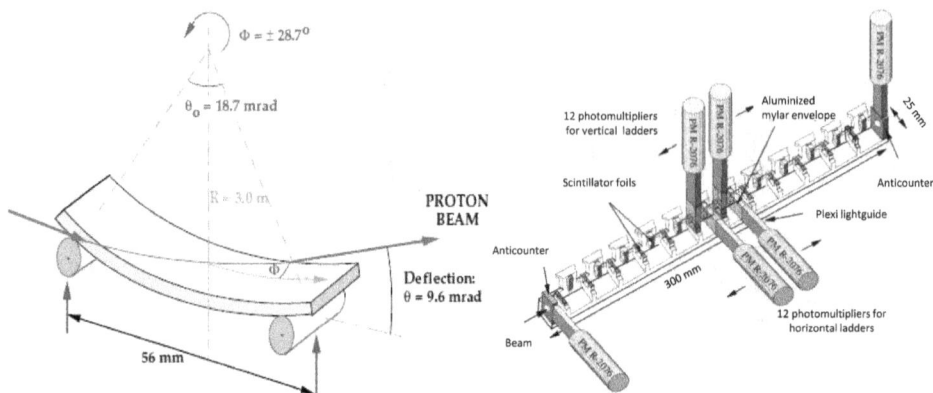

Fig. 7. Left: Schematic view of the bent crystal. Right: Schematic view of the tagger detector.

for the K^0_S beam are derived from the non-interacting protons in the K^0_L target hitting a silicon mono-crystal[31] which is cut to dimensions of $(60, 18, 1.5)$ mm^3 parallel to the (110) crystaline plane. The crystal is bent and mounted on a motorised goniometer that allows a precise positionning to select, deflect and guide along the crystalline planes the desired amount of 3×10^7 ppp (Fig. 7(left)). The channelling properties of a bent crystal produce a clean beam with well defined emittance, deflected into the desired direction by applying the equivalent of a 14.4 T $\cdot$ m bending power in a short length of 6 cm, and containing only a 2×10^{-5} fraction of protons without the presence of a heavy collimator system. The transmitted proton beam traverses a tagging station which registers precisely the passage time of each proton, and is then transported through a series of deflecting magnets to the K^0_S target, located 6 m upstream from the beginning of the fiducial decay region and 72 mm above the K^0_L beam axis. At the aperture of the final collimator, the K^0_S beam goes through an anti-counter used to precisely define the beginning of the K^0_S fiducial region, like in NA31. The K^0_S beam direction is tuned so as to overlap the K^0_L one at the entrance of the main detectors, 120 m downstream. Both K^0_S and K^0_L beams are transported in vacuum inside a large evacuated tank, closed off at its end by a thin Kevlar window to separate the evacuated decay region from the detectors.

4.2. The tagger

The tagging detector placed in the path of the proton beam selected by the bent crystal consists of 24 scintillators mounted in a carbon-fibre structure, alternating in horizontal and in vertical directions[32] and staggered so as to share the beam profile among different counters. For the incoming protons, the scintillators offer in each direction a geometrical overlap of $50\,\mu$m between successive detectors to reduce possible small misalignment issues (Fig. 7(right)). Despite the radiation hardness of the material the effect of 10^{13} protons striking the detector in 100 days caused a 50% decrease of the emitted light. This was monitored and corrected for by the appropriate change of the photomultiplier high voltage. After digitisation by a 8-bit 1 GHz flash-ADC the data are transferred to a ring buffer and are extracted in case of a trigger. The tagger showed excellent performance, offering a reconstructed time resolution of 140 ps and a separation of two close pulses down to 5 ns.

4.3. *The liquid Krypton calorimeter*

The liquid Krypton calorimeter consists of $\sim$13000 cells of $2\,cm \times 2\,cm$ cross-section, defined by Cu–Be–Co ribbons of $40\,\mu m \times 18\,mm \times 125\,cm$ forming longitudinal projective towers pointing to the centre of the decay region and immersed in a bath of $\sim$10 m^3 of liquid Krypton.[33] Each cell is made out of two cathodes and one central anode. Liquid Krypton was adopted as active medium because its radiation length (4.7 cm) is short enough to allow building a compact homogeneous calorimeter while being still affordable. The intrinsic stability of the ionisation signal and the small Moliere radius, ensuring compact lateral shower size, are further assets. A high voltage of 3 kV was applied to generate the drift field. Preamplifiers and calibration system are directly installed at the downstream end of the anodes in the cold volume, minimising the noise and allowing fast charge transfer. The ionisation signal is shaped outside the cryostat and digitised asynchronously by a 10 bit 40 MHz flash ADC. A four gain-switching scheme driven by the signal pulse-height, covers the total dynamic range from 3 MeV to 50 GeV. The readout is restricted to active cells grouped into clusters, by applying zero-suppression using dedicated algorithms.

Energy and time are reconstructed using a digital filter method[34] applied to the three samples around the maximum. Clusters of 3×3 cells are formed around the most energetic cell. The barycentre of the energy deposition is used as estimator of the cluster position and the time of the most energetic cell is considered as time of the cluster. After cell inter-calibration using Ke3 decays, E/p studies result in an energy resolution of $3.2\%/\sqrt{E\,(GeV)}$, with additional contributions of 90 MeV from the electronic noise and 0.42% from the constant term. The energy response is measured to be linear within 0.1% between 5 and 100 GeV. The position resolution is better than 1 mm in both directions and the time resolution is 500 ps.

4.4. *The spectrometer*

Charged decays are triggered by a scintillator hodoscope. The tracks are reconstructed from the hits measured in the four octagon-shaped wire chambers (DCH), two located before and two after the dipole magnet. Each chamber contains eight planes of grounded sense wires tilted by 45° with respect to each other that guarantees a high detection efficiency with sufficient redundancy.[35] All anode wires are instrumented with amplifiers and TDC circuits. The magnet integral field of 0.883 Tm between the second and the third chamber induces a transverse momentum kick of 265 MeV/c in the horizontal direction. The track momentum is reconstructed with a resolution measured in electron beams to be $\sigma(p)/p = 0.48\% \oplus 0.009 \times p\%$ (p in GeV/c), where the first term is due to multiple scattering in the DCH and the surrounding Helium gas, and the second one comes from the position accuracy of the chamber hits. The resolution of the reconstructed kaon mass is $2.5\,\mathrm{MeV}/c^2$. For two-track events, the decay vertex is computed from hits in the two upstream chambers, with a longitudinal (transverse) resolution of 50 cm (2 mm) and a spread of 7 mm on the closest distance of approach.

After the 1999 run, an implosion of the beam tube caused severe damages to the drift chambers. New chambers were installed for the 2001 data taking.

4.5. *The NA48 trigger and data acquisition systems*

The goal of the NA48 trigger is to reduce the 500 kHz rate of particles hitting the detector to a few kHz of accepted events, with minimal inefficiencies and dead time. The first level of the pipelined charged trigger is based on signals in the hodoscope, hit multiplicity in the wire chambers and an energy threshold in the electromagnetic calorimeter. The second level computes the track coordinates, the proper decay time and the invariant mass using positions of the hits in the first, second and fourth chamber.

A pipelined design was also used for the neutral trigger. Using the analogue sums made on 2×8 calorimetric cells in both, horizontal and vertical projections, the kaon energy, the number of photons, their arrival time and the proper decay time are estimated every 25 ns. The trigger requires less than six peaks in each projection to reject $K_L^0 \to 3\pi^0$ decays, a total energy greater than 50 GeV and a decay vertex reconstructed at less than $5\tau_S$ from the final collimator.

The trigger decisions from the sub-detectors are assembled by the trigger supervisor system that defines the event timestamp relative to the 40 MHz clock used to synchronise all the detectors. Event fragments are built up to complete events in on-site PC farms and are transmitted to the CERN computing centre via a Gigabit optical link with a speed of about 10 MB/s. There, raw data are stored on disk and in parallel monitored, reconstructed and further selected in an offline PC farm.[36]

4.6. *The NA48 analysis*

This section provides a description of the analysis of the statistically most precise data period, 1998–1999. Final results obtained combining all the periods are given at the end.

4.6.1. *The neutral decays*

Clusters are chosen in the energy range from 3 to 100 GeV and must be well separated from each other. The 4-cluster time is computed as an energy-weighted average using the two most energetic cells in each cluster. Each cluster time must

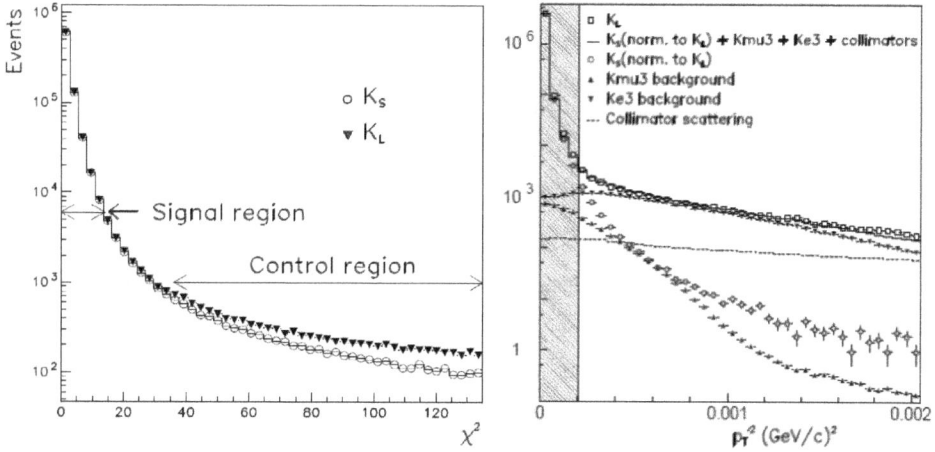

Fig. 8. Left: Distribution of the χ^2 distribution for K_L^0 and K_S^0 events decaying into $2\pi^0$. The signal and control regions are indicated in the figure. Right: Distribution of $P_T^{2'}$ for the charged decays, with all the various contributions.

be compatible within 5 ns with the average time. The longitudinal vertex position is computed from the energies and the positions of the four clusters assuming a kaon decay. The $m_{\gamma\gamma}$ values obtained from the 2-by-2 photon associations are combined in a χ^2 discriminator. Out of the three possible pairings, the one with the lowest χ^2 is kept. The main background in the $2\pi^0$ decay channel results from unidentified $K_L^0 \rightarrow 3\pi^0$ with undetected or merged photons, resulting in only four measured clusters. The background is studied in a control region defined as $36 < \chi^2 < 135$ and subtracted for each energy bin from the signal region, defined as $\chi^2 < 13.5$ (see Fig. 8(left)). The effect of the subtracted background on the double ratio is $(-5.9 \pm 2.0) \times 10^{-4}$. The large reduction with respect to NA31 is due to both the better $m_{\gamma\gamma}$ resolution and the lower overall background acceptance within the significantly shorter decay volume.

The energy scale is adjusted comparing the upstream edge of the reconstructed vertex distribution of $K_S^0 \rightarrow 2\pi^0$ candidates with the AKS counter position. Additional checks are done using π^0 and η decays into two photons, produced in special runs where a π^- beam hits a target placed in different known positions. The agreement of the reconstructed and the nominal target positions was found to be within 1 cm. Comparing the energy scales obtained with two or more distant targets gives constraints on the energy non-linearity.

4.6.2. *The charged decays*

Tracks are reconstructed using the hits and the drift time information. These tracks must come from a common vertex, have momenta greater than 10 GeV and be at least 12 cm away from the centre of each DCH. The tracks must be within the acceptance of the LKr and of the muon vetoes to allow proper particle identification. The reconstructed kaon energy is computed from the momentum ratio of the two tracks, and their relative opening angle θ, assuming a $K \rightarrow \pi^+\pi^-$ decay (see Eq. (10)). The vertex reconstruction is controlled comparing the nominal position of the AKS counter to the upstream edge of the reconstructed spectrum of $K_S^0 \rightarrow \pi^+\pi^-$ decays.

Three-body decays are the main background in the K_L^0 sample. Ke3 decays are rejected by an E/p test applied to both tracks. Removing events with in-time hit in the muon vetoes suppresses the Kμ3 background. Further suppression of both modes is obtained by demanding the two-tracks invariant mass to be compatible with the kaon mass. The final discrimination is done using the transverse missing momentum P_T^2 constructed in a way to be independent of the vertex resolution. Figure 8 right shows the P_T^2 distribution. The signal region is defined as $P_T^2 < 0.0002$ (GeV/$c)^2$. The tails are populated with remaining Ke3 and Kμ3 events, as verified with Ke3 and Kμ3 enriched samples, respectively. The background in the signal region corresponds to a residual fraction of 10.1×10^{-4} Ke3 and 6.2×10^{-4} Kμ3. It is evaluated in each of the twenty energy bins. The K_S^0 sample is also contaminated by $\Lambda \rightarrow \pi^- p$ decays which are suppressed by exploiting the characteristic asymmetric momenta of the two tracks.

All requirements against the backgrounds are applied to both K_S^0 and K_L^0 samples in order to symmetrise the losses affecting genuine 2π decays.

4.6.3. *Corrections: Tagging inefficiency and dilution*

The separation of the K_S^0 and K_L^0 data samples is based on the time coincidence of the tagger with the LKr (for neutral decays) and the charged hodoscope (for charged decays).

As the tagger signal is used for both charged and neutral decays the uncertainties related to its time measurement are symmetric. Two mistagging effects remain:

• The first type of mistagging occurs when a K_S^0 event is labelled as K_L^0 because the reconstructed detector time falls outside the coincidence window. This inefficiency can be measured independently in the charged mode, by selecting K_S^0 decays from the reconstructed vertical position of the decay vertex. The inefficiency, α_{SL}, is the fraction of events lying outside the coincidence window of ± 2 ns (Fig. 9(left)). To measure the inefficiency in the neutral mode, the neutral decays (K_S^0 and K_L^0 into $2\pi^0$ and $3\pi^0$) with a photon converted into an electron–positron pair are considered. On an event-by-event basis, the charged time computed from the electron and positron hits in the hodoscope, is compared with the neutral time, reconstructed as the average time of the photons.

Overall, the mistagging probability α_{SL} has been found to be the same for charged and neutral decays within an uncertainty of $\pm 0.5 \times 10^{-4}$.

• The second type, denoted α_{LS}, concerns K_L^0 events assigned to the K_S^0 sample because of an accidental coincidence with the tagger. This effect, called "dilution",

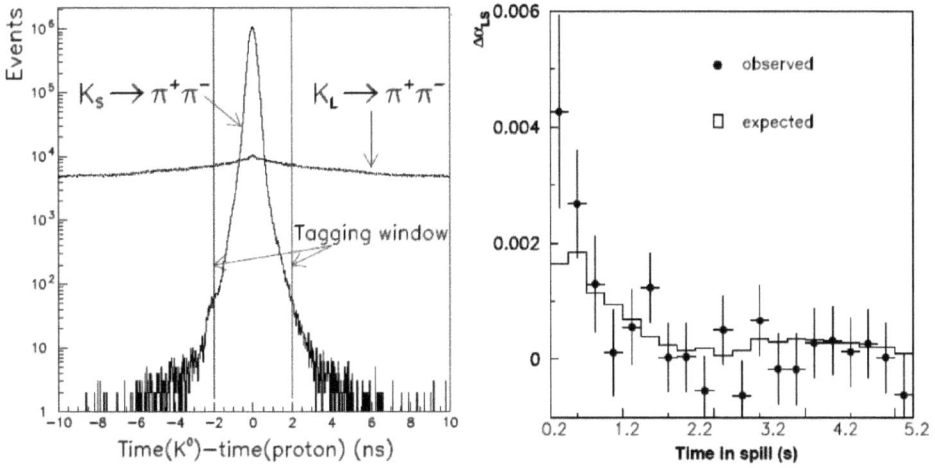

Fig. 9. Left: Time coincidence with the tagger for K_S^0 and K_L^0 charged decays, identified by the vertical position of the reconstructed vertex. The tagging window of ± 2 ns is indicated. Right: variation of $\Delta\alpha_{LS}$ within the time in the spill for the 2001 data taking period. The measured shapes are in good agreement with the expectations from the overlay method. A larger effect is observed at the beginning of the burst where the instantaneous intensity is higher.

is assessed by measuring the probability for a K_L^0 event to be accompanied by a tagged proton within a ± 2 ns window located in the sidebands before or after the event time. This probability is as large as 10%. An additional correction is applied for the small intensity difference between in-time and sideband windows, estimated using independently identified K_L^0 samples: charged decays chosen by their vertex to be K_L^0 and $3\pi^0$ neutral events. The measured value is $\Delta\alpha_{LS} = \alpha_{LS}^{00} - \alpha_{LS}^{+-} = (4.3 \pm 1.4\,(\text{stat}) \pm 1.0\,(\text{syst})) \times 10^{-4}$. The origin of the difference between charged and neutral dilution has been identified to be due to the higher sensitivity of the charged mode to accidentals, at both the trigger and reconstruction levels. This value has been confirmed by the overlay method and varies within the SPS spill as the instantaneous intensity (Fig. 9(right)).

4.6.4. *Corrections: Beam activity, scattering and acceptance*

Most of the accidental activity originates from the high-intensity K_L^0 beam. The concurrent collection of the four modes minimises the sensitivity of the result to this effect. Overlaying the data with specific triggers recorded by the beam monitors in proportion to the intensity, allows to estimate the residual effect. The net (losses-gains) measured effect is larger in the charged mode with respect to the neutral one by $(1.4 \pm 0.7)\%$ (Fig. 10(left)). Because the intensity variations turn out to be similar in both beams within $\pm 1\%$, this effect cancels in the double ratio. The overall uncertainty from accidental effects on the double ratio is $\Delta RR = \pm 4.2 \times 10^{-4}$.

Scattered kaons appear as tails in the distribution of the radius of the centre-of-gravity for neutral and charged events. For K_L^0 these tails are dominated by events

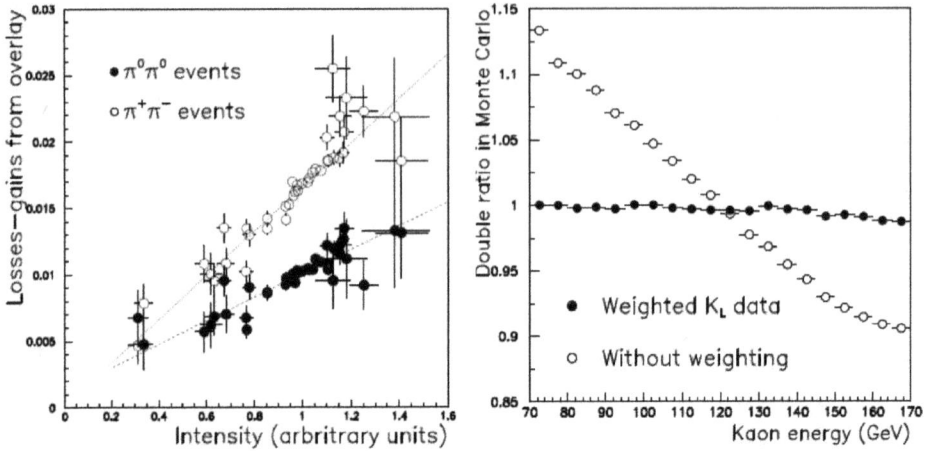

Fig. 10.   Left: Estimated net accidental effect for charged and neutral decays. Right: Monte Carlo acceptance correction in case of unweighted and weighted K_L^0 events.

which have suffered double scattering, resulting in a correction of $-(9.6 \pm 2.0) \times 10^{-4}$ to account for the difference between the losses in the two decay modes.

Despite the convergence of the two beams towards the LKr centre and the use of a short common decay region for the measurement, the different lifetimes of K_L^0 and K_S^0 would imply a large acceptance correction on the double ratio, as shown in Fig. 10(right). To cancel the bulk of this correction, a weighting factor is applied to each K_L^0 event as a function of its proper time resulting into very similar decay spectra for the two beams. This reduces the Monte Carlo correction to $(26.7 \pm 4.1(\text{stat}) \pm 4.0(\text{syst})) \times 10^{-4}$ at the cost of an increased statistical uncertainty.

4.7.   NA48 results

The result is computed in 20 bins of kaon energy in the range 70–170 GeV. In each bin the double ratio is estimated using the numbers of K_S^0 and proper-time weighted K_L^0 charged and neutral decays and applying the various corrections estimated per bin. Table 5 shows the effect of the various corrections on the raw double ratio, together with their uncertainties for the $1998 + 1999$ result.[37]

The final double ratio is extracted by averaging the 20 numbers with a logarithmic estimator.

The overall NA48 statistics and results for the three run periods are given in Table 6. The combined final result of the experiment is $(14.7 \pm 2.2) \times 10^{-4}$,[38] confirming the observation of direct CP violation with a significance of 6.6 standard deviations.

Table 5 Corrections and systematic uncertainties on the double ratio for the 1998 + 1999 data sample.

	in 10^{-4}
$\pi^+\pi^-$ trigger inefficiency	-3.6 ± 5.2
AKS inefficiency	$+1.1 \pm 0.4$
Reconstruction $\pi^0\pi^0$	0 ± 5.8
Reconstruction $\pi^+\pi^-$	$+2.0 \pm 2.8$
Background $\pi^0\pi^0$	-5.9 ± 2.0
Background $\pi^+\pi^-$	$+16.9 \pm 3.0$
Beam scattering	-9.6 ± 2.0
Accidental tagging	$+8.3 \pm 3.4$
Tagging inefficiency	0 ± 3.0
Acceptance	$+26.7 \pm 4.1 \pm 4.0$
Accidental activity	0 ± 4.4
Long term variations of K_S^0/K_L^0	0 ± 0.6
Total	35.9 ± 12.6

Table 6 Number of selected events after background subtraction and corrected for mistagging for the three data taking periods. The K_L^0 statistics are given without lifetime weighting. The corresponding results on ϵ'/ϵ are also given together with their uncertainties.

	1997	1998 + 1999	2001	Combined
Nb of $K_L \to \pi^0\pi^0$ ($\times 1000$)	489	3290	1546	5325
Nb of $K_S \to \pi^0\pi^0$ ($\times 1000$)	975	5209	2159	8343
Nb of $K_L \to \pi^+\pi^-$ ($\times 1000$)	1071	14453	7136	22660
Nb of $K_S \to \pi^+\pi^-$ ($\times 1000$)	2087	22221	9605	33913
ϵ'/ϵ ($\times 10^{-4}$)	18.5	15.0	13.7	14.7
Stat error ($\times 10^{-4}$)	4.5	1.7	2.5	1.4
Syst error ($\times 10^{-4}$)	5.8	2.1	1.9	1.7
Total error ($\times 10^{-4}$)	7.3	2.7	3.1	2.2

The KTeV experiment at Fermilab published its final result in 2011 giving a combined value for ϵ'/ϵ of $(19.2 \pm 2.1) \times 10^{-4}$,[39] in fair agreement with both NA31 and NA48 results.

5.   Concluding Remarks

5.1.   *The world average of ϵ'/ϵ*

The comparison of the experimental results from CERN and Fermilab is shown in Fig. 11. The world average computed by the Particle Data Group is $(16.8 \pm 2.0) \times 10^{-4}$, where the quoted error is inflated by 1.4 to take into account the dispersion of results. The existence of a direct component in CP Violating amplitudes in the kaon system is experimentally established at a level of 8.4 standard deviations.[40]

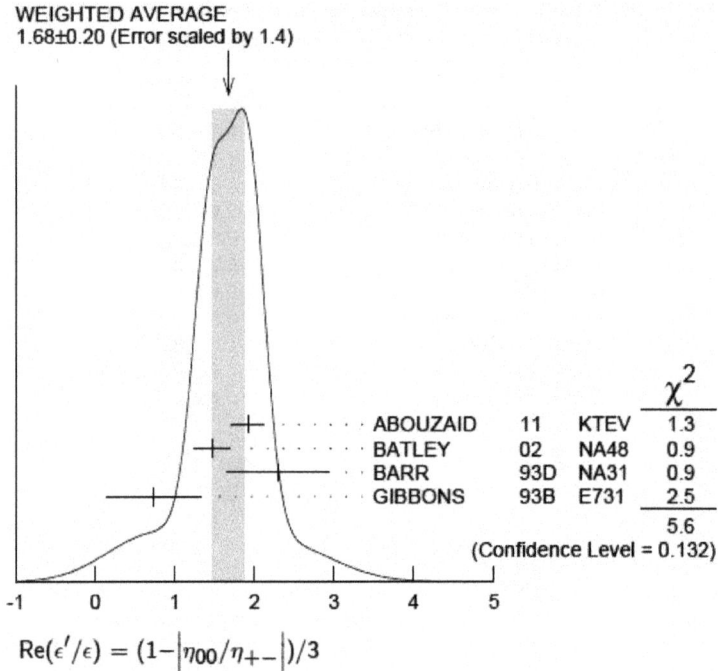

Fig. 11.   Final measurements of ϵ'/ϵ made by the most precise experiments, E731, NA31, KTeV and NA48 together with the average value. The figure is from the 2014 Particle Data Group.[40]

Using formulas (3)–(8), this fundamental result can be expressed as:

$$\frac{\Gamma(K^0 \to \pi^+\pi^-) - \Gamma(\bar{K}^0 \to \pi^+\pi^-)}{\Gamma(K^0 \to \pi^+\pi^-) + \Gamma(\bar{K}^0 \to \pi^+\pi^-)} = 2\mathrm{Re}\ \epsilon' = (5.3 \pm 0.6) \times 10^{-6},$$

$$\frac{\Gamma(K^0 \to \pi^0\pi^0) - \Gamma(\bar{K}^0 \to \pi^0\pi^0)}{\Gamma(K^0 \to \pi^0\pi^0) + \Gamma(\bar{K}^0 \to \pi^0\pi^0)} = -4\mathrm{Re}\ \epsilon' = (-10.6 \pm 1.2) \times 10^{-6}$$

illustrating the occurence of direct CP violation.

5.2.   CP violation in kaons: A portal to heavy meson systems

Since its unexpected discovery in 1964, CP violation opened new horizons in experiments and it has been searched for in heavier meson systems. $D^0\bar{D}^0$ and $B^0\bar{B}^0$ offer a rich phenomenology, which has been exploited in the last 20 years by dedicated experiments at e^+e^- colliders (CESR, LEP, PEP-2, KEK-B) and at $p\bar{p}$ colliders (Tevatron, LHC). Given the high mass of these mesons, many final states are allowed so that the decay widths of the "heavy" and "light" mass eigenstates are much closer than in the kaon system.

In the D system, oscillations due to mixing have been observed and measured recently,[41] with $x = (0.41 \pm 0.14)\%$ and $y = (0.63 \pm 0.07)\%$, where $x = \frac{\Delta M}{\Gamma}$ and

$y = \frac{\Delta\Gamma}{\Gamma}$ with ΔM and $\Delta\Gamma$ being the mass and decay width difference of the two masse eigenstates and $\bar{\Gamma}$ the average decay width. No CP violation was found, in agreement with SM expectations.

The $B^0\bar{B}^0$ phenomenology is much richer. For the B_d^0, $x = 0.774 \pm 0.006$ while y, expected very small in the SM, is experimentally compatible with zero. The first observation of CP violation was made by Babar and Belle[42, 43] in the ψK_S final state where it is due to interference between mixing and decay amplitudes. Direct CP violation has been observed in several final states. For example, the measured asymmetry between $B^0 \to K^+\pi^-$ and $\bar{B}^0 \to K^-\pi^+$ was first observed by the LHCb experiment[44] with a value of -0.082 ± 0.013, widely different from the direct CP violation in the kaon system.

For B_s^0 it took quite a long time before oscillations were observed, because ΔM_s is large compared to Γ_s resulting into oscillations with a very short wave length, beyond the capabilities of early vertex detectors. The first observation was made by CDF.[45] Today this is well established with $x_s = 26.85 \pm 0.13$ and $y_s = 0.137 \pm 0.012$. CP violation in the $B_s^0\bar{B}_s^0$ system has been established as well.[46]

On the rare decay side, a step forward was the observation of the $B_s \to \mu\mu$ mode which has recently been measured with a rate of $(2.9 \pm 0.7) \times 10^{-9}$,[47] compatible with the accurate SM prediction.[48] With additional data this decay will become sensitive to new physics.

5.3. *CP violation in kaons: A portal for theory*

The observation of CP violation fueled spectacular ideas in theory. The CKM formalism and the Standard Model were largely inspired by this discovery. The non-conservation of this symmetry is now a well established phenomenon in the weak decays of hadrons, parametrised by the CKM matrix elements. CP violation was demonstrated by Sakharov[49] in 1967 to be one of the three necessary conditions for baryogenesis. While it is tempting to associate the observed CP violation to the corresponding Sakharov's condition, it is commonly admitted that its magnitude is far too small to play this role. But sources of CP violation other than the CKM matrix are possible. In Higgs-boson mediated transitions with multiple Higgs bosons, CP violation can appear; the value of ϵ in the kaon system imposes however strong constraints on this possibility.[50] Finally, CP violation is also possible and currently being searched for in the neutrino sector with particular emphasis now that $\sin^2 2\theta_{13}$ is known with good precision[51] and has a sizeable value ($\simeq 0.09$).

5.4. *The legacy of CERN kaon experiments*

CERN experiments have had a leading role in the discovery of the direct CP violation component, although the competition with Fermilab was essential. Both NA31 and NA48 groups have improved continuously the beams, the detectors and the analysis methods to better control the systematic uncertainties. In their hunt for precise measurements, the groups developed original tools and methods that are nowadays widely used: the overlay procedure to account for accidental effects, the multiple sampling of calorimetric signals for fast and precise readout, the use of online and offline computer farms, the remote data control, the use of crystal channelling for beam selection and transport, etc.

In parallel with the ϵ'/ϵ data takings, NA31 and NA48 had both undertaken a wide program of studies of rare kaon decays, with original measurements and first observations of several channels.[52-54]

The important contribution of these experiments to the worldwide physics landscape has been awarded by prestigious prizes, given — in particular — to the NA31 collaboration for the first observation of the direct CP component, published in the 1988 paper.[23]

Today, the theoretical precision of ϵ'/ϵ is still unsatisfactory and cannot be usefully confronted with the measurements. Hopes for more accurate predictions are based on ongoing lattice QCD computations.[55]

In the meanwhile, other kaon decay channels are used to push further the limits of the Standard Model. The NA48/2 Collaboration has accumulated in 2003–2004 about $\simeq 4 \times 10^9 K^\pm \rightarrow \pi^\pm \pi^+ \pi^-$ and $\simeq 10^8 K^\pm \rightarrow \pi^\pm \pi^0 \pi^0$ decays in order to study the Dalitz plot of the final states, where direct CP violation could induce an asymmetry. The null result obtained by this experiment is consistent with Standard Model expectations.[56]

Even more interesting would be the study of the very rare decay $K^0 \rightarrow \pi^0 \nu \bar{\nu}$, not yet observed, for which the calculated branching ratio $(2.4 \pm 0.1) \times 10^{-11}$, together with the elusive final state raise extreme experimental challenges. The precision of the SM prediction is however a strong motivation for a dedicated experiment. Several proposals were made in the past; the only one still alive is the E14 proposal at JPARC. Also very clean theoretically, but not dominated by a CP-violating transition, is the $K^+ \rightarrow \pi^+ \nu \bar{\nu}$ mode with an expected rate of about 10^{-10}. Also experimentally challenging, this channel allows a precise measurement of the $|V_{td}|$ CKM parameter. A dedicated experiment, NA62,[57] installed in the EHN2 area at the CERN SPS, had its first run in 2014 and expects to collect in two years about 80 $K^+ \rightarrow \pi^+ \nu \bar{\nu}$ decays. Results from these two very rare decay channels would allow an independent and accurate determination of the Unitarity triangle,[58] probing at the same time physics scenario beyond the Standard Model.

Acknowledgments

The authors are grateful to Ivan Mikulec and Arthur Schaffer for the careful reading and their suggestions. The technical expertise of Catherine Bourge and Bruno Mazoyer contributed efficiently to the preparation of this document.

References

1. J. H. Christenson, J. W. Cronin, V. L. Fitch, R. Turlay, Evidence for the 2π decay of the K_2^0 meson, *Phys. Rev. Lett.* **13**, 138 (1964).
2. S. Bennet, *et al.*, Semileptonic asymmetry, *Phys. Rev. Lett.* **19**, 993 (1967).
3. L. Wolfenstein, Violation of CP invariance and the possibility of very weak interactions, *Phys. Rev. Lett.* **13**, 562 (1964).
4. W. Ochs, πN *Newsletter* **3**, 25 (1991).
5. M. Kobayashi and K. Maskawa, CP violation in the renormalizable theory of weak interactions, *Prog. Theor. Phys.* **49**, 652 (1973).
6. C. Jarlskog, Commutator of the Quark mass matrices in the Standard Electroweak model and a measure of maximal CP nonconservation, *Phys. Rev. Lett.* **55**, 1039 (1985).
7. J. Ellis, M. K. Gaillard and D. V. Nanopoulos, Left-handed currents and CP violation, *Nucl. Phys. B* **109**, 213 (1976).
8. F. J. Gilman and M. B. Wise, The $\Delta = 1/2$ rule and violation of CP in the six-quark model, *Phys. Lett. B* **83**, 83 (1979).
9. B. Guberina and R. Peccei, Quantum Chromodynamics effects and CP violation in the Kobayashi-Maskawa model, *Nucl. Phys. B* **163**, 289 (1980).
10. M. Ciuchini, E. Franco, G. Martinelli, L. Reina and L. Silvestrini, An upgraded analysis of epsilon prime over epsilon at the next to leading order, *Z. Phys. C* **68**, 239 (1995).
11. A. J. Buras, M. Jamin and M. E. Lautenbacher, A 1996 analysis of the CP violating ratio ϵ'/ϵ, arXiv: hep-ph/9608365 (1996).
12. J. M. Flynn and L. Randall, The electromagnetic penguin contribution to epsilon prime over epsilon for large top quark mass, *Phys. Lett. B* **224**, 221 (1989).
13. G. Buchalla, A. J. Buras and M. K. Harlander, The anatomy of epsilon prime over epsilon in the Standard Model, *Nucl. Phys. B* **337**, 313 (1990).
14. M. Holder *et al.*, On the decay $K_L \rightarrow \pi^0\pi^0$, *Phys. Lett. B* **40**, 141 (1972).
15. M. Banner *et al.*, Measurement of $|\eta^{00}/\eta^{+-}|$, *Phys. Rev. Lett.* **28**, 1597 (1972).
16. J. H. Christenson *et al.*, Measurement of the phase and amplitude of η^{00}, *Phys. Rev. Lett.* **43**, 1209 (1979).
17. J. K. Black *et al.*, Measurement of the CP-Nonconservation parameter ϵ'/ϵ, *Phys. Rev. Lett.* **54**, 1628 (1985).
18. R. H. Bernstein *et al.*, Measurement of ϵ'/ϵ in the neutral kaon system, *Phys. Rev. Lett.* **54**, 1631 (1985).
19. M. Woods *et al.*, First results on a new measurement of ϵ'/ϵ in the neutral kaon system, *Phys. Lett. B* **206**, 169 (1988).
20. K. L. Gibbons *et al.*, Measurement of the CP-violation parameter $Re(\epsilon'/\epsilon)$, *Phys. Lett.* **70**, 1203 (1993).
21. G. Barr *et al.*, The beam and detector for a high precision measurement of the CP violation in neutral-Kaon decays , *Nucl. Instr. Meth. A* **268**, 116 (1988).
22. G. D. Barr *et al.*, A large area transition radiation detector, *Nucl. Inst. Meth. A* **294**, 465 (1990).

23. H. Burkhardt *et al.*, First Evidence for direct CP violation, *Phys. Lett. B* **206**, 169 (1988).

24. G. D. Barr *et al.*, A new measurement of direct CP violation in the neutral kaon system, *Phys. Lett. B* **317**, 233 (1993).

25. J. S. Bell and J. Steinberger, in *Proceedings of the Oxford Int. Conf. on Elementary Particles*, 1965, (Oxford University Press, 1966), pp. 195–222.

26. R. Carosi *et al.*, A measurement of the phases of the CP-violating amplitudes in $K \to \pi\pi$ decays and a test of CPT invariance, *Phys. Lett. B* **237**, 303 (1990).

27. E. Abouzaid *et al.*, Precise measurement of direct CP-violation, CPT symmetry, and other Parameters in the neutral kaon system, *Phys. Rev. D* **83**, 092001 (2011).

28. A. Angelopoulos *et al.*, First direct observation of time-reversal non-invariance in the neutral-kaon system, *Phys. Lett. B* **444**, 43 (1998).

29. G. D. Barr *et al.*, Proposal for a precision Measurement of ϵ'/ϵ in CP Violating $K^0 \to 2\pi$ decays, CERN/SPSC/90-22 (1990).

30. V. Fanti *et al.*, The beam and detector for the NA48 neutral kaon CP violation experiment at CERN, *Nucl. Instr. Meth. Phys. Res. A* **574**, 433–471 (2007).

31. N. Doble *et al.*, A novel application of bent crystal channeling to the production of simultaneous particle beams, *Nucl. Instr. Meth. Phys. Res. B* **119**, 181–191 (1996).

32. P. Grafstrom *et al.*, A proton tagging detector for the NA48 experiment, *Nucl. Instr. Meth. Phys. Res. A* **344**, 487–491 (1994).

33. G. D. Barr, *et al.*, Performance of an Electromagnetic Liquid Krypton Calorimeter based on a Ribbon Electrode Tower Structure, *Nucl. Instr. Meth. A* **370**, 507 (1994).

34. W. E. Cleland and E. G. Stern, Signal processing considerations for liquid ionisation calorimeters in a high rate environment , *Nucl. Instr. Meth. A* **338**, 467 (1994).

35. D. Bederede *et al.*, High resolution drift chambers for the NA48 experiment at CERN, *Nucl. Instr. Meth. Phys. Res. A* **367**, 88–91 (1995).

36. A. Peters *et al.*, The NA48 online and offline PC farms, *CHEP 2000* (2000).

37. A. Lai *et al.*, A precise measurement of the direct CP violation parameter Re(ϵ'/ϵ), *Eur. Phys. J. C* **22**, 231–254 (2001).

38. J. R. Batley *et al.*, A precision measurement of direct CP violation in the decay of neutral kaons into two pions, *Phys. Lett. B* **544**, 97–112 (2002).

39. E. Abouzaid *et al.*, Precise measurements of direct CP violation, CPT symmetry and other parameters of the neutral Kaon system, *Phys. Rev. D* **83**, 092001 (2011).

40. Particle Data Group, CP Violation in the quark sector, *Chin. Phys. C* **38**, 945 (2014).

41. HFAG: Charm Physics Parameters, http://www.slac.stanford.edu/xorg/hfag/charm/ FPCP14 (2014).

42. B. Aubert *et al.*, Observation of large CP violation in the neutral B-meson system, *Phys. Rev. Lett.* **87**, 091801 (2001).

43. K. Abe *et al.*, Observation of CP violation in the B^0-meson system, *Phys. Rev. Lett.* **87**, 091801 (2001).

44. R. Aaij *et al.*, First evidence of Direct CP violation in charmless two-body decays of B_s^0 mesons, *Phys. Rev. Lett.* **108**, 201601 (2012).

45. A. Abulencia *et al.*, Measurement of the $B_s - \bar{B}_s$ Oscillation Frequency, *Phys. Rev. Lett.* **97**, 062003 (2006).

46. R. Aaij *et al.*, First observation of CP violation in the decays of B_s mesons, *Phys. Rev. Lett.* **110**, 221601 (2013).

47. V. Khachatryan *et al.*, Observation of the rare $B_S^0 \to \mu^+\mu^-$ decay from the combined analysis of CMS and LHCb data, submitted to *Nature*, arXiv:1411.4413 (2014).

48. C. Bobeth *et al.*, $B_{s,d}^0 \to l^+;^-$ in the Standard Model with reduced theoretical uncertainty, *Phys. Rev. Lett.* **112**, 101801 (2014).

49. A. D. Sakharov, Violation of CP invariance, C asymmetry and baryon asymmetry of the Universe, *JETP Lett.* **6**, 24 (1967).

50. I. I. Bigi and A. I. Sanda, Possible corrections to the KM ansatz: right-handed currents and non-minimal Higgs dynamics, *CP-Violation, Cambridge Monographs on Particle Physics* **28**, 362 (2009).

51. A. B. Balantekin *et al.*, Spectral measurement of electron antineutrino oscillation amplitude and frequency at Daya Bay, *Phys. Rev. Lett.* **112**, 061801 (2013).

52. K. Kleinknecht *et al.*, Results on rare decays of neutral kaons from NA31 experiment at CERN, *Frascati Phys. Ser.* **3**, 377–398 (1994).

53. M. Lenti *et al.*, Kaon and Hyperon rare decays by the NA48 experiment at CERN, hep-exp/0411088 (2004).

54. E. Mazzucato *et al.*, Recent results on CP Violation and rare decays by the NA48 experiment at CERN, *HEP-MAD-2007*, 210 (2007).

55. T. Blum *et al.*, The K$\to (\pi\pi)_{I=2}$ decay amplitude from Lattice QCD, arXiv: 1111.1699 (2011).

56. R. Batley *et al.*, Search for direct CP violating charge asymmetries in the $K^\pm \to \pi^+\pi^+\pi^-$ and $K^\pm \to \pi^0\pi^+\pi^0$ decays, *Eur. Phys. J. C* **52**, 875–891 (2007).

57. E. Cortina Gil *et al.*, NA62 Technical Design Document, NA62-10-07 (2010).

58. K. A. Olive *et al.* (Particle Data Group), The CKM quark-mixing matrix, *Chin. Phys. C* **38**, 090001 (2014), p. 214.

Measurements of Discrete Symmetries in the Neutral Kaon System with the CPLEAR (PS195) Experiment

Thomas Ruf

CERN, CH-1211, Geneva 23, Switzerland

thomas.ruf@cern.ch

The antiproton storage ring LEAR offered unique opportunities to study the symmetries which exist between matter and antimatter. At variance with other approaches at this facility, CPLEAR was an experiment devoted to the study of T, CPT and CP symmetries in the neutral kaon system. It measured with high precision the time evolution of initially strangeness-tagged K^0 and $\overline{K}^0$ states to determine the size of violations with respect to these symmetries in the context of a systematic study. In parallel, limits concerning quantum-mechanical predictions (EPR paradox, coherence of the wave function) or the equivalence principle of general relativity have been obtained. This article will first discuss briefly the unique low energy antiproton storage ring LEAR followed by a description of the CPLEAR experiment, including the basic formalism necessary to understand the time evolution of a neutral kaon state and the main results related to measurements of discrete symmetries in the neutral kaon system. An excellent and exhaustive review of the CPLEAR experiment and all its measurements is given in Ref. 1.

1. The Low Energy Antiproton Ring

The Low Energy Antiproton Ring (LEAR)[2, 3] decelerated and stored antiprotons for eventual extraction to the experiments located in the South Hall. It was built in 1982 and operated until 1996, when it was converted into the Low Energy Ion Ring (LEIR), which provides lead-ion injection for the Large Hadron Collider (LHC). Under the LEAR programme, four machines — the Proton Synchrotron (PS), the Antiproton Collector (AC), the Antiproton Accumulator (AA), and LEAR — worked together to collect, cool and decelerate antiprotons for use in experiments. Protons accelerated to $26\,\text{GeV}/c$ by the PS created antiprotons in collisions with a fixed target. A magnetic spectrometer selected the emerging antiprotons ($3.6\,\text{GeV}/c$) and injected them into the AC. Here they stayed for 4.8 s to reduce their momentum spread by means of stochastic cooling before being stored for a long time in the AA. Whenever the LEAR machine was ready to take a shot ($\approx 5 \times 10^9$) of $\overline{p}$, the AA released a part of its stack to the PS, where the $\overline{p}$'s were decelerated to $609\,\text{MeV}/c$, injected into LEAR, and stochastically cooled down for another 5 min to a momentum spread of $\sigma_p/p = 10^{-3}$. This was followed by electron cooling, resulting in a relative momentum spread of only 5×10^{-4}. LEAR had been equipped with fast and ultra-slow extraction systems, the latter being used for the CPLEAR experiment providing a rate of 1 MHz antiprotons in spills of about 1 h. The last part of extraction line comprised two horizontal and two vertical bending magnets followed by a quadrupole doublet to align and focus the beam on the target in the centre of the detector. The size of the beam spot on the target had a FWHM of about 3 mm.

2. The CPLEAR Experimental Method

CPLEAR made use of charge-conjugate particles K^0 and $\overline{K}^0$ produced in $\overline{p}p$ collisions at rest, with a flavour of strangeness different for particles (K^0) and antiparticles ($\overline{K}^0$):

$$\overline{p}p \rightarrow \frac{K^- \pi^+ K^0}{K^+ \pi^- \overline{K}^0}. \tag{1}$$

The conservation of strangeness in the strong interaction dictates that a K^0 is accompanied by a K^-, and a $\overline{K}^0$ by a K^+. Hence, the strangeness of the neutral kaon at production was tagged by measuring the charge sign of the accompanying charged kaon and was therefore known event by event. If the neutral kaon subsequently decayed to $e\pi\nu$, its strangeness could also be tagged at the decay time by the charge of the decay electron. Indeed, in the limit that only transitions with $\Delta S = \Delta Q$ take place, neutral kaons decay to e^+ if the strangeness is positive at the decay time and to e^- if it is negative. For each initial strangeness, the number of neutral kaon decays was measured as a function of the decay time τ. These numbers, $N_f(\tau)$ and $\overline{N}_f(\tau)$ for a non-leptonic final state f, or $N_\pm(\tau)$ and $\overline{N}_\mp(\tau)$ for an $e\pi\nu$ final state, were combined to form asymmetries, thus dealing mainly with ratios between measured quantities. However, the translation of measured numbers of events into decay rates requires acceptance factors which do not cancel in such asymmetries (a), residual background (b), and regeneration effects (c) to be taken into account.

(a) The major effect arises from the strangeness tagging of the neutral kaon state with the help of detecting and identifying the charge of the accompanying $K^\mp \pi^\pm$ track-pair at the production vertex, and at the decay vertex with the $e^\mp \pi^\pm$ track-pair. Small misalignments of detector components result in different momentum dependent efficiencies for reconstructing positively and negatively charged particles. This effect can be mitigated by changing frequently the polarity of the solenoid magnet, few times per day. A second charge dependent effect arises from different interaction probabilities of particles and antiparticles with the detector material made of matter and not of antimatter. These differences are determined in bins of the kinematics phase-space with large statistics samples of $K^0 \rightarrow \pi^+ \pi^-$ at short decay times for the accompanying $K^\mp \pi^\pm$ track-pair, and with calibration data for the $e^\mp \pi^\pm$ track-pair obtained in a beam at the Paul-Scherrer-Institute (PSI) cyclotron.

(b) The background events mainly consist of neutral kaon decays to final states other than the signal. Since to a high degree of accuracy the amount of background is the same for initial K^0 and $\overline{K}^0$ the contribution cancels in the numerator but not in the denominator of any asymmetry: thus diluting any asymmetry. Their contributions are obtained by Monte Carlo simulations and taken into account by the fits to the asymmetries.

(c) The regeneration probabilities of K^0 and $\overline{K}^0$ propagating through the detector material are not the same, thus making the measured ratio of initial $\overline{K}^0$ and K^0 decay events at time τ different from that expected in vacuum. The effect is called regeneration, since it also leads to the creation of K_S particles when a beam of K_L particles propagates through material, which does not happen in vacuum. A dedicated experimental setup had been used to improve the knowledge on regeneration amplitudes, magnitudes and phases, in the momentum range relevant for the CPLEAR experiment.[4] The effect is being corrected for by applying a weight to each K^0 ($\overline{K}^0$) event equal to the ratio of the decay probabilities for an initial K^0 ($\overline{K}^0$) propagating in vacuum and through the detector.

3.   The CPLEAR Detector

The detector specifications were based on the following essential experimental requirements:

- A very efficient charged kaon identification to separate the signal from the (very) large number of multi-pion annihilation channels.
- To distinguish between the various neutral kaon decay channels.
- To measure the decay proper time between 0 and $\approx$ 20 K_S mean lives. At the highest K^0 momentum measured in our experiment (750 MeV/c), the K_S mean decay length is 4 cm. This sets the size of the cylindrical decay volume to a radius of $\approx$ 60 cm.
- To minimise material to keep the regeneration corrections small, resulting for example in the use of a pressurised hydrogen target instead of liquid hydrogen target.
- To acquire a large number of events, which required both a high rate sophisticated trigger and data acquisition system (1 MHz annihilation rate) and a large geometrical coverage.

Since the antiprotons annihilate at rest, the particles are produced isotropically, thus the detector had a typical near-4π geometry. The whole detector was embedded in a (3.6 m long, 2 m diameter) warm solenoidal magnet which provided a 0.44 T uniform field. The general layout of the CPLEAR experiment is shown in Fig. 1; a comprehensive description of the detector is found in Ref. 5. The incoming

(a)

(b)

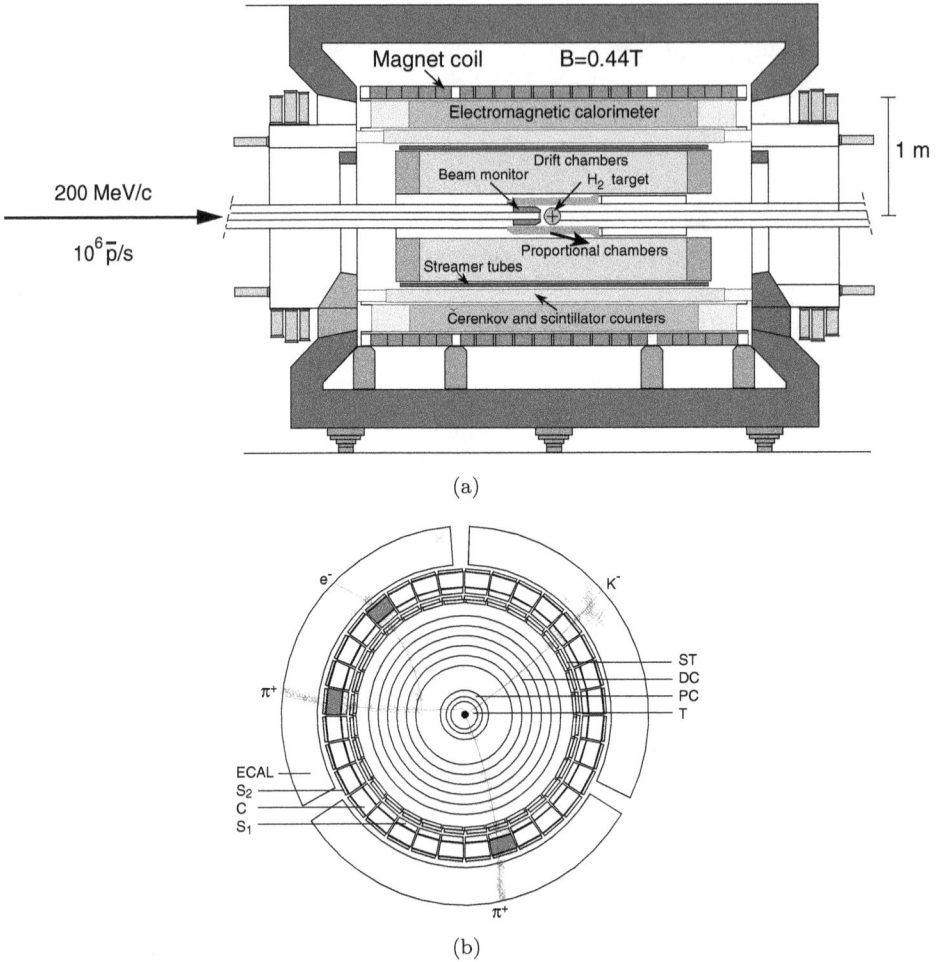

Fig. 1.   CPLEAR detector: (a) longitudinal view and (b) transverse view and display of an event, $\bar{p}p$ (not shown) $\rightarrow K^-\pi^+K^0$ with the neutral kaon decaying to $e^-\pi^+\bar{\nu}$. The view (b) is magnified twice with respect to (a) and does not show the magnet coils and outer detector components. In both views the central region refers to the early data taking without PC0.

antiprotons were stopped in a pressurised hydrogen gas target. For data taken up to mid-1994 the target was a sphere of 7 cm radius at 16 bar pressure. After that date it was replaced by a 1.1 cm radius cylindrical target at 27 bar pressure. A series of light-weighted cylindrical tracking detectors provided information about the trajectories of charged particles in order to determine their charge signs, momenta and positions. There were two proportional chambers (9.5 and 12.7 cm in radius, measuring $r\Phi$), six drift chambers (from 25 to 60 cm, measuring $r\Phi, z$) and two layers of streamer tubes (for a fast z determination within 600 ns). The total material in the target and tracking chambers amounted to $\approx 1\%$ equivalent radiation length (X_0). After trackfit, the spatial resolution was better than $350\,\mu$m in r and

$r\Phi$, and 2 mm in z with sufficient good momentum resolution ($\Delta p/p$ between 5% and 10%).

The tracking detectors were followed by the particle-identification detector (PID), which carried out the charged-kaon identification. The PID comprised a threshold Cherenkov detector, which was mainly effective for K/π separation above 350 MeV/c momentum, and scintillators which measured the energy loss (dE/dx) and the time of flight of charged particles. The PID recognised in ≈ 60 ns the presence of a charged kaon out of a background 250 times higher. The Cherenkov threshold was 300 MeV/c for pions and 700 MeV/c for kaons. The PID was also used to separate electrons from pions below 350 MeV/c.

The outermost detector was a lead/gas sampling calorimeter (ECAL) used to detect the photons produced in π^0 decays. It consisted of 18 layers of 1.5 mm lead converters and high-gain tubes, the latter sandwiched between two layers of pick-up strips ($\pm 30°$ with respect to the tubes), for a total of 64000 readout channels. The design criteria of the calorimeter were mainly dictated by the required accuracy on the reconstruction of the $K^0 \rightarrow 2\pi^0$ or $3\pi^0$ decay vertices. The calorimeter provided e/π separation at higher momenta ($p > 300$ MeV/c) complementary to the PID.

The high annihilation rate and the small value of the branching ratio for the signal reaction ($\approx 2 \times 10^{-3}$) made it necessary to develop a sophisticated trigger and data acquisition system to limit the amount of recorded data and to minimise the dead-time of the experiment. A set of hardwired processors (HWP) was specially designed to reject unwanted events fast and efficient, by providing a full event reconstruction in a few microseconds. The decisions were based on fast recognition of the charged kaon (using the PID hit maps), the number and topology of the charged tracks, the particle identification (using energy-loss, time-of-flight and Cherenkov light response) and kinematic constraints, as well as the number of showers in the ECAL. The overall rejection factor of the trigger was about 10^3, allowing a read-out rate of ≈ 450 events per second at an average beam rate of 800 kHz.

In order to control the bias introduced by the trigger selections, it was essential to confirm the primary $K\pi$ pairs found by the trigger with the primary $K\pi$ pairs found by the offline reconstruction. This matching procedure was achieved by running the trigger simulation on the selected events, requiring the event to pass the trigger criteria with which the data were written, and rejecting events where the trigger and offline reconstruction disagreed on primary tracks. In addition, minimum-bias data, requiring only the coincidence between an incoming $\bar{p}$ signal and a signal in one of the scintillators, were collected at least three times a day, thus providing a representative set of the overall data to be used for calibration purposes and trigger studies.

The CPLEAR detector was fully operational between 1992 and 1996, collecting a total number of antiprotons equal to 1.1×10^{13}. The recorded data of 12 Tbytes consisted of nearly 2×10^8 decays of strangeness-tagged neutral kaons, of which

7×10^7 decays are to $\pi^+ \pi^-$ with a decay time greater than $1\tau_S$, 1.3×10^6 to $e\pi\nu$, 2×10^6 to $\pi^0 \pi^0$, 5×10^5 to $\pi^+ \pi^- \pi^0$, and 1.7×10^4 to $\pi^0 \pi^0 \pi^0$. With these data CPLEAR achieved a number of results on the discrete symmetries in the neutral kaon system,[6-12] and measured other related quantities.[4, 13-15] The large statistics of decays to $\pi^+\pi^-$ allowed testing the equivalence principle of general relativity,[16] by looking at possible annual, monthly and diurnal modulations of the CP violation parameter η_{+-} caused by variations in astrophysical potentials. With a slightly modified setup originally introduced to measure neutral kaon forward scattering cross-sections in carbon,[17] CPLEAR was also able to perform an Einstein–Podolski–Rosen-type experiment.[18] Combining several CPLEAR measurements enabled tests of quantum mechanics,[19] setting limits on parameters describing the possible evolution of pure states into mixed states, sensitive to physics at ultra-high energies, as well as precise determinations of mass and lifetime differences between K^0 and its antiparticle $\overline{K}^0$ using the unitarity relation.[20, 21] Some of these measurements will be discussed in more detail in the following sections after a short introduction into the formalism of the time evolution of neutral kaon states.

4. Phenomenology of the Neutral Kaon System

4.1. *Time evolution*

In the absence of any strangeness-violating interaction, the stationary states $|K^0 \rangle$ and $|\overline{K}^0 \rangle$ of a K^0-meson and $\overline{K}^0$-meson respectively, are mass eigenstates of the strong and electromagnetic interactions and of strangeness S:

$$(\mathcal{H}_{st} + \mathcal{H}_{em})|K^0 \rangle = m_0|K^0 \rangle \qquad (\mathcal{H}_{st} + \mathcal{H}_{em})|\overline{K}^0 \rangle = m_0|\overline{K}^0 \rangle, \qquad (2)$$

$$S|K^0 \rangle = +|K^0 \rangle \qquad S|\overline{K}^0 \rangle = -|\overline{K}^0 \rangle. \qquad (3)$$

Since the strangeness-violating interaction $\mathcal{H}_{wk}$ is much weaker than the strong and electromagnetic interaction, perturbation theory can be applied (Wigner–Weisskopf approach).[22, 23] The time evolution of the neutral kaon wave function is then described by the following differential equation:[24, 25]

$$i\frac{d}{d\tau} \begin{pmatrix} K^0 (\tau) \\ \overline{K}^0 (\tau) \end{pmatrix} = \Lambda \begin{pmatrix} K^0 (\tau) \\ \overline{K}^0 (\tau) \end{pmatrix} = (M - \frac{i}{2}\Gamma) \begin{pmatrix} K^0 (\tau) \\ \overline{K}^0 (\tau) \end{pmatrix}, \qquad (4)$$

where Λ can be split into two Hermitian matrices M and Γ, called mass and decay matrices respectively. The matrix elements of Λ are given by:

$$\Lambda_{ij} = m_0\delta_{ij} + \langle i|\mathcal{H}_{wk}|j\rangle + \sum_f \mathcal{P} \left(\frac{\langle i|\mathcal{H}_{wk}|f\rangle\langle f|\mathcal{H}_{wk}|j\rangle}{m_0 - E_f} \right)$$

$$- i\pi \sum_f \langle i|\mathcal{H}_{wk}|f\rangle\langle f|\mathcal{H}_{wk}|j\rangle\delta(m_0 - E_f), \qquad (5)$$

where $\mathcal{P}$ stands for the principal part and the indices $i, j = 1$ and $i, j = 2$ correspond to K^0 and $\overline{K}^0$ respectively. They can be calculated within the Standard Model, although with large uncertainties because of non-perturbative effects. The same formalism applies to the two neutral B-meson systems (B_d and B_s) where the matrix elements of Λ can be calculated rather reliable due to the much larger mass of the b-quark compared to the s-quark. Since no direct K^0–$\overline{K}^0$ transition exists within the Standard Model, the second term of Eq. (5) vanishes. We use the following parametrisation of Λ with eight real and positive parameters:

$$
\Lambda = \begin{pmatrix} m_{K^0} & M_{12}e^{i\varphi_M} \\ M_{12}e^{-i\varphi_M} & m_{\overline{K}^0} \end{pmatrix} - \frac{i}{2} \begin{pmatrix} \Gamma_{K^0} & \Gamma_{12}e^{i\varphi_\Gamma} \\ \Gamma_{12}e^{-i\varphi_\Gamma} & \Gamma_{\overline{K}^0} \end{pmatrix},
\tag{6}
$$

where m_{K^0} and $m_{\overline{K}^0}$ are equal to the masses, and $1/\Gamma_{K^0}$ and $1/\Gamma_{\overline{K}^0}$ to the lifetimes of the K^0 and $\overline{K}^0$ states respectively.

The time evolution of initially pure K^0 and $\overline{K}^0$ states is given by

$$
\begin{pmatrix} K^0(\tau) \\ \overline{K}^0(\tau) \end{pmatrix} = T(\tau) \begin{pmatrix} K^0(0) \\ \overline{K}^0(0) \end{pmatrix}
\tag{7}
$$

with

$$
T(\tau) = \begin{pmatrix} f_+(\tau) + \dfrac{\Lambda_{22} - \Lambda_{11}}{\Delta\lambda} f_-(\tau) & -2\dfrac{\Lambda_{21}}{\Delta\lambda} f_-(\tau), \\[3mm] f_+(\tau) - \dfrac{\Lambda_{22} - \Lambda_{11}}{\Delta\lambda} f_-(\tau) & -2\dfrac{\Lambda_{12}}{\Delta\lambda} f_-(\tau) \end{pmatrix},
\tag{8}
$$

$$
f_\pm(\tau) = \frac{e^{-i\lambda_S\tau} \pm e^{-i\lambda_L\tau}}{2},
$$

$$
\lambda_{L,S} = m_{L,S} - \frac{i}{2}\Gamma_{L,S} = \frac{\Lambda_{11} + \Lambda_{22}}{2} \pm \sqrt{\frac{(\Lambda_{22} - \Lambda_{11})^2}{4} + \Lambda_{12}\Lambda_{21}},
$$

$$
\Delta\lambda = \lambda_L - \lambda_S = \sqrt{(\Lambda_{22} - \Lambda_{11})^2 + 4\Lambda_{12}\Lambda_{21}},
$$

where λ_S and λ_L are the eigenvalues of the matrix Λ. The corresponding eigenvectors are given by:

$$
|K_S\rangle = \frac{e^{i\varphi_S}}{\sqrt{1 + |r_S|^2}} \left(r_S|K^0\rangle + |\overline{K}^0\rangle \right),
$$

$$
|K_L\rangle = \frac{e^{i\varphi_L}}{\sqrt{1 + |r_L|^2}} \left(r_L|K^0\rangle + |\overline{K}^0\rangle \right)
\tag{9}
$$

with arbitrary phases φ_S, φ_L and

$$
r_S = \frac{2M_{12}}{\Lambda_{22} - \Lambda_{11} - \Delta\lambda},
\tag{10}
$$

$$
r_L = \frac{2M_{12}}{\Lambda_{22} - \Lambda_{11} + \Delta\lambda}.
$$

4.2. *Discrete symmetries*

CP and CPT transformations change a stationary K^0 state into a $\overline{K}^0$ state and vice versa, whereas a T transformation does not alter the states except for an arbitrary phase:

$$CP\,|K^0\,\rangle = \qquad e^{i\phi_{CP}}|\overline{K}^0\,\rangle, \quad CP\,|\overline{K}^0\,\rangle = \qquad e^{-i\phi_{CP}}|K^0\,\rangle,$$

$$T\,|K^0\,\rangle = \qquad e^{i\phi_T}|K^0\,\rangle, \quad T\,|\overline{K}^0\,\rangle = \qquad e^{i\overline{\phi}_T}|\overline{K}^0\,\rangle, \tag{11}$$

$$CPT\,|K^0\,\rangle = e^{i(\phi_{CP}+\phi_T)}|\overline{K}^0\,\rangle, \, CPT\,|\overline{K}^0\,\rangle = e^{i(-\phi_{CP}+\overline{\phi}_T)}|K^0\,\rangle.$$

By requiring $CPT\,|K^0\,\rangle = T\,CP\,|K^0\,\rangle$, it follows for the phases ϕ_i that

$$2\phi_{CP} = \overline{\phi}_T - \phi_T\,. \tag{12}$$

If Λ is invariant under T, CPT or CP transformations, the following conditions must be satisfied:

$$\begin{aligned}
T &\ : |\Lambda_{12}| = |\Lambda_{21}|, \\
CPT &: \Lambda_{11} = \Lambda_{22}, \\
CP &\ : |\Lambda_{12}| = |\Lambda_{21}| \quad \text{and} \quad \Lambda_{11} = \Lambda_{22}.
\end{aligned} \tag{13}$$

It is convenient to introduce the following T and CPT violation parameters:

$$\varepsilon_T \equiv \sin\left(\varphi_{SW}\right) \frac{|\Lambda_{12}|^2 - |\Lambda_{21}|^2}{\Delta\Gamma\Delta m} e^{i\varphi_{SW}}, \tag{14}$$

$$\delta \equiv \cos\left(\varphi_{SW}\right) \frac{\Lambda_{22} - \Lambda_{11}}{\Delta\Gamma} e^{i(\varphi_{SW}+\pi/2)} \tag{15}$$

with $\varphi_{SW} = \operatorname{atan}\left(2\Delta m/\Delta\Gamma\right)$. The lifetime difference is found to be about twice the mass difference $\Delta m \equiv m_L - m_S$ and therefore $\varphi_{SW} \approx 45°$. Assuming small T and CPT violation, the time evolution of initially-pure strangeness states can be rewritten as

$$|K^0\,(\tau)\rangle = [f_+(\tau) - 2\delta f_-(\tau)]\,|K^0\,\rangle + (1 - 2\varepsilon_T)\,e^{-i\varphi_\Gamma}\,f_-(\tau)|\overline{K}^0\,\rangle, \tag{16}$$

$$|\overline{K}^0\,(\tau)\rangle = [f_+(\tau) + 2\delta f_-(\tau)]\,|\overline{K}^0\,\rangle + (1 + 2\varepsilon_T)\,e^{i\varphi_\Gamma}\,f_-(\tau)|K^0\,\rangle. \tag{17}$$

Additional violations of discrete symmetries may occur in the decay of particles, either

1 through the interference of a decay amplitude with the oscillation amplitude, i.e. the phase of the decay amplitude is different from φ_Γ,
2 through the interference of two decay amplitudes with different weak phases,
3 or through direct CPT in a decay amplitude.

The neutral kaon system is rather special compared to the other neutral meson systems (D^0, B_d^0 and B_s^0), in the sense that due to the low mass of the kaon, the number of different final states is rather limited. This enables a rather complete systematic study of CP in the neutral kaon system. And in addition, one decay amplitude ($K^0 \to \pi\pi$, $I = 0$) dominates over all other decay amplitudes. This makes the effects of 1 and 2 very small in the kaon system while they are dominating in the B systems.

In case, there is one amplitude contributing to the instant decay of a neutral kaon to a final state f, this can then be described for K^0 and $\overline{K}^0$ by the amplitudes $\mathcal{A}_f$, $\overline{\mathcal{A}}_{\overline{f}}$,

$$\mathcal{A}_f = \langle f|\mathcal{H}_{\mathrm{wk}}|K^0\rangle, \quad \overline{\mathcal{A}}_{\overline{f}} = \langle \overline{f}|\mathcal{H}_{\mathrm{wk}}|\overline{K}^0\rangle,$$

$$= (A_f + B_f)\,e^{i\delta_f}, \quad \overline{\mathcal{A}}_{\overline{f}} = \left(A_f^* - B_f^*\right)e^{i\delta_f}. \tag{18}$$

The amplitudes A_f and B_f are CPT symmetric and antisymmetric, respectively, δ_f is a strong phase describing a possible final state interaction. Alternatively we can express the rates in terms of K_S and K_L decay amplitudes, which is common in case f is a CP eigenstate:

$$\mathcal{A}_{fS} = \langle f|\mathcal{H}_{\mathrm{wk}}|K_S\rangle, \quad \mathcal{A}_{fL} = \langle f|\mathcal{H}_{\mathrm{wk}}|K_L\rangle. \tag{19}$$

It is then possible for example to calculate the time dependent decay rates into $f = \pi^+\pi^-$ as:

$$R_{K^0 \to \pi\pi}(\tau) = B\left[e^{-\tau/\tau_S} + |\eta_{+-}|^2 e^{-\tau/\tau_L} + 2|\eta_{+-}|e^{-\overline{\Gamma}\tau}\cos(\Delta m\tau - \phi_{+-})\right],$$
$$\tag{20}$$
$$R_{\overline{K}^0 \to \pi\pi}(\tau) = \overline{B}\left[e^{-\tau/\tau_S} + |\eta_{+-}|^2 e^{-\tau/\tau_L} - 2|\eta_{+-}|e^{-\overline{\Gamma}\tau}\cos(\Delta m\tau - \phi_{+-})\right]$$

with

$$\eta_{+-} = \frac{\langle f|\mathcal{H}_{\mathrm{wk}}|K_L\rangle}{\langle f|\mathcal{H}_{\mathrm{wk}}|K_S\rangle}, \quad \overline{B}/B = [1 + 4\Re(\varepsilon_T + \delta)]. \tag{21}$$

Since the $\pi^+\pi^-$, $\pi^0\pi^0$ final states are governed by isospin $I = 0$ and $I = 2$ amplitudes, with $|A_2/A_0| \approx 0.045$,[26] the different contributions to η_{+-} and η_{00} are given by:

$$\eta_{+-} = \varepsilon + \varepsilon' \quad \text{and} \quad \eta_{00} = \varepsilon - 2\varepsilon' \tag{22}$$

$$\text{with} \quad \varepsilon = \varepsilon_T + \delta + i\Delta\phi + \Delta A,$$

where $\varepsilon_T + \delta$ represent T and CPT respectively in mixing, ε' direct CP through interference of $I = 0$ and $I = 2$ amplitudes, $\Delta\phi$ CP through interference between mixing and $I = 0$ decay amplitude, and ΔA represents CPT in the dominating $I = 0$ amplitude.

4.3. Measurement of CP violation in the decay to $\pi^+\pi^-$

The CPLEAR measurement of the decay rate asymmetry Fig. 2) shows that large rate differences between K^0 and $\overline{K}^0$ occur between 8 and 16 K_S lifetimes, despite

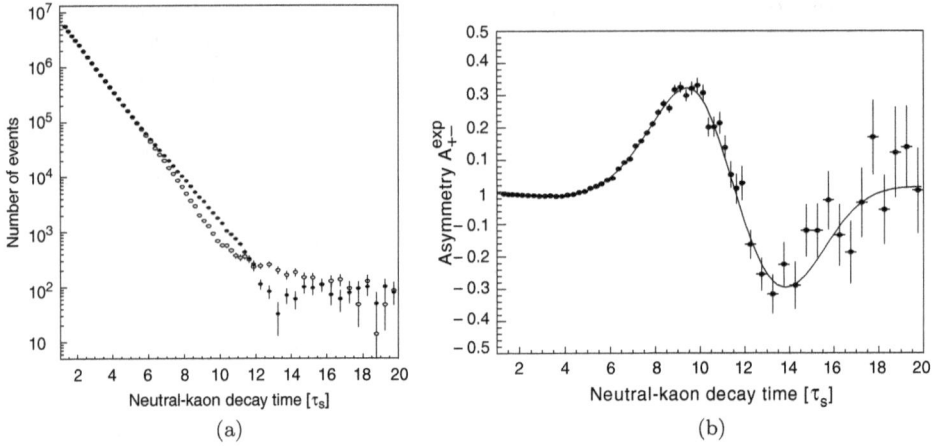

Fig. 2. Decay to $\pi^+\pi^-$: (a) The measured decay rate (acceptance corrected and background subtracted) as a function of the decay time τ, separately for K^0 (open circle) and $\overline{K}^0$ (black circle). (b) The data points (black circle) are the measured time dependent decay-rate asymmetry $A_{+-}(\tau)$. The continuous curve is the result of the best fit (Eq. (23)).

the fact that $|\eta_{+-}|$ is only about $\approx 2.3 \times 10^{-3}$. The measured decay rates need to be corrected for charge asymmetric detection efficiencies of the accompanying primary $K^\pm\pi^\mp$ pair, which is done using the high statistics data of the $\pi^+\pi^-$ mode at short decay times, where contributions due to $\mathcal{CP}$ described by η_{+-} is known with sufficient accuracy. However, this only allows one to determine $w = [1+4\Re(\varepsilon_T+\delta)]\xi$, with ξ describing the detector effects. The experimental measured asymmetry then becomes:

$$A^{+-}(\tau) = \frac{R(\overline{K}^0 \to \pi^+\pi^-) - k*w\,R(K^0 \to \pi^+\pi^-)}{R(\overline{K}^0 \to \pi^+\pi^-) + k*w\,R(K^0 \to \pi^+\pi^-)}$$

$$= -2\frac{|\eta_{+-}|e^{\frac{1}{2}(1/\tau_S - 1/\tau_L)\tau}\cos(\Delta m \times \tau - \phi_{+-})}{1 + [|\eta_{+-}|^2 + \mathrm{Bck}(\tau)]e^{(1/\tau_S - 1/\tau_L)\tau}} \qquad (23)$$

with $\mathrm{Bck}(\tau)$ describing the residual background contributions mainly from semileptonic decays, k a free parameter of the fit accounting for the statistical uncertainties in the normalisation weights and for the correlations between the magnitudes of these weights and the fitted $\mathcal{CP}$-violation parameters. Using the 1998 PDG average values for Δm, τ_S and τ_L,[27] the final CPLEAR result is:

$$|\eta_{+-}| = (2.264 \pm 0.023_{\mathrm{stat}} \pm 0.026_{\mathrm{syst}} \pm 0.007_{\tau_S}) \times 10^{-3},$$

$$\phi_{+-} = 43.19^\circ \pm 0.53^\circ{}_{\mathrm{stat}} \pm 0.28^\circ{}_{\mathrm{syst}} \pm 0.42^\circ{}_{\Delta m}. \qquad (24)$$

The improved precision in the value of the phase ϕ_{+-} had been an important ingredient for setting a limit to a possible $\mathcal{CPT}$ violating K^0–$\overline{K}^0$ mass difference, see Section 4.5.

4.4. *Direct measurements of the T and CPT violation parameters*

Semileptonic decays of neutral kaons have the distinctive feature that the charge of the lepton tags the strangeness at the time of the decay ($K^0 \to \pi^- l^+ \nu$ and $\overline{K}^0 \to \pi^0 l^- \overline{\nu}$). Within the standard model, $\Delta S = \Delta Q$ violating decays ($K^0 \to \pi^0 l^- \overline{\nu}$ and $\overline{K}^0 \to \pi^- l^+ \nu$) are expected to be heavily suppressed (10^{-7}; Ref. 28) and have not been observed so far, only upper limits have been measured. This allows one to measure for example, very precisely the oscillation frequency of an originally K^0 state to change to a $\overline{K}^0$ state,[13] and moreover observe directly T violation by measuring the rate asymmetry between a K^0 decaying as $\overline{K}^0$ and its T-conjugated process, $\overline{K}^0$ decaying as K^0.[10]

In the absence of $\Delta S = \Delta Q$ violating processes, the time dependent decay rate asymmetry A_T measures directly the difference in magnitude of the off-diagonal elements of Λ without any assumption about the smallness of CP and the magnitude of $\Delta \Gamma$:

$$A_T(\tau) = \frac{R(\overline{K}^0 \to K^0)(\tau) - R(K^0 \to \overline{K}^0)(\tau)}{R(\overline{K}^0 \to K^0)(\tau) + R(K^0 \to \overline{K}^0)(\tau)} = \frac{|\Lambda_{12}|^2 - |\Lambda_{21}|^2}{|\Lambda_{12}|^2 + |\Lambda_{21}|^2} = 4\Re(\varepsilon_T).$$

(25)

With $\Delta S = \Delta Q$ violating processes, three more parameters related to the semileptonic decay amplitudes appear in the formalism of semileptonic decay rate asymmetries: $\Re(y)$ describing direct CPT violation in the $\Delta S = \Delta Q$ allowed decay, (x_+) CP violating and CPT conserving and (x_-) CPT violating contributions through $\Delta S = \Delta Q$ violating amplitudes. For a detailed definition see Ref. 1, Section 2.2. A_T then becomes:

$$A_T(\tau) = 4\Re(\varepsilon_T) - 2\Re(x_- + y)$$

$$+ 2\frac{\Re(x_-)\left(e^{-(1/2)\Delta\Gamma\tau} - \cos(\Delta m \tau)\right) + \Im(x_+)\sin(\Delta m \tau)}{\cosh(\frac{1}{2}\Delta\Gamma\tau) - \cos(\Delta m \tau)}$$

$$\longrightarrow 4\Re(\varepsilon_T) - 2\Re(x_- + y) \quad \text{for } \tau \gg \tau_S.$$

(26)

In addition, correcting for charge depending detector asymmetries affecting the detection of the accompanying primary particles ($K^\pm \pi^\mp$) using the $\pi^+ \pi^-$ data at early lifetimes, yields an additional contribution to the asymmetry of $2\Re(\varepsilon_T + \delta)$. Using high precision measurements of the semileptonic decay asymmetry,[27] $\delta_l = 2\Re(\varepsilon_T + \delta - y - x_-) = (3.27 \pm 0.12) \times 10^{-3}$, this results in:

$$A_T{}^{\text{exp}}(\tau) = 4\Re(\varepsilon_T) - \Re(x_- + y)$$

$$+ 2\frac{\Re(x_-)\left(e^{-(1/2)\Delta\Gamma\tau} - \cos(\Delta m \tau)\right) + \Im(x_+)\sin(\Delta m \tau)}{\cosh(\frac{1}{2}\Delta\Gamma\tau) - \cos(\Delta m \tau)}.$$

(27)

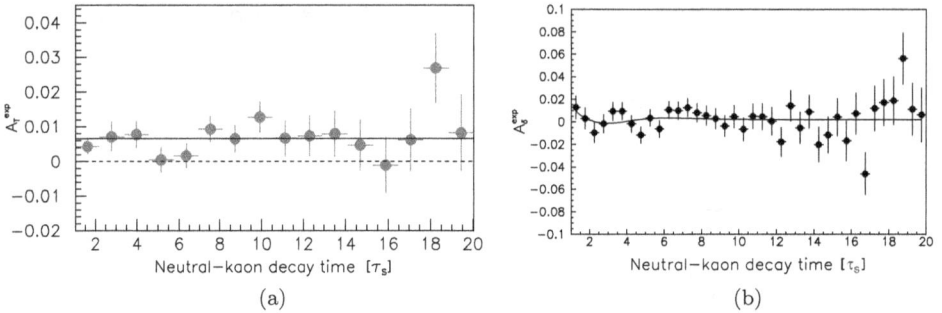

Fig. 3.　(a) Experimental demonstration of T-violation: the asymmetry A_T^{exp} versus the neutral-kaon decay time (in units of τ_S). The positive values show that a $\overline{\mathrm{K}}^0$ develops into a K^0 with higher probability than does a K^0 into a $\overline{\mathrm{K}}^0$. The solid line represents the fitted average $\langle A_T^{\mathrm{exp}} \rangle = (6.6 \pm 1.3) \times 10^{-3}$. (b) The experimentally measured $\mathcal{CPT}$ violating asymmetry A_δ. The solid line represents the result of the fit.

In the original publication of the A_T asymmetry,[10] an assumption about $\mathcal{CPT}$ invariance in the semileptonic decay amplitudes was made when fitting the experimental data (Fig. 3) resulting in:

$$4\Re(\varepsilon_T) = (6.2 \pm 1.4_{\mathrm{stat}} \pm 1.0_{\mathrm{syst}}) \times 10^{-3},$$

$$\Im(x_+) = (1.2 \pm 1.9_{\mathrm{stat}} \pm 0.9_{\mathrm{syst}}) \times 10^{-3} \tag{28}$$

having observed directly for the first time $\mathcal{T}$ violation at work. Compiling the CPLEAR data together with other world averages for some of the neutral kaon parameters, together with the Bell–Steinberger (or unitarity) relation,[20] constraints the quantity $\Re(x_- + y)$ to be within $(-0.2 \pm 0.3) \times 10^{-3}$ confirming the assumption that the possible contribution to A_T^{exp} from $\mathcal{CPT}$-violating decay amplitudes is negligible. Until today (2014), this is the only direct observation of $\mathcal{T}$ violation in the mixing of neutral mesons.

A similar asymmetry can be constructed for the case of $\mathcal{CPT}$, for simplicity assume absence of $\Delta S = \Delta Q$ violating processes:

$$A_{\mathrm{CPT}}(\tau) = \frac{R(\overline{\mathrm{K}}^0 \to \overline{\mathrm{K}}^0)(\tau) - R(\mathrm{K}^0 \to \mathrm{K}^0)(\tau)}{R(\overline{\mathrm{K}}^0 \to \overline{\mathrm{K}}^0)(\tau) + R(\mathrm{K}^0 \to \mathrm{K}^0)(\tau)}$$

$$= 2\Re(y) + 4\frac{\Re(\delta)\sinh\left(\frac{1}{2}\Delta\Gamma\tau\right) + \Im(\delta)\sin(\Delta m\tau)}{\cosh(\frac{1}{2}\Delta\Gamma\tau) + \cos(\Delta m\tau)}. \tag{29}$$

Including $\Delta S = \Delta Q$ violating contributions and correcting for the primary charge asymmetry as before with the 2π data, if finally turns out that a direct measurement of $\mathcal{CPT}$ can be obtained by combining the two asymmetries 26 and

29 to become:

$$
\begin{aligned}
A_\delta(\tau) &= A_{\text{CPT}}^{\text{exp}}(\tau) + A_{\text{T}}^{\ \text{exp}}(\tau) \\
&= 4\Re(\delta) + 4\frac{\Re(\delta)\sinh(\frac{1}{2}\Delta\Gamma\tau) + \Im(\delta)\sin(\Delta m\tau)}{\cosh(\frac{1}{2}\Delta\Gamma\tau) + \cos(\Delta m\tau)}
\end{aligned}
$$

$$
- 4\frac{\Re(x_-)\cos(\Delta m\tau)\sinh(\frac{1}{2}\Delta\Gamma\tau) - \Im(x_+)\sin(\Delta m\tau)\cosh(\frac{1}{2}\Delta\Gamma\tau)}{[\cosh(\frac{1}{2}\Delta\Gamma\tau)]^2 - [\cos(\Delta m\tau)]^2}
$$

(30)

$$
\longrightarrow 8\Re(\delta) \quad \text{for } \tau \gg \tau_{\text{S}}.
$$

The final fit results are:

$$
\Re(\delta) = (3.0 \pm 3.3_{\text{stat}} \pm 0.6_{\text{sys}}) \times 10^{-4},
$$

$$
\Im(\delta) = (-1.5 \pm 2.3_{\text{stat}} \pm 0.3_{\text{sys}}) \times 10^{-2},
$$

$$
\Re(x_-) = (0.2 \pm 1.3_{\text{stat}} \pm 0.3_{\text{sys}}) \times 10^{-2},
$$

$$
\Im(x_+) = (1.2 \pm 2.2_{\text{stat}} \pm 0.3_{\text{sys}}) \times 10^{-2}.
$$

4.5. $\mathcal{T}$ and $\mathcal{CPT}$ parameters constrained by the unitarity relation

As mentioned earlier, the neutral kaon system is unique in the sense that due to the rather limited number of final states, Eq. (5) can directly be used as a constraint by summing up all relevant final states. By improving the precision of the three-pion decay rates[8, 9] and measuring precisely the semileptonic decay rates,[12] CPLEAR made possible the determination of many parameters of the neutral kaon system with unprecedented accuracy. Rewriting Eq. (5) in the $K_S - K_L$ basis, we derive the well known Bell–Steinberger relation[29, 30] relating all decay channels of neutral kaons to the parameters describing $\mathcal{T}$ and $\mathcal{CPT}$ non-invariance in the neutral kaon mixing:

$$
\Re(\varepsilon_{\text{T}}) - i\Im(\delta) = \frac{1}{2(i\Delta m + \overline{\Gamma})} \times \sum \langle f|\mathcal{H}_{\text{wk}}|K_L\rangle^* \langle f|\mathcal{H}_{\text{wk}}|K_S\rangle .
$$

(31)

The sum on the right-hand side of the above equation can be written as

$$
\sum \langle f|\mathcal{H}_{\text{wk}}|K_L\rangle^* \langle f|\mathcal{H}_{\text{wk}}|K_S\rangle = \sum (\text{BR}^S_{\pi\pi}\Gamma_S\eta_{\pi\pi}) + \sum (\text{BR}^L_{\pi\pi\pi}\Gamma_L\eta^*_{\pi\pi\pi})
$$

$$
+ 2\left[\Re(\varepsilon_{\text{T}}) - \Re(y) - i\left(\Im(x_+) + \Im(\delta)\right)\right]\text{BR}^L_{l\pi\nu}\Gamma_L .
$$

Here BR stands for branching ratio, the upper index refers to the decaying particle and the lower index to the final state and l denotes electrons and muons. The radiative modes, like $\pi^+\pi^-\gamma$, are essentially included in the corresponding parent modes. Channels with BR^S_f (or $\text{BR}^L_f \times \Gamma_L/\Gamma_S$) $< 10^{-5}$ do not contribute to Eq. (31) within the accuracy of the CPLEAR analysis. Using data from CPLEAR together with the most recent world averages (1998) for some of the neutral kaon parameters, the following result is being obtained:[20]

$$\Re(\varepsilon_T) = (164.9 \pm 2.52_{\text{stat}} \pm 0.1_{\text{sys}}) \times 10^{-5},$$

$$\Im(\delta) = (2.4 \pm 5.02_{\text{stat}} \pm 0.1_{\text{sys}}) \times 10^{-5},$$

$$\Re(\delta) = (2.4 \pm 2.72_{\text{stat}} \pm 0.6_{\text{sys}}) \times 10^{-4},$$

which establishes unambiguously $\mathcal{T}$ violation at the level of 65σ and sets stringent limits on $\mathcal{CPT}$ in mixing but also in various decay amplitudes, for more results see Ref. 20. The unitarity relation in the K^0–$\overline{K}^0$ basis can also be used to derive a limit on the phase difference between Γ_{12} and the dominating $I = 0$ decay amplitude in the $\pi\pi$ mode. In the neutral B-system, such a phase difference which corresponds to the interference of mixing and decay amplitudes is the dominating source for $\mathcal{CP}$ violation. In the kaon system, it is small $\Delta\Phi = \frac{1}{2}[\varphi_\gamma - \arg(A_0^*\overline{A}_0)] = (-1.2 \pm 8.5) \times 10^{-6}$.[31]

The $\mathcal{CPT}$ theorem,[32-34] which is based on general principles of the relativistic quantum field theory, states that any order of the triple product of the discrete symmetries $\mathcal{C}$, $\mathcal{P}$ and $\mathcal{T}$ should represent an exact symmetry. The theorem predicts, among other things, that particles and antiparticles have equal masses and lifetimes. With the above results for $\Re(\delta)$ and $\Im(\delta)$ and using Eq. (15), it is straightforward to obtain:

$$\Gamma_{K^0 K^0} - \Gamma_{\overline{K}^0 \overline{K}^0} = (3.9 \pm 4.2) \times 10^{-18}\,\text{GeV},$$

$$M_{K^0 K^0} - M_{\overline{K}^0 \overline{K}^0} = (-1.5 \pm 2.0) \times 10^{-18}\,\text{GeV}, \tag{32}$$

with a correlation coefficient of -0.95. In contrast to earlier compilations, for example Ref. 27, the CPLEAR results are free of any prejudice of $\mathcal{CPT}$ invariance in decay amplitudes. Assuming $\mathcal{CPT}$ invariance in all decays, the precision on the mass difference $(-0.7 \pm 2.8) \times 10^{-19}\,\text{GeV}$ improves by about one order of magnitude. These are still the best limits for a mass difference between particles and antiparticles, thanks to the small value of $\Delta m = 3.484 \times 10^{-12}\,\text{MeV}$ which works like a magnification glass. In the neutral B-systems, the mass differences are ≈ 100 and ≈ 300 larger for the B_d and B_s respectively compared to the kaon system and therefore B-systems are less sensitive to $\mathcal{CPT}$ effects.

4.6.　*Measurements related to basic principles*

In the last section, I would like to discuss three CPLEAR results[16, 18, 19] which are related to the basics of Quantum Mechanics (QM) and general relativity.

4.6.1.　*Probing a possible loss of QM coherence*

All results discussed so far are based on a framework of QM of closed systems, solutions of Eq. (4) are pure states and evolve as such in time. Some approaches to quantum gravity[35] suggest that topologically non-trivial space–time fluctuations (space–time foam, virtual black-holes) entail an intrinsic, fundamental information loss, and therefore induce transitions from pure to mixed states,[36] and define the arrow of time. In the K^0–$\overline{K}^0$ system such a behavior can be described by a phenomenological ansatz using a 2×2 density matrix ρ, which obeys

$$\dot{\rho} = -i[\Lambda \rho - \rho \Lambda^\dagger] + \delta/\Lambda\rho \qquad (33)$$

where Λ is the 2×2 matrix of Eq. (6), and the term $\delta\!\!\!/\Lambda\,\rho$ induces a loss of quantum coherence in the observed system. In the case of the neutral kaon system, if the conservation of energy and strangeness are assumed, the open-system equation (33) introduces[36] three $\mathcal{CPT}$-violation parameters α, β and γ. Before CPLEAR, existing measurements of $\mathcal{CP}$ violation in the mixing of neutral kaons could have been solely explained by these $\mathcal{CPT}$-violation parameters. Having measured decay-rate asymmetries over a large range of lifetimes ($\sim 20\tau_S$) for the $\pi^+\pi^-$ and $e\pi\nu$ decay channels together with the constraint of $|\eta_{+-}|$ and δ_l measured at long lifetimes,[27] enabled CPLEAR to obtain 90% CL limits, $\alpha < 4.0 \times 10^{-17}\,\text{GeV}$, $\beta < 2.3 \times 10^{-19}\,\text{GeV}$ and $\gamma < 3.7 \times 10^{-21}\,\text{GeV}$, to be compared with a possible order of magnitude of $\mathcal{O}(m_K^2/m_{\text{Planck}}) = 2 \times 10^{-20}\,\text{GeV}$ for such effects if relevant for our universe.

4.6.2. *Testing the non-separability of the $K^0\overline{K}^0$ wave function*

For this measurement, pairs of $K^0\overline{K}^0$ were selected being produced simultaneously in the reaction:

$$\overline{p}p \to K^0\overline{K}^0 . \qquad (34)$$

Depending on the angular momentum between the K^0 and $\overline{K}^0$, the wave function describing the time evolution of the two entangled states is either symmetric or antisymmetric with respect of changing $K^0 \leftrightarrow \overline{K}^0$. However, it turns out that in 93% of the cases,[37] the wave function is antisymmetric with $J^{PC} = 1^{--}$:

$$\langle\Psi(0,0)| = \frac{1}{\sqrt{2}}[\langle K^0|_a\langle\overline{K}^0|_b - \langle\overline{K}^0|_a\langle K^0|_b] \qquad (35)$$

Switching on the time evolution of the neutral kaons, Eq. (8), and separating into combinations of unlike and like-strangeness at time t_a and t_b yields:

$$\langle\Psi(t_a,t_b)|_{K^0\overline{K}^0} \propto T_{11}(t_a)T_{22}(t_b) - T_{12}(t_a)T_{21}(t_b),$$

$$\langle\Psi(t_a,t_b)|_{\overline{K}^0\overline{K}^0} \propto T_{21}(t_a)T_{22}(t_b) - T_{22}(t_a)T_{21}(t_b).$$

From which follows the prediction of QM, independent of any $\mathcal{CP}$ in the mixing of neutral kaons: at equal times, the probability to observe the two states with equal strangeness goes to zero.

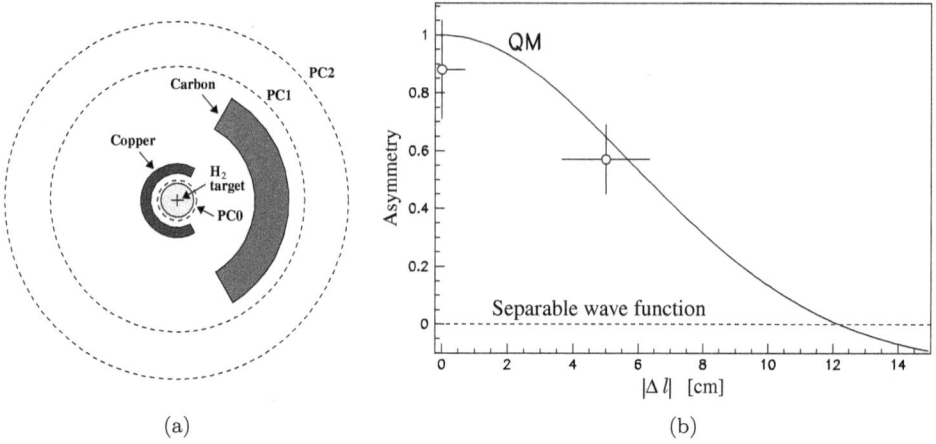

Fig. 4.   (a) Conceptual sketch of the experiment (see text); (b) asymmetry of the measured $\Lambda K^{\pm}$ yields after background subtraction. The two points show the long distance correlation of the entangled kaons, in agreement with quantum mechanics.

The speciality of this measurement, the strangeness is monitored by strong interaction in two absorbers near the target, see Fig. 4, via the observation in the same event, at two different times, of a Λ and a K^{+} (unlike strangeness) or a Λ and a K^{-} or two Λ (like strangeness). The tagging via strong interaction bypasses any potential complications arising from $\Delta S = \Delta Q$ violating neutral meson decays. The asymmetries of the yields for unlike- and like-strangeness events were measured for two experimental configurations C(0) and C(5), see Fig. 4(a), corresponding to ≈ 0 and $1.2\tau_S$ proper time differences between the two strangeness measurements, or path differences $|\Delta l|$ of ≈ 0 and $5\,\mathrm{cm}$. As shown in Fig. 4(b), these asymmetries are consistent with the values predicted from QM, and therefore consistent with the non-separability hypothesis of the $K^0\overline{K}^0$ wave function. The non-separability hypothesis is also strongly favoured by the yield of $\Lambda\Lambda$ events.

4.6.3.   *Test of the equivalence principle for particles and antiparticles*

With the large statistics of $\pi^{+}\pi^{-}$ decays, CPLEAR had been able to search for possible annual, monthly and diurnal modulations of the observables $|\eta_{+-}|$ and ϕ_{+-} that could be correlated with variations in astrophysical potentials. No such correlations were found within the CPLEAR accuracy.[16] Data were analyzed assuming effective scalar, vector and tensor interactions, with the conclusion that the principle of equivalence between particles and antiparticles holds to a level of $(6.5, 4.3, 1.8)\times10^{-9}$, respectively, for scalar, vector and tensor potential originating from the Sun with a range much greater than the Earth–Sun distance. Figure 5 shows a compilation of the upper limits on $|g - \bar{g}|_J$, the gravitational coupling difference between K^0 and $\overline{K}^0$, as a function of the interaction range r_J where $J = 0, 1, 2$ for scalar, vector and tensor potential, respectively.

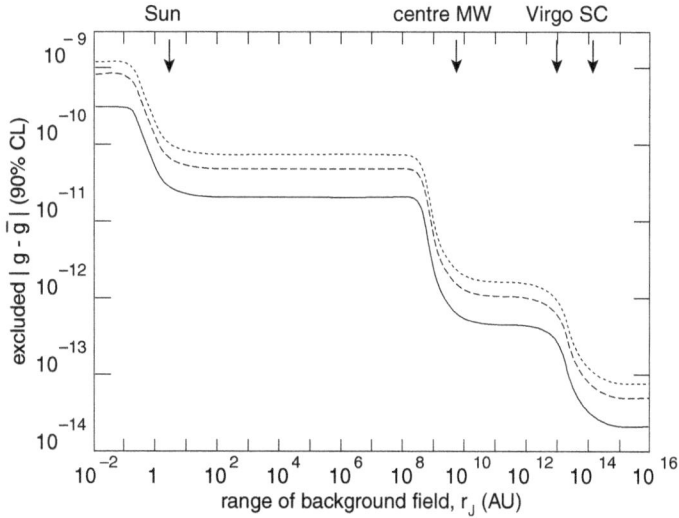

Fig. 5.   Limits on the gravitational coupling difference between K^0 and $\overline{K}^0$, $|g - \overline{g}|_J$, obtained from the measured K^0–$\overline{K}^0$ mass difference as a function of the effective interaction range r_J, with $J = 0, 1, 2$ for scalar, vector and tensor potential, respectively. Labels along the top indicate the distances to several astronomical bodies (Milky Way: MW; Shapley supercluster: SC) measured in Astronomical Units (AU). The curves are upper limits shown separately for tensor (solid line), vector (dashed line) and scalar (dotted line) interactions.

5.   Conclusion

To summarise, CPLEAR had been a nice small size experiment studying with unprecedented precision violations of discrete symmetries ($\mathcal{T}$, $\mathcal{CPT}$ and $\mathcal{CP}$) in the neutral kaon systems and addressing fundamental physics questions ranging from a possible breakdown of quantum coherence of the wave function to the equivalence principle of general relativity. Thanks to the idea of using flavour tagged neutral kaon "beams".

References

1. A. Angelopoulos *et al.*, Physics at CPLEAR, *Physics Reports* **374**, 165 270 (2003).
2. S. Baird *et al.*, in *Proc of the 1997 Particle Accelerator Conference*, Vancouver, M. Comyn *et al.* (ed.) (IEEE, Piscataway, 1998), p. 982.
3. CERN, (2014). https://home.web.cern.ch/about/accelerators/low-energy-antiproton-ring.
4. A. Angelopoulos *et al.*, Measurement of the neutral kaon regeneration amplitude in carbon at momenta below 1-GeV/c, *Phys. Lett. B* **413**, 422 (1997).
5. R. Adler *et al.*, *Nucl. Instrum. Methods A* **379**, 76 (1996).
6. A. Apostolakis *et al.*, A detailed description of the analysis of the decay of neutral kaons to $\pi^+\pi^-$ in the cplear experiment, *Eur. Phys. J. C* **18**, 41 (2000).

7. A. Angelopoulos *et al.*, Measurement of the CP violation parameter n_{00} using tagged $\overline{K}^0$ and K^0, *Phys. Lett. B* **420**, 191 (1998).

8. A. Angelopoulos *et al.*, The neutral kaons decays to $\pi^+\pi^-\pi^0$: a detailed analysis of the CPLEAR data, *Eur. Phys. J. C* **5**, 389 (1998).

9. A. Angelopoulos *et al.*, Search for CP violation in the decay of tagged $\overline{K}^0$ and K^0 to $\pi^0\pi^0\pi^0$, *Phys. Lett. B* **425**, 391 (1998).

10. A. Angelopoulos *et al.*, First direct observation of time-reversal non-invariance in the neutral-kaon system, *Phys. Lett.* **444**, 43 (1998).

11. A. Angelopoulos *et al.*, A determination of the CPT violation parameter $\mathrm{Re}(\delta)$ from the semileptonic decay of strangeness-tagged neutral kaons, *Phys. Lett.* **444**, 52 (1998).

12. A. Angelopoulos *et al.*, T-violation and CPT-invariance measurements in the CPLEAR experiment: a detailed description of the analysis of neutral-kaon decays to $e\pi\nu$, *Eur. Phys. J. C* **22**, 55 (2001).

13. A. Angelopoulos *et al.*, Measurement of the K_L-K_S mass difference using semileptonic decays of tagged neutral kaons, *Phys. Lett.* **444**, 38 (1998).

14. A. Apostolakis *et al.*, Measurement of the energy dependence of the form factor f_+ in K_{e3}^0 decay, *Phys. Lett.* **473**, 186 (2000).

15. A. Apostolakis *et al.*, Measurement of the energy dependence of the form factor f_+ in K_{e3}^0 decay, *Phys. Lett.* **473**, 186 (2000).

16. A. Apostolakis *et al.*, Tests of the equivalence principle with neutral kaons, *Phys. Lett.* **452**, 425 (1999).

17. W. Fetscher *et al.*, Regeneration of arbitrary coherent neutral kaon states: A new method for measuring the K^0-$\overline{K}^0$ forward scattering amplitude, *Z. Phys. C* **72**, 543 (1996).

18. A. Apostolakis *et al.*, An epr experiment testing the non-separability of the $\overline{K}^0$ K^0 wave function, *Phys. Lett. B.* **422**, 339 (1998).

19. A. Angelopoulos *et al.*, Test of CPT symmetry and quantum mechanics with experimental data from CPLEAR, *Phys. Lett.* **364**, 239 (1995).

20. A. Apostolakis *et al.*, Determination of the T and CPT violation parameters in the neutral kaon system using the Bell-Steinberger relation and data from CPLEAR, *Phys. Lett. B* **456**, 297–303 (1999). doi: 10.1016/S0370-2693(99)00483-9.

21. A. Angelopoulos *et al.*, K^0-$\overline{K}^0$ mass and decay-width differences: CPLEAR evaluation, *Phys. Lett.* **471**, 332 (1999).

22. V. Weisskopf and E. Wigner, Over the natural line width in the radiation of the harmonius oscillator, *Z. Phys.* **65**, 18–29 (1930). doi: 10.1007/BF01397406.

23. V. Weisskopf and E. P. Wigner, Calculation of the natural brightness of spectral lines on the basis of Dirac's theory, *Z. Phys.* **63**, 54–73 (1930). doi: 10.1007/BF01336768.

24. S. Treiman and R. Sachs, Alternate modes of decay of neutral K mesons, *Phys.Rev.* **103**, 1545–1549 (1956). doi: 10.1103/PhysRev.103.1545.

25. T. Lee, R. Oehme, and C.-N. Yang, Remarks on possible noninvariance under time reversal and charge conjugation, *Phys. Rev.* **106**, 340–345 (1957). doi: 10.1103/PhysRev.106.340.

26. T. J. Devlin and J. O. Dickey, Weak hadronic decays: $K \rightarrow 2\pi$ and $K \rightarrow 3\pi$, *Rev. Mod. Phys.* **51**, 237 (1979). doi: 10.1103/RevModPhys.51.237.

27. C. Caso *et al.*, Review of particle physics, *Phys. J. C* **3**, 1 (1998).

28. C. Dib and B. Guberina, Almost forbidden $\Delta Q = -\Delta S$ processes, *Phys. Lett. B* **255**, 113–116 (1991). doi: 10.1016/0370-2693(91)91149-P.

29. J. S. Bell and J. Steinberger, Weak interactions of kaons, in *Proc. of the Oxford International Conference on Elementary Particles*, R. G. Moorhouse *et al.* (eds.), (Oxford University Press, 1966), p. 195.

30. K. R. Schubert, B. Wolff, J. Chollet, J. Gaillard, M. Jane, *et al.*, The phase of η_{00} and the invariances CPT and T, *Phys. Lett. B* **31**, 662–665 (1970). doi: 10.1016/0370-2693(70)90029-8.

31. T. Ruf, Status of CP and CPT violation in the neutral kaon system, in *Proc of the 16th International Conference on Physics in Collision*, Mexico City, Mexico, 19–21 June 1996 (1996).

32. R. S. B. House, ed. *Time Reversal in Field Theory*, Vol. 231, (1955). doi: 10.1098/rspa.1955.0189.

33. R. Jost, A remark on the C.T.P. theorem, *Helv. Phys. Acta.* **30**, 409–416 (1957).

34. G. Luders, Proof of the TCP theorem, *Annals of Physics* **2**, 1–15 (1957). doi: 10.1016/0003-4916(57)90032-5.

35. S. Hawking, The unpredictability of quantum gravity, *Commun. Math. Phys.* **87**, 395–415 (1982). doi: 10.1007/BF01206031.

36. J. R. Ellis, J. Hagelin, D. V. Nanopoulos, and M. Srednicki, Search for violations of quantum mechanics, *Nucl. Phys. B* **241**, 381 (1984). doi: 10.1016/0550-3213(84)90053-1.

37. R. Adler *et al.*, Experimental measurement of the $K_S K_S / K_S K_L$ ratio in antiproton annihilations at rest in gaseous hydrogen at 15 and 27 bar, *Phys. Lett. B* **403**, 383–389 (1997). doi: 10.1016/S0370-2693(97)00489-9.

An ISR Discovery: The Rise of the Proton–Proton Cross-Section

Ugo Amaldi

Technische Universität München, Arcisstraße 21, D-80333 Munich, Germany
TERA Foundation, Via Puccini 11, 28100 Novara, Italy
ugo.amaldi@cern.ch

The Intersecting Storage Rings (ISR) were the first hadron collider ever built, providing proton–proton collisions at centre-of-mass energies as high as 62 GeV, almost five times larger than any previous accelerator. When in 1971 the ISR began operation the Regge-pole approach dominated and the proton–proton total cross-section was expected to have already reached a finite asymptotic value. However, ISR experiments found that the cross-section was rising by 10% between 22 and 62 GeV, while the interaction radius was increasing by 5%, a trend that continues up to the hundred times larger energies available at the Large Hadron Collider. In order to accurately measure the total and elastic cross-sections, new experimental methods — uniquely adapted to the environment of a hadron collider — had to be developed; they are described in the central part of this paper, which closes with a review of the data obtained at the LHC since they put in a wider perspective the forty years old ISR results.

1. Hadron–Hadron Cross-Sections at the Beginning of the 1970s

The first unexpected result produced at the Intersecting Storage Rings (ISR) was the discovery, in 1973, that the proton–proton total cross-section was not constant over the newly opened energy range. Today it is difficult to describe and explain the surprise and scepticism with which the news of the "rising total cross-section" was received by all knowledgeable physicists. Among the many episodes, I vividly recall what Daniele Amati told me while walking out of the CERN Auditorium after the seminar of March 1973 in which I had described the results obtained independently by the CERN–Rome and Pisa–Stony Brook Collaborations: "Ugo, you must be wrong, otherwise the pomeron trajectory would have to cut the axis above 1!"

Nowadays, all those who still care about the pomeron know the phenomenon, find it normal and accept the explanations of this fact given by the experts. However, at that time the Reggeon description of all small-angle hadronic phenomena was the only accepted dogma since it could explain the main experimental results in strong interaction physics:

(i) The tendency of the total cross-sections of all hadron–hadron collisions to become energy independent, as shown in Fig. 1:[1]

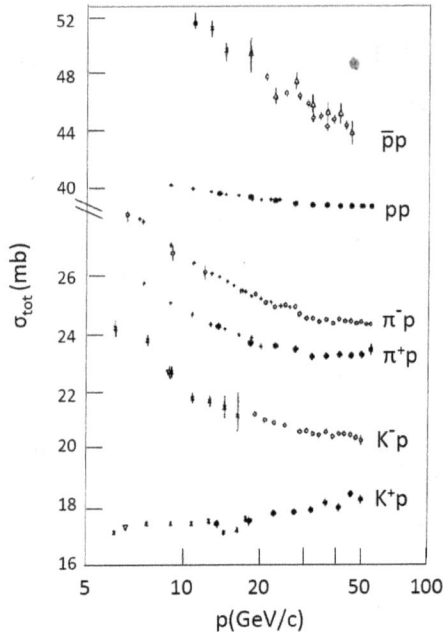

Fig. 1. The total cross-sections σ_{tot} (measured in the early 1970s at the Serpukhov 70 GeV synchrotron and at lower-energy accelerators) plotted versus the laboratory momentum p of the proton.[1]

(ii) The "shrinking" of the forward differential cross-sections when the collision energy was increasing,[2] which meant that the forward proton–proton differential elastic cross-section at small centre-of-mass angles θ_{cm} (i.e. at small momentum transfers $q = cp_{\text{cm}}\sin\theta_{\text{cm}}$, usually measured in GeV) is proportional to $\exp(-Bq^2)$, with a slope parameter B that *increased* with the centre-of-mass energy, indicating, through the uncertainty principle, that the proton–proton interaction radius *increased* as $\sqrt{B}$.

The regime in which all the total cross-sections would become energy independent was called "asymptopia", and theorists and experimentalists alike were convinced that the ISR would demonstrate that the total proton–proton cross-section, which slightly decreases in the Serpukhov energy range (Fig. 1), would tend to a constant of about 40×10^{-27} cm^2 (40 mb), thus confirming the mainstream interpretation of all hadronic phenomena, the Regge model.

2.   The Theoretical Framework

In the 1960s the forward differential cross-sections had found a universally accepted interpretation in terms of the collective effect of the exchanges of all the particles,

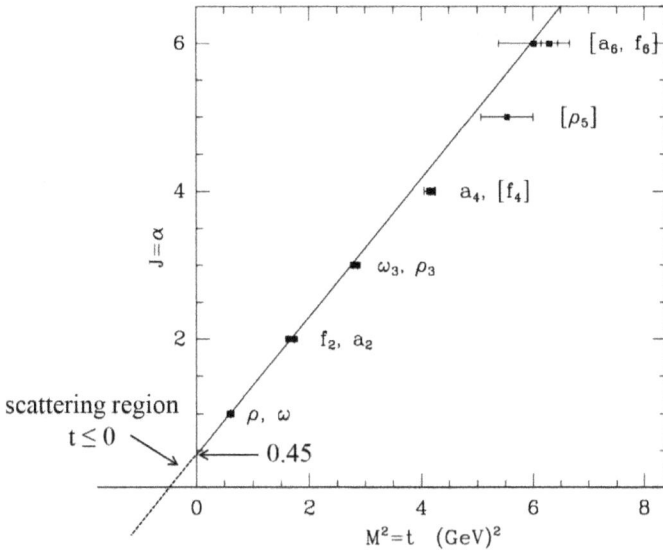

Fig. 2. The present situation of the Chew–Frautschi plot shows that the Regge trajectory containing the ρ meson (mass = 770 MeV) is practically linear up to very large masses.

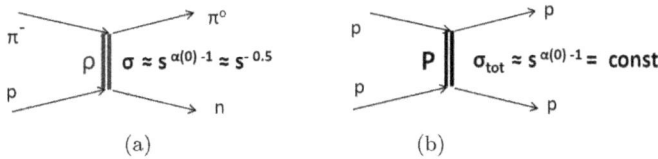

Fig. 3. (a) The main contribution to the pion charge-exchange phenomenon is the exchange of the ρ trajectory. (b) In the Regge model, the exchange of a pomeron trajectory is the dominant phenomenon in all high-energy elastic collisions.

which, in the mass^2–spin plane, lay on a "Regge trajectory". The present knowledge of the ρ trajectory is represented in Fig. 2.[3]

The exchange of the ρ trajectory dominates the charge-exchange cross-section of Fig. 3(a). By using the usual parameter $s = E_{cm}^2$, where E_{cm} is the centre-of-mass energy, the recipes of the Regge model give a cross-section that varies as $s^{\alpha(t=0)-1}$ (Fig. 3).

Since in Fig. 2 $\alpha(0) \approx 0.45$, the charge-exchange cross-section of Fig. 3(a) was predicted to vary roughly as $s^{-0.5} = 1/E_{cm}$.

In the 1960s, the experimental confirmation of this prediction — see for instance Ref. 4 — was one of the strongest arguments in favour of the Regge description of the scattering of two hadrons. Such a description is still used because these phenomena cannot be computed with quantum chromodynamics.

As shown in Fig. 3(b), in the Regge approach the proton–proton *elastic* scattering process was also described by the exchange of a trajectory, the "pomeron",

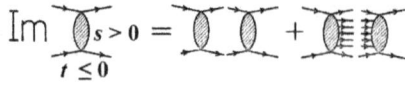

$$\mathrm{Im}\ \underset{\substack{s>0\\ t\le 0}}{\overbrace{}} = \overbrace{} + \overbrace{}$$

$$4\pi\,\mathrm{Im}\,f(t)/k = G_{\mathrm{el}}(t) + G_{\mathrm{in}}(t).$$

$$\mathbf{For}\ t = 0\!:\ 4\pi\,\mathrm{Im}\,f(0)/k = \sigma_{el}(s) + \sigma_{inel}(s) = \sigma_{tot}(s)$$

Fig. 4.   The graphical representation of the unitarity relation, at a given s and for $t \le 0$, explains the definition of the elastic and inelastic overlap integrals $G_{\mathrm{el}}(t)$ and $G_{\mathrm{in}}(t)$ with $k = p/\hbar$.

which — given the fact that σ_{tot} is proportional to $s^{\alpha(t=0)-1}$ — had to have an "intercept" $\alpha_{\mathrm{P}}(t = 0) = 1$ to be consistent with an energy-independent total cross-section. For this reason, at the beginning of the 1970s, the so often heard "asymptopia" and "the pomeron intercept is equal to 1" were used as different ways of saying the same thing.

Since there were no particles belonging to the pomeron trajectory, its slope could be fixed only by measuring the t dependence of the forward elastic proton–proton cross-section, which could be described by the simple exponential $\exp(-B|t|)$ with a "slope" B that *increased* with the centre-of-mass energy. The accepted slope of the pomeron trajectory was $\alpha'_{\mathrm{P}}(0) \approx 0.25\,\mathrm{GeV}^{-2}$.

In parallel with this "t-channel" description, other theorists, working on the "s-channel description", were deriving rigorous mathematical consequences from the fundamental properties of the S-matrix, which describes the scattering processes: unitarity, analyticity and crossing.

Unitarity of the S-matrix implies that one can compute the imaginary part of the forward scattering amplitude $\mathrm{Im}\ f(t)$ by taking the product of a scattering amplitude and its conjugate and summing them over all possible intermediate states, as graphically depicted in Fig. 4.

The sum is made up of two contributions, which are called "elastic and inelastic overlap integrals" $G_{\mathrm{el}}(t)$ and $G_{\mathrm{in}}(t)$. In the forward direction, i.e. for $t = 0$, the overlap integrals reduce to the elastic and inelastic cross-sections, and the unitarity relation gives the "optical theorem", which states that the imaginary part of the forward scattering amplitude equals the total cross-section σ_{tot}, except for a factor $k/4\pi$, which depends on the definition chosen for the amplitude itself.

The figure and the formulae indicate that hadron–hadron forward elastic scattering ($t = 0$) is determined by the amplitudes of both elastic and inelastic reactions. When the collision energy is large, there are many open inelastic channels, the incoming wave is absorbed and the elastic scattering amplitude is dominated by its imaginary part, which is the "shadow" of the elastic and inelastic processes. In such a *diffraction phenomenon*, the ratio $\rho = \mathrm{Re}(f)/\mathrm{Im}(f)$ between the real and imaginary parts of the elastic amplitude is small, so that, in the expression for the forward elastic cross-section deduced from the optical

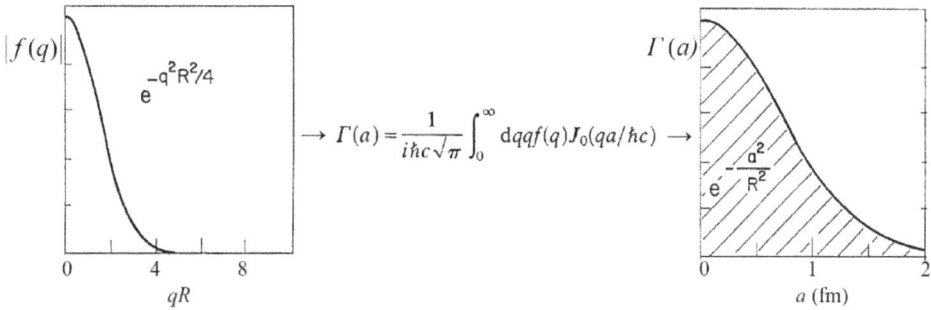

Fig. 5.   A Gaussian real elastic profile function corresponds to an imaginary scattering amplitude that decreases exponentially with $q^2 = |t|$. (In the integral, J_0 is the Bessel function of order zero.)

theorem,

$$\left(\frac{d\sigma_{el}}{d|t|}\right)_{t=0} = \frac{(1+\rho^2)\sigma_{tot}^2}{16\pi} \quad \text{with } \rho = \frac{\mathrm{Re}\, f(0)}{\mathrm{Im}\, f(0)}$$

where the term ρ^2 is of the order of a few percent.

The unitarity equation expressed in term of the variable $q = (-t)^{1/2}$ can also be written as a function of the complementary variable, the impact parameter a in the plane perpendicular to the momenta of the colliding particles. By applying the transformation written in Fig. 5 to the scattering amplitude $f(q)$, one can compute the "profile function" $\Gamma(a)$ as a function of a.

By applying the transformation of Fig. 5 to the three terms of the unitarity relation of Fig. 4, one obtains

$$2\mathrm{Re}\,\gamma(a) = |\Gamma(a)|^2 + G_{in}(a) \quad \text{with } 0 \leq \Gamma(a) \leq 1, \quad 0 \leq G(a) \leq 1.$$

This equation shows how, in the diffraction limit, i.e. when the scattering amplitude is essentially imaginary because ρ is small, the "profile function" $\Gamma(a)$ is real and the "inelastic overlap integral" $G_{in}(a)$ determines the elastic "profile function", $\Gamma(a) = 1 - \sqrt{[1 - G_{in}(a)]}$, and vice versa. The two inequalities to the right express the limits imposed by unitarity.

If for $a \leq R$ the absorption is complete, i.e. $G_{in}(a \leq R)] = 1$ and $\Gamma(a \leq R)] = 1 - \sigma_{el} = \sigma_{in} = \pi R^2$ and $\sigma_{tot} = \sigma_{el} + \sigma_{in} = 2\pi R^2$. Thus a ratio $\sigma_{el}/\sigma_{tot} = 0.50$ is a clear sign of the fact that the "black disk" model has to be adopted.

Combining unitarity with analyticity and crossing, in the 1960s three important theorems had been demonstrated.

- The *Pomeranchuk theorem*[5] states that, in the limit $s \to \infty$, the hadron–hadron and the antihadron–hadron cross-sections become equal.

- According to the *Froissart–Martin theorem*[6, 7] the total cross-section must satisfy the bound

$$\sigma_{\text{tot}} \leq C \ln^2\left(\frac{s}{s_0}\right) \approx 60 \,\text{mb} \, \ln^2\left(\frac{s}{s_0}\right)$$

where the numerical value $C = \pi(\hbar/m_\pi)^2$ is determined by the mass of the pion, which is the lightest particle that can be exchanged between the two colliding hadrons, and s_0 is usually taken equal to 1 GeV^2.

- Finally, the *Khuri–Kinoshita theorem*[8] relates the energy dependence of ρ with the energy dependence of the total cross-section by stating that, if σ_{tot} increases with energy, ρ passes from small negative values to positive values. This is a consequence of the "dispersion relations", which connect the real part of the forward elastic amplitude with some appropriate energy integrals of the total cross-section. Khuri and Kinoshita showed that, if σ_{tot} follows the Froissart–Martin bound and increases proportionally to $\ln^2 s$, for $s \to \infty$ the ratio ρ is *positive* and tends to zero from above as $(\ln s)^{-1}$.

3. Three ISR Proposals

In March 1969, the ISR Committee received three proposals that are relevant to the subjects discussed in this paper.

The title of the proposal by the Pisa group (signed by G. Bellettini, P. L. Braccini, R. R. Castaldi, C. Cerri, T. Del Prete, L. Foà, A. Menzione and G. Sanguinetti) was "Measurements of the p–p total cross-section".[9] Giorgio Bellettini presented orally the proposal to the ISR Committee.

Two of the figures of the proposal are reproduced in Fig. 6. The very large scintillator hodoscopes would detect the outgoing particles and count the total number of events. Moreover, small-angle telescopes would detect forward elastic events in order to estimate the number of elastic events not recorded because the protons, scattered at small angles, would be lost in the ISR vacuum chamber.

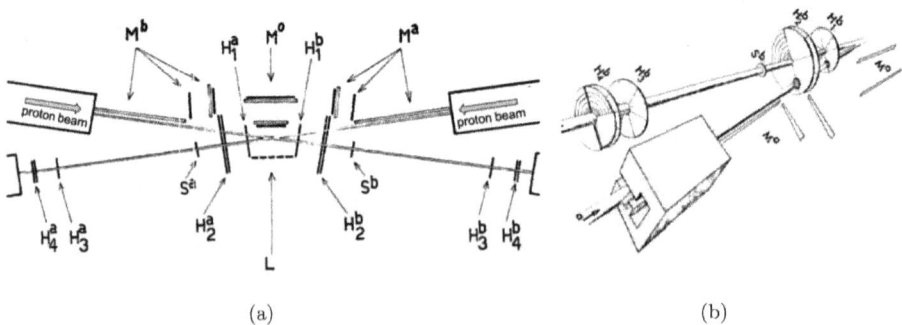

(a)   (b)

Fig. 6.   The initial proposal by the Pisa group to measure the proton–proton total cross-section.

At a collider, in order to measure any cross-section it is necessary to determine the "luminosity" L. In the case of a beam of parallel particles that cross at an angle Φ, the only important spatial variable is the vertical one y. Given the normalised vertical distributions of the two beams, $\rho_1(y - y_0)$ and $\rho_2(y)$, which are displaced vertically by y_0, the luminosity is proportional to the two currents and depends upon the crossing angle of the beams according to the formula:

$$L(y_0) = \underbrace{\frac{I_1 I_2}{ce^2 \tan(\phi/2)}} \quad \underbrace{\int \rho_1(y - y_0)\rho_2(y)dy}$$

$$R(y_0) = K \times \sigma \times \quad \text{(overlap integral)}.$$

To obtain the luminosity, the Pisa group proposed to measure ρ_1 and ρ_2 separately with the two sets of spark chambers indicated in Fig. 6 with the letters M^o, M^a and M^b, and then to compute numerically the beam overlap integral.

The problem of measuring the ISR luminosity was amply debated during 1968 and various proposals to do so by *separated measurements* of the vertical distributions were put forward by Darriulat and Rubbia,[10] Rubbia,[11] Schnell,[12] Steinberger[13] and Onuchin.[14] Another method proposed in different forms by Cocconi,[15] di Lella[16] and Rubbia and Darriulat[17] was based on the detection of the two protons scattered at angles smaller than about 1 mrad, where the *known* Coulomb elastic scattering cross-section dominates.

All the proposals requiring the separate measurements of the vertical distributions of the two beams were superseded by a very simple observation made by Simon Van der Meer.[18] He remarked that the cross-section σ_M of a particular type of event (detected by a set of monitor counters surrounding the interaction region) can be obtained by measuring the rate of the monitor events $R_M(y_0)$ as a function of the distance y_0 between the centres of the two beams, which are moved vertically in small and precisely known steps.

Since in the integral $I_{VdM} = \int R_M(y_0)dy_0$ the double integral over dy_0 and dy — implicit in $R_M(y_0)$ — equals 1, because ρ_1 and ρ_2 are normalised, the cross-section of the monitor counters is given by $\sigma_M = I_{VdM}/K$ and the cross-section σ corresponding to any other rate R is simply obtained as

$$\sigma = \frac{R}{R_M}\sigma_M = \frac{I_{VdM}}{K}\frac{R}{R_M}.$$

The magnets needed to precisely displace the two beams vertically were installed in the ISR, and since then the Van der Meer method has been used to measure proton–proton luminosities at all colliders.

Figure 7 shows the apparatus built by what became the Pisa–Stony Brook Collaboration after the Pisa group joined with the Stony Brook group led by Guido Finocchiaro and Paul Grannis.

Coulomb scattering and its interference with nuclear scattering was the focus of the proposal "The measurement of proton–proton differential cross-section in the

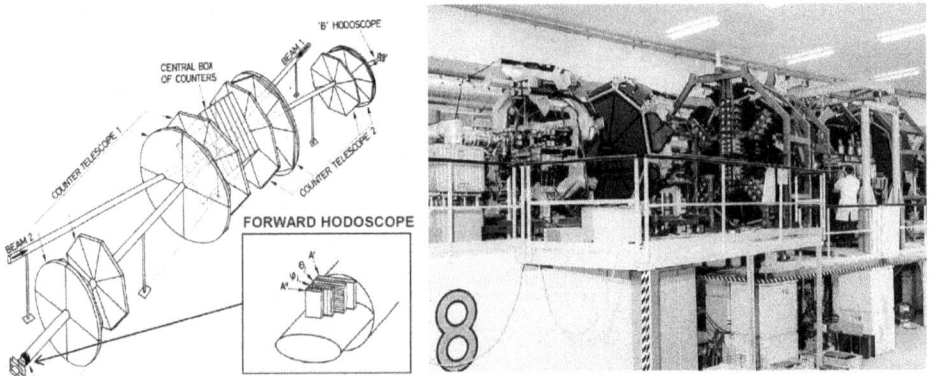

Fig. 7. In the final detector built by the Pisa–Stony Brook Collaboration, forward telescopes were used to measure elastic scattering events at small angles.

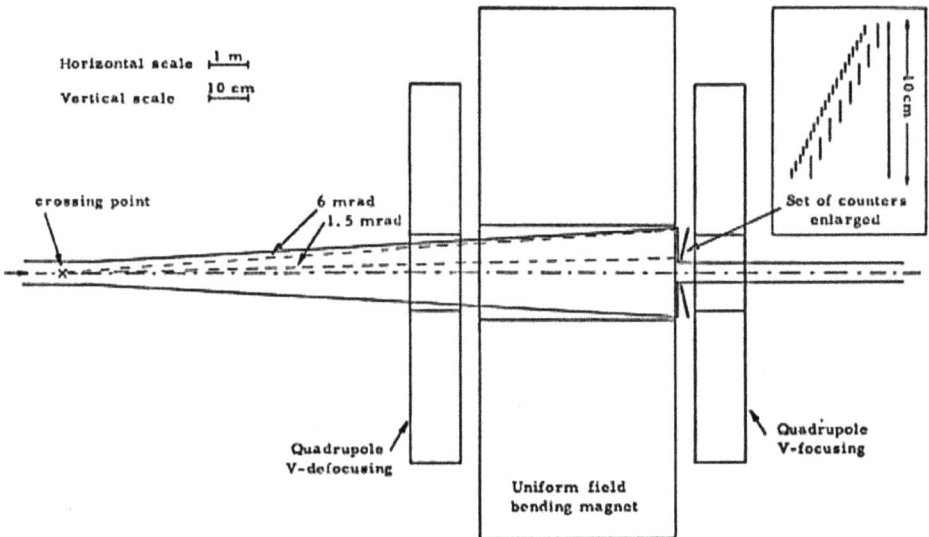

Fig. 8. In the first proposal, two quadrupoles and one magnet focused the protons and bent them so as to measure protons scattered down to 1.5 mrad.

angular region of Coulomb scattering at the ISR"[19] presented by Giorgio Matthiae on behalf of the Rome–Sanità group. The proposal was signed by U. Amaldi, R. Biancastelli, C. Bosio, G. Matthiae and P. Strolin; Strolin, at the time was an ISR engineer. The apparatus (shown in Fig. 8) required a modification of the ISR vacuum pipe, with new magnets to be installed on each beam. A few months later, in an addendum to the proposal, the authors wrote: "In discussions with the specialists of the machine (R. Calder and E. Fischer) we found a simple way for allocating the detectors near the beam, which does not imply a modification of the standard parts of the vacuum chamber."

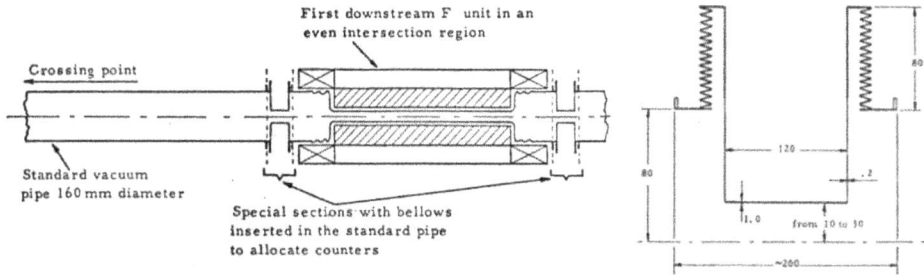

Fig. 9. In the 1969 proposal there were four movable sections on each beam and the forward-scattered protons were detected by a coincidence between counters located upstream and downstream of the first ISR magnet.

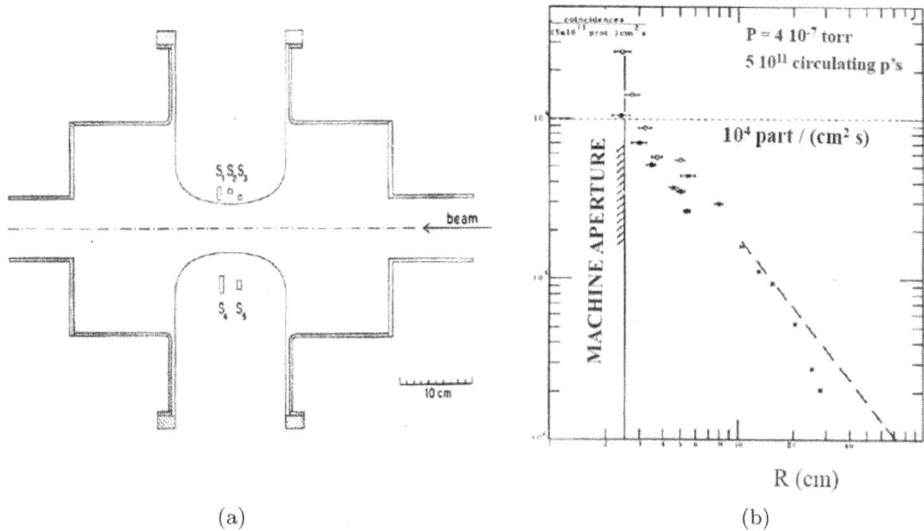

(a)   (b)

Fig. 10. Special section of the PS that allowed the measurement (a) of the rate detected by scintillators placed very close to a circulating beam formed by 5×10^{11} protons (b).

The proposal (Fig. 9) foresaw getting the bottoms of the movable sections as close as 10 mm to the beam with the bottom of the movable sections, as described many years before by Larry Jones.[20] This was a daring operation and many people worried so much that, in an ISR meeting, Carlo Rubbia said: "Your scintillators will give light as bulbs!"

To counter the criticisms, in 1970 a test was performed at the CERN PS to check whether one could install scintillation counters very close to a circulating proton beam (Fig. 10). Eifion Jones participated in the planning and in the tests — in which the PS beam was moved towards the scintillators. Previously Hyams and Agoritsas had performed similar measurements.[21]

Fig. 11.   Paolo Strolin describes to Sacha Skrinsky (Novosibirsk) the ACHGT experiment, which detected with magnetostrictive spark chambers the protons scattered between 30 and 100 mrad.

The memorandum sent to the ISR Committee[22] concluded that, down to a few millimetres from the beam, the rate to be found at the ISR would have been sufficiently low to allow the Coulomb experiment (Fig. 11).

The ISR movable sections of the vacuum chamber soon became known as "Roman pots", which was the translation of the expression *"les pots de Rome"* invented by the French draftsman whom we visited, regularly travelling from Rome to Geneva and who, under the direction of Franco Bonaudi, transformed our rough sketches into construction drawings.

In October 1970, the ISR Committee took various decisions on pending experiments. Following it, the CERN group of Giuseppe Cocconi, Alan Wetherell, Bert Diddens and Jim Allaby wrote the Committee a memorandum, which said: "At the meeting of the ISRC on 14 October, it was concluded that there is no way to fit the proposed experiment on deep inelastic scattering into the present ISR experimental programme. As a result we have decided, on their invitation, to collaborate with the Rome group (U. Amaldi *et al.*) on the small-angle scattering experiment."

For the final experiment, the newly formed CERN–Rome Collaboration decided to retain only the four movable sections located *in front* of the first ISR magnet, a decision that simplified the experiment and its interactions with the accelerator.

In the same ISR Committee meeting in which the Pisa and the Rome experiments were presented, Carlo Rubbia described the third proposal by the CERN–Genoa–Torino group (P. Darriulat, C. Rubbia, P. Strolin, K. Tittel, G. Diambrini, I. Giannini, P. Ottonello, A Santroni, G. Sette, V. Bisi, A. Germak, C. Grosso. The title of the proposal was "Measurement of the elastic scattering cross-section at the ISR".[23] The apparatus of Fig. 12 was made of two parts such that "the whole angular range from 1 mrad to about 100 mrad can be covered. The very

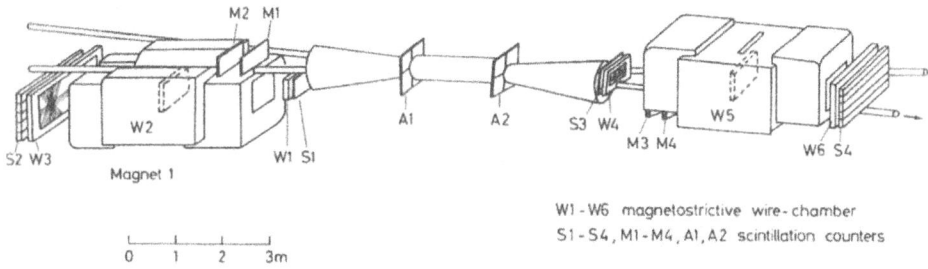

Fig. 12. The two septum magnets of the ACHGT Collaboration have been used to measure the forward elastic cross-section.

small-angle events (in the Coulomb region) are detected by a two-arm spectrometer sharing the first four magnets with the storage ring system. The larger-angle events are momentum-analysed with a pair of magnets that do not perturb the circulating beams."

After many discussions, the ISR Committee decided to approve only the system made of two septum magnets installed in the intersection regions and to leave the detection of elastic scattering in the Coulomb region to the scintillators mounted in the Roman pots. Since then, Carlo Rubbia has described the ISR experimental programme as "key-hole physics".

After the approval, Rubbia's Collaboration was joined by the Aachen and Harvard groups and became the Aachen–CERN–Harvard–Genoa–Torino (ACHGT) Collaboration.

The two elastic scattering experiments were mounted in interaction regions I6 of the ISR (Fig. 12), while interaction region I8 was assigned to the total cross-section experiment.

4. First Results on Elastic Scattering and Total Cross-Sections

The slope of the forward elastic cross-section was the easiest measurement to perform. The 1971 results,[24, 25] reported in Fig. 13, confirmed the behaviour first found at the PS and confirmed at Serpukhov: in the range $30 \leq s \leq 3000 \, \mathrm{GeV}^2$, the elastic slope B is linear in $\ln s$, in agreement with the description based on pomeron exchange, and in the ISR energy range ($23 \leq \sqrt{s} \leq 62 \, \mathrm{GeV}$, i.e. $550 \leq s \leq 3800 \, \mathrm{GeV}^2$) B increases by about 10%, which corresponds to a 5% increase of the proton–proton interaction radius.

In the Regge description

$$B = B_0 + 2 \, \alpha'_{\mathrm{P}}(0) \, \ln\left(\frac{s}{s_0}\right)$$

and the dashed line of Fig. 11 corresponds to $\alpha'(0) = 0.28 \, \mathrm{GeV}^{-2}$, confirming what was already known from lower-energy data: the pomeron slope at $t = 0$ is definitely smaller than the slope $\alpha'_\rho(0) \approx 1 \, \mathrm{GeV}^{-2}$ of the ρ trajectory (Fig. 2).

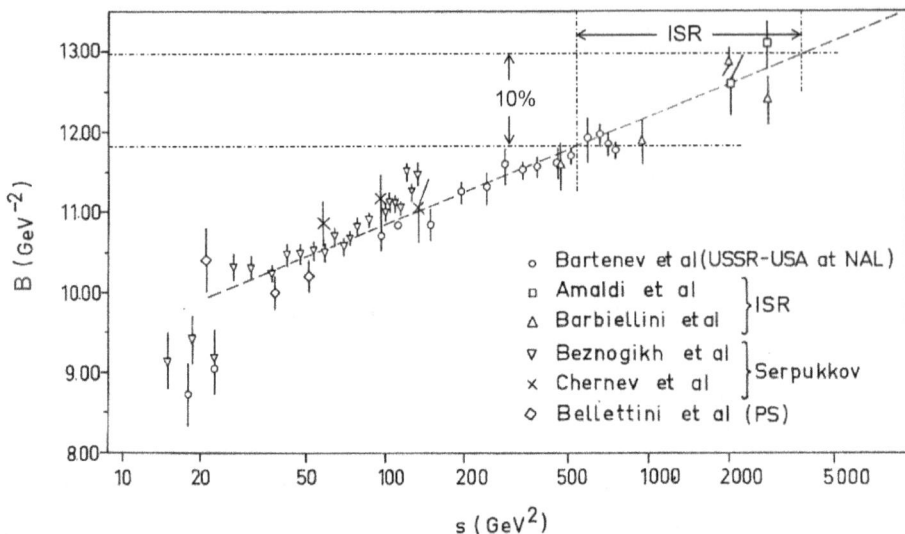

Fig. 13. The data available in 1971 for $-t \leq 0.12\,\mathrm{GeV}^2$ and the results of the measurement performed in 1972 at NAL (Fermilab).[26] The dashed line shows that, over a very large energy range, the t-width (which is equal to $1/B$) of the forward elastic peak decreases as the inverse of $(a + b \ln s)$.

In 1972, the ACHGT Collaboration reported the experimental distributions plotted in Fig. 14, which show that

(i) The forward elastic cross-section has a variation of slope at $|t| \approx 0.16\,\mathrm{GeV}^2$;[27]

(ii) The deep diffraction minimum located at $|t| \approx 1.4\,\mathrm{GeV}^2$ is the energy-dependent deepening of the structure observed at lower energies.[28]

However, the real surprise came with the measurements of the total cross-section done by the Pisa–Stony Brook Collaboration, with the apparatus of Fig. 7, and by the ACHGT and the CERN–Rome Collaborations, by measuring the forward elastic cross-section and using the optical theorem.

This method, which, as far as I know, was not considered before the ISR start-up, was pioneered in 1971 by ACHGT:[29] the hadron–hadron forward elastic cross-section (measured outside the Coulomb peak with the Van der Meer method) is extrapolated to zero angle to obtain $(d\sigma/dt)_0$ and the optical theorem is applied to obtain

$$\sigma_{\mathrm{tot}} = \frac{\sqrt{16\pi(d\sigma/dt)_0}}{(1 + \rho^2)}.$$

In the autumn of 1972 the three collaborations were competing to be the first to measure the total proton–proton cross-section. I remember very vividly that period, because I was the one performing the analysis of the CERN–Rome data.

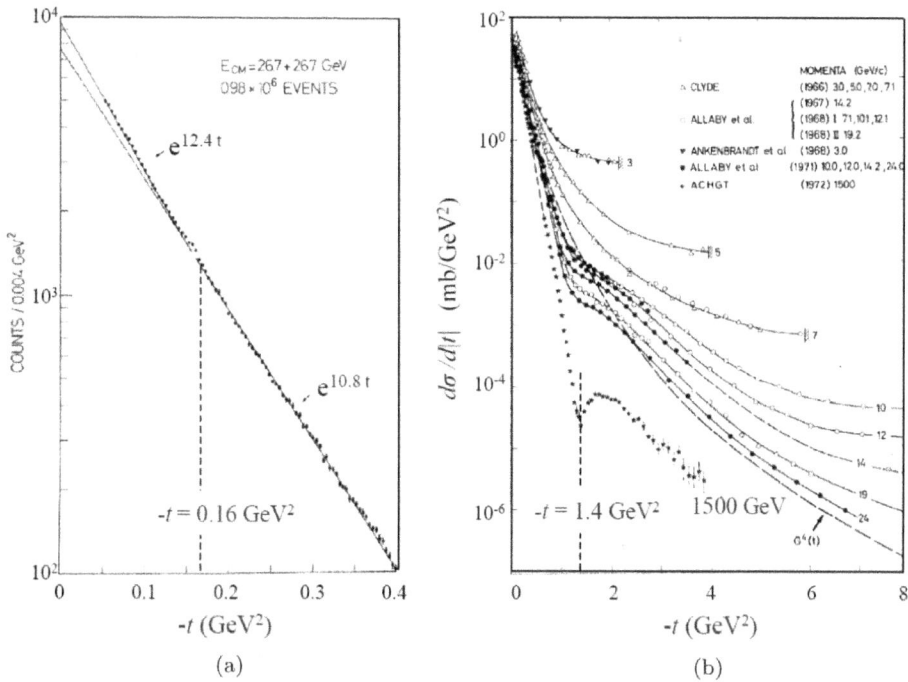

Fig. 14. First measurements by the ACHGT Collaboration of proton–proton elastic scattering (a) in the forward region[28] and (b) at large momentum transfer.[29]

The confusing status of the measurements in October 1972 is presented in Fig. 15, which I prepared for the invited talk I gave in September 1973, at the II Aix en Provence International Conference on Elementary Particles.[30]

The Conference session of September 12, 1973 — in which I presented the rising cross-section data — was the most momentous one I ever contributed to. Daniele Amati gave the first talk on "Strong interaction theory"; he started the presentation by placing on the overhead projector a transparency with a hand-made Chilean flag because the night before the Pinochet *coup d'état* had overthrown Allende's government. Then Alan Mueller spoke on "High multiplicity reactions", I presented "Elastic scattering and low multiplicities" and Steven Weinberg discussed "Recent progress in gauge theories of weak, electromagnetic and strong interactions". Finally Paul Musset described — in "Neutrino interactions" — the neutral current events discovered at CERN by Gargamelle; the applause never ended. In his Nobel speech Abdus Salam said, "At the Aix-en-Provence meeting, that great and modest man, Lagarrigue, was also present and the atmosphere was that of a carnival — at least this is how it appeared to me."

Figure 15 shows that in fall 1972 the Pisa–Stony Brook and CERN–Rome Collaborations had an indication of the rising cross-section, while AGHGT was

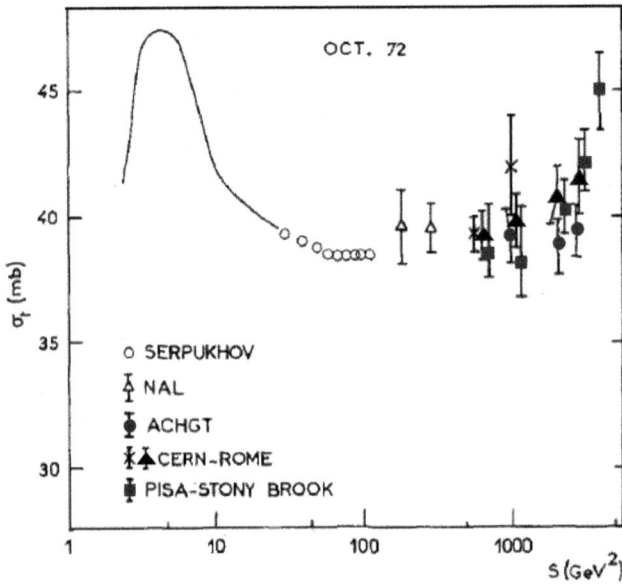

Fig. 15.   Status of the total cross-section measurements in October 1972.[31] The points by the CERN–Rome Collaboration were obtained with the luminosity measured with both the Van der Meer method and Coulomb scattering.

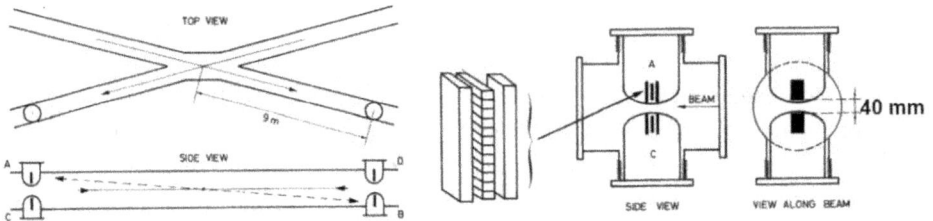

Fig. 16.   The 1972 telescope systems of the CERN–Rome Collaboration[32] were used (i) to obtain the ISR luminosity using the Coulomb scattering events and (ii) to measure ρ.

finding no energy dependence; this negative result was publicised in many seminars and for many months the difference with the other two was hotly debated.

In February 1972, the CERN–Rome Collaboration had published the first measurement of the ratio ρ between the real and imaginary parts of the forward scattering amplitude and of the total cross-section using Coulomb scattering as normalisation.[31] The measurement could only be performed at the two lowest ISR energies because, with the apparatus of Fig. 16, the minimum scattering angle was fixed at about 2.5 mrad by the background rate due to the beam halo. Thus at the highest ISR energies, after completion of the stacking process in the two ISR rings,

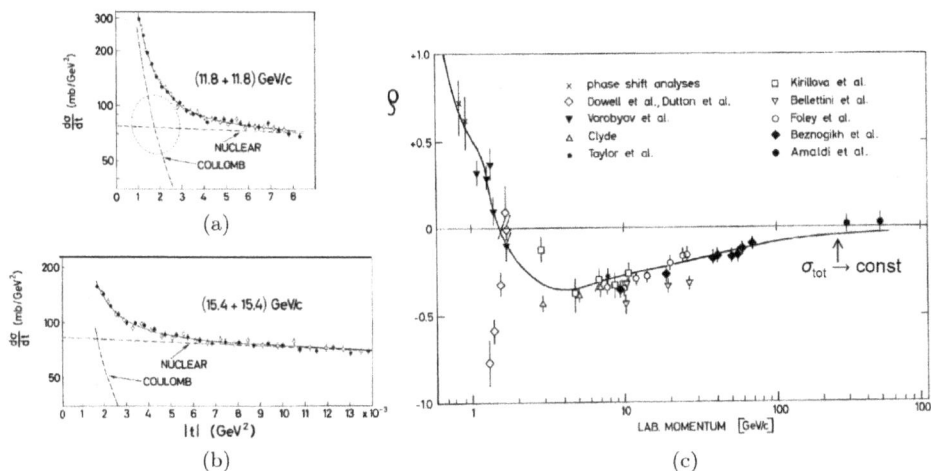

Fig. 17. The first measurements of the real part of the forward scattering amplitude were performed at the two lowest ISR energies.[32]

the pots could not be moved close enough to the beams to reach the t-range where the Coulomb scattering amplitude is as large as the nuclear one.

The measured differential cross-sections are shown in Figs. 17(a) and 17(b).

The t-dependence of the Coulomb amplitude is well known, because it is due to large-impact-parameter collisions of two point-like charges, is essentially real and decreases proportionally to $1/t$. In the t range indicated by the dashed ellipse, the nuclear amplitude varies little and its (small) real part interferes with the Coulomb amplitude, which is well known, being due to an electromagnetic phenomenon. The ratio ρ can thus be obtained by a fit to the very precise data.

The results of this first experiment are shown as full dots in Fig. 17(c). The two data points indicated that ρ was becoming positive in the ISR energy range which, because of the Khuri–Kinoshita theorem, was a signal of the rise of the total proton–proton cross-section. The error bars are large, but within the Collaboration we knew that the indication was stronger than it appeared because, after many discussions, the experimental errors were *doubled* to be on the safe side in the first paper reporting the result of a new delicate experiment.

The CERN–Rome and Pisa–Stony Brook data — presented at CERN in my 1973 seminar and published shortly after[32, 33] — definitely demonstrated that

(i) The proton–proton total cross-section increases by about 10% in the ISR energy range (Fig. 18(a)),

(ii) The elastic cross-section (computed by integrating the measured differential cross-section) increases by the about same amount, so that in the full ISR energy range the ratio is $\sigma_{el}/\sigma_{tot} \approx 0.17$, while it decreases monotonically at lower energies.

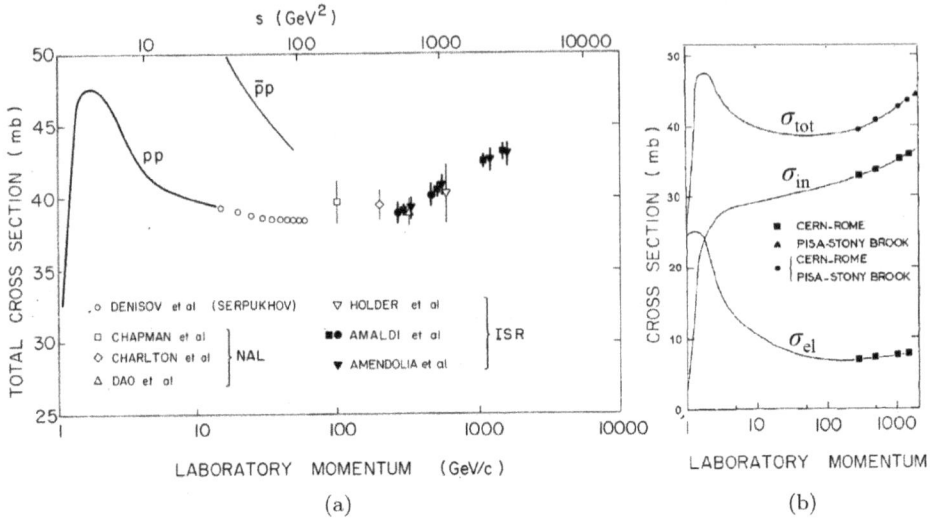

Fig. 18.   (a) The proton–proton total cross-section increases for laboratory momenta larger than 300 GeV/c ($s > 500$ GeV^2). (b) The inelastic cross-section was computed by subtraction: $\sigma_{in} = \sigma_{tot} - \sigma_{el}$.

Since our final paper was ready before the one of the Pisa–Stony Brook Collaboration, we waited a couple of weeks so that the two papers could be published one below the other in the same issue of *Physics Letters*.

The *constant* ratio $\sigma_{el}/\sigma_{tot} \approx 0.17$ and the 10% *increase* of the proton–proton forward slope are easily attributed to the combination of an *energy-independent* value of the inelastic overlap integral $G_{in}(0)$ and of the profile function $\Gamma(0) = 1 - \sqrt{[1 - G_{in}(0)]}$, with an interaction radius that *increases* by 5% in the ISR energy range. In this simple model — called "geometrical scaling" — the shape of $G_{in}(a)$ does not change with the collision energy.

The inelastic cross-section is four times larger than the elastic cross-section and increases roughly proportionally to $s^{0.04}$ from about 50 MeV/c to the maximum ISR energy (Fig. 18(b)). Looking at the three curves of this figure, it appears that the shallow minimum of the total proton–proton cross-section $\sigma_{tot} = \sigma_{in} + \sigma_{el}$ around $s = 100$ GeV^2 is a consequence of the *continuously* rising inelastic cross-section which, through unitarity, seems to drive the increase of the elastic cross-section.

If the energy dependence of the high-energy total cross-section is fitted with the formula of the Froissart–Martin bound, one obtains

$$\sigma_{tot} \cong \left[38.4 + 0.5 \ln\left(\frac{s}{s_0}\right)^2 \right] \text{ mb},$$

where $\sqrt{s_0} = 140$ GeV.[32] Since the coefficient 0.5 mb is much smaller than the limiting value predicted by the Froissart–Martin bound, the very good fit obtained with $\ln(s/s_0)^2$ is probably uncorrelated with the bound itself.

As I said, at the time most experts were convinced of the constancy of the cross-sections at high energies, with two important exceptions. In 1952, Werner Heisenberg had published a paper that described pion production in proton–proton collisions as a shock wave problem governed by a non-linear equation and deduced a $\ln^2 s$ dependence of the cross-section.[34] The model proposed by H. Cheng and T. T. Wu[35] is much more sophisticated because it is based on quantum field theory, specifically on a massive version of quantum electrodynamics. After the announcement of the ISR results, the model was reconsidered and fitted to the experimental data by Cheng, Walker and Wu.[36]

The CERN seminar of March 1973 and, soon after, the two publications made a certain impression also outside the physics community, so much so that I was invited to write an article for *Scientific American*. In spring and summer 1973 this took me a lot of time since the editor was following very closely the writing of the text and the production of the figures. The article was published in September 1973[37] after a drastic cut of the part of the article containing the impact parameter description of the ISR collision. As a replacement, I introduced the quantity "average opaqueness" $O = 2\sigma_{el}/\sigma_{tot}$, which in wave mechanics is $O = 1$ for a black disk, and showed with a figure how O decreases at low energies and becomes roughly constant ($O \approx 0.35$) in the whole ISR energy range.

I may add that letters and telex exchanges were needed to convince the editor to insert the 29 names of the members of the CERN–Rome and Pisa–Stony Brook Collaborations, a request that in the past *Scientific American* — as they told me — had always refused because "they are too many and the readers are not interested". At that time a collaboration of 20 scientists were considered to be very large and papers in molecular biology were signed by 2–3 authors.

5. Second-Generation Experiments

In the years 1974–1978, three experiments brought more precise data. The first one was performed by the Annecy–CERN–Hamburg–Heidelberg–Vienna Collaboration that used the Split Field Magnet to accurately measure the elastic cross-section up to $-t = 12$ $(GeV/c)^2$.[38] It was observed that the minimum at $-t = 1.4$ $(GeV/c)^2$ deepens around $E_{CM} = 30\,GeV$ and fills up at larger energies (Fig. 19(a)). It was interesting to remark that the deepest minimum happens at the same energies at which the forward real part is practically zero (Fig. 21(b)), possibly indicating that the fill-up at higher energy is due to a non-zero real part of the large-angle scattering amplitude.

In 1973 the CERN–Rome and Pisa–Stony Brook Collaborations proposed to the ISR Committee a joint experiment that would be done in new Roman pots installed — with more precise hodoscopes — in intersection region I8 where the Pisa–Stony Brook apparatus was located. Figure 20 shows the overall apparatus.

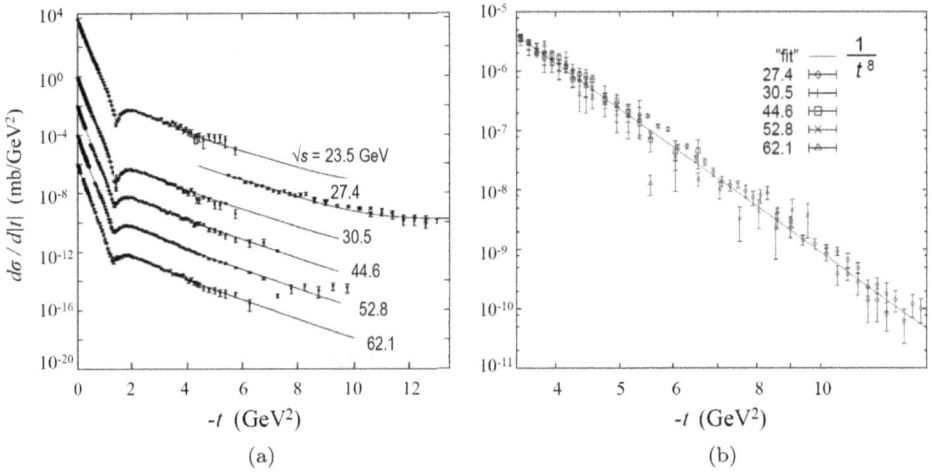

Fig. 19. (a) The elastic differential cross-sections at large momentum transfers plotted on different vertical scales.[38] (b) The elastic cross-section is energy independent and decreases as $1/t^8$.[39]

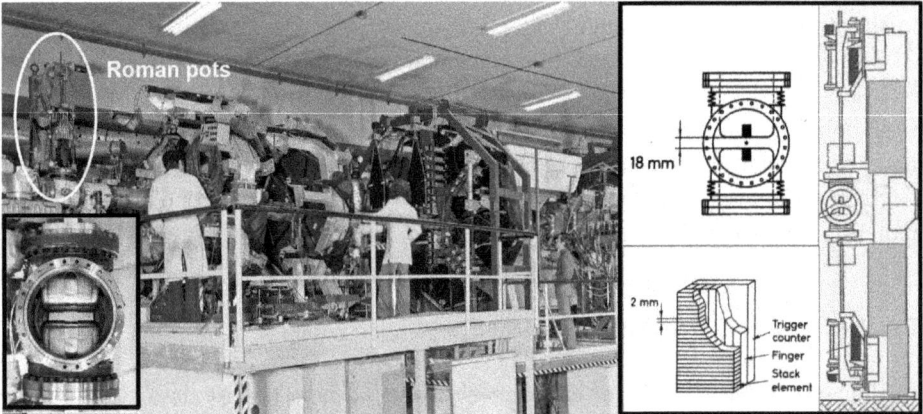

Fig. 20. The CERN–Rome–Pisa–Stony Brook apparatus and its Roman pots.

As the inset to Fig. 20 shows, the four pots — two per side — had very thin and flat windows, which allowed the pots — and the new systems of "finger" scintillators they contained — to be moved much closer to the circulating proton beams than in the previous experiment, once the beam stacking process was completed. The set-up also allowed a much more accurate measurement of the distance between the edges of the two hodoscopes located one on top of the other. I well remember Giuseppe Cocconi and the NIKHEF PhD student Jheroen Dorenbosch spending long hours to improve — through accurate position measurements — the knowledge of the momentum transfer q. (It can be mentioned that in 1977 one of the CERN–Rome scintillation hodoscope was requested by the National Museum of History and Technology in Washington to be shown to the public.)

The combination of the two detectors opened the way to the application of the new method for measuring total cross-sections. This is based on the measurement of (i) the total number of inelastic events N_{in}, measured by the Pisa–Stony Brook detector in a given run, which is, after small corrections due to the unavoidable losses, proportional to σ_{tot} and (ii) the extrapolated forward rate $(dN/dt)_0$, measured by the CERN–Rome hodoscopes, which is proportional to σ_{tot}^2. Because of the optical theorem, σ_{tot} is proportional to the ratio $(dN_{el}/dt)_0/N_{tot}$, where $(dN/dt)_0$ is the extrapolated forward number of events and $N_{tot} = N_{in} + N_{el}$ is the total number of inelastic and elastic events, computed by integrating the differential rate dN_{el}/dt.

$$\sigma_{tot} = \frac{16\pi}{(1+\rho^2)} \frac{(dN_{el}/dt)_0}{N_{el} + N_{in}}$$

The combined results of the three methods are plotted in Fig. 21(a).[40] (It is worth noting that the ratio ρ is small and contributes a negligible error to σ_{tot}).

The CERN–Rome measurements of the real part of the forward amplitude[41, 42] — obtained with the improved Roman pots of Fig. 20 — are plotted in Fig. 21(b). The curves of the two figures have been obtained by taking into account the dispersion relation that connects the forward real parts to energy integrals of the total cross-sections. The physical content of the complicated mathematics can be understood by stating that, at high energies, ρ becomes roughly proportional to the *logarithmic* derivative of the total cross-section, $d\sigma_{tot}/d(\ln s)$. This fits with

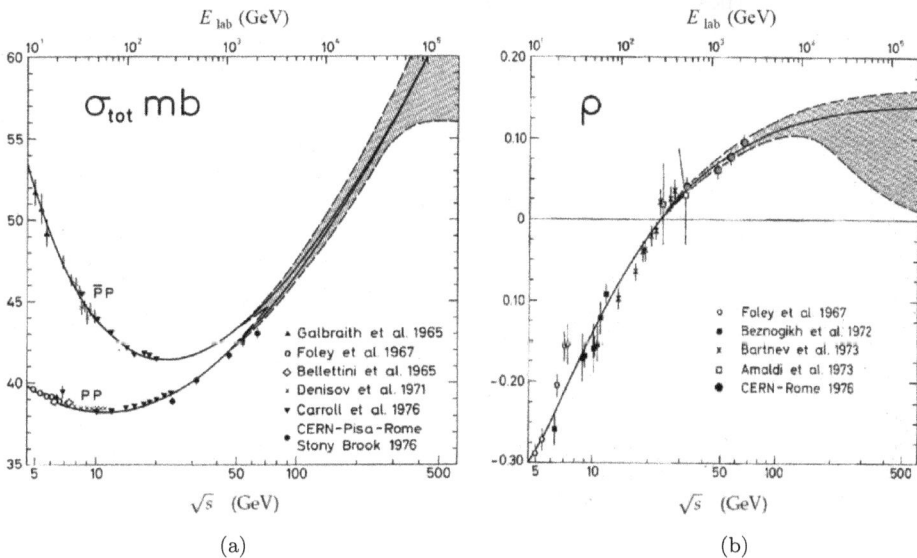

Fig. 21.   The curves are fitted to the energy dependence of the total cross-sections and the forward real part, and are based on the analyticity properties of the scattering matrix.[45, 46]

the Khuri–Kinoshita theorem, which states that $\rho \to \pi \ln s$ for a cross-section that increases proportionally to $\ln^2 s$ — and explains why precise measurements of ρ at $\sqrt{s} \approx 50 \, \text{GeV}$ determine the total cross-section up to about $500 \, \text{GeV}$. (A discussion of this rough argument can be found in Ref. 43.)

This was the first experiment in which the measured ratio ρ was used to obtain information on the energy dependence of the total cross-section at energies much larger than those available.

The CERN–Rome fit[41] gives a total cross-section that increases as $\ln(s/s_0)^\gamma$ with $\gamma = 2.1 \pm 0.1$ and $s_0 = 1 \, \text{GeV}$. As in the first generation experiments, the exponent coincides, with a smaller error, with the limiting value of the Froissart–Martin bound. This fact was confirmed by a second experiment performed just before the demise of the ISR, when the availability of the CERN Antiproton Accumulator allowed a measurement of the real part of the antiproton–proton forward scattering amplitude.[44] The CERN–Louvain-la-Neuve–Northwestern–Utrecht Collaboration used the apparatus of the CERN–Rome Collaboration and inherited its techniques: I remember Jheroen Dorenbosch and myself passing to Martin Bloch the codes that we had developed over the years.

6. Overlap Integrals in the ISR Energy Range

To understand the significance of these results, let us step back the definition of the profile function $\Gamma(a)$ and the inelastic overlap integral $G_{\text{in}}(a)$. In 1980, from all the measured elastic differential elastic cross-sections Klaus Schubert and myself have computed these quantities in a much-quoted article.[45]

Figure 22(a) shows that the profile function is Gaussian-like and completely different from that of Fig. 22(b),[43] which describes a "black disk" having a radius

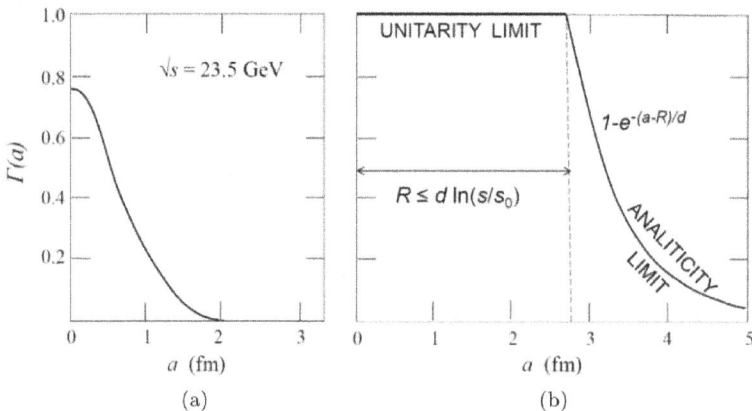

Fig. 22.   At ISR energies the profile function is far from saturating the unitarity and analyticity constraints that define the Froissart–Martin bound. The length d, which is determined by the pion mass, fixes the constant C that multiplies $\ln^2(s/s_0)$ in the Froissart–Martin bound.

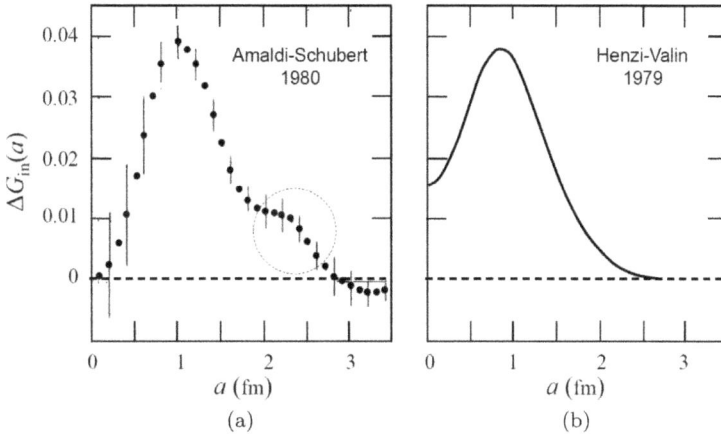

Fig. 23. The impact parameter dependence of variation of the inelastic overlap integral in the ISR energy range is the best way to understand the significance of the rising p–p cross-sections.

proportional to $\ln(s/s_0)$ and a grey periphery of constant width, as needed to saturate the Froissart–Martin bound. (It is worthwhile recalling that this is the high-energy behaviour predicted by the Cheng and Wu massive quantum electrodynamics model.[35, 36])

In the already quoted original works of the 1960s,[6, 7] C was proven to be equal to $\pi(\hbar/m_\pi)^2 \approx 60\,\mathrm{mb}$, but in a 2009 paper[46] André Martin derived for the *inelastic* cross-section the new limit $C = \pi(\hbar/2m_\pi)^2$, which is four times smaller and corresponds to $d = \hbar/[(2\sqrt{2})m_\pi] \approx 0.5\,\mathrm{fm}$. Still the new constant is thirty times larger than the best fit to the experimental data.

I now consider the *increase* $\Delta G_{in}(a)$ of the inelastic overlap integral over the ISR energy range. In 1973 I presented such an analysis in Aix en Provence, concluding that the increase of the proton–proton cross-section is a peripheral phenomenon,[30] a conclusion reached at the same time by others.[47, 48]

This is confirmed by Fig. 23(a), which is the result of the analysis performed with Klaus Schubert.[45] The novelties brought by this analysis were the *direct* calculation of $G_{in}(a)$ from the experimental data in the ISR energy range ($23\,\mathrm{GeV} \leq \sqrt{s} \leq 62\,\mathrm{GeV}$) and the careful estimates of statistical and systematic errors.

It is worth mentioning that the physical origin of the bump of $\Delta G_{in}(a)$ at $a = 2.3\,\mathrm{fm}$, first noted in Ref. 45, is not yet known.

Figure 23(b) displays the results of the analysis by Henzi and Valin,[49] who used a different approach by first fitting the differential cross-sections with analytical functions and then computing $G_{in}(a)$.

We can see that the shadow of the inelastic channels increases by $\Delta G_{in} = 0.04$ at 1 fm, which confirms the peripheral nature of the phenomenon. At $a = 0$ the two analyses are compatible when the errors are properly taken into account and indicate that $\Delta G_{in}(0)$ is less than three times smaller than $\Delta G_{in}(a = 1\,\mathrm{fm})$. It could even be zero, since small impact parameters imply large momentum transfers, and in this region the analytical fits to the cross-section[49] are not perfect, a problem that is not encountered when the experimental data are used directly, as done in Ref. 45.

As mentioned above, the fitted exponent of the logarithmic increase of σ_{tot} is 2, with a very small error. We can now answer the question: is this fact connected with the exponent 2 predicted by the Froissart–Martin bound? The answer must be negative, because the overlap integral of Fig. 22(a) is very different from that of Fig. 22(b), but the coincidence is so puzzling that, without understanding, the expression "qualitative saturation of the Froissart–Martin bound" was introduced and much used.

In synthesis, the ISR measurements of elastic scattering, total cross-section highlighted an unexpected state of affairs: with increasing collision energy, the proton–proton opacity at zero impact parameter does not decrease — as predicted by the "classical" pomeron exchange model — but remains roughly constant.

7. The ISR "Small-Angle Physics" Seen from Higher Energies

In forty years, the energy of hadron–hadron colliders has passed from $\sqrt{s} = 30\,\mathrm{GeV}$, the ISR minimum value, to the $\sqrt{s} = 8000\,\mathrm{GeV}$ available at the LHC in 2012. A review of the results obtained at this high-energy frontier is beyond the scopes of this paper; however, before closing, some remarks concerning the energy evolution of the main phenomena discussed in the previous sections may be useful.

Figure 24 reproduces the data obtained at the CERN antiproton–proton collider, at the Tevatron and, recently, at the Large Hadron Collider by the TOTEM Collaboration.[50, 51] ATLAS and CMS have published similar data.[52, 53] It is seen that the low energy trend continues and the rough cosmic ray data (see, for instance, Ref. 54) are in agreement with the precise results obtained at LHC.

To detect the protons scattered at very small angles the TOTEM collaboration has located its Roman pots at hundreds of meters from the interaction point, in a high-beta interaction region, used — as proposed for the ISR by Darriulat and Rubbia in Ref. 17 — as a magnifying lens.

The slope of the forward elastic cross-section continues to increase up to $2000\,\mathrm{GeV}$ (Fig. 25(a)).

However the TOTEM value (in red) is surprisingly larger than the prediction of the fit to lower energy data. By excluding it from the fit, the slope of the pomeronslope of the pomeron trajectory is in agreement with the value obtained at lower energies: $\alpha_{P}{\prime}(0) = 0.25\,\mathrm{GeV}^{-2}$.

An overall fit to the total cross-section of Fig. 24 and to the ρ-parameter of Fig. 25(b) gives for the exponent of $\ln(s)$ the value $\gamma = 2.23 \pm 0.15$,[55] in agreement with the value obtained in Ref. 45: $\gamma = 2.1 \pm 0.1$. This is a confirmation of the fact that the ISR value has nothing to do with the Froissart–Martin bound.

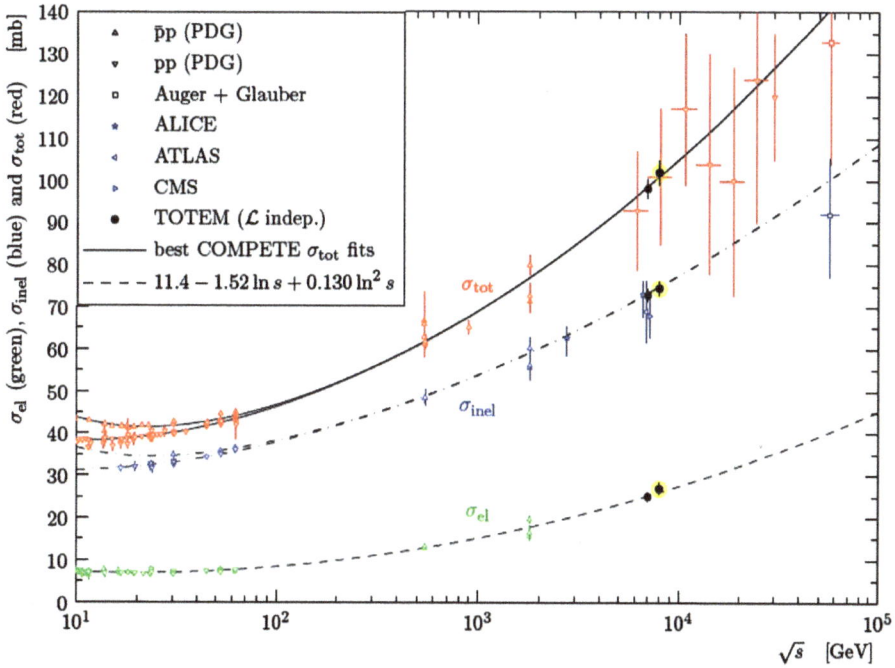

Fig. 24. Summary of the available data on the total, inelastic and elastic cross-sections. The TOTEM points are in black.

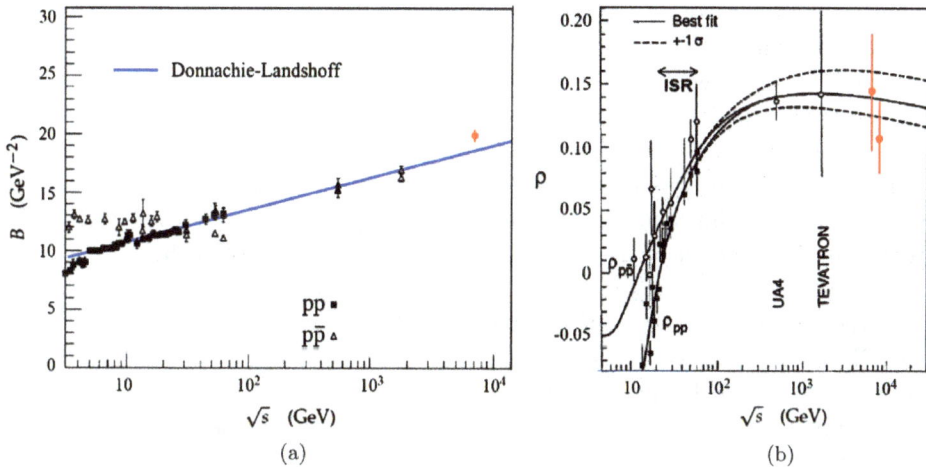

Fig. 25. The slope B of forward elastic cross-section shrinks in an enormous energy range: $30 \le \sqrt{s} \le 7000$ GeV and the ratio between the imaginary and real part of the forward amplitude has a shallow maximum around 1000 GeV. The TOTEM results are in red.

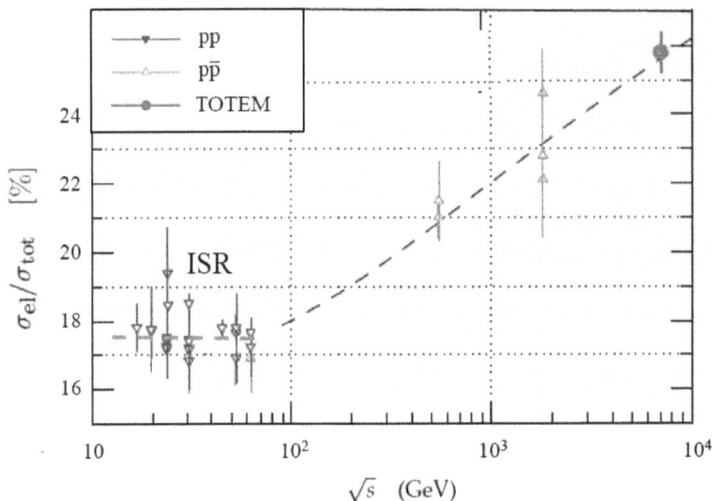

Fig. 26.   Above the ISR energies the ratio σ_{el}/σ_{tot} is a linear function of the centre-of-mass energy.

By comparing the error bars of Fig. 25(b) with the one-sigma band — defined by a fit to the lower energy data — one can conclude that the precision of the measurement has to be improved by at least a factor of 3 in order to derive useful constraints on the behaviour of the total cross-section at energies much larger than the one available at LHC, as it was done in the 1970s at the ISR (Fig. 21).

Figure 26 shows that the constant value $\sigma_{el}/\sigma_{tot} \approx 0.175$ — an indication of what was called "geometrical scaling" — is valid only in the ISR energy range.

Approximate geometrical scaling implies that in this energy region the central inelastic overlap integral $G_{in}(0)$ is *almost constant* while the effective proton–proton interaction radius increases so that the total cross-section increases. Before 1973, in the framework of the Pomeron model with intercept equal 1, most theorists were instead predicting a *decreasing* $G_{in}(0)$ so to exactly compensate the increasing proton–proton radius and produce an energy independent total cross-section. This is the physical content of the unexpected result obtained at the ISR.

Since the ratio σ_{el}/σ_{tot} increases above 100 GeV, it does not come as a surprise that the central inelastic overlap integral $G_{in}(0)$, in passing from the ISR to the CERN proton–antiproton collider, increases, as shown by Henzi and Valin,[56] at variance from what happens in the ISR energy range (Fig. 27).

In summary, the s-channel description based on a *purely peripheral* increase of the inelastic overlap integral may be valid in the ISR energy range but certainly it is not at higher energies. From this wider point of view the much discussed "geometrical scaling" of the 1970s is a transition regime of the restricted energy region where the total proton–proton cross-section begins its rise.

I conclude this discussion of the s-channel description of high-energy scattering by recalling that Henzi and Valin gave a descriptive title to their 1983 paper:[56]

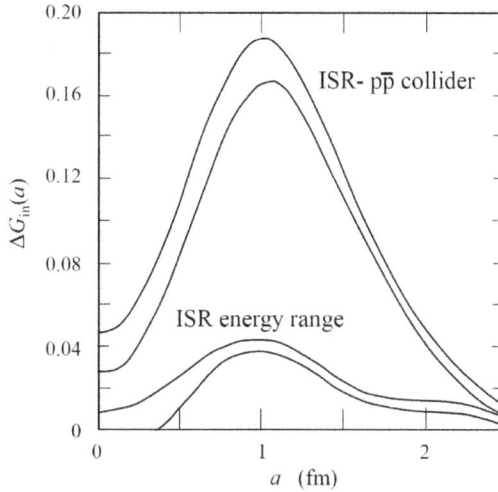

Fig. 27.　$G_{in}(0)$ increases in the energy range $\sqrt{s} = 53\text{--}550\,\text{GeV}$[55] while from 23 to 62 GeV $G_{in}(0)$ is zero within the errors.[45, 49] The bands represent the estimated error bars.

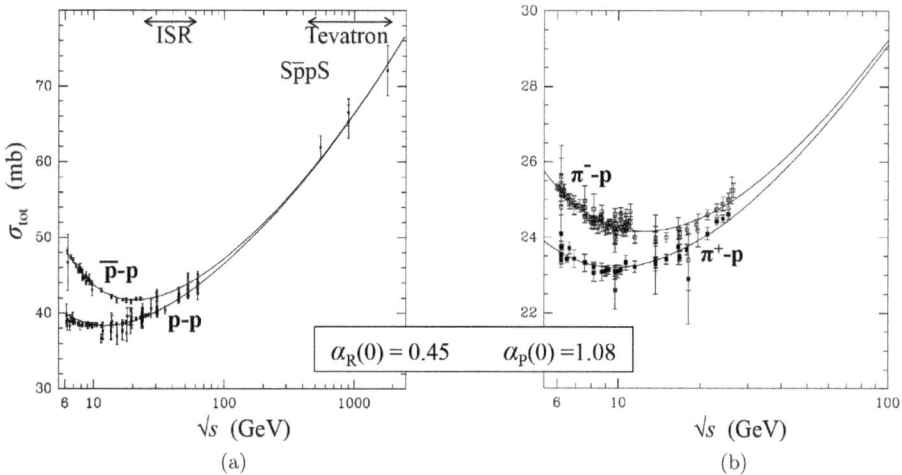

Fig. 28.　The figure shows the fits obtained by A. Donnachie and P. Landshoff to the best-measured total cross-sections.[3] The intercept of the Reggeon trajectory ($\alpha_R(0) = 0.45$) is in good agreement with the value derived from the masses of the particles belonging to it (Fig. 2).

"Towards a blacker, edgier and larger proton". Moreover in 2015, to summarise their overall fit to all the available data, in Ref. 57 Martin Block *et al.* wrote: "The cross-sections approach a black disk limit asymptotically. The approach to the limit is, however, very slow: A 'black disk' of logarithmically growing radius is supplemented by a soft 'edge' whose properties are in invariant with energy."

This vision is contrasted by the *t*-channel descriptions of the energy dependence of hadron–hadron cross-sections.

In 1992, Donnachie and Landshoff wrote all the hadron–hadron total cross-sections as the sum $\sigma_{\text{tot}} = Xs^\varepsilon + Ys^{-\eta}$ of two powers, the first due to pomeron exchange and the second to the exchange of the trajectory of Fig. 2.[3] Figure 28 shows the experimental points and the fitted curves for the four best-measured channels. They state their conclusion in the following terms: "The fact that all cross-sections rise with energy at the same rate s^ε makes it unnatural to attribute the rise to some intrinsic property of the hadrons involved. It is unhelpful to adopt a geometrical approach and to talk of hadrons becoming bigger and blacker as the energy increases. Rather the rise is a property of something that is exchanged, the pomeron, and this is why the rise is universal. Our conclusions are in accord with the recent important results from UA8 at the CERN collider, which indicate that the pomeron does have a rather real existence: it can hit hadrons hard, break them up and knock most of their fragments sharply forward."

In the fit of Fig. 28 the standard pomeron intercept is at $\alpha(0) = 1.08$ but the authors warn the reader that the exponent $\varepsilon = 0.08$ (appearing in the energy dependence s^ε of the total cross-sections) is a little less than $\alpha(0) - 1$ because of multiple pomeron exchange.

This shows that twenty years ago a debate between the followers of the *s*-channel and the *t*-channel approaches was going on. And it is still alive, as indicated by a paper by Donnachie and Landshoff[58] who in 2011 — forty years after the first ISR physics runs — have analysed the data produced at the LHC by the TOTEM Collaboration coming to the conclusion that their picture is still valid but a term has to be added due to the "hard pomeron" observed in electron–proton collisions at HERA by ZEUS and H1.[59]

In 1973 Daniele Amati could not accept a pomeron intercept above 1, even if his Aix en Provence talk started with the following words: "Despite of the title of this session, there is no theory of strong interactions. Our hadron world is complex and we lack a dynamical theory that could allow us to understand and calculate its properties".[60] Forty years later an analysis of all hadron–hadron cross-sections lead to the conclusion that the soft pomeron intercept is at 1.0926 ± 0.0016,[61] a very precise number; still we are not able to compute it from quantum chromodynamics, the well-established fundamental theory of strong interactions.

8. Concluding Remarks

It is often said that the ISR did not have the detectors needed to discover fundamental phenomena made accessible by its large and new energy range. This is certainly true for "high-momentum-transfer physics", which, since the end of the 1960s, became a main focus of research, but the statement does not apply to the field that is the subject of this paper — elastic and total cross-sections — and to diffraction dissociation, for which the interested reader is referred to Ref. 62.

In fact, looking back to the results obtained at the ISR by the experiments aimed at measuring total cross-sections and small-angle scattering and particle production, one can safely say that the detectors were very well suited to the tasks and performed much better than foreseen.

As far as the results are concerned, in this particular branch of hadron-hadron physics, very precise measurements were performed, new phenomena were discovered, unexpected scaling laws were found and the first detailed studies of that still elusive concept which goes under the name "pomeron", were performed.

Moreover, some precision techniques and methods have had a lasting influence: since then all colliders had and have their Roman pots, and the different methods developed at the ISR for measuring the luminosity are still in use.

"Small-angle physics" is not very fashionable today but it gives a lot of satisfaction to those who accurately labour around it and, in addition, has a great merit: it requires a very close collaboration among machine physicists and experimentalists, an invaluable gift that as experimentalists we enjoyed for the first time at a wonderful collider, the Intersecting Storage Rings.

References

1. S. P. Denisov *et al.*, Total cross-sections of π^+, K^+ and p on protons and deuterons in the momentum range $15\,\text{GeV}/c$ to $60\,\text{GeV}/c$, *Phys. Lett. B* **36**, 415–421 (1971).
2. A. N. Diddens *et al.*, High-energy proton scattering, *Phys. Rev. Lett.* **9**, 108–111 (1962).
3. A. Donnachie and P. V. Landshoff, Total cross-sections, *Phys. Lett. B* **296**, 227–232 (1992).
4. V. N. Bolotov *et al.*, Negative pion charge exchange scattering on protons in the momentum range 20–$50\,\text{GeV}/c$, *Nucl. Phys. B* **73**, 365–386 (1974).
5. I. Ya. Pomeranchuk, Equality of the nucleon and antinucleon total interaction cross-section at high energies, *Sov. Phys. JETP* **7**, 499–501 (1958).
6. M. Froissart, Asymptotic behavior and subtractions in the Mandelstam representation, *Phys. Rev.* **123**, 1053–1057 (1961).
7. A. Martin, Unitarity and high-energy behaviour of scattering amplitudes, *Phys. Rev.* **129**, 1432–1436 (1963).
8. N. N. Khuri and T. Kinoshita, Real part of the scattering amplitude and the behavior of the total cross-section at high energies, *Phys. Rev. B* **137**, 720–729 (1965).
9. G. Bellettini *et al.*, Measurement of the p–p total cross-section, CERN/ISRC/69-12, 14 March 1969.
10. P. Darriulat and C. Rubbia, On beam monitoring for ISR experiments, ISR User's Meeting, CERN Internal Report, 10–11 June 1968.
11. C. Rubbia, Report to the ISR user's meeting, ISR User's Meeting, CERN Internal Report, 10–11 June 1968.

12. W. Schnell, A mechanical beam profile monitor for the ISR, CERN Internal Report, ISR User's Meeting, CERN Int. Rep., ISR-RF/68-19, 22 April 1968.

13. J. Steinberger, Suggestions for the luminosity measurement of the ISR, ISR User's Meeting, CERN Internal Report, 10–11 June 1968.

14. A. P. Onuchin, Suggestions for the luminosity measurements at the ISR, CERN Internal Report, NP/68/26.

15. G. Cocconi, An absolute calibration of the ISR luminosity, CERN Internal Report, NP/67/436, 1967.

16. L. di Lella, Elastic proton–proton scattering with the ISR, CERN ISR User's Meeting, CERN Internal Report, 10–11 June 1968.

17. C. Rubbia and P. Darriulat, High beta interaction region: A magnifying lens for very small angle proton–proton scattering, CERN Internal Report, 1968.

18. S. Van der Meer, Calibration of the effective beam height in the ISR, CERN Internal Report, ISR-PO/68-31, 18 June 1968.

19. U. Amaldi *et al.*, The measurement of proton–proton differential cross-section in the angular region of Coulomb scattering at the ISR, CERN/ISRC/69–20, 24 March 1969.

20. L. W. Jones, Recent U.S. work on colliding beams, in *Proc. 4th Int. Conf. on High Energy Accelerators*, 21–27 Aug 1963, Dubna, eds. A. A. Kolomensky, A. B. Kusnetsov and A. N. Lebedev (JINR, 1964), pp. 379–390.

21. B. D. Hyams and V. Agoritsas, Background in the ISR, CERN Internal Report, AR/Int. SG/65-29.

22. U. Amaldi, R. Biancastelli, C. Bosio, G. Matthiae, E. Jones and P. Strolin, Report on background measurements at the PS in preparation of the small angle ISR elastic scattering experiment, ISRC 70–25.

23. P. Darriulat *et al.*, Measurement of the elastic scattering cross-section at the ISR, CERN/ISRC/69–19, 16 March 1969.

24. M. Holder *et al.* (ACHGT Collaboration), Observation of small angle proton–proton elastic scattering at 30 GeV and 45 GeV center-of mass energy, *Phys. Lett. B* **35**, 355–360 (1971).

25. U. Amaldi *et al.* (CERN–Rome Collaboration), Measurements of small angle proton–proton elastic scattering at the CERN Intersecting Storage Rings, *Phys. Lett. B* **36**, 504–508 (1971).

26. V. Bartenev *et al.*, Measurement of the slope of the diffraction peak in elastic pp scattering from 50 GeV to 400 GeV, *Phys. Rev. Lett.* **31**, 1088–1091, 1367–1370 (1973).

27. G. Barbiellini *et al.* (ACHGT Collaboration), Small angle proton–proton elastic scattering at very high energies ($460 \, \text{GeV}^2 < s < 2900 \, \text{GeV}^2$), *Phys. Lett. B* **39**, 663–667 (1972).

28. A. Boehm *et al.* (ACHGT Collaboration), Observation of a diffraction minimum in proton–proton elastic scattering at the ISR, *Phys. Lett. B* **49**, 491–495 (1974).

29. M. Holder *et al.* (ACHGT Collaboration), Further results on small angle elastic proton–proton scattering at very high energies, *Phys. Lett. B* **36**, 400–402 (1971).

30. U. Amaldi, Elastic scattering and low multiplicities, in *Proc. Aix en Provence Int. Conf. on Elementary Particles, 6–12 September 1973*, J. Phys. (Paris), Suppl. 10, C1, **34**, 241–260 (1973).

31. U. Amaldi *et al.* (CERN–Rome Collaboration), Measurements of the proton–proton total cross-section by means of Coulomb scattering at the Intersecting Storage Rings, *Phys. Lett. B* **43**, 231–236 (1973).

32. U. Amaldi *et al.* (CERN–Rome Collaboration), The energy dependence of the proton–proton total cros-section for center-of-mass energies between 23 and 53 GeV, *Phys. Lett. B* **44**, 112–118 (1973).

33. R. Amendolia *et al.* (Pisa–Stony Brook Collaboration), Measurements of the total proton–proton cross-section at the ISR, *Phys. Lett. B* **44**, 119–124 (1973).

34. W. Heisenberg, Production of mesons as a shock wave problem, *Z. Phys.* **133**, 65–79 (1952).

35. H. Cheng and T. T. Wu, Limit of cross-sections at infinite energy, *Phys. Rev. Lett.* **24**, 1456–1460 (1970).

36. H. Cheng, J. K. Walker and T. T. Wu, Impact picture of proton–proton, antiproton–proton, pion–proton and kaon–proton elastic scattering from 20 to 5000 GeV, *Phys. Lett. B* **44**, 97–101 (1973).

37. U. Amaldi, Proton interactions at high energies, *Sci. Am.* **299**, 36–44 (1973).

38. E. Nagy *et al.* (Annecy–CERN–Hamburg–Heidelberg–Wien Collaboration), Measurements of elastic proton–proton scattering at large momentum transfer at the CERN Intersecting Storage Rings, *Nucl. Phys. B* **150**, 221–267 (1979).

39. A. Donnachie and P.V. Landshoff, The interest of large-*t* elastic scattering, *Phys. Lett. B* **387**, 637–641 (1996).

40. U. Amaldi *et al.* (CERN–Rome–Pisa–Stony Brook Collaboration), New measurements of proton–proton total cross-sections at the CERN Intersecting Storage Rings, *Phys. Lett. B* **62**, 460–464 (1976).

41. U. Amaldi *et al.* (CERN–Rome Collaboration), The real part of the forward proton–proton scattering amplitude measured at the CERN Intersecting Storage Rings, *Phys. Lett. B* **66**, 390–394 (1977).

42. J. Dorenbosch, The real part of the forward proton–proton scattering amplitude measured at the CERN Intersecting Storage Rings, PhD thesis, University of Amsterdam, The Netherlands (1977).

43. U. Amaldi, Elastic processes at the Intersecting Storage Rings and their impact parameter description, in *Laws of hadronic matter, Proc. Erice School 1973*, ed. A. Zichichi (Academic Press, New York, 1975), pp. 672–741.

44. N. Amos *et al.* (CERN–Louvain-la-Neuve–Northwestern–Utrecht Collaboration), Measurements of small-angle proton-antiproton and proton–proton elastic scattering at the CERN Intersecting Storage Rings, *Nucl. Phys. B* **262**, 689–714 (1985).

45. U. Amaldi and K. Schubert, Impact parameter interpretation of proton–proton scattering from a critical review of all ISR data, *Nucl. Phys. B* **166**, 301–320 (1980).

46. A. Martin, Froissart bound for inelastic cross-sections, *Phys. Rev. D* **80**, (2009) 065013.

47. H. I. Miettinen, s-channel phenomenology of diffraction scattering, in *Proc. Aix en Provence Int. Conf. on Elementary Particles, 6–12 September 1973, J. Phys. (Paris), Suppl. 10, C1,* **34**, (1973), 263–267.

48. R. Henzi, B. Margolis and P. Valin, Energy dependence of factorizable models of elastic scattering, *Phys. Rev. Lett.* **32**, 1077–1080 (1974).

49. R. Henzi, and P. Valin, On elastic proton–proton diffraction scattering and its energy dependence, *Nucl. Phys. B* **148**, 513–573 (1979).

50. C. Augier *et al.*, Predictions on the total cross-section and real part at LHC and SSC, *Phys. Lett. B* **315**, 503–506 (1993).

51. G. Antchev *et al.* (TOTEM Collaboration), Luminosity-independent measurement of the proton–proton total cross-section at $\sqrt{s} = 8$ TeV, *Phys. Rev. Lett.* **111**, 012001 (2013).

52. ATLAS Collaboration, Measurement of the total cross-section from elastic scattering in pp collisions at $\sqrt{s} = 7$ TeV with the ATLAS detector, *Nucl. Phys. B* **889**, 486–548 (2014).

53. CMS Collaboration, Measurements of the inelastic proton–proton cross-section at $\sqrt{s} = 7$ TeV, *Phys. Lett. B* **722**, 5–27 (2013).

54. P. Abreu *et al.*, Pierre Auger Collaboration, Measurement of the proton-air cross-section at $\sqrt{s} = 57$ TeV with the Pierre Auger Observatory, *Phys. Rev. Lett.* **109**, 062002 (2012).

55. M. J. Menon and P. V. R. G. Silva, An updated analysis of the rise of the hadronic total cross-sections at the LHC energy region, *Int. J. Mod. Phys. A* **28**, 1350099 (2013).

56. R. Henzi and P. Valin, Towards a blacker, edgier and larger proton, *Phys. Lett. B* **132**, 443–448 (1983).

57. A. Donnachie and P. V. Landshoff, Elastic scattering at the LHC, arXiv:1112.2485 [hep-ph], 12 December 2011.

58. A. Donnachie and P. V. Landshoff, New data on the hard pomeron, *Phys. Lett. B* **518**, 63–71 (2001).

59. D. Amati, Strong interaction theory, in *Proc. Aix en Provence Int. Conf. on Elementary Particles, 6–12 September 1973, J. Phys. (Paris), Suppl. 10, C1*, **34**, 129–140 (1973).

60. M. J. Menon, P. V. R. G. Silva, A study on analytic parameterization for proton–proton cross-sections and asymptotia, *J. Phys. G: Nucl. Part. Phys. G* **40**, 125001 (2013).

61. U. Amaldi, M. Jacob and G. Matthiae, Diffraction of hadronic waves, *Ann. Rev. Nucl. Sci.* **26**, 385–456 (1976).

Deep Inelastic Scattering
with the SPS Muon Beam

Gerhard K. Mallot[1] and Rüdiger Voss[2]

CERN, CH-1211 Geneva 23, Switzerland
[1] *gerhard.mallot@cern.ch*
[2] *ruediger.voss@cern.ch*

We review results from deep inelastic muon scattering experiments at the SPS which started in 1978, and are still actively pursued today. Key results include the precision measurement of scaling violations and of the strong coupling constant, spin-dependent structure functions, and studies of the internal spin structure of protons and neutrons. These experiments have revealed a wealth of details about the internal structure of nucleons in terms of quarks and gluons.

1. Introduction

At the 14th International Conference in High Energy Physics in Vienna in 1968, SLAC reported for the first time the "scaling" behaviour of the electron–nucleon cross-section in the deep inelastic continuum, and W. K. H. Panofsky remarked that "... theoretical speculations are focused on the possibility that these data might give evidence on the behaviour of point-like, charged structures within the nucleon."[1] Soon after, it was realised that the parton structure of the nucleon discovered by the first electron–nucleon scattering experiments in the deep inelastic regime indeed confirmed the quark model of Gell-Mann[2] and Zweig.[3]

The early SLAC results on the quark–parton structure of the nucleon had a profound impact on the first-generation experimental programme of the CERN Super Proton Synchrotron (SPS), and most notably on the muon and neutrino scattering experiments. Several groups realised the potential of this new machine to extend the landmark SLAC experiments much 'deeper' into the inelastic regime by building a high-intensity, high-energy muon beam. This was the beginning of one of the most prolific fixed-target physics programmes of CERN that started in 1978, soon after the commissioning of the SPS, and is still vigorously pursued today.

This brief review focuses on two central components of this programme which, from a present-day perspective, have had the most lasting impact: (a) the precision measurement of scaling violations for tests of perturbative QCD and measurements of the strong coupling constant, and (b) the measurement of spin-dependent structure functions, the discovery of the 'spin crisis', and comprehensive studies of the spin structure of the nucleon. However, it must not be overlooked that the CERN muon programme has, over the years, produced a wealth of other, sometimes unexpected results. Examples are the discovery of nuclear effects in deep inelastic scattering,[4] the first observation of weak–electromagnetic interference effects in muon scattering, or measurements of charm production.

2.   Beam and Detectors

The SPS muon beam M2[5] was first commissioned in 1978 and is still in operation today, with only minor modifications. It is likely to be the best and most versatile high-energy muon beam ever designed, combining a wide range of momenta up to 300 GeV with high intensities and minimal halo background. The beam has a natural longitudinal polarisation that can be tuned by varying the momentum ratio of decay muons to parent pions, and can reach values up to ≈80%. A high beam polarisation is an essential prerequisite for the measurement of spin-dependent structure functions.

2.1.   *Early detectors*

Two large detectors were built for the first generation of experiments, the NA2 experiment of the European Muon Collaboration (EMC),[6–8] and the NA4 experiment of the Bologna–CERN–Dubna–Munich–Saclay (BCDMS) Collaboration.[9, 10] The two collaborations choose radically different, complementary experimental approaches. The EMC detector was a conventional open-geometry spectrometer built around a large air-gap dipole magnet, instrumented with proportional and drift chambers for particle tracking. The main advantages of this design were an excellent momentum resolution, a large kinematic range, and the ability to partly resolve the hadronic final state of the deep inelastic interaction. A disadvantage was the maximum target length allowed by the spectrometer layout, of order 1 m, which limits the statistical accuracy of many measurements.

In contrast, the BCDMS spectrometer was specifically designed for the inclusive measurement of high-momentum final state muons. It was based on a large, modular toroidal iron magnet of 50 m length instrumented with multiwire proportional chambers. In the centre, the toroid contained a modular target of almost the same length that could be filled with liquid hydrogen or deuterium, or replaced by solid target material. Principal advantages of this design were the enormous luminosity and the excellent muon identification through immediate absorption of the hadronic shower, which could not be resolved by the detector. Another obvious drawback was the comparatively poor momentum resolution due to multiple scattering in the iron magnet, limited to $\Delta p/p \approx 10\%$ over most of the momentum range.

The EMC and BCDMS experiments took data from 1978 until 1985, both with liquid hydrogen, liquid deuterium, and solid nuclear targets. In addition, EMC made first measurements with a polarised solid ammonia target. Whereas the BCDMS spectrometer was subsequently dismantled, the more versatile EMC spectrometer underwent several upgrades, in particular for the later NMC (NA37, 1986–1989) and SMC (NA47, 1992–1996) experiments. The NMC Collaboration (where the N stands for "New") refined and improved the EMC measurements of unpolarised structure functions, with a strong focus on the study of nuclear effects with a variety of heavy targets.[11] The SMC experiment (where the S stands for "Spin", obviously) was devoted exclusively to polarised muon-nucleon scattering with solid butanol, deuterated butanol, and ammonia targets.

2.2. *The COMPASS detector*

The most comprehensive rebuilt of the EMC/NMC/SMC spectrometer was undertaken by the COMPASS (NA58) Collaboration, which today continues the successful tradition of muon scattering at CERN, and still uses some of the original EMC equipment. The COMPASS experiment[12] started taking data in 2002.

Contrary to the one-stage EMC[6] and SMC spectrometers, the COMPASS detector (Fig. 1) is a two-stage magnetic spectrometer with the SM1 and SM2 dipoles. This results in a very large acceptance which is important for semi-inclusive deep inelastic scattering (SIDIS) experiments. Other essential additions and improvements concern the particle identification detectors, the large-acceptance, superconducting target magnet, and last but not least the high rate and data acquisition capabilities, which went up from the order of 100 Hz to 25 kHz.

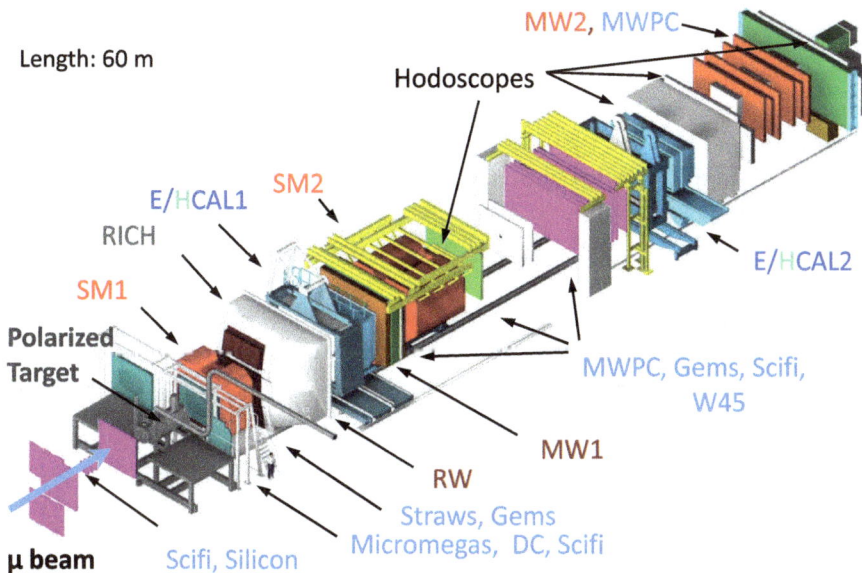

Fig. 1.   Artist's view of the COMPASS spectrometer. For a description see the text.

The COMPASS spectrometer is installed in the M2 muon beam line delivering muons of 160–200 GeV with a polarisation of about 80%. The usable beam intensity is typically 2×10^7/s during a 9.6 s long spill. The repetition rate varies and is typically about 1/40 s. The momentum of each beam muon is measured in the beam momentum station.

Charged particles are tracked in the beam region by scintillating fibre stations (SciFi) and silicon detectors. In the region close to the beam, micromega and gas-electron-multiplier (GEM) gaseous detectors with high rate capabilities are deployed. The backbone of tracking in the intermediate region is formed by multiwire proportional chambers (MWPCs). Finally, the large area tracking is covered by drift chambers (DC, W45) and drift tubes (Straws, RW, MW).

The velocity of charged particles is measured in a ring-imaging Cherenkov detector (RICH), which can separate pions and kaons from 9 GeV up to 50 GeV. The photon detector comprises multianode-photomultiplier tubes and in the periphery MWPCs with photosensitive CsI cathodes. The energy of charged particles is measured in sampling hadron calorimeters (HCAL), while neutral particles, in particular high-energy photons, are detected in electromagnetic calorimeters (ECAL).

Event recording is triggered by the scattered muon, which is "identified" by its ability to traverse thick hadron absorbers located just upstream of the Muon Wall detectors (MW), and detected by various systems of scintillator hodoscopes. The same spectrometer is also used for an experimental programme on hadron spectroscopy using pion, kaon and proton beams.[13]

2.3. *The COMPASS polarised target*

The heart of the experiment is the superconducting polarised target system. It comprises a 2.5 T solenoid and a 0.6 T dipole magnet, a ^{3}He/^{4}He dilution refrigerator originating from SMC, a 70 GHz microwave system for the dynamic nuclear polarisation (DNP), and an NMR system to measure the target polarisation. The target material is cooled down to about 60 mK in frozen spin mode. Irradiated ammonia (NH$_3$) and lithium-6 deuteride (^{6}LiD) were selected as proton and deuteron targets, respectively. Typical polarisations achieved are 85% for protons and 50% for deuterons. The target volume has an overall length of 1.3 m and comprises two or three cells with opposite polarisations. The target spins are rotated typically once per day by rotation of the magnetic field vector. The rotation can be stopped in transverse position for measurements with transverse target polarisation. For such measurements the polarisation is inverted typically once per week by DNP.

3.   Unpolarised Nucleon Structure Functions

Deep inelastic lepton–nucleon scattering is loosely defined as scattering at energy transfers much larger than the parton binding energy in the nucleon, such that the interaction occurs at the parton level and thus probes the internal quark–parton

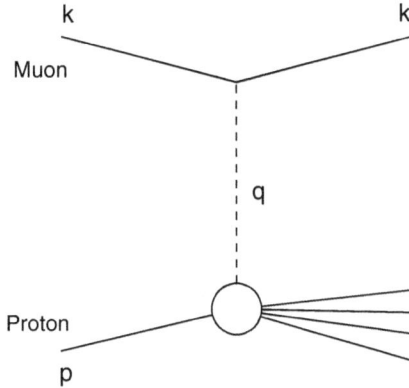

Fig. 2. Deep inelastic muon–nucleon scattering in lowest order.

structure of the target nucleon. Charged lepton scattering can be mediated through neutral-current γ or Z exchange, or through charged current $W^\pm$ exchange. At typical SPS fixed-target energies of a few hundred GeV, it is dominated by single photon exchange (Fig. 2). Consequently, muon scattering at CERN has mostly focused on this channel. The excellent luminosity of the BCDMS spectrometer has allowed for measurements of γ–Z interference[14, 15] which however have been completely superseded by data from HERA, and are not reviewed here.

3.1. *Cross-section and structure functions*

For inclusive scattering where the scattering amplitudes are summed over all possible hadronic final states, the unpolarised cross-section can be written as a function of two independent kinematic variables. One usually chooses two of the following Lorentz invariant variables,

- the squared four-momentum transfer

$$Q^2 = -q^2 = -(k - k')^2 = 4EE' \sin^2 \theta; \tag{1}$$

- the energy transfer to the hadronic system

$$\nu = \frac{p \cdot q}{M} - E' - E; \tag{2}$$

- the Bjorken scaling variable

$$x = \frac{Q^2}{2p \cdot q} = \frac{Q^2}{2M\nu}; \tag{3}$$

- and the scaling variable

$$y = \frac{p \cdot q}{p \cdot k} = \frac{\nu}{E}. \tag{4}$$

In these equations, k, k', p and q are the four-vectors of the initial and final state lepton, the target nucleon, and the exchanged boson. M is the mass of the target nucleon and the lepton mass has been neglected. E, E', and θ are the energies of the incident and scattered lepton, and the lepton scattering angle, in the laboratory frame.

The differential cross-section for unpolarised deep inelastic charged lepton scattering can be written, in the Born approximation, as[16, 17]

$$\frac{d^2\sigma}{dQ^2 dx} = \frac{4\pi\alpha^2}{Q^4}\frac{1}{x}\left[xy^2 F_1(x, Q^2) + \left(1 - y - \frac{Mxy}{2E}\right)F_2(x, Q^2)\right], \tag{5}$$

where α is the electromagnetic coupling constant and $F_1(x, Q^2)$ and $F_2(x, Q^2)$ are the unpolarised structure functions of the nucleon,

$$F_1(x, Q^2) = \frac{1}{2x}\sum_i e_i^2 x q_i(x, Q^2), \tag{6}$$

$$F_2(x, Q^2) = 2x F_1(x, Q^2) = \sum_i e_i^2 x q_i(x, Q^2). \tag{7}$$

In these expressions, $q_i(x, Q^2)$ is the probability distribution of partons of flavour i in the kinematic variables x and Q^2 and the index i runs over the active parton flavours in the nucleon. The SLAC discovery that the structure functions depend, at least approximately, on the dimensionless scaling variable x only,[18, 19]

$$q_i(x, Q^2) \approx q_i(x) \tag{8}$$

— the effect commonly referred to as "scaling" — is interpreted in the Quark–Parton Model (QPM) as elastic scattering on dimensionless, i.e. pointlike scattering centres inside the nucleon. Scaling becomes exact in the Bjorken limit where Q^2, $\nu \to \infty$ at constant x,[20] such that the transverse momentum of partons in the infinite momentum frame of the proton becomes negligible.

3.2. *Scaling violations*

When the muon experiments at the SPS started taking data in 1978, scaling and the QPM were well established experimentally and phenomenologically. The key interest of the experiments shifted soon to the measurement of small deviations from exact scaling behaviour, or *scaling violations*. As an example, the most representative fixed-target measurements of the proton structure function $F_2^p(x, Q^2)$ are shown in Fig. 3.[a] They exhibit a characteristic rise of the structure function with Q^2 at small x, a decrease at large x, and "apparent scaling" at $x \approx 0.15$.

---

[a]The first-generation data from muon scattering at the SPS were plagued by significant disagreements between the EMC and BCDMS results on F_2. The NMC Collaboration later remeasured this structure function with the upgraded EMC spectrometer, and eventually confirmed the BCDMS results.

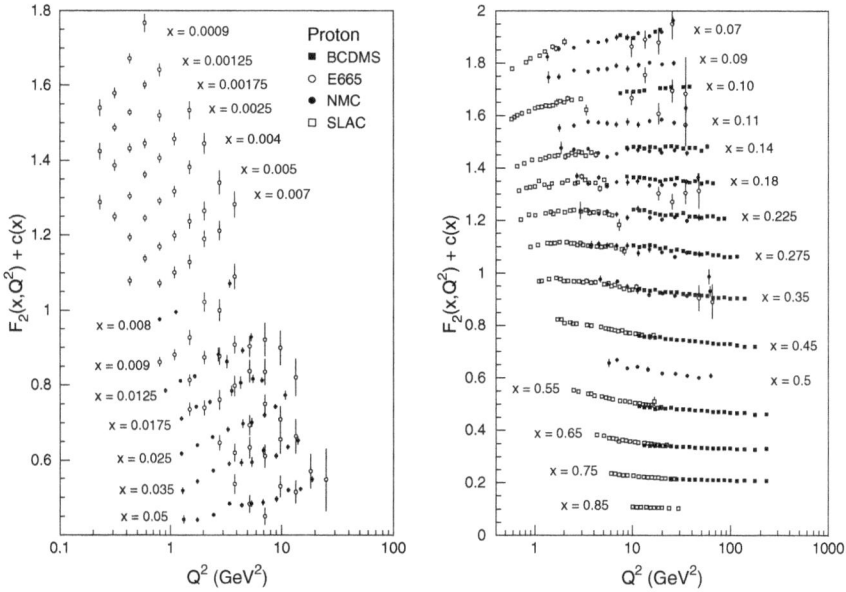

Fig. 3. The proton structure function F_2^p measured in deep inelastic muon scattering by the BCDMS[21] and NMC[22] experiments, shown as a function of Q^2 for bins of fixed x. The CERN data are complemented at small Q^2 by the SLAC electron scattering data,[23] and at small x by muon scattering data from the Fermilab E665 experiment.[24] Only statistical errors are shown. For the purpose of plotting, a constant $c(x) = 0.1i_x$ is added to F_2^p where i_x is the number of the x bin, ranging from 1 ($x = 0.05$) to 14 ($x = 0.0009$) on the left-hand figure, and from 1 ($x = 0.85$) to 15 ($x = 0.007$) on the right-hand figure.

3.3. *Tests of perturbative QCD*

Scaling violations occur naturally in Quantum Chromodynamics since, at large parton momenta x and increasing Q^2, the structure functions are increasingly depleted by hard gluon radiation from quarks; at small x, they are enriched by gluon conversion into low-momentum quark–antiquark pairs. The initial years of experimentation with the SPS muon beam coincided with the emergence of QCD as the universally accepted theory of the strong interaction, and were an active and exciting period of cross-fertilisation of phenomenology and experiments. Precise data on scaling violations turned out to be one of the most powerful tools to test the perturbative branch of the new theory, and allowed for one of the best early measurements of the strong coupling constant.

The Q^2 evolution of the strong coupling constant α_s is controlled by the renormalisation group equation of QCD. The "canonical" — but by no means unique — solution usually adopted for the analysis of deep inelastic data is, in next-to-leading order (NLO),

$$\alpha_s(Q^2) = \frac{4\pi}{\beta_0 \ln(Q^2/\Lambda^2)} \left[1 - \frac{\beta_1}{\beta_0^2} \frac{\ln \ln(Q^2/\Lambda^2)}{\ln(Q^2/\Lambda^2)} \right] \tag{9}$$

where the so-called beta functions are given by

$$\beta_0 = 11 - \frac{2}{3}N_f, \quad \beta_1 = 102 - \frac{38}{3}N_f$$

and N_f is the effective number of active quark flavours in the scattering process. The parameter Λ is the so-called "mass scale" of QCD and has the physical meaning of a typical energy at which the running coupling constant (9) becomes large and the perturbative expansion breaks down. Its value is not predicted by QCD and can only be determined by experiment. Since α_s is the physical observable, the numerical value of Λ depends on N_f and, beyond leading order, on the renormalisation scheme assumed to compute the perturbative QCD expansions.

The Q^2 evolution of the effective quark and gluon distribution is predicted by the Altarelli–Parisi equations,[25]

$$\frac{dq^{NS}(x, Q^2)}{d\ln Q^2} = \frac{\alpha_s(Q^2)}{2\pi} \int_x^1 q^{NS}(t, Q^2) P^{NS}\left(\frac{x}{t}\right) \frac{dt}{t}, \tag{10}$$

$$\frac{dq^{SI}(x, Q^2)}{d\ln Q^2} = \frac{\alpha_s(Q^2)}{2\pi} \int_x^1 \left[q^{SI}(t, Q^2) P_{qq}\left(\frac{x}{t}\right) + C_q g(t, Q^2) P_{qg}\left(\frac{x}{t}\right)\right] \frac{dt}{t}, \tag{11}$$

$$\frac{dg(x, Q^2)}{d\ln Q^2} = \frac{\alpha_s(Q^2)}{2\pi} \int_x^1 \left[g(t, Q^2) P_{gg}\left(\frac{x}{t}\right) + C_g q^{SI}(t, Q^2) P_{gq}\left(\frac{x}{t}\right)\right] \frac{dt}{t}, \tag{12}$$

where SI and NS denote flavour singlet and non-singlet combinations of quark distributions, respectively, g is the gluon distribution, and the C_i are a set of coefficients. P^{NS}, P_{qq}, etc. are so-called splitting functions describing the QCD diagrams which can be calculated in perturbative QCD as power series in α_s.

3.4. *Measurement of the strong coupling constant*

The F_2 measurements of BCDMS at large x and Q^2 with carbon, hydrogen, and deuterium targets were the first high statistics data that yielded a conclusive determination of Λ_{QCD}.[26–28] The original BCDMS fits to the hydrogen and deuterium data were later superseded by a careful analysis by Virchaux and Milsztajn of the combined SLAC and BCDMS hydrogen and deuterium data.[29] Since the SLAC data extend down to four-momentum transfers as low as $Q^2 = 1\,\text{GeV}^2$, these authors make an allowance for non-perturbative "higher twist" contributions to the observed scaling violations at small Q^2. These higher twist effects are mostly due to long-distance final state interactions which are difficult to calculate in perturbative QCD and there is little theoretical prejudice about their kinematical dependence except that they can be expanded into power series in $1/Q^2$.[30] This suggests an *ansatz*

$$F_2(x, Q^2) = F_2^{LT}(x, Q^2)\left[1 + \frac{C_{HT}(x)}{Q^2}\right] \tag{13}$$

where the leading twist structure function F_2^{LT} follows the Altarelli–Parisi equations, and which gives indeed a very satisfactory fit to the data (Fig. 4). The quality

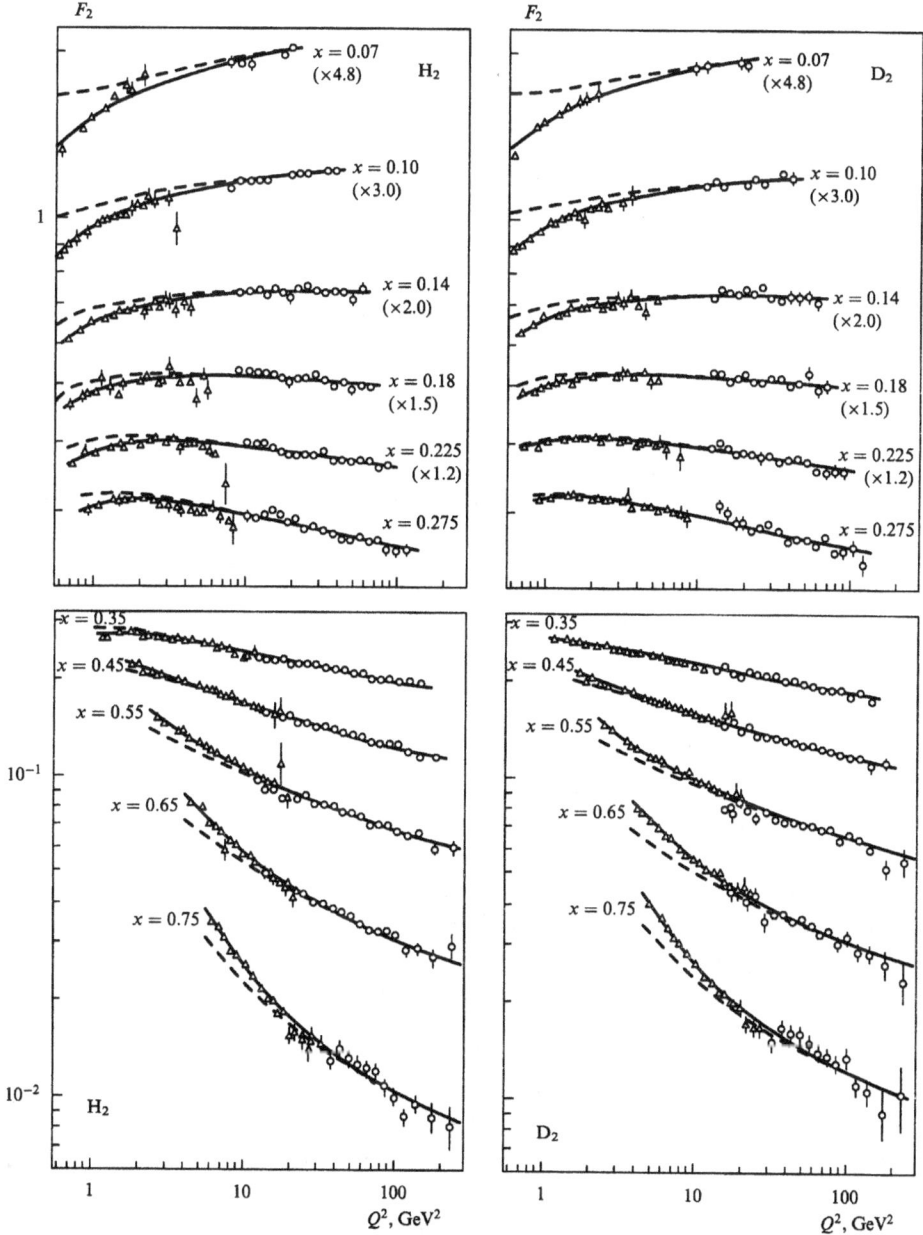

Fig. 4. QCD fit to the SLAC (triangles) and BCDMS (circles) data on $F_2(x, Q^2)$. The dashed line is the purely perturbative fit with the leading twist structure function $F_2^{LT}(x, Q^2)$. The solid line includes the higher twist contribution discussed in the text.

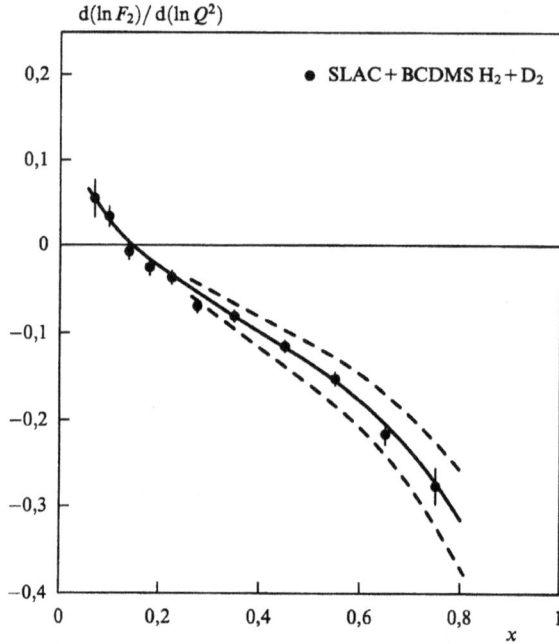

Fig. 5.   Scaling violations $d\ln F_2/d\ln Q^2$ observed in the combined SLAC/BCDMS hydrogen and deuterium data. Errors are statistical only. The solid line is a QCD fit corresponding to $\alpha_s(M_Z^2) = 0.113$; the dashed lines correspond to $\Delta\alpha_s(M_Z^2) = 0.010$.

of the fit is best illustrated in the representation of the "logarithmic slopes" which shows the derivative of the structure function with respect to $\ln Q^2$ as predicted by the Altarelli–Parisi equations, averaged over the Q^2 range of each bin in x (Fig. 5). In this analysis, the higher twist term $C_{HT}(x)$ of Eq. (13) is fitted by a set of constants in each bin of x. These coefficients are compatible with zero for $x < 0.4$, i.e. perturbative QCD can describe scaling violations in this region down to Q^2 as small as $1\,\mathrm{GeV}^2$.

In the same analysis, Virchaux and Milsztajn have also estimated the "theoretical" uncertainty due to the neglect of higher order terms in the perturbative QCD expansions. Their final result for α_s at $Q^2 = M_Z^2$ is

$$\alpha_s(M_Z^2) = 0.113 \pm 0.003 \text{ (exp.)} \pm 0.004 \text{ (theor.)}.$$

A later analysis by Alekhin[31] based on the combined SLAC, BCDMS, and NMC data yielded $\alpha_s(M_Z^2) = 0.1183 \pm 0.0021$ (exp.) ± 0.0013 (theor.). These data still provide significant input to the present world average of α_s from deep inelastic scattering, and are in good agreement with the final combined result from LEP.[32]

As a byproduct, these QCD fits can also provide an estimate of the gluon distribution in the nucleon (Eq. (11)). Since the gluon distribution is strongly peaked at small x, however, this estimate is now superseded by fits to more recent data, in particular data from HERA, with better coverage of this kinematic region.

4. Nucleon Spin and Polarised Deep Inelastic Scattering

One of the last experiments performed by the EMC was in 1985 the measurement of the double-spin cross-section asymmetry for inclusive deep inelastic scattering of longitudinally polarised muons off longitudinally polarised protons. As in the unpolarised case initial measurements had been performed at SLAC at lower energies in a limited kinematic range. While the SLAC measurements were in line with expectations from the QPM, the EMC results showed in the previously unmeasured small-x region ($x < 0.1$) a clear disagreement with potentially dramatic consequences.[33, 34] In the QPM, the nucleon spin is supposed to arise entirely from the quark spins, while in relativistic quark models a contribution of about 60% is expected. However, the EMC result was compatible with zero. Leader and Anselmino conclude in 1988[35] in an article entitled "A crisis in the parton model: where, oh where is the proton's spin?":

(a) Orbital angular momentum may be important; and this is perfectly consistent with what is known about the intrinsic k_T [transverse momentum] of quarks.
(b) The sacrosanct Bjorken sum rule may be broken. A measurement of g_1^n [of the neutron] is clearly now vital!
(c) The experiment may be wrong. Given its fundamental importance it should be redone, ...obviously, with great emphasis on the small x region.

The unexpected result, dubbed the spin crisis, gave birth to many new experiments, including those of the Spin Muon (SMC) and COMPASS Collaborations at CERN. A very fruitful interplay between theory and experiment enrolled opening up a whole field of research now extending to transverse polarisation, transverse-momentum dependent (TMD) and generalised parton distributions (GPD). A comprehensive recent review is presented in Ref. 36 including non-CERN work by HERMES at DESY, Jefferson Lab and the Relativistic Heavy Ion Collider (RHIC) at the Brookhaven National Laboratory not covered here.

4.1. *Longitudinal spin*

The spin of the nucleon of $1/2$ (in units of $\hbar$) can be decomposed into contributions from spins 'Δ' and orbital angular momenta L from both quarks q and gluons g

$$\frac{1}{2} = \frac{1}{2}\Delta\Sigma + \Delta g + L_g + L_q, \tag{14}$$

with

$$\Delta\Sigma = \Delta u + \Delta d + \Delta s + \text{aq.}, \tag{15}$$

where "aq." indicates the corresponding terms for antiquarks. The individual spin contributions from the up, down, and strange quarks to the nucleon spin are given by the first moments Δu, Δd, and Δs of the corresponding helicity distributions $\Delta q_i(x)$

$$\Delta i = \int_0^1 \Delta q_i(x)\, dx \quad \text{with } i = u, d, s, \text{ and antiquarks,} \tag{16}$$

where

$$\Delta q_i(x) = q_i^+(x) - q_i^-(x). \tag{17}$$

Here, the superscripts $+$ and $-$ denote the helicity of the quarks; the gluon helicity distribution $\Delta g(x)$ is defined accordingly. While in the unpolarised case the sum $q_i(x) = q_i^+(x) + q_i^-(x)$ of the number densities of quarks appears, in the polarised case this role is taken by their difference.

The quark helicity distributions $\Delta q_i(x, Q^2)$ in the nucleon can be accessed via the spin-dependent structure function $g_1(x, Q^2)$, which appears in the DIS cross-section. In the QPM the structure function g_1 is given by

$$g_1(x) = \frac{1}{2} \sum_i e_i^2 \Delta q_i(x), \tag{18}$$

where e_i denotes the electric charge of the struck quark (compare Eq. (6)). Like F_1, also g_1 depends on Bjorken x and logarithmically on Q^2.

4.1.1.   *Sum rules*

For the proton the first moment Γ_1 of g_1 can be decomposed into three axial charges: the isovector charge a_3, the octet charge a_8 and the flavour-singlet charge a_0

$$\Gamma_1^p(Q^2) = \int_0^1 g_1^p(x, Q^2)\, dx = \frac{1}{12}\left(a_3 + \frac{1}{3}a_8\right) + \frac{1}{9}a_0. \tag{19}$$

They are given in terms of flavour contributions by

$$a_3 = \Delta u - \Delta d + \text{aq.}, \quad a_8 = \Delta u + \Delta d - 2\Delta s + \text{aq.}, \quad a_0 = \Delta u + \Delta d + \Delta s + \text{aq.} \tag{20}$$

The isovector and isoscalar charges come with Q^2-dependent Wilson coefficients, which are calculable in perturbative QCD and are omitted here. For the Q^2-dependent flavour-singlet axial charge a_0 usually its normalisation-scheme independent value at $Q^2 \to \infty$ is quoted. In the $\overline{\text{MS}}$ renormalisation scheme, a_0 is identical to $\Delta\Sigma$, the sum of all quark spins (Eq. (15)). However, while a_0 is an observable, $\Delta\Sigma$ *per se* is not. The isovector axial charge a_3 is equal to the weak coupling constant $|g_A/g_V|$ measured independently in neutron decay and a_8 is known from hyperon decays assuming SU(3) flavour symmetry. Both are Q^2 independent.

Subtracting from Eq. (19) the corresponding equation for the neutron yields the fundamental Bjorken sum rule[37, 38] which for $Q^2 \to \infty$ reads

$$\Gamma_1^p - \Gamma_1^n = \frac{1}{6}\left|\frac{g_A}{g_V}\right|. \tag{21}$$

The a_0 and a_8 axial charges cancel in the difference of proton and neutron, i.e. when Δu is replaced by Δd and vice versa. This famous sum rule links the first moment of the structure function g_1 (for $Q^2 \to \infty$) to the neutron decay constant and was derived already in 1966 using current algebra. However, Bjorken first dismissed it as a 'worthless' equation, because performing a measurement with a polarised neutron target seemed impossible at the time. Only three years later he reconsidered this statement "in light of the present experimental and theoretical situation". It took until 1992 that the first neutron (deuteron and helium-3) measurements were performed. Earlier proposals to measure the neutron as part of the E130 experiment at SLAC were finally not carried out. At this time the proton results were in line with expectations and thus a neutron measurement was less pressing.

In 1973, Ellis and Jaffe used Eq. (19) to make a prediction[39] for Γ_1 assuming an unpolarised strange sea ($\Delta s = 0$) in which case the singlet and octet axial charges are identical (Eq. (20)). Taking a_8 from hyperon decay constants, they obtained $\Gamma_1^p = 0.185$ and $\Gamma_1^n = -0.023$ for the proton and the neutron, respectively. Unlike the Bjorken sum rule, the Ellis–Jaffe sum rules depend on several assumptions, in particular a vanishing polarisation of strange quarks in the nucleon.

4.2. *Experimental method of the CERN experiments*

The three CERN experiments by the EMC (1985), the SMC (1992–1996) and the COMPASS Collaboration (since 2002) share the same principle. All of them use the M2 beam line providing longitudinally polarisded positive muons with momenta of up to 200 GeV. A polarisation of about 80% was measured by the SMC[40, 41] using two dedicated beam polarimeters. The solid-state polarised target consists of two or three cells with material of opposite polarisations, which are inverted at regular intervals. The open forward spectrometer and the polarised target are described in Sections 2.2 and 2.3.

The experiments measure the DIS cross-section asymmetry for parallel and antiparallel orientation of muon and nucleon spins, taking advantage of the cancellation of several important quantities in the asymmetry: the dominant unpolarised cross-section, the beam flux, the number of target nuclei, and the spectrometer acceptance. From the measured DIS cross-section asymmetry the virtual-photon asymmetry

$$A_1 = \frac{\sigma_{\frac{1}{2}} - \sigma_{\frac{3}{2}}}{\sigma_{\frac{1}{2}} + \sigma_{\frac{3}{2}}} = \frac{g_1 - \frac{Q^2}{\nu^2} g_2}{F_1} \to \frac{g_1}{F_1} \tag{22}$$

is determined taking into account the beam and target polarisations, the fraction of polarisable nucleons in the target material and the depolarisation of the virtual photon with respect to the parent muon. Here $\sigma_{\frac{1}{2}}$ and $\sigma_{\frac{3}{2}}$ are the cross-sections for the absorption of a transversely polarised photon with spin antiparallel and parallel to the spin of the longitudinally polarised nucleon. The contribution of the structure function g_2 is suppressed by Q^2/ν^2 and A_1 is essentially equal to the ratio of the spin-dependent and the spin-averaged structure functions g_1 and F_1.

4.3.   *Experimental results*

4.3.1.   *Sum rules*

The EMC proton result $\Gamma_1^p = 0.126 \pm 0.010 \pm 0.015$ is in clear disagreement with the Ellis–Jaffe prediction of 0.185 ± 0.005. From this the EMC deduced a small axial singlet charge of $a_0 = 0.098 \pm 0.076 \pm 0.113$ and a negative strange quark contribution to the proton spin of $\Delta s + \Delta \bar{s} = -0.095 \pm 0.016 \pm 0.023$.[34] Recent COMPASS results indicate a somewhat larger quark spin contribution of $a_0 = 0.33 \pm 0.03 \pm 0.05$ and a similar strange quark contribution $\Delta s + \Delta \bar{s} = -0.08 \pm 0.01 \pm 0.02$.[43] Still the original EMC conclusion that the quark spins do not account for most of the proton spin holds.

In 1992, the SMC performed the first measurement of the neutron g_1 structure function[44] using a polarised deuteron target and the EMC result for the proton. The measurement revealed a violation of the Ellis–Jaffe sum rule for the neutron and confirmed the Bjorken sum rule for the difference of proton and neutron first moments of g_1 (Eq. (21)). As for the deuteron, the measured x-range was subsequently extended also for the proton[45] down to $x = 0.004$ (for $Q^2 > 1$ GeV^2) confirming essentially the EMC result.

Also in 1992, the ^{3}He experiment E142 at SLAC reported the contradicting findings: a validation of the Ellis–Jaffe sum rule for the neutron and thus a violation of the Bjorken sum rule.[46] Due to the lower beam energy of 19 GeV–26 GeV, E142 had to struggle with large QCD radiative corrections of order $\alpha_s(Q^2)/\pi$ for the Bjorken sum rule. From this Q^2 evolution Ellis and Karliner determined in 1994 the strong coupling constant $\alpha_s(M_Z^2) = 0.122^{+0.005}_{-0.009}$ using corrections up to order $(\alpha_s/\pi)^4$.[47] Applying these corrections, the E142 result turned out to be also compatible with the Bjorken sum rule.

The most recent COMPASS result for the Bjorken integral and for the isosinglet "Ellis–Jaffe" integral $\int_{x_{\min}}^1 (g_1^p + g_1^n)\, dx$ is shown in Fig. 6 as a function of the lower integration limit $x_{\min}$ at $Q^2 = 3$ GeV^2. Note that while for the Bjorken sum there is a large contribution for $x < 0.1$, the contribution from this region to the Ellis–Jaffe sum is negligible. With $a_3 = 1.28 \pm 0.07 \pm 0.010$ compared to the PDG value for $|g_A/g_V| = 1.2723 \pm 0.0023$ the Bjorken sum rule is confirmed at the 10% level.[42]

4.3.2.   *Structure functions and quark helicity distributions*

The spin-dependent structure function data for the proton as obtained from the asymmetry measurements using Eq. (22) are shown in Fig. 7 as a function of x and Q^2. The world data come from COMPASS,[42, 43] SMC,[48] EMC,[34] SLAC,[49–53] HERMES,[54] and Jefferson Lab.[55, 56] The smallest-x data were obtained by the

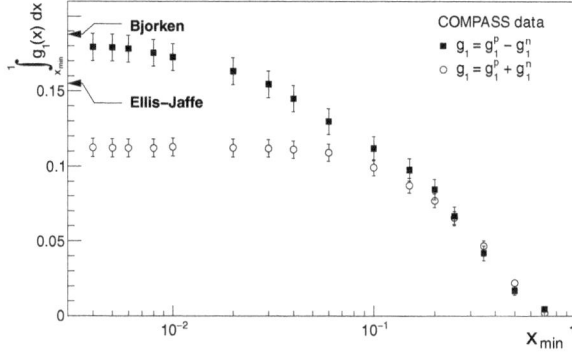

Fig. 6. Convergence of the first moments of $g_1^p \mp g_1^n$ as a function of the lower integration limit x_{min} from the COMPASS proton and deuteron data.[42] The arrows indicate the theoretical expectations. Error bars are statistical only.

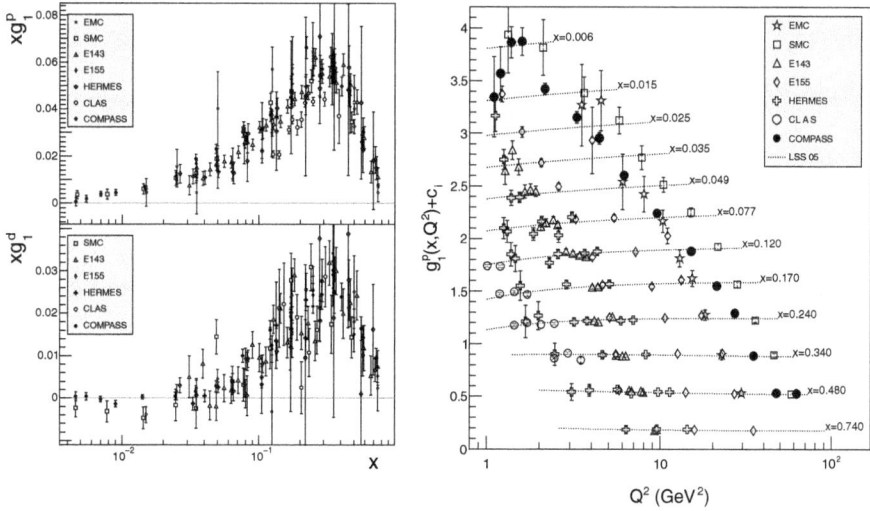

Fig. 7. Left: $xg_1(x, Q^2)$ as function of x with $Q^2 > 1$ GeV^2 for the proton (top) and the deuteron (bottom). Right: Proton data for $g_1(x, Q^2)$ as a function of x and Q^2 with $W > 2.5$ GeV. For clarity the g_1 data for the i-th x bin (starting from $i = 0$) were offset by $c_i = 0.28(11.6 - i)$. Error bars are statistical errors only.

CERN experiments. Similar data exist for the deuteron. HERMES, SLAC and Jefferson Lab also obtained some neutron (^{3}He) data.

Insight into the individual quark and gluon helicity distributions can be gained from semi-inclusive deep inelastic scattering (SIDIS, Fig. 8). The probability for a quark q of flavour i to fragment into a hadron h with energy fraction $z = E_h/\nu$ is described by the fragmentation function $D_i^h(z, Q^2)$. Due to the factorisation theorem, x and z dependences appear as a product of quark distribution and

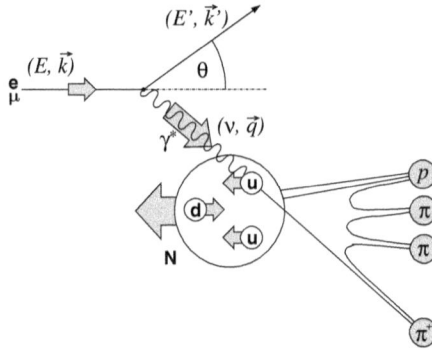

Fig. 8.   Semi-inclusive deep inelastic scattering.

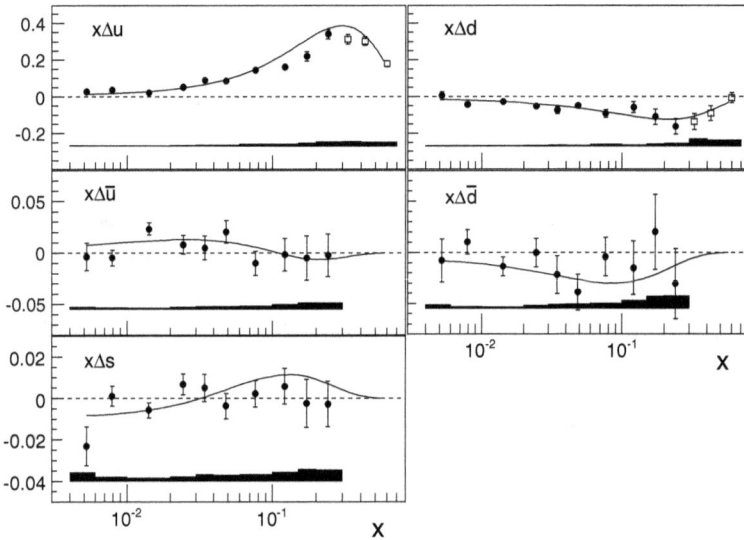

Fig. 9.   Quark helicity distributions from a LO analysis.[58] The bands indicate the systematic uncertainty.

fragmentation functions. Similar to the inclusive asymmetries, one obtains double-spin cross-section asymmetries for the production of a hadron h

$$A_1^h(x, Q^2, z) \simeq \frac{\sum_i e_i^2 \, \Delta q_i(x, Q^2) \, D_i^h(z, Q^2)}{\sum_i e_i^2 \, q_i(x, Q^2) \, D_i^h(z, Q^2)}. \tag{23}$$

An up quark preferentially fragments into a π^+ while a down quark favours fragmenting into a π^-. The difference of favoured and unfavoured fragmentation allows for a flavour separation of the quark helicity distributions. The first leading-order (LO) determination of the valence and nonstrange sea polarisation using this method was made by the SMC.[57] A recent result by COMPASS is shown in Fig. 9. The up quark polarisation is positive and the one of the down quark negative. The strange quark polarisation is slightly positive contrary to the x-integrated result from the first moment of g_1. This issue is still under discussion and may be linked to the uncertainties in the fragmentation functions for strange quarks.

4.3.3. *Gluon helicity distributions*

In 1988, it was shown that the gluon polarisation contributes via the axial anomaly to the singlet axial charge like

$$a_0 = \sum_q \Delta q - 3 \frac{\alpha_s}{2\pi} \Delta g \tag{24}$$

with $\alpha_s \Delta g$ constant, i.e. independent of Q^2.[61, 62] This led to the suggestion that maybe a large positive gluon polarisation would mask the quark spin contribution to the nucleon spin. In order to recover the value of 0.6 for $\Delta\Sigma$, values of $2\hbar$ to $3\hbar$ are required for Δg. This gave birth to the COMPASS Collaboration, which set out to determine the gluon polarisation.

The gluon polarisation can be probed in SIDIS via the gluon–photon fusion process (PGF) $\gamma g \to q\bar{q}$. Particularly interesting is the production of open charm, e.g. of D mesons, because of the absence at small x of charmed quarks in the nucleon. Furthermore, high-p_T hadron pairs and single hadrons can be used to determine the gluon polarisation. The first indication that the gluon polarisation is much smaller then required by the anomaly scenario came 2005 from COMPASS[63] using events with high-p_T hadron pairs at low Q^2. Later results from open charm[64] and events with high-p_T hadron pairs at $Q^2 > 1$ GeV^2 followed.[65] The LO results of such determinations are summarised in Fig. 10. Results from RHIC confirmed the small gluon polarization, but recently also indicated that the gluon still may contribute significantly to the nucleon spin.[66]

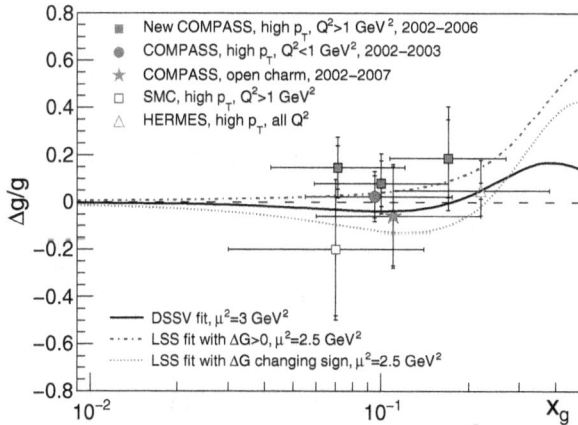

Fig. 10. Gluon polarisation $\Delta g/g$ from LO determinations as function of x_g. The horizontal error bar indicates the x range of the measurement. Also shown are the results from NLO QCD analyses of the world data.[59, 60]

4.3.4. *Global QCD analyses*

Like in the spin-averaged case (see Section 3.3), the Q^2 evolution of the g_1 structure function (Fig. 7) is described by the DGLAP equations.[25] From next-to-leading (NLO) QCD analyses one obtains the individual quark, antiquark and gluon helicity distributions $\Delta q(x, Q^2)$ and $\Delta g(x, Q^2)$. Modern global QCD analyses[59, 60, 66] take into account inclusive DIS and SIDIS data as well as data from polarised pp collisions at RHIC. While the quark distributions are well determined, the gluon distribution still have considerable uncertainties owing to the small Q^2 range of the data for a given x. A polarised electron–ion collider would change this situation dramatically as HERA did in the unpolarised case.

4.4. Transverse spin

4.4.1. *Transversity*

Apart from the spin-averaged (F_1) and spin-dependent structure function (g_1), there is at leading twist a third, chiral-odd structure function h_1 describing the distribution of transverse quark spins in a transversely polarised nucleon

$$h_1(x) = \frac{1}{2} \sum_i e_i^2 \delta q_i(x) \quad \text{with} \quad \delta q_i(x) = q_i^\uparrow(x) - q_i^\downarrow(x), \tag{25}$$

where $q^\uparrow$ and $q^\downarrow$ respectively indicate the number densities of quarks with spin orientation parallel and antiparallel to the transverse nucleon spin. In the nonrelativistic case $h_1(x)$ is equal to $g_1(x)$. This structure function does not contribute to inclusive scattering, because it implies a flip of the quark spin, which is conserved for massless quarks. However, in SIDIS h_1 can be coupled to the chirally odd Collins fragmentation function $\Delta_T D_i^h(z, p_T)$ and thus lead to an azimuthal sine modulation of the cross-section asymmetry in the Collins angle[b] $\phi_{\text{Coll}} = \phi_h + \phi_S + \pi$ with an amplitude of

$$A_{\text{Coll}}(x, z) \sim \frac{\sum_i e_i^2 \, \delta q_i(x) \, \Delta_T D_i^h(z, p_T^h)}{\sum_q e_i^2 \, q_i(x) \, D_i^h(z, p_T^h)}. \tag{26}$$

Here p_T^h denotes the transverse hadron momentum with respect to the virtual photon; ϕ_h and ϕ_S are the azimuthal angles of the hadron and the nucleon spin. Figure 11 (top) shows the Collins asymmetry for the proton as measured for positive and negative hadrons (dominantly pions). Similar measurements exist for identified pions and kaons[67] and from HERMES. The corresponding asymmetries for the deuteron are compatible with zero due to a cancellation of the up and down quark contributions.

Transversity can — instead of to the Collins function — also couple to another chiral-odd fragmentation function, the interference fragmentation function (IFF),

---

[b]Note that some experiments, e.g. HERMES, use a definition of ϕ_{Coll} without adding π.

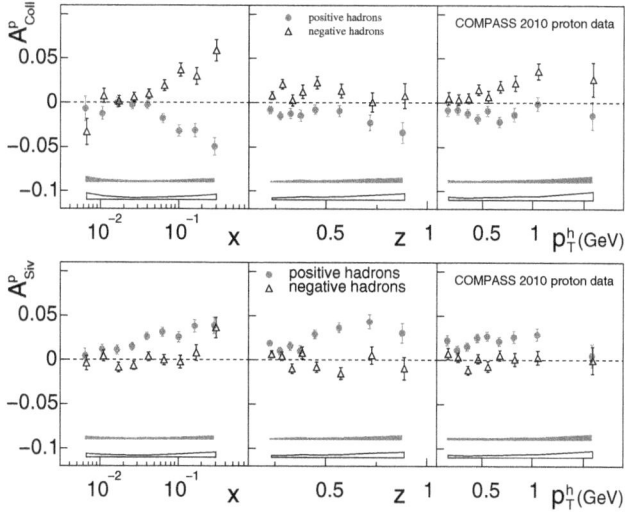

Fig. 11. Collins (top) and Sivers (bottom) asymmetry[68, 69] of the proton for positive and negative hadrons as functions of x, z, and p_T^h from COMPASS. The bands indicate the systematic uncertainty.

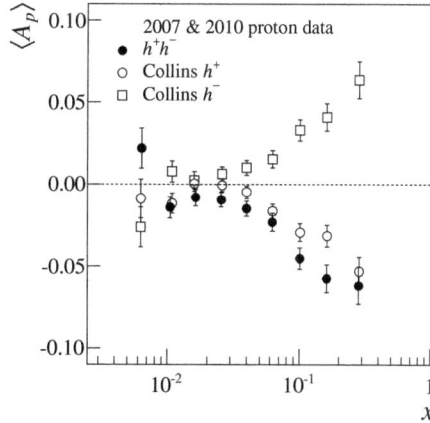

Fig. 12. Collins asymmetries for positive (lower open circles) and negative hadrons (upper open circles) and 2-hadron asymmetries (full circles) as function of x.[68, 70]

which generates a pair of oppositely charged hadrons. The similarity of the Collins asymmetry (for positive hadrons) and the 2-hadron asymmetry suggests that a common mechanism is at work in both cases (Fig. 12).

Phenomenological determinations of the transversity structure function[71] have been made using data from COMPASS, HERMES, and Belle. At Belle, the Collins and IFF fragmentation function have been measured in e^+e^- collisions. These analyses show a positive transverse polarisation of the up quarks and a negative one for down quarks in a transversely polarised nucleon.

4.4.2. *Transverse-momentum-dependent parton distributions*

The PDFs discussed so far do not explain the strong transverse single-spin asymmetries observed in both hadron and DIS reactions indicating significant spin-orbit coupling in the nucleon associated with the quark transverse momentum k_T. Transverse-momentum-dependent (TMD) distributions allow for such a k_T dependence in addition to the one on the longitudinal momentum fraction x carried by the parton. In QCD there are eight leading-twist transverse-momentum-dependent parton distributions listed below.

N \ q	U	L	T
U	f_1		$h_1^\perp$
L		g_1	$h_{1L}^\perp$
T	$f_{1T}^\perp$	$g_{1T}^\perp$	h_1 $h_{1T}^\perp$

Here U, L, and T stand for unpolarised, longitudinally polarised, and transversely polarised nucleons (rows) and quarks (columns), respectively. Upon integration over k_T the TMD PDFs in the diagonal (in bold) yield the usual structure functions $F_1(x)$, $g_1(x)$, and $h_1(x)$, while all off-diagonal TMD PDFs vanish.

The best known TMD PDF is the Sivers function $f_{1T}^\perp$ which describes the distribution of unpolarised quarks in a transversely polarised nucleon. It couples to the standard unpolarised fragmentation function D_i^h and causes an azimuthal asymmetry in $\sin\phi_{\mathrm{Siv}}$, where the Sivers angle is defined by $\phi_{\mathrm{Siv}} = \phi_h - \phi_S$. Figure 11 (bottom) shows the Sivers asymmetry of the proton for positive and negative hadrons. For positive hadrons a clear positive asymmetry is visible, in particular at larger x and z. The Boer–Mulders function $h_1^\perp$ describes the distribution of transversely polarised quarks in an unpolarised nucleon and couples to the Collins fragmentation function. As the Sivers function, the Boer–Mulders function is odd under time reversal and only nonvanishing due to initial or final-state interactions. For these functions only a restricted universality is predicted implying a change of sign when going from SIDIS to Drell–Yan (DY) reactions

$$f_{1T}^\perp|_{\mathrm{SIDIS}} = -f_{1T}^\perp|_{\mathrm{DY}} \quad \text{and} \quad h_1^\perp|_{\mathrm{SIDIS}} = -h_1^\perp|_{\mathrm{DY}}. \tag{27}$$

An experimental test of this prediction is an important next step in spin physics.

COMPASS has the unique possibility to measure with the same spectrometer the sign change of the T-odd Sivers and Boer–Mulders PDFs in the upcoming, first-ever polarised Drell–Yan experiment with a pion beam planned for 2015.

4.4.3. *Generalised parton distributions*

The role of the orbital angular momentum in the nucleon is still unclear and the only known access to this quantity is via generalised parton distribution functions (GPD),[72] which correlate longitudinal momentum and transverse spatial degrees of freedom. They enter in the cross-sections for deeply virtual Compton scattering and hard exclusive meson production. COMPASS started to study theses processes and in 2016/17 will have GPD runs with a 2.5 m long liquid hydrogen target.

5. Conclusions

Deep inelastic muon scattering at the SPS has been a unique success story for more than 35 years now, and has grown into the most long-lived experimental programme of CERN. It has had a fundamental impact on the present-day understanding of the innermost structure of hadronic matter, and has been an important and fruitful testing ground for the Quark–Parton Model and for Quantum Chromodynamics, which it helped establishing as the universally accepted theory of the strong interaction of quarks and gluons. It has even outlived the HERA electron–proton collider programme at DESY, which had some of its major roots in the CERN muon experiments.

The question whether deep inelastic lepton scattering has a future after COMPASS and HERA cannot be answered today. Only the discovery of a substructure of quarks or leptons would warrant the investment in a major new programme; by colliding protons from the LHC — or a future hadron collider of even higher energy — with a new, high-energy electron beam, CERN would be well equipped to repeat the successes of its deep inelastic scattering programmes at a new energy frontier.

References

1. W. Panofsky, Electromagnetic Interactions: Low q^2 Electrodynamics: Elastic and Inelastic Electron (and Muon) Scattering, in *Proc. 14th Int. Conf. on High-Energy Physics, Vienna* (1968).
2. M. Gell-Mann, A Schematic Model of Baryons and Mesons, *Phys. Lett.* **8**, 214–215 (1964). doi: 10.1016/S0031-9163(64)92001-3.
3. G. Zweig, An SU(3) model for strong interaction symmetry and its breaking (Version 2), CERN-TH-412 (1964).
4. J. Aubert *et al.*, The ratio of the nucleon structure functions F_2 for iron and deuterium, *Phys. Lett. B* **123**, 275 (1983). doi: 10.1016/0370-2693(83)90437-9.
5. R. Clifft and N. Doble, Proposed Design of a High-Energy, High Intensity Muon Beam for the SPS North Experimental Area (1974), CERN/LAB. II/EA/74-2, CERN/SPSC/74-12.

6. J. Aubert *et al.*, A Large Magnetic Spectrometer System for High-Energy Muon Physics, *Nucl. Instrum. Meth.* **179**, 445–466 (1981). doi: 10.1016/0029-554X(81)90169-5.

7. J. Aubert *et al.*, A Detailed Study of the Proton Structure Functions in Deep Inelastic Muon-Proton Scattering, *Nucl. Phys. B* **259**, 189 (1985). doi: 10.1016/0550-3213(85)90635-2.

8. J. Albanese *et al.*, The Vertex and Large Angle Detectors of a Spectrometer System for High-energy Muon Physics, *Nucl. Instrum. Meth.* **212**, 111 (1983). doi: 10.1016/0167-5087(83)90682-8.

9. D. Bollini *et al.*, A High Luminosity Spectrometer for Deep Inelastic Muon Scattering Experiments, *Nucl. Instrum. Meth.* **204**, 333 (1983). doi: 10.1016/0167-5087(83)90063-7.

10. A. Benvenuti *et al.*, An Upgraded Configuration of a High Luminosity Spectrometer for Deep Inelastic Muon Scattering Experiments, *Nucl. Instrum. Meth. A* **226**, 330 (1984). doi: 10.1016/0168-9002(84)90045-7.

11. P. Amaudruz *et al.*, Precision measurement of the structure function ratios $F_2(\text{He})/F_2(\text{D})$, $F_2(\text{C})/F_2(\text{D})$ and $F_2(\text{Ca})/F_2(\text{D})$, *Z. Phys. C* **51**, 387–394 (1991). doi: 10.1007/BF01548560.

12. P. Abbon *et al.*, The COMPASS experiment at CERN, *Nucl. Instrum. Meth. A* **577**, 455–518 (2007). doi: 10.1016/j.nima.2007.03.026.

13. P. Abbon *et al.*, The COMPASS Setup for Physics with Hadron Beams, hep-ex/1410.1797 (2014).

14. A. Argento *et al.*, Electroweak Asymmetry in Deep Inelastic Muon-Nucleon Scattering, *Phys. Lett. B* **120**, 245 (1983). doi: 10.1016/0370-2693(83)90665-2.

15. A. Argento *et al.*, Measurement of the Interference Structure Function $xg_3(x)$ in Muon–Nucleon Scattering, *Phys. Lett. B* **140**, 142, (1984). doi: 10.1016/0370-2693(84)91065-7.

16. F. Halzen and A. D. Martin, *Quarks and Leptons* (Wiley, 1984). ISBN 9780471887416.

17. R. Roberts, *The Structure of the Proton: Deep Inelastic Scattering* (Cambridge Monographs on Mathematical Physics), (Cambridge University Press, 1990).

18. E. D. Bloom *et al.*, High-Energy Inelastic *e-p* Scattering at 6 Degrees and 10 Degrees, *Phys. Rev. Lett.* **23**, 930–934 (1969). doi: 10.1103/PhysRevLett.23.930.

19. M. Breidenbach *et al.*, Observed Behavior of Highly Inelastic Electron-Proton Scattering, *Phys. Rev. Lett.* **23**, 935–939 (1969). doi: 10.1103/PhysRevLett.23.935.

20. J. Bjorken, Asymptotic Sum Rules at Infinite Momentum, *Phys. Rev.* **179**, 1547–1553 (1969). doi: 10.1103/PhysRev.179.1547.

21. A. Benvenuti *et al.*, A High Statistics Measurement of the Proton Structure Functions $F_2(x, Q^2)$ and R from Deep Inelastic Muon Scattering at High Q^2, *Phys. Lett. B* **223**, 485 (1989). doi: 10.1016/0370-2693(89)91637-7.

22. M. Arneodo *et al.*, Measurement of the proton and the deuteron structure functions F_2^p and F_2^d, *Phys. Lett. B* **364**, 107–115 (1995). doi: 10.1016/0370-2693(95)01318-9.

23. L. Whitlow, E. Riordan, S. Dasu, S. Rock, and A. Bodek, Precise measurements of the proton and deuteron structure functions from a global analysis of the SLAC deep inelastic electron scattering cross-sections, *Phys. Lett. B* **282**, 475–482 (1992). doi: 10.1016/0370-2693(92)90672-Q.

24. M. Adams *et al.*, Proton and deuteron structure functions in muon scattering at 470 GeV, *Phys. Rev. D* **54**, 3006–3056 (1996). doi: 10.1103/PhysRevD.54.3006.

25. G. Altarelli and G. Parisi, Asymptotic Freedom in Parton Language, *Nucl. Phys. B* **126**, 298 (1977). doi: 10.1016/0550-3213(77)90384-4.

26. A. Benvenuti *et al.*, Test of QCD and a Measurement of Λ From Scaling Violations in the Nucleon Structure Function $F_2(x, Q^2)$ at High Q^2, *Phys. Lett. B* **195**, 97 (1987). doi: 10.1016/0370-2693(87)90892-6.

27. A. Benvenuti *et al.*, Test of QCD and a Measurement of Λ From Scaling Violations in the Proton Structure Function $F_2(x, Q^2)$ at High Q^2, *Phys. Lett. B* **223**, 490 (1989). doi: 10.1016/0370-2693(89)91638-9.

28. A. Benvenuti *et al.*, A High Statistics Measurement of the Deuteron Structure Functions $F_2(x, Q^2)$ and R From Deep Inelastic Muon Scattering at High Q^2, *Phys. Lett. B* **237**, 592 (1990). doi: 10.1016/0370-2693(90)91231-Y.

29. M. Virchaux and A. Milsztajn, A Measurement of α_s and higher twists from a QCD analysis of high statistics F_2 data on hydrogen and deuterium targets, *Phys. Lett. B* **274**, 221–229 (1992). doi: 10.1016/0370-2693(92)90527-B.

30. R. K. Ellis, W. Furmanski, and R. Petronzio, Unraveling Higher Twists, *Nucl. Phys. B* **212**, 29 (1983). doi: 10.1016/0550-3213(83)90597-7.

31. S. I. Alekhin, Combined analysis of SLAC-BCDMS-NMC data at high x: α_s and high twists, hep-ph/9907350 (1999).

32. K. Olive *et al.* (Review of Particle Physics), *Chin. Phys. C* **38**, 090001 (2014). doi: 10.1088/1674-1137/38/9/090001.

33. J. Ashman *et al.*, A Measurement of the Spin Asymmetry and Determination of the Structure Function g_1 in Deep Inelastic Muon-Proton Scattering, *Phys. Lett. B* **206**, 364 (1988). doi: 10.1016/0370-2693(88)91523-7.

34. J. Ashman *et al.*, An Investigation of the Spin Structure of the Proton in Deep Inelastic Scattering of Polarized Muons on Polarized Protons, *Nucl. Phys. B* **328**, 1 (1989). doi: 10.1016/0550-3213(89)90089-8.

35. E. Leader and M. Anselmino, A Crisis in the Parton Model: Where, Oh Where Is the Proton's Spin?, *Z. Phys. C* **41**, 239 (1988). doi: 10.1007/BF01566922.

36. C. A. Aidala *et al.*, The Spin Structure of the Nucleon, *Rev. Mod. Phys.* **85**, 655–691 (2013). doi: 10.1103/RevModPhys.85.655.

37. J. Bjorken, Applications of the Chiral U(6) × (6) Algebra of Current Densities, *Phys. Rev.* **148**, 1467–1478 (1966). doi: 10.1103/PhysRev.148.1467.

38. J. Bjorken, Inelastic Scattering of Polarized Leptons from Polarized Nucleons, *Phys. Rev. D* **1**, 1376–1379 (1970). doi: 10.1103/PhysRevD.1.1376.

39. J. R. Ellis and R. L. Jaffe, A Sum Rule for Deep Inelastic Electroproduction from Polarized Protons, *Phys. Rev. D* **9**, 1444 (1974). doi: 10.1103/PhysRevD.10.1669.2, 10.1103/PhysRevD.9.1444

40. B. Adeva *et al.*, Measurement of the polarization of a high-energy muon beam, *Nucl. Instrum. Meth. A* **343**, 363–373 (1994). doi: 10.1016/0168-9002(94)90213-5.

41. D. Adams *et al.*, Measurement of the SMC muon beam polarization using the asymmetry in the elastic scattering off polarized electrons, *Nucl. Instrum. Meth. A* **443**, 1–19 (2000). doi: 10.1016/S0168-9002(99)01017-7.

42. M. Alekseev *et al.*, The Spin-dependent Structure Function of the Proton g_1^p and a Test of the Bjorken Sum Rule, *Phys. Lett. B* **690**, 466–472 (2010). doi: 10.1016/j.physletb.2010.05.069.

43. V. Y. Alexakhin *et al.*, The Deuteron Spin-dependent Structure Function g_1^d and its First Moment, *Phys. Lett. B* **647**, 8–17, (2007). doi: 10.1016/j.physletb.2006.12.076.

44. B. Adeva *et al.*, Measurement of the spin dependent structure function $g_1(x)$ of the deuteron, *Phys. Lett. B* **302**, 533–539 (1993). doi: 10.1016/0370-2693(93)90438-N.

45. D. Adams *et al.*, Measurement of the spin dependent structure function $g_1(x)$ of the proton, *Phys. Lett. B* **329**, 399–406 (1994). doi: 10.1016/0370-2693(94)90793-5.

46. P. Anthony *et al.*, Determination of the neutron spin structure function, *Phys. Rev. Lett.* **71**, 959–962 (1993). doi: 10.1103/PhysRevLett.71.959.

47. J. R. Ellis and M. Karliner, Determination of α_S and the nucleon spin decomposition using recent polarized structure function data, *Phys. Lett. B* **341**, 397–406 (1995). doi: 10.1016/0370-2693(95)80021-O.

48. B. Adeva *et al.*, Spin asymmetries A_1 and structure functions g_1 of the proton and the deuteron from polarized high-energy muon scattering, *Phys. Rev. D* **58**, 112001 (1998). doi: 10.1103/PhysRevD.58.112001.

49. P. Anthony *et al.*, Deep inelastic scattering of polarized electrons by polarized He-3 and the study of the neutron spin structure, *Phys. Rev. D* **54**, 6620–6650 (1996). doi: 10.1103/PhysRevD.54.6620.

50. K. Abe *et al.*, Precision determination of the neutron spin structure function g_1^n, *Phys. Rev. Lett.* **79**, 26–30 (1997). doi: 10.1103/PhysRevLett.79.26.

51. K. Abe *et al.*, Measurements of the proton and deuteron spin structure functions g_1 and g_2, *Phys. Rev. D* **58**, 112003, (1998). doi: 10.1103/PhysRevD.58.112003.

52. P. Anthony *et al.*, Measurement of the deuteron spin structure function $g_1^d(x)$ for 1 $(\text{GeV}/c)^2 < Q^2 < 40$ $(\text{GeV}/c)^2$, *Phys. Lett. B* **463**, 339–345 (1999). doi: 10.1016/S0370-2693(99)00940-5.

53. P. Anthony *et al.*, Measurements of the Q^2 dependence of the proton and neutron spin structure functions g_1^p and g_1^n, *Phys. Lett. B* **493**, 19–28 (2000). doi: 10.1016/S0370-2693(00)01014-5.

54. A. Airapetian *et al.*, Precise determination of the spin structure function g_1 of the proton, deuteron and neutron, *Phys. Rev. D* **75**, 012007 (2007). doi: 10.1103/PhysRevD.75.012007.

55. X. Zheng *et al.*, Precision measurement of the neutron spin asymmetry A_1^N and spin flavor decomposition in the valence quark region, *Phys. Rev. Lett.* **92**, 012004 (2004). doi: 10.1103/PhysRevLett.92.012004.

56. K. Dharmawardane *et al.*, Measurement of the x- and Q^2-dependence of the asymmetry A_1 on the nucleon, *Phys. Lett. B* **641**, 11–17 (2006). doi: 10.1016/j.physletb.2006.08.011.

57. B. Adeva *et al.*, Polarization of valence and nonstrange sea quarks in the nucleon from semiinclusive spin asymmetries, *Phys. Lett. B* **369**, 93–100 (1996). doi: 10.1016/0370-2693(95)01584-1.

58. M. Alekseev *et al.*, Quark helicity distributions from longitudinal spin asymmetries in muon–proton and muon–deuteron scattering, *Phys. Lett. B* **693**, 227–235 (2010). doi: 10.1016/j.physletb.2010.08.034.

59. D. de Florian *et al.*, Extraction of Spin-Dependent Parton Densities and Their Uncertainties, *Phys. Rev. D* **80**, 034030 (2009). doi: 10.1103/PhysRevD.80.034030.

60. E. Leader, A. V. Sidorov, and D. B. Stamenov, Determination of Polarized PDFs from a QCD Analysis of Inclusive and Semi-inclusive Deep Inelastic Scattering Data, *Phys. Rev. D* **82**, 114018 (2010). doi: 10.1103/PhysRevD.82.114018.

61. G. Altarelli and G. G. Ross, The Anomalous Gluon Contribution to Polarized Leptoproduction, *Phys. Lett. B* **212**, 391 (1988). doi: 10.1016/0370-2693(88)91335-4.

62. A. Efremov and O. Teryaev, Spin Structure of the Nucleon and Triangle Anomaly, *Nucl. Phys.* (1988).

63. E. Ageev *et al.*, Gluon polarization in the nucleon from quasi-real photo-production of high-p_T hadron pairs, *Phys. Lett. B* **633**, 25–32 (2006). doi: 10.1016/j.physletb.2005.11.049.

64. C. Adolph *et al.*, Leading and Next-to-Leading Order Gluon Polarization in the Nucleon and Longitudinal Double Spin Asymmetries from Open Charm Muopro-duction, *Phys. Rev. D* **87** (5), 052018, (2013). doi: 10.1103/PhysRevD.87.052018.

65. C. Adolph *et al.*, Leading order determination of the gluon polarisation from DIS events with high-p_T hadron pairs, *Phys. Lett. B* **718**, 922–930 (2013). doi: 10.1016/j.physletb.2012.11.056.

66. D. de Florian *et al.*, Evidence for polarization of gluons in the proton, *Phys. Rev. Lett.* **113**, 012001 (2014). doi: 10.1103/PhysRevLett.113.012001.

67. C. Adolph *et al.* Collins and Sivers asymmetries in muon production of pions and kaons off transversely polarised proton, hep-ex/1408.4405 (2014).

68. C. Adolph *et al.*, Experimental investigation of transverse spin asymmetries in muon-p SIDIS processes: Sivers asymmetries, *Phys. Lett. B* **717**, 383–389 (2012). doi: 10.1016/j.physletb.2012.09.056.

69. C. Adolph *et al.*, Experimental investigation of transverse spin asymmetries in muon-p SIDIS processes: Collins asymmetries, *Phys. Lett. B* **717**, 376–382 (2012). doi: 10.1016/j.physletb.2012.09.055.

70. C. Adolph *et al.*, A high-statistics measurement of transverse spin effects in dihadron production from muon-proton semi-inclusive deep inelastic scattering, *Phys. Lett. B* **736**, 124–131 (2014). doi: 10.1016/j.physletb.2014.06.080.

71. M. Anselmino *et al.*, Simultaneous extraction of transversity and Collins functions from new SIDIS and e^+e^- data, *Phys. Rev. D* **87**, 094019 (2013). doi: 10.1103/PhysRevD.87.094019.

72. X.-D. Ji, Gauge-Invariant Decomposition of Nucleon Spin, *Phys. Rev. Lett.* **78**, 610–613 (1997). doi: 10.1103/PhysRevLett.78.610.

Revealing Partons in Hadrons:
From the ISR to the SPS Collider

Pierre Darriulat[1] and Luigi Di Lella[2]

[1] *VATLY, INST, 179 Hoang Quoc Viêt, Cau Giay, Ha Noi, Viêt Nam*
[2] *Università di Pisa, Physics Department, Largo Bruno Pontecorvo 3, 56127 Pisa, Italy*
pierre.darriulat@gmail.com

Our understanding of the structure of hadrons has developed during the seventies and early eighties from a few vague ideas to a precise theory, Quantum Chromodynamics, that describes hadrons as made of elementary partons (quarks and gluons). Deep inelastic scattering of electrons and neutrinos on nucleons and electron–positron collisions have played a major role in this development. Less well known is the role played by hadron collisions in revealing the parton structure, studying the dynamic of interactions between partons and offering an exclusive laboratory for the direct study of gluon interactions. The present article recalls the decisive contributions made by the CERN Intersecting Storage Rings and, later, the proton–antiproton SPS Collider to this chapter of physics.

1. Preamble

In the mid-sixties, when the ISR were being born, the idea that hadrons could be composite particles was still far from being generally accepted. Summer school lectures were giving as much weight to bootstrap ideas[1] as to the newly born quark model.[2] We remember a seminar by C. N. Yang[3] at CERN, just before the ISR first collisions, introducing the concept of limiting fragmentation, which we were religiously listening to in the hope that it could give us an idea of what to expect from our imminent exploration of the high energy territory. In spite of the spectacular success of Gell-Mann's eightfold way, the quark model had to face two very strong counter-arguments: the failure of many quark search experiments to find any hint for fractional charges and the apparent incompatibility of the quark model with Fermi–Dirac statistics. Indeed, we did not know about colour, nor about the peculiar behavior of the strong force to get weaker at short distances. The light would come from SLAC at the very end of the decade, with deep inelastic electron scattering soon followed by SPEAR and its harvest of revolutionary results.

If hadrons are composite, it should be possible to understand hadron collisions in terms of interactions and rearrangements of the constituents, the so-called partons, and, in particular, to eject one of them, as in nuclear physics with $(p,2p)$ or (p,pn) reactions. It is indeed possible, but it took a decade to reach this goal. Hadrons are very different from nuclei, which can be qualitatively described classically in this context. For two main reasons: one is that hadron masses are much larger than parton masses, making the picture fully relativistic; the other is the increase of the strength of the strong force with distance, making it impossible to eject an isolated parton: as it is pulled apart from its parent hadron, the field of the strong force in between takes such high values that quark–antiquark pairs are produced in the form of mesons that accompany the ejected parton. To identify such a collection of hadrons as the filiation of the parent parton among the host of other hadrons produced in the collision is only possible when they fly close enough together to form what is called a jet. In what follows, we try to recall how this was achieved at CERN, from the ISR to the proton–antiproton SPS collider, between the early seventies and the mid-eighties. Most of it is borrowed from two earlier papers of ours.[4,5]

2. The ISR as a Gluon Collider

2.1. *Introduction*

It so happens that the lifetime of the ISR, roughly speaking the seventies, coincides with a giant leap in our understanding of particle physics. However, it is honest to say that, to first order, there is no causal relation between the two. Yet, those of us who have worked at the ISR remember these times with the conviction that we were not merely spectators of the ongoing progress, but also — admittedly modest — actors. The ISR contribution, it seems to us, is too often unjustly forgotten in the accounts that are commonly given of the progress of particle physics during this period. We shall try to present arguments of relevance to this issue in what we hope to be as neutral and unbiased way as possible. We restrict the scope of the presentation to large transverse momentum processes, or equivalently to the probing of the proton structure at short distances. This, however, is not much of a limitation, as the ISR did not significantly contribute to the progress achieved in the weak sector.

Each individual has his own vision of the past and history can merely be an attempt at collecting all such visions into as coherent as possible a story. In physics, this is particularly true when discoveries and new ideas occur at a rapid pace, as was the case in the seventies. Each of us remembers a seminar, a discussion at coffee, the reading of a particular article, or another event of this kind as a milestone in his own understanding of the new ides. Reading accounts by Steve Weinberg,[6] David Gross,[7] Gerard 't Hooft[8] or Jerry Friedman[9] of how they remember this period is particularly instructive in this respect. The same kind of disparity that exists between the visions of different individuals also occurs between the visions of different science communities. In particular, during the seventies, the e^+e^-, neutrino, fixed target and ISR communities had quite different perceptions of the progress that was being achieved. It is therefore useful to recall briefly the main events in this period.

2.2. The main milestones

When Vicky Weisskopf, in December 1965, in his last Council session as CERN Director-General obtained approval for the construction of the ISR, there was no specific physics issue at stake, which the machine was supposed to address; its only justification was to explore the *terra incognita* of higher centre-of-mass energy collisions (to our knowledge, since then, all new machines have been proposed and approved with a specific physics question in mind, which they were supposed to answer). The strong interaction was perceived as a complete mystery. The eightfold way, today understood as the approximate $SU(3)$ flavour symmetry associated with interchanges of u, d and s quarks, was not believed to have significant consequences in the dynamics of the strong interaction. The fact that no free quark had been found in spite of intensive searches, and that states such as Δ^{++}, with spin-parity $3/2^+$, could not be made of three identical spin $1/2$ u quarks without violating Fermi statistics, were discouraging such interpretations.

The first hint to the contrary came in 1968–1969 at SLAC[9] with the discovery of an important continuum in the deep inelastic region of electron proton scattering. The 2-mile linear accelerator had started operation the preceding year and the experimental program, using large spectrometers, extended over several years. From the very beginning, experimenters and theorists were in close contact, feeding each other with new data and new ideas, starting with Bjorken's ideas on scaling[10] and Feynman's ideas on partons,[11] both early advocates of a proton structure consisting of point-like constituents. However, one had to wait until 1972 for the case for a quark model to become strong: by then, scaling had been established; the measurement of a small R value (the ratio of the absorption cross sections of transverse and longitudinal virtual photons) had eliminated competitors such as the then popular Vector Dominance Model; deuterium data had been collected allowing for a comparison between the proton and neutron structure functions; a number of sum rules had been tested; evidence for the quarks to carry but a part of the proton longitudinal momentum had been obtained; the first neutrino deep inelastic data from Gargamelle had become available.[12] By the end of 1972, the way was paved for Gross, Wilczek and Politzer[13] to conceive the idea of asymptotic freedom and its corollary, infrared slavery, explaining why one could not see free quarks. By the end of 1973, the connection with non-Abelian gauge theories had been established and the "advantages of the colour-octet gluon picture", including the solution of the Fermi statistics puzzle, had been presented by Fritzsch, Gell-Mann and Leutwyler.[14] QCD was born and, by 1974, was starting to be accepted by the whole community as *the* theory of the strong interaction. It took another three to four years for it to come of age.

By mid-1972, SPEAR, the Stanford electron–positron collider, had begun operation. In November 1974, it shook the physics community with what has since been referred to as a revolution: the discovery of the Ψ going hand in hand with the simultaneous discovery of the J at Brookhaven. It immediately exploited its ability to produce pure quark–antiquark final states to measure the number of colours. However, there were so many things happening in the newly available energy domain (opening of the naked charm channels, crowded charmonium spectroscopy, production of the τ lepton) that it took some time to disentangle their effects and to understand what was going on. By the end of the decade, scaling violations had been studied both in neutrino interactions and in electron–positron annihilations (DORIS had started operation in Hamburg two years after SPEAR). QCD had reached maturity and the only puzzling questions that remained unanswered, the absence of a CP violating phase and our inability to handle the theory at large distances, are still with us today.

2.3. *What about the ISR?*

The above account of the progress of particle physics in the seventies, while following the standard folklore, does not even mention the name of the ISR. Being asked whether he was aware of the results obtained at the ISR and whether they had an impact on the development of QCD, David Gross answered:[15] *"Every one was aware of the qualitative phenomena observed in hadronic physics at large p_T, which were totally consistent with simple scattering ideas and parton model ideas [...] The tests were not as clean as in deep inelastic scattering, the analysis was more difficult and deep inelastic scattering was much cleaner in the beginning of perturbative QCD [...] Parton ideas did not test QCD at all, they simply tested the idea that there were point-like constituents but not the dynamics."* His answer illustrates well the way in which the ISR were generally perceived: a collider that was shooting Swiss watches against each other, as Feynman once jokingly described. Yet, some theorists followed closely what the ISR were producing; paradoxically, Feynman was one of them, Bjorken was another.

David Gross could have returned the question to us: *"How aware were you, the ISR community, of the experimental progress at SLAC and of the new ideas in theory?"* The first name that comes to mind in answer to this question is that of Maurice Jacob. Maurice had spent a sabbatical at Stanford where, together with Sam Berman, he had written a seminal paper on point-like constituents and large transverse momentum production.[16] Back at CERN, he organised a lively series of discussions between ISR experimenters and theorists that proved to be extremely successful in permeating our community with the progress in deep inelastic scattering and, later, in electron–positron collisions. At that time, our community was small enough to fit in the ISR auditorium. Maurice was gifted with an unusual talent to make theoretical ideas accessible to us. We all remember these seminars as a most profitable experience that brought coherence and unity in our community. For this reason, it makes sense to talk about a common ISR culture. In particular, by 1972, we were aware of the basic parton ideas and of the picture of large transverse momentum production factorised in three steps (Fig. 1): singling

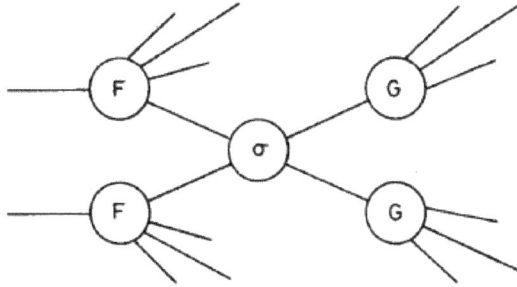

Fig. 1. Parton model picture of high p_T hadron interactions. One parton of each of the incident hadrons (structure function F) experiences a binary collision (σ) and the outcoming partons fragment into hadrons (fragmentation function G).

out a parton in each proton, making them interact (how, was not clear) in a binary collision and letting the final state partons fragment into hadrons. There were a few papers[11,16–21] in support of such a picture which most of us had read and which were our basic reference. Yet, in these early days, there was a typical delay of at least six months between SLAC and us for a new idea to be digested. There was even more delay, for most of us, to digest the subtle development of non-Abelian gauge theories: we only heard of them from our theorist friends.

Table 1 lists leading order diagrams involving quarks or gluons. A simple glance at it illustrates the originality of the ISR: gluons contribute to leading order. In electron–proton annihilations and deep inelastic scattering, gluons contribute to next to leading order only, in the form of radiative corrections associated with a bremsstrahlung gluon radiated from a quark line. This does not mean that such gluon contributions are unimportant: the scaling violations which they induce have been one of the most powerful tools in the development of our understanding of QCD. But, at the ISR, gluons not only contribute to leading order but indeed dominate the scene: in the low x regime characteristic of the ISR, collisions involving gluons, either gluon–gluon or quark–gluon, account for most of the high p_T cross-section. Gluon interactions being a privileged domain of the ISR, and gluons having been the last component of the theory to be understood and digested, it seems difficult to argue that the ISR have played but a minor role. The more so when one considers that the ISR had exclusive access to the three and four gluon vertices, a specific expression of QCD as a non Abelian gauge theory.

2.4. *Large transverse momentum: Inclusive production data*

In 1972–1973, three ISR teams[22–24] announced the observation of an unexpectedly copious pion yield at large transverse momenta (Fig. 2), orders of magnitude above a (traditionally called naive) extrapolation of the exponential distribution observed at low p_T values, $\sim \exp(-6p_T)$. "Unexpectedly" is an understatement. The whole ISR experimental program had been designed under the assumption that all hadrons

Table 1   Leading order processes involving quarks or gluons. The symbols $><$ and $][$ stand for s and t channel exchange, respectively. The last column gives the coupling constants, the number of structure functions (F) and the number of fragmentation functions (G). Couplings are written α_n for $\alpha/(\sin\theta_W \cos\theta_W)^2$ and α_{ch} for $\alpha/\sin\theta_W^2$ with θ_W being the Weinberg angle. Processes involving gluons in the initial state are shaded.

		Electron–positron annihilations	
1		$e^+e^- > \gamma < q\bar{q}$	$\alpha^2 G^2$
		Deep inelastic electron scattering	
2		$eq]\,\gamma\,[eq$	$\alpha^2 FG$
		Deep inelastic neutrino scattering	
3	Neutral currents	$\nu q]\,Z\,[\nu q$	$\alpha_n^2 FG$
4	Charged currents	$\nu q]\,W\,[lq$	$\alpha_{ch}^2 FG$
		Proton–proton collisions (ISR)	
5	Drell–Yan	$q\bar{q} > \gamma < l^+l^-$	$\alpha^2 F^2$
6	Direct photons	$q\bar{q}]\,q\,[\gamma g$	$\alpha\alpha_s F^2 G$
7		$qg]\,q\,[\gamma q$	
8	Large p_T hadrons	$qq]\,g\,[qq$	$\alpha_s^2 F^2 G^2$
9		$qq]\,q\,[gg$	
10		$q\bar{q} > g < gg$	
11		$q\bar{q} > g < q\bar{q}$	
12		$qg]\,q\,[qg$	
13		$qg]\,g\,[qg$	
14		$qg > q < qg$	
15		$gg > g < q\bar{q}$	
16		$gg > g < gg$	
17		$gg]\,q\,[qq$	
18		$gg]\,g\,[gg$	
19		$gg >< gg$	

would be forward produced. The best illustration was the Split Field Magnet, meant to be the general multipurpose detector at the ISR. No experiment was equipped with very large solid angle good quality detectors at large angle. This first discovery was opening the ISR to the study of large transverse momentum production and was providing a new probe of the proton structure at short distances. That was the good side of it. But it also had a bad side: the background that had been anticipated in the search for new particles had been strongly underestimated and such searches were becoming much more difficult than had been hoped for.

Bjorken scaling was found to apply, in support of the parton picture, but the index of the p_T power law was twice as higher than the value expected from point-like constituents, 8 rather than 4. Precisely, the π^0 inclusive invariant cross-section was of the form $p_T^{-n} \exp(-kx_T)$ where $x_T = 2p_T/\sqrt{s}$, $n = 8.24 \pm 0.05$ and $k = 26.1 \pm 0.5$. The impact of this result was quite strong and brought into fashion

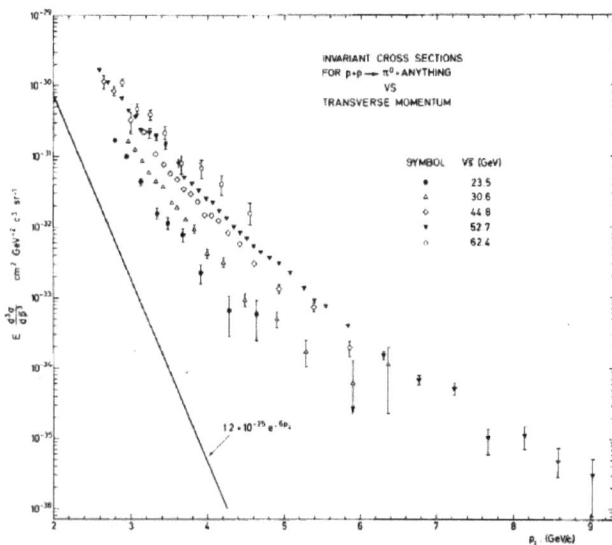

Fig. 2. Early inclusive π^0 cross-section[24] giving evidence for copious production at high p_T well above the exponential extrapolation of lower energy data.

the so-called constituent interchange model.[25] The idea was to include mesons in addition to quarks among the parton constituents of protons: deep inelastic scattering would be blind to such mesons because of their form factor but hadron interactions would allow for quark rearrangements such as $\pi^+ + d \rightarrow \pi^0 + u$. The cross-section was then predicted to be of the form $p_T^{-2(n-2)}(1 - x_T)^{2m-1}$ at large values of x_T, where n stands for the number of "active quark lines" taking part in the hard scattering and m stands for the number of "passive" quark lines wasting momentum in the transitions between hadrons and quarks. The model, that correctly predicted the power 8 measured at the ISR, had many successes but did not stand the competition with early QCD models that were starting to be developed.

Such an example is illustrated in Fig. 3, giving evidence for important quark–gluon and gluon–gluon contributions[26] beside the quark–quark term. By then, the inclusive production of charged pions, kaons, protons and antiprotons as well as η mesons had been studied at the ISR, and at Fermilab where a π^- beam had also been used, providing decisive evidence in favour of QCD. It was then understood that the p_T power law was evolving to p_T^{-4} at high values of x_T, which, however, were only accessible, in practice, to larger centre-of-mass energy collisions. The successes of the constituent interchange models were then relegated to the rank of "higher twist corrections" to the leading order perturbative regime. Between 1973 and 1978, inclusive high p_T single hadron production in hadron collisions had given exclusive contributions to the establishment of QCD as the theory of the strong interaction in a domain where other experiments — deep inelastic scattering and electron–positron annihilations — could not contribute: that of short distance collisions involving

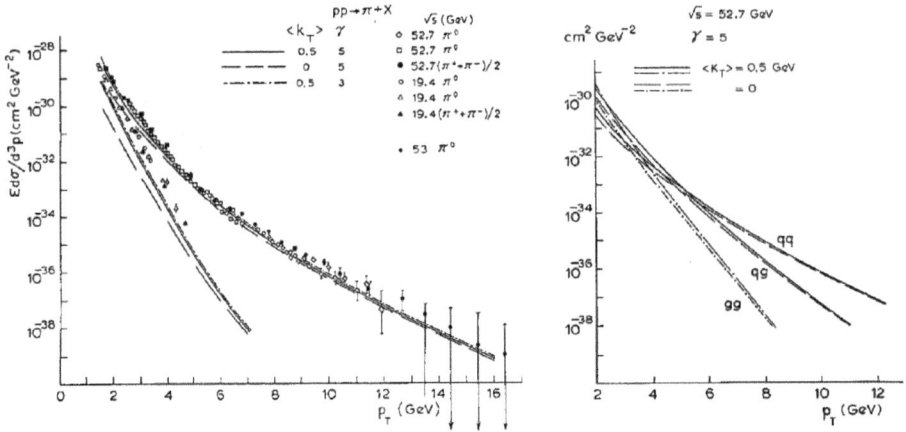

Fig. 3. A typical QCD fit[26] to inclusive pion data (left) and the relative contributions of quark–quark, quark–gluon and gluon–gluon diagrams (right).

gluons to leading order of the perturbative expansion. In this domain, the data collected at the CERN ISR — at the higher centre-of-mass energies — and at Fermilab — with a variety of beams and targets — nicely complemented each other. As the results were confirming the validity of QCD, and as there were so many important events happening elsewhere in physics, people tended to neglect or forget these important contributions.

2.5. *Event structure and jets*

The early evidence in favour of the parton picture encouraged studies of the global event structure and, in particular, experiments aiming at the detection of the hadron jets into which the hard scattered partons were supposed to fragment. Unfortunately, none of the existing ISR detectors was matched to the task. In March 1975, a large magnetic detector serving precisely this purpose had been proposed to the ISR Committee but had been rejected in October of the same year. The proposal had been reiterated with various amendments. It was enjoying the support of the ISR community, of a working party that had been appointed to assess "the need for a new magnetic facility at the ISR", with Nino Zichichi in the chair, and of the ISR Committee. It was definitively turned down two weeks later by the Research Board. Meanwhile, step by step, the existing ISR experiments had upgraded their set ups as well as they could but one had to wait until 1982, with the Axial Field Spectrometer in I8 and the Superconducting Solenoid in I1 to see detectors having large calorimeter coverage. When the ISR closed down in 1984, a rich set of important results had been obtained by these two groups,[27] with two-jet events (Fig. 4, left) dominating the scene for transverse energies in excess of 35 GeV;[28] but the CERN proton–antiproton collider, which had published its first jets in 1982,[29] had already taken the limelight away from the ISR.

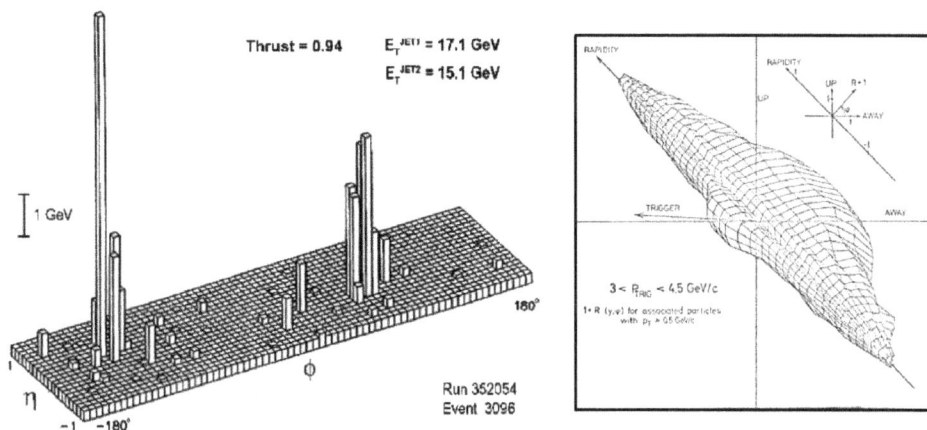

Fig. 4.   Left: A lego plot from the AFS experiment showing the two-jet structure that dominates at larger transverse energies.[28] Right: Longitudinal phase space density (relative to minimum bias events) associated with a single particle trigger at 90° (see text).

There is no doubt that the lack of proper instrumentation has been a major handicap for the ISR in their contribution to the physics of hard collisions. More support from the CERN management would probably have made it possible to gain two precious years. Retrospectively, it is difficult to estimate how much of a negative impact the approval of a new large facility at the ISR would have had on the high priority CERN programs, LEP and the proton–antiproton collider, where quark and gluon jets could be studied in optimal conditions: in comparison, the ISR were quite marginal. Moreover, the ISR beam geometry, with a crossing angle of 15° implying large vacuum chambers, was making the design of a 4π detector difficult. Seen from today, nearly forty years later, our frustration was certainly understandable and legitimate, but the decision of the CERN management now sounds more reasonable than it did then.

Between 1973 and 1978, several ISR experiments had completed studies of the event structure and the evidence for hard jets in the final state, already clear in 1976,[30] had strengthened. Figure 4 (right) shows the longitudinal phase space density of charged particles produced in a hard scattering collision. It is an average of data collected by the British French Collaboration using a charged particle trigger at 90° and momentum analysing in the Split Field Magnet the charged particles produced in association. Particle densities are normalised to those obtained in minimum bias collisions. Several features are visible: diffraction is suppressed at large rapidities, a "same-side" jet is present alongside the trigger and "away-side jets", at opposite azimuth to the trigger, cover a broad rapidity range. A difficulty inherent to the study of hard hadron collisions is the presence of a so-called "underlying event" which contains the fragments of the spectator partons that do not take part in the hard collision. This is at variance with electron–positron annihilations where all hadrons are fragments of hard partons and, to a lesser extent, with deep inelastic scattering where most of the information is carried by the structure functions. It implies a transverse momentum threshold, half a GeV to one GeV, below which a particle cannot be unambiguously identified as being a fragment of a hard scattered parton. At ISR energies, it is a serious limitation.

A second difficulty, resulting from the lack of proper calorimeter coverage in the first decade of ISR operation, was the so-called "trigger bias". Since the hard parton scattering cross-section has a much steeper p_T dependence than has the fragmentation process, it is very likely for a particle of a given p_T to be the leading fragment of a rather soft jet. This distortion of the "same-side" jet fragmentation creates an asymmetry between it and the "away-side" jet, which makes it more difficult to compare their properties. For this reason, an ideal experiment should trigger on the total transverse energy E_T using calorimetric devices. Numerous studies of the "same-side" correlations have been performed at the ISR, establishing early that they were not the result of resonance production but of a jet fragmentation characterised by a limited transverse momentum around the jet axis.

Evidence for an excess of particles at opposite azimuth to the trigger had been obtained very early and it had soon been recognised that it was due to a collimated jet produced at a rapidity which was different from event to event. The "away-side" jet multiplicity could then be measured and compared with that of quark jets observed in deep inelastic and electron–positron annihilations (Fig. 5, left). ISR jets being dominantly gluon jets, one could expect to see a difference but the p_T range accessible to the ISR was still too low to reveal significant differences in the fragmentation functions of quark and gluon jets (Fig. 5, right).

In electron–positron collisions, the first evidence for quark jets came from SPEAR in 1975[31] and the first evidence for gluon jets came from PETRA in 1979–1980.[32] The former were 4 GeV quark jets, PETRA's gluon jets were typically 6 GeV; ISR jets — mostly gluon jets — were at least 10 GeV. The e^+e^- data were analysed in terms of event shapes: sphericity, oblateness, thrust, triplicity, etc. There was no doubt that, without any theoretical preconception, the evidence for ISR jets was stronger than the evidence for quark jets at SPEAR in 1975 and the evidence for gluon jets at PETRA in 1979–1980; the ISR physicists who studied large transverse momentum production were rightly feeling frustrated with the relative lack of public recognition given to their data compared with the enthusiasm generated by the SPEAR and PETRA results. The worse sceptics were to be found in the fixed target community where too low values of the centre-of-mass energy prevented jets to be revealed.

Part of the imbalance in the reception given to ISR data compared with SPEAR and PETRA data was subjective: the analysis of ISR data was complicated, which for many meant "was not clean". But, one must recognise that a good part was objective. First because the SPEAR and PETRA detectors were better fit to such studies and second because the beauty of the SPEAR results came from two important features which gave strong support to the quark jet hypothesis: the

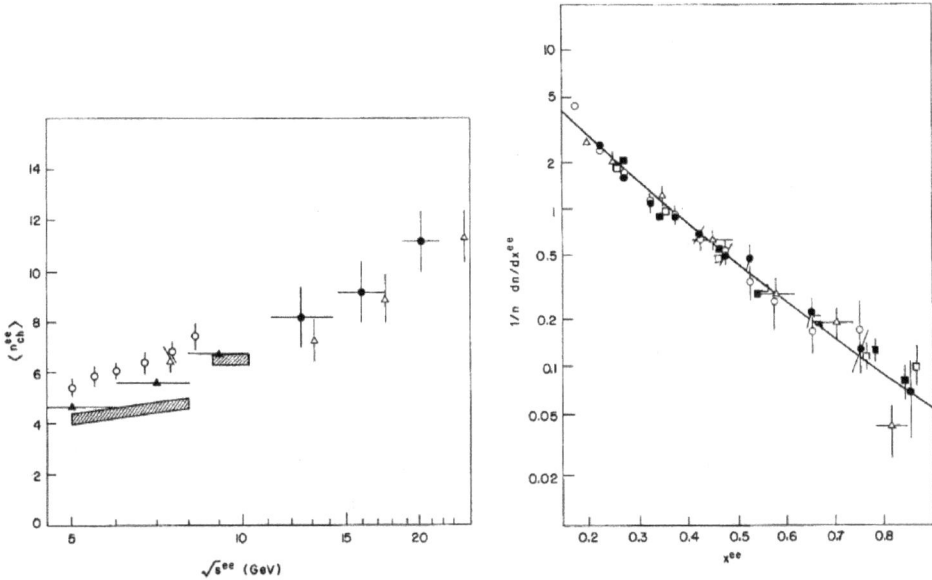

Fig. 5. Left: Mean charge multiplicity of hadron jets as a function of the equivalent e^+e^- energy as measured at SPEAR and DORIS (cross-hatched rectangles), at PETRA (open triangles), in neutrino deep inelastic scattering (full triangles) and in high p_T hadronic interactions at the ISR (open circles). Right: Jet fragmentation functions measured in different processes (triangles are for neutrino deep inelastic scattering, circles for high p_T hadronic interactions at the ISR and the solid line for e^+e^- interactions).

azimuthal distribution of the jet axis displayed the behaviour expected from the known beam polarisation and its polar angle distribution obeyed the $1 + \cos^2 \theta$ law expected in the case of spin 1/2 partons. At PETRA, by the mid-eighties, all four experiments had presented clear evidence for gluon bremsstrahlung, including convincing comparisons with QCD predictions. At the ISR, the complexity of the physics processes at stake was undoubtedly much larger than at electron–positron colliders, making it difficult to devise decisive QCD tests independent from what had been learned at other accelerators. But, once again, ISR data were exploring elementary processes which were not accessible to other accelerators and were shown to nicely fit in a coherent QCD picture embedding deep inelastic as well as e^+e^- annihilation results. This was clearly an independent and essential contribution to the validation of QCD.

2.6. *Direct photons*

In addition to hadron jets, other production mechanisms revealed the parton structure of the colliding protons, such as the production of leptons, heavy flavours[33] and direct photons. The latter was soon recognised to be a particularly simple process: its comparison with QCD predictions could be expected to be instructive. It proceeds either by a quark–antiquark pair in the initial state radiating a photon and a gluon in the final state or by a Compton-like interaction between a quark and a gluon producing a quark and a photon. In both cases, the photon is produced alone, without high p_T companions, and its transverse momentum is balanced by a hadron jet. At the ISR, the Compton diagram dominates: the study of direct photon production should provide information on the gluon structure function as well as a measurement of α_s, the quark fragmentation being borrowed from e^+e^- data. In the first half of the decade, pioneering measurements have established the existence of a signal and identified backgrounds, the main source being π^0 and η decays sending one of the two decay photons alongside their own momentum. At the end of the decade, clear signals were observed[34,35] and a series of measurements followed, which, together with fixed target data, provided a very successful laboratory for QCD. Once again, hadronic interactions, both on fixed target machines and at the ISR, had made use of their unique ability to study gluon collisions and to give essential contributions to the study of the strong interaction in the QCD perturbative regime.[36]

2.7.   *The ISR legacy*

We hope that this brief review of ISR contributions to the new physics that was born in the seventies, and specifically to QCD becoming the theory of the strong interaction, has convinced the reader that they were more than a mere test of the idea that there were point-like constituents inside the proton.[37] Together with hard hadron interactions on fixed target machines, they made optimal use of their exclusive property to study the gluon sector of QCD to leading order. The ISR had the privilege of a higher centre-of-mass energy, fixed target machines had the privilege of versatility: their respective virtues nicely complemented each other. Many factors have contributed to the relative lack of recognition which has been given to ISR physics results: the absence, for many years, of detectors optimised for the study of hard processes, the fact that the weak sector, which during the decade was the scene of as big a revolution as the strong sector, was completely absent from the ISR landscape and, may be most importantly, the fact that hard hadron collisions imply complex processes which may seem "dirty" to those who make no effort to study them in detail.

   We, who worked at the ISR, tend not to attach much importance to this relative lack of recognition because for us, their main legacy has been to have taught us how to make optimal use of the proton–antiproton collider, which was soon to come up. They had given us a vision of the new physics and of the methods to be used for its study which turned out to be extremely profitable. They had played a seminal role in the conception of the proton–antiproton collider experiments, they were the first hadron collider ever built in the world, and they were the machine where a generation of physicists learned how to design experiments on hadron colliders. We tend to see the ISR and the proton-antiproton colliders, both at CERN and at

Fermilab, as a lineage, father and sons, the success of the latter being inseparable from the achievements of the former.

3. Jets at the SPS Collider

3.1. *Introduction*

The SPS collider produced its first collisions in July 1981.[38] It owed its existence to the determination of Carlo Rubbia and his team, gathering together many outstanding competences, including in particular Simon Van der Meer's decisive contribution on stochastic cooling. The motivation of such an effort was to be first to produce and detect the weak bosons that were predicted at the time to be accessible to proton–antiproton collisions at 540 GeV centre-of-mass energy with production cross-sections at nanobarn scale. To this aim, a general purpose 4π detector, UA1, had been designed and constructed. It included a central tracking chamber embedded in a magnetic field and surrounded by calorimetry (Figs. 6 and 7, left). While the performance of the tracker was at the cutting edge of current technology, the constraints imposed on the overall design by the magnetic field implied a rather coarse calorimeter design. A second, much cheaper detector, UA2, had been conceived with the idea to compete with and complement UA1 on only part of the weak boson physics without being constrained by a requirement of universality (Figs. 6 and 7, right). Its ambition being limited to the detection

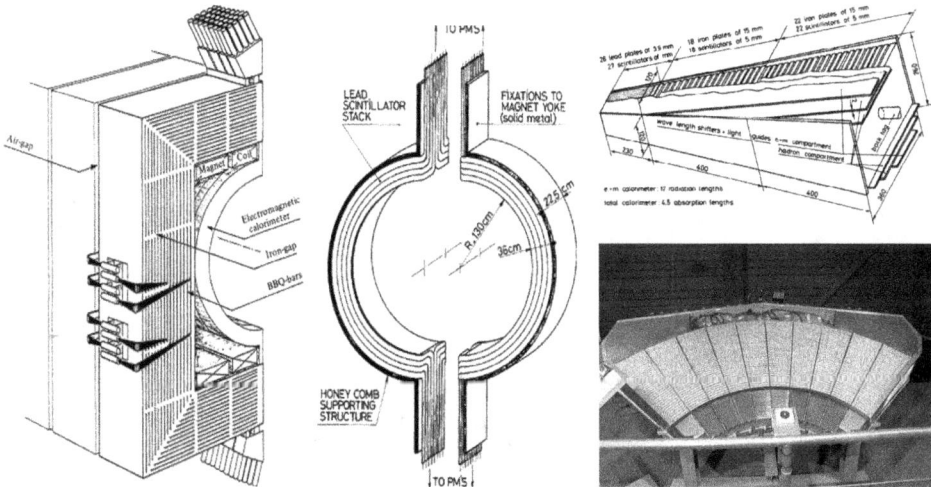

Fig. 6. Left: schematic view of the UA1 hadronic calorimeter, showing two half modules of the magnet yoke instrumented with iron scintillator sandwiches. Centre: schematic view of the UA1 electromagnetic calorimeter, showing a pair of lead-scintillator "gondolas" surrounding the central tracker. Right, up: schematic arrangement of one of the 240 projective cells of the UA2 central calorimeter. Right, down: an azimuthal sector of the central UA2 calorimeter (orange slice) during assembly.

Fig. 7.   Overall views of the UA1 (left) and UA2 (right) detectors.

of electrons and hadron jets, it could afford having no magnetic field at the price of giving up muon detection. At variance with UA1 it was equipped with a small central tracker surrounded by calorimeters optimised for the task, of a design making full use of the lessons that had been learned at the ISR and with better energy and angular resolutions than UA1.

3.2.   *Evidence for jet production*

For this reason, the first experiment to obtain clear evidence for jet production in hadron collisions using a method free from trigger bias was UA2.[29] In the first collider run, while its azimuthal coverage was not yet complete — it was missing a 60° wedge — it detected a sample of high transverse energy jet pairs standing above an underlying event of particles having only some $0.4\,\mathrm{GeV}/c$ transverse momentum on average.[39,40] This result marked the end of the doubts shed by fixed target experiments[41,42] on the ISR claim to have evidence for the production of hadron jets. Following the UA2 observation, jets were soon observed also by UA1[43] after some hesitation: in the February 1982 issue of Physics Today,[44] a report on first preliminary results of UA1 states that "... the anomalously high total transverse energy appears generally to be distributed quite uniformly among the particles emerging in all azimuthal directions. Clean parton-model jets will be much more elusive in hadron–hadron scattering than in e^+e^- collisions." The UA2 detector included a total-absorption calorimeter covering the full azimuth over the polar angle interval $40° < \theta < 140°$. This calorimeter[45] was subdivided into 240 independent cells, each subtending the interval $\Delta\theta \times \Delta\varphi = 10° \times 15°$. For each event it was possible to measure the total transverse energy $\sum E_T$, defined as $\sum E_T = \sum_i E_i \sin\theta_i$ where E_i is the energy deposited in the ith cell, θ_i is the polar angle of the cell centre, and the sum extends to all cells. The observed $\sum E_T$

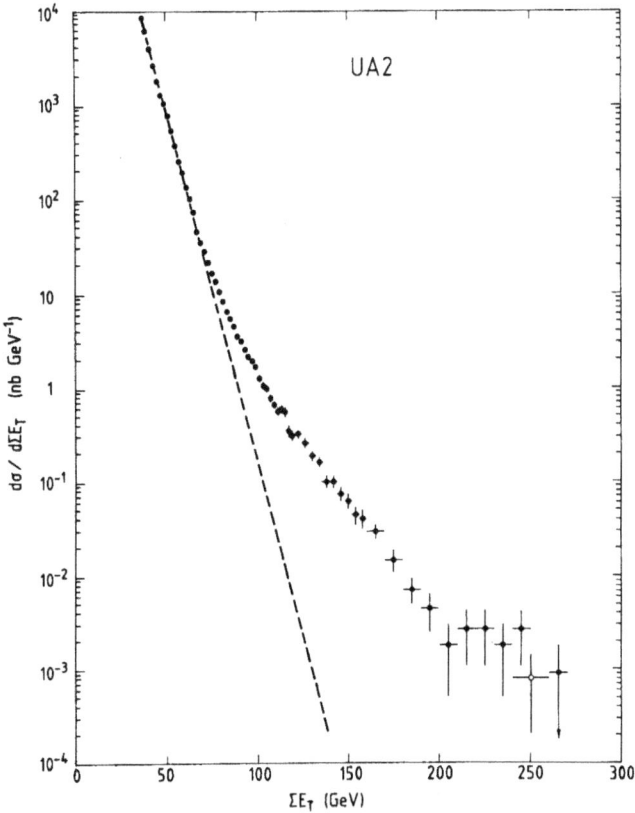

Fig. 8.   Distribution of the total transverse energy $\sum E_T$ observed in the UA2 central calorimeter.

distribution[46] (see Fig. 8) shows a clear departure from the exponential when $\sum E_T$ exceeds 60 GeV.

In order to study the pattern of energy distribution in the events, energy clusters were constructed by joining all calorimeter cells sharing a common side and containing at least 0.4 GeV. In each event, these clusters were then ranked in order of decreasing transverse energies ($E_T^1 > E_T^2 > E_T^3 > \cdots$). Figure 9 (left) shows the mean value of the fractions $h_1 = E_T^1/\sum E_T$ and $h_2 = (E_T^1 + E_T^2)/\sum E_T$ as a function of $\sum E_T$. Their behaviour reveals that, when $\sum E_T$ is large enough, a very substantial fraction of it is shared on average by two clusters with roughly equal transverse energies (an event consisting of only two clusters with equal transverse energies would have $h_1 = 0.5$ and $h_2 = 1$).

The azimuthal separation $\Delta\varphi_{12}$ between the two largest clusters is shown in Fig. 9 (right) for events with $\sum E_T > 60$ GeV and $E_T^1, E_T^2 > 20$ GeV. A clear peak is observed at $\Delta\varphi_{12} = 180°$, indicating that the two clusters are coplanar with the beam direction.

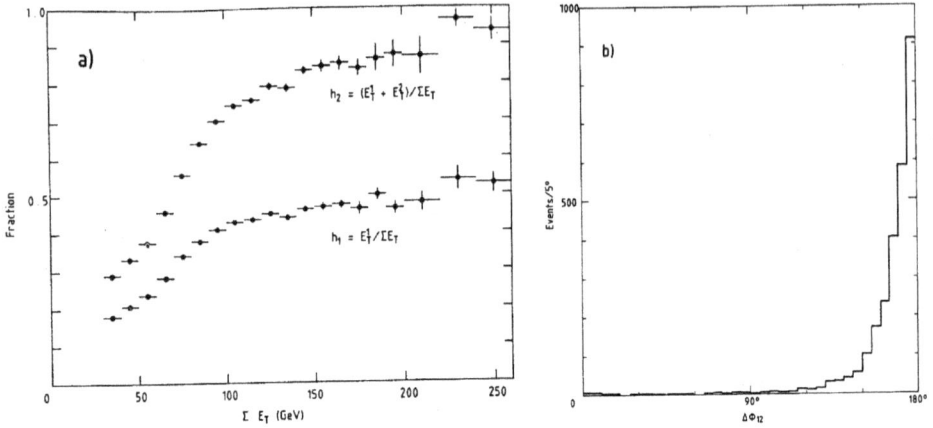

Fig. 9. Left: Mean value of the fraction h_1 (h_2) of the total transverse energy $\sum E_T$ contained in the cluster (in the two clusters) having the largest E_T, as a function of $\sum E_T$. Right: Azimuthal separation between the two largest E_T clusters in events with $\sum E_T > 60$ GeV and E_T^1, $E_T^2 > 20$ GeV.

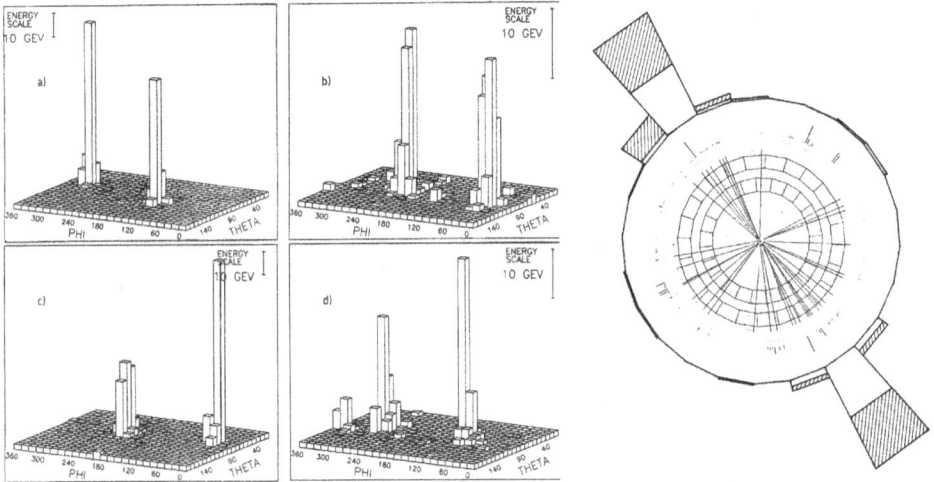

Fig. 10. Left: Four typical transverse energy distributions for events with $\sum E_T > 100$ GeV in the $\theta - \varphi$ plane. Each bin represents a cell of the UA2 calorimeter. Right: Projection of a typical two-jet event perpendicular to the beams in the UA2 detector. The heights of the trapezoids are proportional to transverse energy. The open and shaded areas represent the energy depositions in the electromagnetic and hadronic sections of the calorimeter, respectively.

The emergence of two-cluster structures in events with large $\sum E_T$ is even more dramatically illustrated by inspecting the transverse energy distribution over the calorimeter cells. Figure 10 (left) shows such a distribution for four typical events having $\sum E_T > 100$ GeV. The transverse energy appears to be concentrated within two (or, more rarely, three) small angular regions. These energy clusters are associated with collimated multiparticle systems (jets), as shown in Fig. 10 (right) which displays the reconstructed charged particle tracks in these events (there is no magnetic field in the central region of the UA2 detector, so all tracks appear straight).

3.3.   *Theoretical interpretation*

Jet production in hadronic collisions is interpreted in the framework of the parton model as hard scattering among the constituents of the incident hadrons. Since the incident proton and antiproton contain quarks, antiquarks and gluons, there are several elementary subprocesses that contribute to jet production. For each subprocess the scattering cross-section, calculated to first order in the strong coupling constant α_s is given by the expression

$$\frac{d\sigma}{d\cos\theta^*} = \frac{\pi\alpha_s^2}{2\hat{s}}|M|^2 \tag{1}$$

where θ^* is the scattering angle and $\hat{s}$ the square of the total energy in the centre-of-mass of the two partons; M is the matrix element, which is itself a function of $\hat{s}$ and θ^*. Explicit expressions for $|M|^2$ have been calculated.[47] They show that subprocesses involving initial gluons, such as gg and qg scattering, are dominant whenever the gluon density in the incident proton (or antiproton) is comparable to that of the quarks (or antiquarks).

   The cross-section for inclusive jet production as a function of the jet p_T and angle of emission θ can be calculated to leading order in α_s as a sum of convolution integrals:[48]

$$\frac{d^2\sigma}{dp_T d(\cos\theta)} = \frac{2\pi p_T}{\sin^2\theta} \sum_{A,B} \int dx_1 dx_2 F_A(x_1) F_B(x_2)$$

$$\times \delta\left(p_T - \frac{\sqrt{\hat{s}}}{2}\sin\theta^*\right)\alpha_s^2 \sum_f \frac{|M(AB\to f)|^2}{\hat{s}} \tag{2}$$

where F_A and F_B are structure functions describing the densities of partons A and B in the incident hadrons, Q^2 is the square of the four-momentum transfer in the subprocess, and the sum extends over all initial partons types A, B, and all possible final states f. The structure functions depend on Q^2: they are measured in deep inelastic lepton–nucleon scattering experiments ($Q^2 \leq 20\,\text{GeV}^2$) and extrapolated to the Q^2 range of interest (up to $10^4\,\text{GeV}^2$ at the energy of the proton–antiproton collider) according to the predicted QCD evolution.[49]

   At the energy of the proton–antiproton collider, jets with p_T around $30\,\text{GeV}/c$ produced near 90° arise from hard scattering of partons with relatively small values of x ($x < 0.1$). In this region gluon jets are expected to dominate, both because there are many gluons in the nucleon at small x and because subprocesses involving initial gluons have large cross-sections. This is in contrast with e^+e^- collisions, where the production of quark jets dominates hadronic final states.

A number of uncertainties affect the comparison between predicted cross-section and experimental data. The most obvious is that Eq. (2) predicts the yield of high-p_T massless partons, whereas the experiments measure hadronic jets with a total invariant mass of several GeV. The relation between the parton p_T and the measured cluster transverse energy E_T is usually determined with the help of QCD-inspired simulations in which the outgoing partons evolve into jets according to a specific hadronisation model, and the detector response to hadrons is taken into account. An important uncertainty in the theoretical predictions arises from the Q^2 extrapolation of the structure functions, especially those describing the gluons. Finally, in addition to the statistical errors, the data are also affected by a number of systematic effects, such as uncertainties in the calorimeter energy scale and detector acceptance. These effects amount typically to an overall uncertainty of $\pm 50\%$ in the measured jet yields. Altogether, a comparison between the theoretical predictions and the experimental results is only possible to an accuracy not greater than a factor of 2.

Figure 11 (left) shows the inclusive jet production cross-section around $\theta = 90°$, as measured by UA1[43] and UA2[46] during the first physics runs of the proton–antiproton collider. Also shown is a band of QCD predictions[48,50] with a width that illustrates the theoretical uncertainties. The agreement between data and theory is remarkable, especially because the theoretical curves are not a fit to the data but represent absolute predictions made before the data became available.

Subsequent improvements in the collider luminosity and progress in theory are illustrated in Fig. 11 (right), where the inclusive jet production cross-section for the central region, as measured by UA2 in 1988–89[23] is compared with a QCD prediction based on more refined structure functions.[24]

3.4. Angular distribution of parton–parton scattering

The study of the jet angular distribution in two-jet events provides a way to measure the angular distribution of parton–parton scattering, and can therefore be considered as the analogue of Rutherford's experiment in QCD. We can write

$$\frac{d^3\sigma}{dx_1 dx_2 d\cos\theta^*} = \sum_{A,B} \frac{F_A(x_1)}{x_1} \frac{F_B(x_2)}{x_2} \sum_{C,D} \frac{d\sigma(AB \to CD)}{d(\cos\theta^*)} \qquad (3)$$

where $F_A(x_1)$ [$F_B(x_2)$] is the structure function describing the density of parton A [B] within the incident hadrons, and the sum extends to all subprocesses $AB \to CD$. Then, if the total transverse momentum of the two-jet system is zero, or very much smaller than the transverse momentum of each jet, it is possible to determine

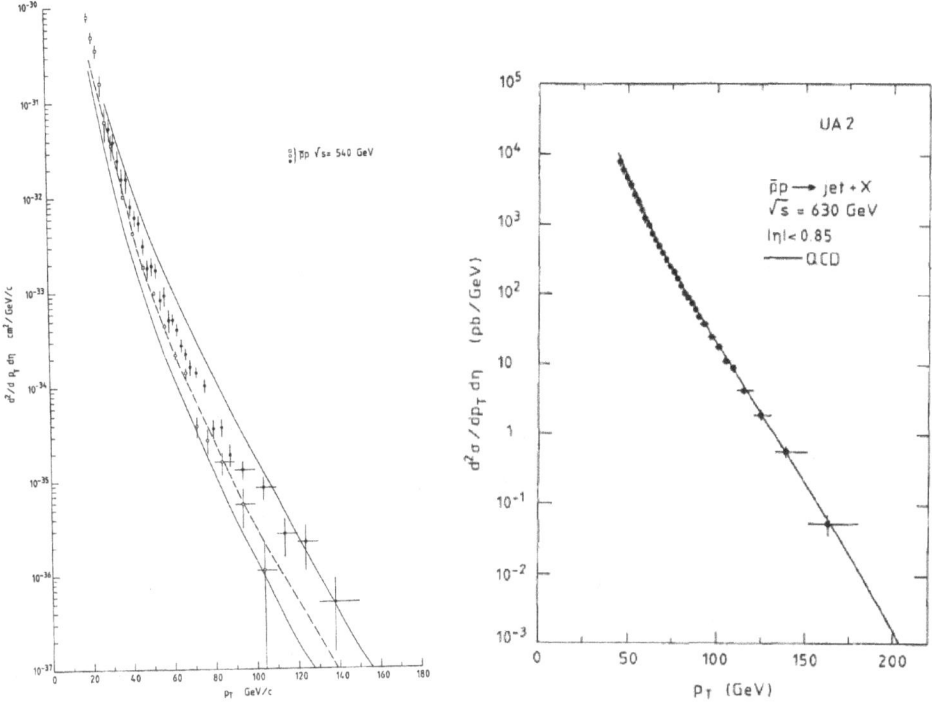

Fig. 11.   Left: Early collider measurements of the cross-section for inclusive jet production around $\theta = 90°$, as a function of the jet p_T. Full circles: UA2[46]; open circles and squares: UA1.[43] The dashed curve represents a theoretical prediction.[48] The two full curves define a band of QCD predictions.[50] Right: Inclusive jet cross-section in the central region ($|\eta| < 0.85$, where $\eta = -\ln \tan \theta/2$), as measured by UA2 in 1988–89.[51] The curve represents a QCD prediction.[52]

simultaneously for each event the momentum fractions x_1, x_2 carried by the two incident partons and their scattering angle θ^*.

Equation (3) may at first sight appear hopeless in view of the many terms involved. However, in the case of proton-antiproton collisions the dominant sub-processes are $gg \to gg$, $qg \to qg$ (or $\bar{q}g \to \bar{q}g$), and $q\bar{q} \to q\bar{q}$, which to a very good approximation have the same $\cos \theta^*$ dependence. Equation (3) can then be approximately factorised as

$$\frac{d^3\sigma}{dx_1 dx_2 d\cos\theta^*} = \left[\frac{1}{x_1}\sum_A F_A(x_1)\right]\left[\frac{1}{x_2}\sum_A F_B(x_2)\right]\frac{d\sigma}{d(\cos\theta^*)}. \tag{4}$$

If $d\sigma/d(\cos\theta^*)$ is taken to be the differential cross-section for gluon–gluon elastic scattering, which to leading order in QCD has the form

$$\frac{d\sigma}{d(\cos\theta^*)} = \frac{9\pi\alpha_s^2}{16 x_1 x_2 s}\frac{(3+\cos^2\theta^*)^3}{(1-\cos^2\theta^*)^2} \tag{5}$$

where s is the square of the proton–antiproton total centre-of-mass energy, then it becomes possible to write

$$\sum_A F_A(x) = g(x) + \frac{4}{9}[q(x) + \bar{q}(x)] \tag{6}$$

where $g(x)$, $q(x)$ and $\bar{q}(x)$ are the gluon, quark, and antiquark structure functions of the proton, respectively. The factor $4/9$ in Eq. (6) reflects the relative strength of the quark–gluon and gluon–gluon couplings in QCD.

The term $d\sigma/d(\cos\theta^*)$ in Eq. (5) contains a singularity at $\theta^* = 0$ with the familiar Rutherford form $\sin^{-4}(\theta^*/2)$ which is typical of gauge vector boson exchange. In the subprocesses $gg \to gg$ and $qg \to qg$ (or $\bar{q}g \to \bar{q}g$) it arises from the three-gluon vertex. It is also present in the subprocess $q\bar{q} \to q\bar{q}$, but in this case it would be present in an Abelian theory as well, as for e^+e^- scattering in QED.

Figure 12 (left) shows the $\cos\theta^*$ distribution measured by UA1[53] for jets with $p_T > 20\,\mathrm{GeV}/c$. Both data and theoretical curves for the three dominant subprocesses are normalised to 1 at $\cos\theta^* = 0$. The UA2 results[54] are shown in Fig. 12 (right), where they are compared with the $\cos\theta^*$ distribution predicted by QCD with no approximation (the UA2 data cover only the range $|\cos\theta^* < 0.6$ because of the limited polar-angle interval covered by the UA2 calorimeter). Both sets of data agree with QCD expectations, and they clearly show the

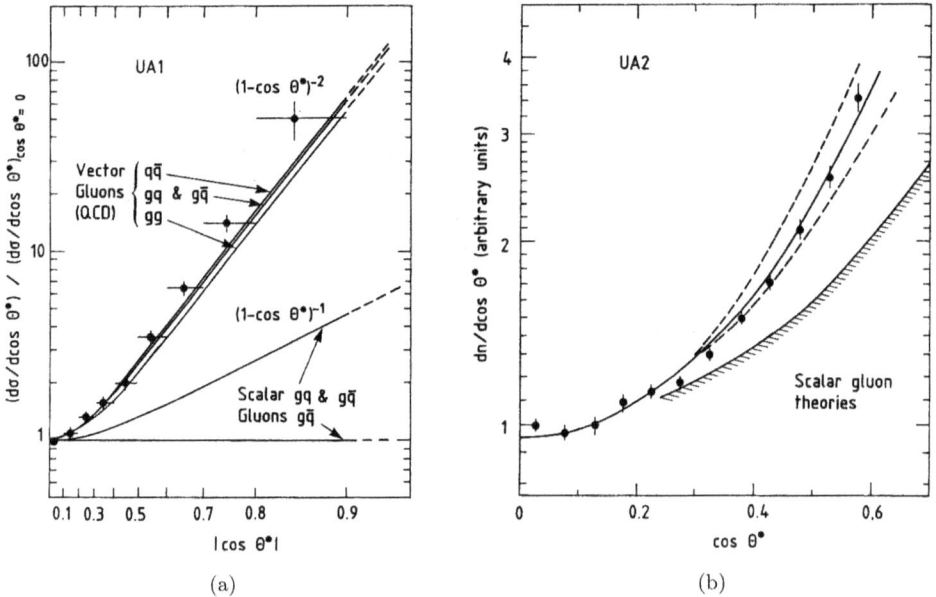

Fig. 12. (a) Distribution of $\cos\theta^*$ for hard parton scattering as measured by UA1,[53] normalised to 1 at $\cos\theta^* = 0$. (b) Distribution of $\cos\theta^*$ for hard parton scattering as measured by UA2.[54] All QCD subprocesses lie in the area between the two dashed curves. The full line is the QCD prediction, normalised to the data.

increase towards the forward direction expected from the Rutherford singularity. For historical reasons, Fig. 12 displays also expectations from theories with scalar gluons, in strong disagreement with the data.

3.5. *Determination of the proton structure function*

The effective structure function $F(x)$ (see Eq. (6)) can also be extracted from the analysis of two-jet events. Figure 13 shows the function $F(x)$ as determined by UA1[53] and UA2.[54] In addition to the statistical errors there is a systematic uncertainty of $\sim$50% in the overall normalisation which reflects theoretical uncertainties associated with the absence of higher-order terms. Also shown in the figure are curves representing the function $g(x) + (4/9)[q(x) + \bar{q}(x)]$ as expected from fits to neutrino and antineutrino deep inelastic scattering data.[55] The collider results agree with the behaviour expected at the large Q^2 values typical of the collider experiments ($Q^2 \approx 2000\,\mathrm{GeV}^2$). They show directly the very large gluon density in the proton at small x values.

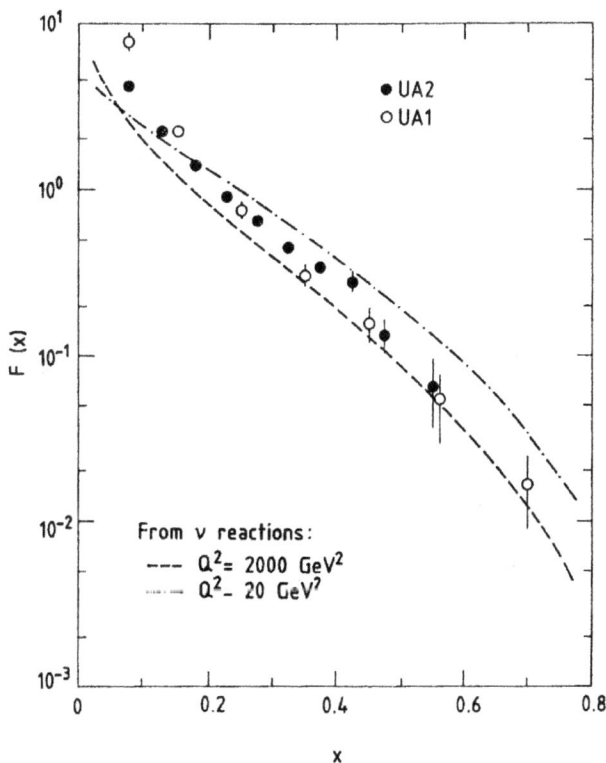

Fig. 13. Effective structure function measured from two-jet events.[53,54] The dashed lines are obtained from deep inelastic neutrino scattering experiments.[55]

3.6. *Direct photon production*

Direct photon production at high p_T is expected to result from the subprocesses $qg \rightarrow q\gamma$, $\bar{q}g \rightarrow \bar{q}\gamma$, or $q\bar{q} \rightarrow g\gamma$. It was first observed at the ISR, where the explored p_T range did not exceed ~ 10 GeV. The cross-section is expected to be proportional to the product $\alpha_s\alpha$ and thus it is two to three orders of magnitude smaller than the cross-section for jet production at the same $\boldsymbol{p}_T$ value.

This process has the great advantage that the photon p_T is not affected by fragmentation effects, resulting in experimental uncertainties which are considerably smaller than those obtained in the measurement of the jet cross-section. The production of high p_T jets is, however, a large source of background: hadron jets often contain one or more π^0 (or η) mesons which decay into very asymmetric photon pairs or narrow photon pairs that are not resolved by the calorimeters. This background has a cross-section much larger than the direct photon signal. The latter, however, results in isolated electromagnetic clusters, whereas the background from hadronic jets is accompanied by jet fragments, so that an "isolation requirement" is very effective in reducing the contamination of the signal sample. The residual contamination from high p_T isolated π^0 (or η) mesons is measured in UA2 and subtracted on a statistical basis by considering the fraction of photons that initiate showers in a 1.5 radiation length thick lead converter located in front of the calorimeter.

The UA2 measurement of direct photon production[56] is shown in Fig. 14, which displays the invariant differential cross-section as a function of the photon p_T. The data are in good agreement with a next-to-leading order QCD calculation.[57]

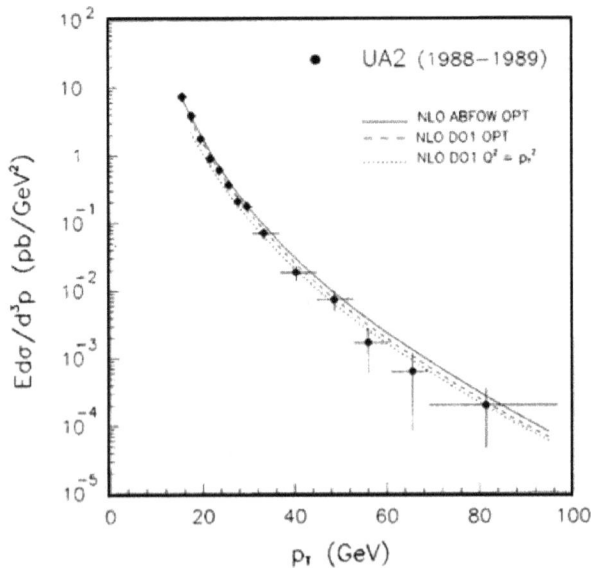

Fig. 14. Invariant differential cross-section for direct photon production.[56] The curves represent QCD predictions[57] for different sets of structure functions.

The proton–antiproton collider has been a powerful laboratory for many other mechanisms than jet and direct photon production, which have given access to precise and decisive QCD tests, such as the production of weak bosons and of heavy flavours. Their presentation is beyond the scope of the present article.

3.7. *Total transverse momentum of the two-jet system*

If the two partons that undergo hard scattering have no initial p_T, the total transverse momentum of the final two-jet system, P_T, should be equal to zero. In reality, this does not happen because the incident partons have a small "primordial" transverse momentum, and, furthermore, both incident and outgoing partons may radiate gluons.

Experimentally, P_T is determined from the sum of two large and approximately opposite two-dimensional vectors p_{T1} and p_{T2}, and it is therefore sensitive to instrumental effects such as the calorimeter energy resolution and incomplete jet containment due to edge effects in the detector. These effects can be made small by considering only the component of P_T, P_η, parallel to the bisector of the angle defined by p_{T1} and p_{T2}.

Figure 15 shows the distribution of P_η, as measured by UA2.[54] The data are in good agreement with a QCD prediction[58] illustrated by the curve. In QCD, gluon radiation by a gluon $(g \to gg)$, which occurs because of the three-gluon vertex, has a rate 9/4 times higher than that of $q \to qg$, and prediction based on the assumption that gluons radiate like quarks disagree with the data (Fig. 15). Since gluon jets dominate in the p_T range explored at the collider, we can consider the good agreement between the data and the theoretical prediction as further evidence in favour of a QCD description of high-p_T jet production.

3.8. *Multijet final states*

Three-jet final states were first observed in e^+e^- annihilations to hadrons.[59] They were interpreted as an effect of gluon radiation by the outgoing quark or antiquark. Such an effect is also expected in the case of hadron collisions, where, however, gluons can be radiated not only by the outgoing high-p_T partons, but also by the incident partons and at the parton scattering vertex as well.

At tree level, the QCD matrix element for two-to-three parton scattering processes have been calculated by several authors.[60] Under the assumption of massless partons, the final-state configuration, at fixed centre-of-mass energy $\hat{s}$, is specified by four independent variables. Two variables are required to specify how the available energy is shared between the three final-state partons, and two variables serve to fix the orientation of the three-jet system with respect to the axis defined by the colliding beams (we do not consider the overall azimuthal angle, which is irrelevant because the incident beams are not polarised). The most commonly used variables are z_1, z_2, z_3 (the energies of the outgoing partons scaled such that

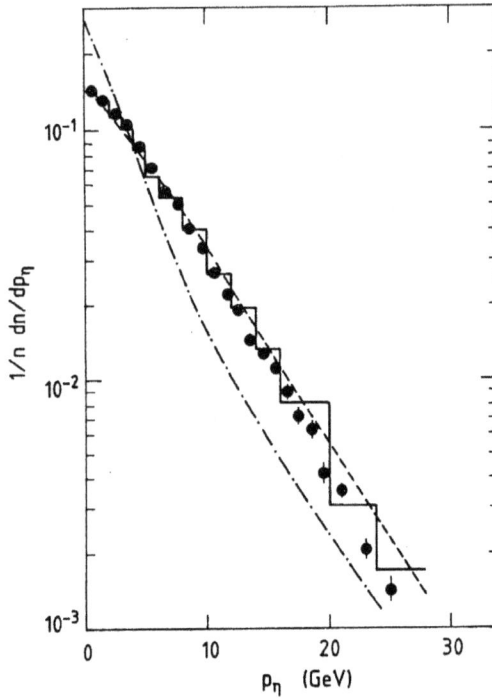

Fig. 15. Distribution of the component P_η of the total transverse momentum of the two-jet system, as measured by UA2.[54] The dashed line is a QCD prediction.[58] The dashed-dotted line is the same prediction, but assumes that gluons radiate as quarks. The histogram is the standard QCD prediction with the detector effects taken into account.

$z_1 + z_2 + z_3 = 2$ and ordered such that $z_1 > z_2 > z_3$); θ_1, the angle of parton 1 with respect to the beam axis; and ψ, the angle between the plane containing partons 2 and 3, and the plane defined by parton 1 and the beam axis.

The UA2 analysis of three-jet events uses variables defined by $x_{ik} = (m_{ik})^2/\hat{s}$, where m_{ik} is the invariant mass of any two of the three jets. The three x_{ik} variables are simply related to the z_i as follows: $x_{12} = 1 - z_3$; $x_{13} = 1 - z_2$; and $x_{23} = 1 - z_1$. They satisfy the constraint $x_{12} + x_{13} + x_{23} = 1$. The three-jet scatter plot in the x_{12}, x_{23} plane measured by UA2[61] is shown in Fig. 16 (left). The absence of events at small x_{23} is due to the inability to resolve jets at small angle to each other, and the absence of events at large x_{12} is due to the requirement that all three jet p_T values exceed $10\,\mathrm{GeV}/c$. The increase in event density with decreasing x_{23} for fixed x_{12} reflects the tendency of final-state gluon radiation to be produced at small angle to the radiating parton. The projections of the scatter plot onto the x_{12} and x_{23} axes are also shown. The data are in acceptable agreement with the leading order QCD predictions but inconsistent with phase space distributions.

The three-jet angular distributions ($\cos\theta_1$ versus ψ) measured by UA1[62] are shown in Fig. 16 (right). The distribution of $\cos\theta_1$ shows a pronounced

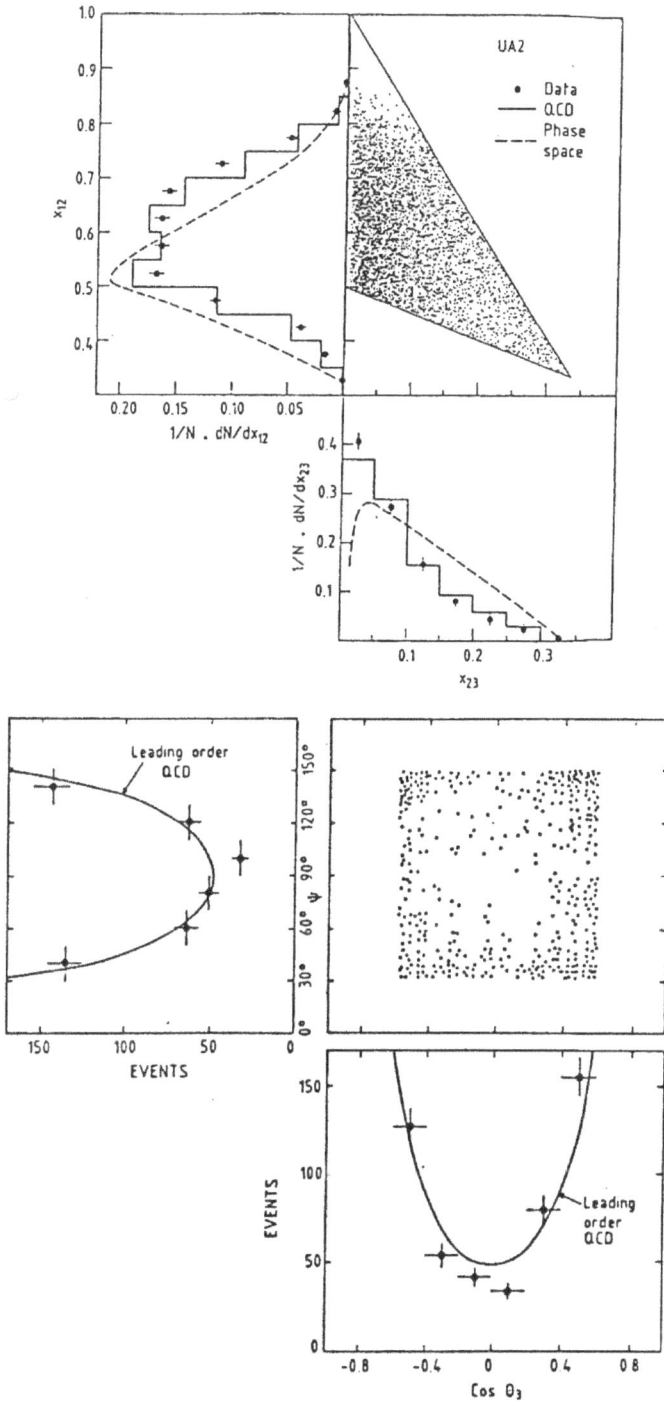

Fig. 16. Top: Three-jet scatter plot x_{12} versus x_{23}, as measured by UA2.[61] Bottom: Three-jet angular distribution, as measured by UA1.[62]

forward–backward peaking, which is qualitatively similar to the one observed in two-jet events. The $|\psi|$ distribution shows that the configuration in which jets 2 and 3 lie close the plane defined by jet 1 and the beam axis ($|\psi| \approx 30°$ or $150°$) are preferred relative to configurations for which $|\psi| \approx 90°$. This effect reflects the tendency of initial-state gluon radiation to be produced at small angles to the incoming partons. The projection of the scatter plot onto the $\cos\theta_1$ and $|\psi|$ axes are also shown, together with the theoretical curves calculated from the leading-order QCD formulae neglecting scale breaking effects. The data are in fair agreement with these predictions. It has been shown[63] that the inclusion of scale breaking effects in the theoretical calculations improves the agreement with experiment.

3.9.   *Conclusion*

One of the first results from the CERN collider was the observation of clear, uncontroversial jets in hadronic collisions. This result had been long awaited and had a very significant impact on the field of particle physics. It was the successful culmination of years of experimental effort, carried over from the CERN ISR and elsewhere, on a difficult and subtle experimental problem. It certainly ranks among the most important collider discoveries, not only because it provided by far the most spectacular evidence at the time of the physical reality of the partons inside the proton, but also because it opened the door to many quantitative studies of jet related phenomena which followed, first at the CERN collider, a few years later at the higher energy collider at Fermilab and now at the LHC. All these studies have amply confirmed the interpretation of these phenomena in terms of parton–parton scattering, as described by perturbative QCD.

References

1. G. F. Chew, *High Energy Physics*, Les Houches (Gordon & Breach, 1965), p. 187.
2. R. H. Dalitz, *High Energy Physics*, Les Houches (Gordon & Breach, 1965), p. 251.
3. J. Benecke, T. T. Chou, C. N. Yang and E. Yen, *Phys. Rev.* **188**, 2159 (1969).
4. P. Darriulat, *Pontif. Acad. Sci. Scr. Varia* **119**, 109–119 (2011); CERN-2012-004, pp. 63 (2012), arXiv:1206.4131 [physics.acc-ph].
5. L. Di Lella, *Phys. Rep.* **403–404**, 147–164 (2004).
6. S. Weinberg, Electroweak Reminiscences, in *History of Original Ideas and Basic Discoveries in Particle Physics*, eds. H. B. Newman and T. Ypsilantis, NATO ASI series, B352 (Plenum Press, 1994) p. 27; The Making of the Standard Model, in *Prestigious Discoveries at CERN*, eds. R. Cashmore, L. Maiani and J.-P. Revol (Springer, 2003), p. 9.
7. D. J. Gross, Asymptotic freedom, Confinement and QCD, in *History of Original Ideas and Basic Discoveries in Particle Physics*, eds. H. B. Newman and T. Ypsilantis, NATO ASI series, B352 (Plenum Press, 1994), p. 75.

8. G. 't Hooft, Gauge Theory and Renormalization, in *History of Original Ideas and Basic Discoveries in Particle Physics*, eds. H. B. Newman and T. Ypsilantis, NATO ASI series, B352 (Plenum Press, 1994), p. 37; The Creation of Quantum Chromodynamics, in *The Creation of Quantum Chromodynamics and the Effective Energy*, World Scientific Series in XXth century Physics, Volume 25, ed. L. N. Lipatov (World Scientific, 2001).

9. J. I. Friedman, Deep Inelastic Scattering Evidence for the Reality of Quarks, in *History of Original Ideas and Basic Discoveries in Particle Physics*, eds. H. B. Newman and T. Ypsilantis, NATO ASI series, B352 (Plenum Pres, 1994), p. 725.

10. J. D. Bjorken, private communication to the MIT-SLAC group, 1968, and *Phys. Rev.* **179**, 1547 (1969).

11. R. P. Feynman, *Phys. Rev. Lett.* **23**, 1415 (1969); in *Proceedings of the IIIrd International Conference on High Energy Collisions* (Gordon and Breach, New York, 1969).

12. D. H. Perkins, in *Proceedings of the XVIth International Conference on High Energy Physics, Chicago and NAL*, Vol. 4 (1972), p. 189.

13. D. J. Gross and F. Wilczek, *Phys. Rev. Lett.* **30**, 1343 (1973); *Phys. Rev. D* **8**, 3633 (1973) and *Phys. Rev. D* **9**, 980 (1974).

14. H. Fritzsch, Murray Gell-Mann and H. Leutwyler, *Phys. Lett.* **47B**, 368 (1973).

15. See: *History of Original Ideas and Basic Discoveries in Particle Physics*, eds. H. B. Newman and T. Ypsilantis, NATO ASI series B352 (Plenum Press, 1994), p. 140.

16. S. M. Berman and M. Jacob, *Phys. Rev. Lett.* **25**, 1683 (1970).

17. J. D. Bjorken and E. A. Paschos, *Phys. Rev.* **185**, 1975 (1968).

18. S. D. Drell, D. J. Levy and T. M. Yan, *Phys. Rev.* **187**, 2159 (1969).

19. S. M. Berman, J. D. Bjorken and J. Kogut, *Phys. Rev. D* **4**, 3388 (1971).

20. J. Kuti and V. F. Weisskopf, *Phys. Rev. D* **4**, 3418 (1971).

21. S. D. Drell and T. M. Yan, *Phys. Rev. Lett.* **25**, 316 (1970).

22. B. Alper *et al.*, *Phys. Lett. B* **44**, 521 (1973).

23. M. Banner *et al.*, *Phys. Lett. B* **44**, 537 (1973).

24. F. W. Büsser *et al.*, *Phys. Lett. B* **46**, 471 (1973), and *Proc. 16th Int. Conf. on High Energy Phys.* (Chicago and NAL, 1972).

25. For a review, see D. Sivers, R. Blankenbecler and S. J. Brodsky, *Phys. Rep. C* **23**, 1 (1976).

26. A. P. Contogouris, R. Gaskell and S. Papadopoulos, *Phys. Rev. D* **17**, 2314 (1978).

27. T. Akesson *et al.*, *Phys. Lett. B* **118**, 185 (1982) and 193; *Phys. Lett. B* **123**, 133 (1983); *Phys. Lett. B* **128**, 354 (1983); *Z. Phys. C* **25**, 13 (1984); *Z. Phys. C* **30**, 27 (1986); *Z. Phys. C* **32**, 317 (1986); *Z. Phys. C* **34**, 163 (1987).
A. L. S. Angelis *et al.*, *Phys. Lett. B* **126**, 132 (1983); *Nucl. Phys. B* **244**, 1 (1984); *Nucl. Phys. B* **303**, 569 (1988).

28. C. W. Fabjan and N. McCubbin, *Physics athe CERN Intersecting Storage Rings (ISR) 1978–1983*, *Phys. Rep.* 403–404 165–175 (2004).

29. M. Banner *et al.*, *Phys. Lett. B* **118**, 203 (1982); J.-P. Repellin for the UA2 collaboration, *J. Phys. Colloques* **43**, C3–571, 578 (1982).

30. P. Darriulat, in *Proc. XVIIIth Int. Conf. on High Energy Phys.*, Tbilisi, USSR, 1976, ed. N. N. Bogolioubov *et al.*, JINR Dubna, 1977; *Large Transverse Momentum Hadronic Processes*, *Ann. Rev. Nucl. Part. Sci.* **30** 159 (1980).

31. G. Hanson *et al.*, *Phys. Rev. Lett.* **35**, 1609 (1975).
32. J. G. Branson, Gluon Jets, in *History of Original Ideas and Basic Discoveries in Particle Physics*, eds. H. B. Newman and T. Ypsilantis, NATO ASI series, B352 (Plenum Press, 1994), p. 101.
33. D. Drijard *et al.*, *Phys. Lett. B* **81**, 250 (1979); *Phsy. Lett. B* **85**, 452 (1979).
34. M. Diakonou *et al.*, *Phys. Lett. B* **87**, 292 (1979).
35. A. L. S. Angelis *et al.*, *Phys. Lett. B* **94**, 106 (1980).
36. T. Ferbel and W. M. Molzon, *Rev. Mod. Phys.* **56**, 181 (1984); L. F. Owens, *Rev. Mod. Phys.* **59**, 485 (1987); P. Aurenche *et al.*, *Nucl. Phys. B* **297**, 661 (1988) and *Eur. Phys. C* **9**, 107 (1999).
37. General reviews can be found in:
    G. M. Giacomelli and M. Jacob, Physics at the CERN-ISR, *Phys. Rep.* **55**, 1 (1979).
    L. van Hove and M. Jacob, Highlights of 25 Years of Physics at CERN, *Phys. Rep.* **62** (1980).
    Maurice Jacob, in *A Review of Accelerator and Particle Physics at the CERN Intersecting Storage Rings*, CERN 64–13 21—81 (1984).
38. C. Rubbia, *Phys. Rep.* **239**, 241 (1994).
39. G. Arnison *et al.*, *Phys. Lett. B* **118**, 167 (1982).
40. M. Banner *et al.*, *Phys. Lett. B* **122**, 322 (1983).
41. C. DeMarzo *et al.*, *Nucl. Phys. B* **211**, 375 (1983).
42. B. Brown *et al.*, *Phys. Rev. Lett.* **49**, 7117 (1982).
43. G. Arnison *et al.*, *Phys. Lett. B* **123**, 115 (1983); *Phys. Lett. B* **132**, 214 (1983).
44. B. M. Schwarzchild, CERN SPS now running as 540-GeV p̄p collider, *Physics Today* **35**(2), 17 (1982).
45. A. Beer *et al.*, *Nucl. Instrum. Methods A* **224**, 360 (1984).
46. P. Bagnaia *et al.*, *Z. Phys. C* **20**, 117 (1983); *Phys. Lett. B* **138**, 430 (1984).
47. B. L. Combridge, J. Kripfganz and J. Ranft, *Phys. Lett. B* **70**, 234 (1977).
48. R. Horgan and M. Jacob, *Nucl. Phys. B* **179**, 441 (1981).
49. J. F. Owens *et al.*, *Phys. Rev. D* **17**, 3003 (1979); R. Baier *et al.*, *Z. Phys. C* **2**, 265 (1983); F. E. Paige and S. D. Protopopescu, in *Proc. 1982 DPF Summer Study on Elem. Part. Phys. and Future Facilities, Snowmass, Colorado*, eds. R. Donaldson, R. Gustafson and F. Paige (AIP, 1982), p. 471.
50. N. G. Antoniou *et al.*, *Phys. Lett. B* **128**, 257 (1983); Z. Kunszt and E. Pietarinen, *Phys. Lett. B* **132**, 453 (1983); B. Humpert, *Z. Phys. C* **27**, 257 (1985).
51. J. Alitti *et al.*, *Phys. Lett. B* **257**, 232 (1991).
52. E. Eichten *et al.*, *Rev. Mod. Phys.* **56**, 579 (1984); *Rev. Mod. Phys.* **58**, 1065(E) (1986).
53. G. Arnison *et al.*, *Phys. Lett. B* **136**, 294 (1984).
54. P. Bagnaia *et al.*, *Phys. Lett. B* **144**, 283 (1984).
55. H. Abramowicz *et al.*, *Z. Phys. C* **12**, 289 (1982); *Z. Phys. C* **13**, 199 (1982); *Z. Phys. C* **17**, 283 (1983); F. Bergsma *et al.*, *Phys. Lett. B* **123**, 269 (1983).
56. J. Alitti *et al.*, *Phys. Lett. B* **263**, 544 (1991).
57. P. Aurenche *et al.*, *Phys. Lett. B* **140**, 87 (1984); *Nucl. Phys. B* **297**, 661 (1988).
58. M. Greco, *Z. Phys. C* **26**, 567 (1985).
59. For a review see P. Söding and G. Wolf, *Ann. Rev. Nucl. Part. Sci.* **31**, 231 (1981).

60. Z. Kunszt, *Nucl. Phys. B* **164**, 45 (1980); T. Gottschalk and D. Sivers, *Phys. Rev. D* **21**, 102 (1980).
61. A. Appel *et al.*, *Z. Phys. C* **30**, 341 (1986).
62. G. Arnison *et al.*, *Phys. Lett. B* **158**, 494 (1985).
63. E. J. Buckley, Ph.D. Thesis, *Rutherford Appleton Laboratory Thesis* 029 (1986) (unpublished).

Properties of Antiprotons and Antihydrogen, and the Study of Exotic Atoms

Michael Doser

CERN, EP-1211 Geneva, Switzerland
michael.doser@cern.ch

The study of exotic atoms, of antiprotons and of antihydrogen atoms provides many windows into the investigation of fundamental symmetries, of interactions between particles and nuclei, of nuclear physics and of atomic physics. This field appeared at CERN simultaneously with the first accelerators, and has advanced over the decades in parallel with improvements and advances in its infrastructure.

1. Introduction

Since the earliest days of CERN, antiprotons and exotic atoms have formed a central part of its experimental program. Complementary to particle physics approaches, studies of exotic atoms have played an important role in probing particle properties (particle mass determinations), in studying the strong interaction at relatively large distances of several fm (in pionic or antiprotonic atoms, for example), in investigating nuclear physics (through measurements of atomic transitions of pionic or kaonic atoms comprised of different nuclei) or nuclear radii (with muonic atoms), or in tests of fundamental symmetries (measurements of mass, charge and magnetic moments of antiprotons in antiprotonic helium).

At the same time, antiprotons which were used initially as a workhorse in the study of mesonic states, before techniques to accumulate and store them were developed in the 1970s and 1980s, have since then been studied at very high precision, both as individual particles in traps, as well as in the form of atoms consisting completely of antimatter. Here too, the goals are similar: search for violation of symmetries like CPT through precision spectroscopy of antihydrogen atoms or precision comparisons of mass, charge and magnetic moment between antiprotons and protons, and tests of the weak equivalence principle through measurements of the gravitational interaction of (neutral) antihydrogen atoms.

The study of these exotic systems does not quite follow the general history of CERN: while accelerators at CERN became ever more powerful over the decades, the energies required to produce the sufficiently long-lived particles (muons, pions, kaons and antiprotons) that form the building blocks of exotic atoms are modest; even today, the same Proton Synchrotron (PS, Fig. 1) that produced the first antiprotons at CERN in 1960,[1] shortly after its start-up in 1959 still

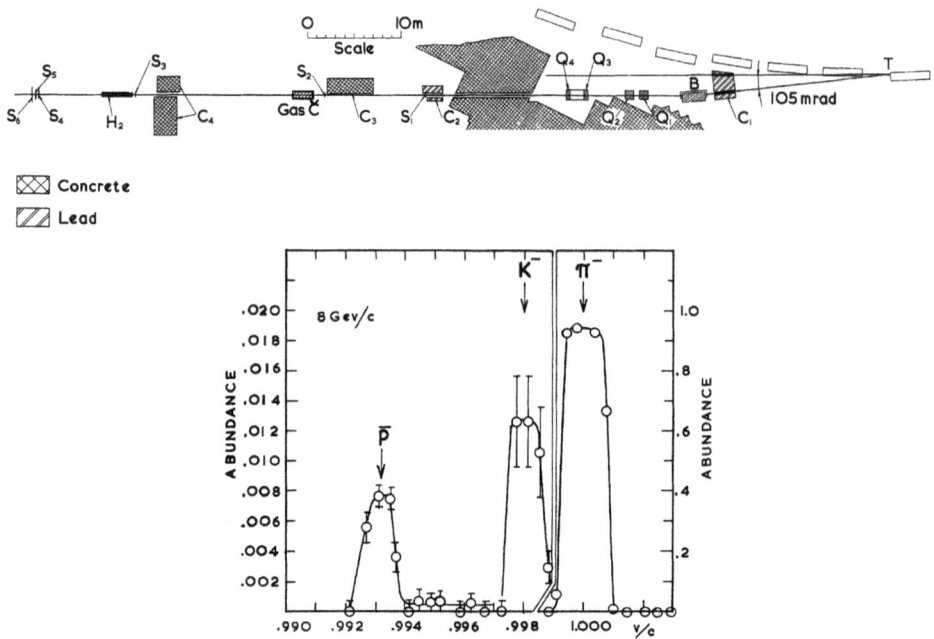

Fig. 1.   Top: First antiproton beam line at CERN (1960). The dashed line to the right of the shielding blocks is the Proton Synchrotron.[1] Bottom: beam composition in this beam line.[1]

continues to provide the antiprotons that form the heart of experiments on them or on antihydrogen atoms, and muonic and pionic atoms are now studied at high-intensity (but low energy) accelerators outside of CERN. Nevertheless, technical developments at CERN have played a defining role in the study of exotic atoms and of antiprotons. In particular, experiments relying on antiprotons have only become possible through the invention of stochastic cooling, the construction of a dedicated storing and cooling accelerator infrastructure (the Antiproton Accumulator (AA) and the Antiproton Collector (AC)), the development of antiproton trapping techniques in the 1980s at CERN's dedicated antiproton experimental facility LEAR (low energy antiproton ring) from 1982 to 1996, and — since 2000 — of the transformation of the AC into the unique Antiproton Decelerator (AD) facility, which hosts all existing experiments worldwide that require trapped antiprotons.

2.   Pionic, Muonic, Kaonic and other Exotic Atoms

Negatively charged particles with a sufficiently long lifetime (relative to the atomic processes involved with capture and decay of the resultant atoms to deeply bound states) offer a window to study atomic physics processes (Rydberg states, cascades, binding energies, lifetimes), but also nuclear physics processes: the deeply bound states' energy levels and lifetimes are affected by strong-interaction effects, which in turn provide the opportunity to study nuclear forces at large distances ("nuclear stratosphere") as well as isotope-related nuclear deformations. As the capture and subsequent de-excitation process occurs on time-scales of ps ~ ns,[2, 3] muons, pions, kaons, antiprotons, but also shorter-lived baryons, such as Σ^-, or even potentially Ξ^- and Ω^-, can form exotic relatively long-lived atoms (although Ξ^- or Ω^- exotic atoms have to date not been observed).

The first such measurements at CERN started at the SC in 1961: observation of 2p–1s transitions in muonic atoms[4] and subsequently in pionic atoms.[5]

With the advent of intense K^- beams in the late 1960s, counter experiments on hypernuclei became possible. While the focus was mainly on hypernuclear continuum states, the study of kaonic atoms also took place at CERN: shift and width of the 1s level of the K^-p atom occupied a number of experiments at the PS,[6] leading also to the first observations, and then studies, of Σ^- exotic atoms from 1970 onward.[7] In the same year, also the first X-ray transitions of antiprotonic atoms were detected and investigated by the same group.[8]

Although earlier observations of exotic atoms had been made, the contemporaneous development of solid-state detectors — by allowing high resolution γ-ray detection — made precision spectroscopy possible, and with it precise determinations of energies (and thus of minute energy shifts), of natural line-widths and of intensities for a large range of transitions and nuclei possible.

2.1. *Atomic physics*

Through their unusual composition, but also through the formation process, the physics involved in de-excitation, and the much smaller radii of the ground state levels (and thus a strongly enhanced proximity to the nucleus), exotic atoms are an ideal test bed for aspects of atomic physics that are significantly more difficult to probe in conventional atoms.

Leaving aside the more complex highly excited exotic atoms, where the heavy negative particle shell radius is larger than the lowest electron Bohr radius, the exotic atom can be considered hydrogen-like, with bound state energy eigenvalues of

$$E(n,l) = \frac{m_x}{m_e} E_0(n,l) \tag{1}$$

with m_x the reduced mass of the exotic atom with the negative particle, m_e the electron mass, and E_0 the energy levels of the ordinary atom. The resulting Bohr radius is consequently (since the mass ratio for even the lightest such negative particle — the muon — is ~ 200) comparable to or even smaller than the radius of the nucleus. For an exotic atom with nuclear charge Z and containing a negative particle of mass m_x, the fine structure splitting is given by:

$$\Delta E = (\mu_D + 2\mu_a) \frac{(Z\alpha)^4}{2n^3} \frac{m_x c^2}{l(l+1)} \tag{2}$$

with μ_D the Dirac moment and μ_a the anomalous magnetic moment.[9] Measurement of the transition energies (proportional to the mass of the negative particle) and of the fine structure splitting (proportional to its magnetic moment) thus provide access to the negative particles properties. Depending on whether the negative particle is a spin-$\frac{1}{2}$ particle ($\mu^-, \bar{p}, \Sigma^-$) or a meson (π^-, K^-), different corrections need to be applied. Hadronic negative particles will be sensitive to the distribution of hadronically interacting particles in the nucleus; leptonic probes will be sensitive to the (related) charge distribution in the nucleus, once their Bohr radius is sufficiently close to the nuclear surface to interact with it.

2.1.1. *Formation processes*

The formation of exotic atoms, described first by Fermi and Teller,[10] and subsequently detailed by Ponomarev,[11] consists of a series of steps: the negative meson (or baryon) slows down from relativistic velocities to velocities of the order of atomic electron velocities; at this point, the negative particle is captured into a highly excited (Rydberg) state, with principal quantum number dependent on the particle mass and ejects an electron. The capture cross-section is related to the overlap between the wave functions of the particle and the atomic electron[11] so the heavy particle will initially populate atomic states with radii close to that of the (ejected) electrons. The observation of metastable antiprotonic atoms allows probing this process: the PS205 experiment at LEAR (see Section 4.1) was able to follow the trapping of antiprotons into the large-n and large-l metastable states of neutral antiprotonic helium. The principal quantum number n_0 of these initial states is expected to be

$$n_0 \sim (m_x/m_e)^{1/2} \sim 40 \tag{3}$$

where m_x is the reduced mass of the exotic atom with the negative particle, and m_e the electron mass. The first observed laser-induced transition in antiprotonic helium[12] corresponds to a transition $(n, l) = (39, 35) \rightarrow (38, 34)$, in agreement with the expected population of high-n states.

2.1.2. *De-excitation processes*

The formed exotic atoms have several possible de-excitation pathways: radiative transitions, Auger effect, or Stark transitions $(n, l) \rightarrow (n, l - 1)$ through collisions with other atoms. The details of the de-excitation depend on many parameters (target density, type of negative particle, etc.) and the final states populated depend sensitively on these (for kaonic atoms, for example, see Ref. 13). Measurements of cascade times for K^-,[14] for Σ^{-}[15] or for antiprotons,[16, 17] as well as searches for X-ray transitions into the lowest-lying s- and p-states in pionic, kaonic, Σ^- and antiprotonic atoms, were carried out in a number of experiments at CERN, and confirm the details of the cascade calculations, in particular with regards to the importance of Stark transitions, which dramatically shorten the cascade time, and populate high-n, low-l states in the course of the cascade. A consequence of this is that in bubble chamber experiments, antiproton annihilation occurs almost exclusively from the s-wave, with implications on the distribution of final states produced in these annihilations.

2.2.   Particle parameters

Masses, charges and magnetic moments of the negatively-charged particles in exotic atoms determine the transition energies. Compared to measurements based on kinematics ($\pi \rightarrow \mu\nu$, or $K^- \rightarrow \pi^+\pi^-\pi^-$), spectroscopic measurements can reach higher precision (although calibration issues are of concern).

The first precision measurement of the pion mass via 4f–3d X-ray lines in Ca and Ti exotic atoms using a crystal spectrometer[18] took place at Berkeley in 1967, and the same technique was also employed there in kaonic atoms (4f–3d X-ray lines in Cl) to provide the first spectroscopic determination of the negative kaon mass.[19] Subsequent exotic-atom determinations of mass or magnetic moments of K^-,[20] $\bar{p}$[9] and Σ^-[21] were carried out at Brookhaven's AGS from 1975. However, the advent of improved detection techniques (a higher efficiency Ge(Li) detector), improvements in the accuracy and reliability of calculations of corrections to the energy levels of mesonic atoms (allowing also to choose less sensitive transitions), and better calibration lines allowed an improved measurement of the pion mass to be carried out at CERN's SC in 1971 by Backenstoss *et al.*,[22] allowing them also to provide the first upper limit to the mass of the ν_μ of less than 1 MeV. The start-up of the PS at CERN was also immediately used by the same group to carry out a 4-fold improved determination of the K^- mass via kaonic X-rays in Au and Ba.[23]

The special case of measurements of the antiproton's mass, charge and magnetic moment in antiprotonic helium (which affords far higher precision than is possible when measuring energies of X-rays emitted during a cascade) is dealt with in Sections 4.1 and 4.2.

2.3.   The strong interaction

Depending on the type of the negatively-charged particle in the exotic atoms, strong interaction effects may affect the lowest-lying bound states, shifting their energy level, modifying the lifetime of the state, and changing the transition probabilities (and thus the transition intensities) with respect to a pure QED reference value. A number of probes are also sensitive to the presence of a diffuse neutron halo at the nuclear surface.

A detailed understanding of the strong interaction became possible in part through improved understanding of the K^-N and π^-N interactions as probed in the deeply-bound pionic and kaonic atoms, as the strong interaction will induce an energy shift and broadening of these states. Pionic 2p–1s transitions in Be, ^{10}B, C, N, ^{16}O, ^{18}O, F and Na[5] at CERN's SC were among the first to have sufficient sensitivity to detect the 1s level shifts and widths and to establish the presence of isospin dependence (Fig. 2).

The same group also carried out the first measurements in a kaonic X-ray transition of the natural linewidth and energy shift due to the strong kaon-nuclear interaction,[6, 27] which were instrumental in ruling out a number of mechanisms proposed to explain an increase of the K^- absorption in nuclear matter, among them the need for an attractive real part in the K-nucleus optical potential (rather than the repulsive real part suggested by K–N scattering), or an extended neutron halo above the nucleus. A number of subsequent investigations have clarified this point: the presence of a nuclear resonance (the $\Lambda(1405)$ s-wave resonance in the isospin $I = 0$ channel) strongly affects this amplitude; the sub-threshold K–p interaction dominates over the K–n interaction.

A rather recent (late 1990s) experiment at CERN to probe QCD at very low energies studied fully-exotic hydrogen-like atoms in which both participants are either pions or kaons. The DIRAC experiment at the PS focused first on observing such atoms (produced through electromagnetically-interacting, kinematically-matched opposite-charge pions produced in 24 GeV p–N interactions), and then determining the lifetime of $\pi^+\pi^-$ atoms. The corresponding measurement[29] of this lifetime of $\tau_{2\pi} = (3.15^{+0.20}_{-0.19}(\text{stat})^{+0.20}_{-0.18}(\text{sys}))$ fs represents a very sensitive test of Chiral Perturbation Theory (ChPT). Similar πK atoms probe the more general 3-flavour SU(3) structure of low-energy hadronic interactions that is not accessible to $\pi\pi$ atoms. The first observation of these atoms[30] only allows setting a lower limit of 0.8 fs (at 90% CL) for these atoms whose lifetime is predicted by ChPT to be 3.7 ± 0.4 fs.[31]

2.4. _Nuclear physics_

Shape and charge distributions of the nucleus can be probed through an accurate measurement of the energy and, to a lesser degree, the relative intensities of X-rays in muonic atoms. In a series of measurements beginning in 1961,[4] Backenstoss and collaborators — using the μ-channel of the SC, which provided an intense and rather pure beam of muons — carried out a systematic exploration of nuclei. Electrical charge distributions of nuclei are best measured through muonic X-ray transitions and electron scattering experiments. The lowest levels of the muonic atom provide the highest sensitivity to the charge distribution, and this sensitivity increases with increasing Z. Often using the 2p–1s transition as their workhorse to determine the energy shifts and intensities of the transitions, this group deduced nuclear charge distribution parameters for, e.g. spherical nuclei from Cl to Pb.[32, 33] This method is of course also applicable to deformed nuclei, such as Sm, Eu, Tb, Ho, Hf, Ta, W, Os, Ir, Th, ^{233}U, ^{235}U, ^{238}U, ^{239}Pu.[34] Also light ($Z < 17$) nuclei were probed via this transition. However, due to insufficient resolution at that time, only one parameter of the charge distribution (the r.m.s. radius) could be derived for higher transitions,

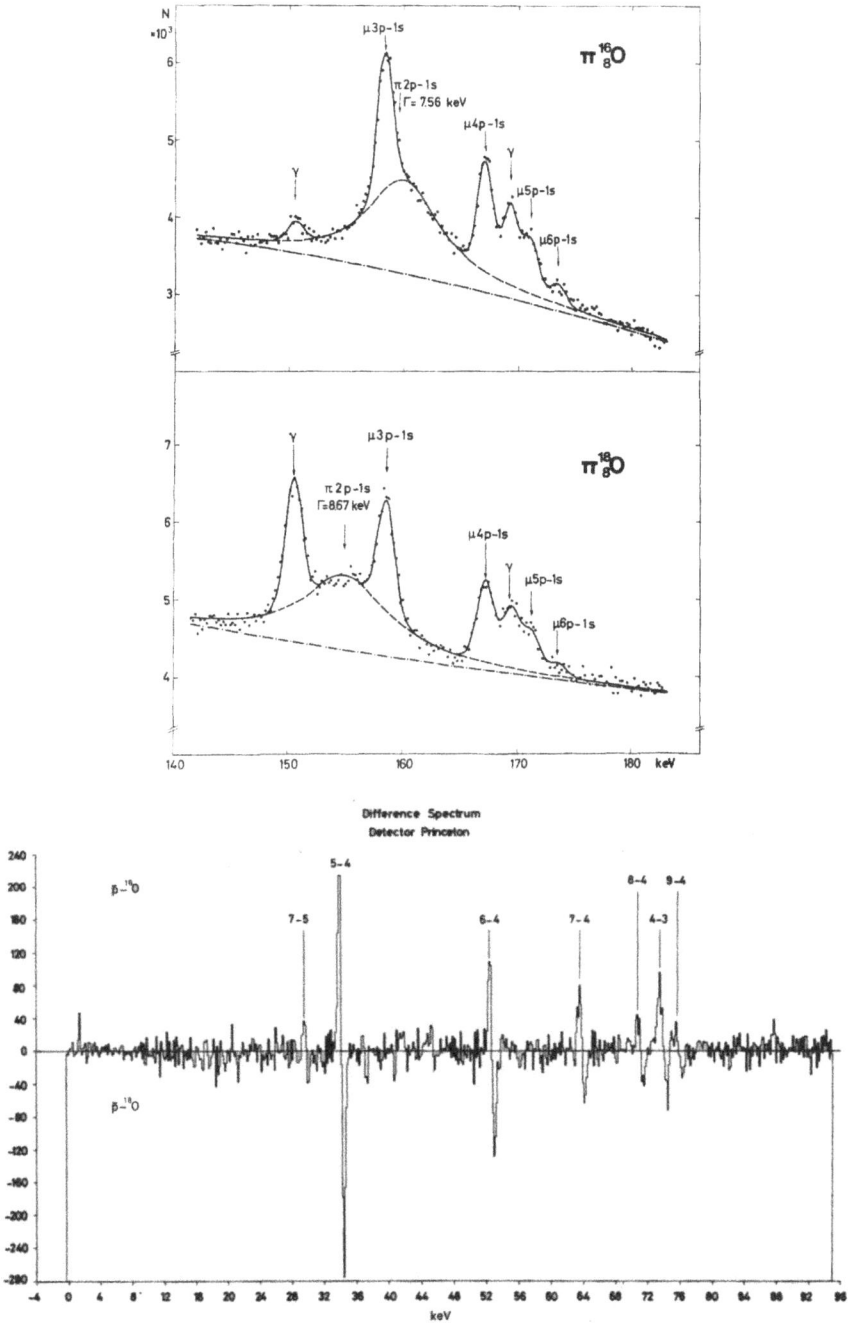

Fig. 2. Isotopic shifts in exotic atoms. Top: Pionic 2p–1s transitions in ^{16}O and ^{18}O.[5] The broad pionic 2p–1s transition and background muonic atomic transitions are visible (data published in 1967). Bottom: Antiprotonic oxygen: difference spectrum of $\bar{p}-^{16}$O and $\bar{p}-^{18}$O[26] (data published in 1978).

e.g. 3d–2p. Since this parameter could be determined from the X-ray energy of the transition almost independently of the model of the charge distribution it was used as a proxy for the size of the charge distribution. The precision on the r.m.s. radii of the probed nuclei was surpassed by electron scattering experiments only for the lightest ($Z < 8$) nuclei.

Another way to probe the nuclear periphery relies on antiprotonic atoms. Neutron density distributions can be sampled in (heavy) nuclei by correlating measurements of their antiprotonic X-ray cascade with a radiochemical determination of the same nuclei after they have been exposed to antiproton capture and annihilation (and are consequently one mass unit lighter). The density distributions can be described via a two-parameter Fermi distribution which defines a half-density radius and a diffuseness parameter for both protons and neutrons. In a series of measurements, the PS209 experiment at LEAR investigated a range of 34 different nuclei, from ^{40}Ca, to ^{238}U via both techniques. If proton distributions are constrained to the values from electron scattering or muonic-X-ray measurements, then the neutron density distribution is best reproduced in terms of a half-density radius compatible with that of the proton, but a significantly larger diffuseness[36] in the case of neutron-rich nuclei.

While current research on exotic atoms at CERN is mainly focused on antiprotonic atoms, muonic exotic atoms with ever improved experimental precision continue to play an essential role elsewhere, sensitive perhaps even to physics beyond the standard model.[37] One such area is that of muonic hydrogen. A 7σ discrepancy separates the determination of the proton radius obtained through a measurement of the Lamb shift in muonic hydrogen,[38, 39] at the CREMA experiment at PSI from that obtained from electron scattering and spectroscopic studies of atomic hydrogen and deuterium. This discrepancy — barring experimental errors — can only be accounted for by modifications at a very deep level (QED, lepton universality, change in the Rydberg constant),[37] which illustrates the topicality of research on exotic atoms even today.

3. Antiprotonic Atoms and Protonium

The simplest antiprotonic atom is protonium, a bound state of an antiproton with a proton. These atoms are formed by slowing down antiprotons (by ionisation loss) in hydrogen. Once the antiproton is (almost) at rest, it will replace the electron of the hydrogen atom, in an orbit whose radius is that of the first Bohr radius of hydrogen, approximately 5.3×10^{-9} cm. The protonium thus formed will have a large angular momentum, l, and a principal quantum number, n, of about $(m_p/2m_e)^{1/2}$ ($\sim$30), where m_p and m_e are the mass of the proton and of the electron.[40] During de-excitation, (n, l) quantum numbers are reshuffled: for formation of protonium in liquid hydrogen, Desai[40] showed that annihilation of the proton–antiproton system would take place essentially exclusively from states with zero angular momentum (albeit with possibly high principal quantum number). This prediction depends on the likelihood of Stark mixing, however, and so is strongly density dependent: protonium formation and cascades in low density hydrogen (requiring of course low-momentum antiprotons) would be expected to contribute annihilations from p (or higher angular momentum) states.

Of course, antiprotonic atoms with heavier nuclei are also possible. The special case of antiprotonic helium will be discussed below, as measurement of transitions between its meta-stable states allow precision determinations of the antiproton's mass, charge and magnetic moment. In general however, antiprotonic atoms are short-lived, and are ideal probes of strong interaction effects. Atomic physics with antiprotons at CERN began with the first observation of antiprotonic atoms in 1970,[8] also providing the first spectroscopic measurement of the antiproton mass. Figure 3 shows the X-ray spectrum produced by stopping 14×10^6 antiprotons in a Tl target. Comparison between the measured and calculated transitions allowed the authors to give a 68% CL upper limit on any mass difference between protons and antiprotons of ($|m_p - m_{\bar{p}}| < 0.5 \, \text{MeV}$), a relative precision of 5×10^{-4}. Only a consistency check on the equality of the magnetic moment of the proton and of the antiproton could be provided, the (limited) accuracy of the measurements precluding any quantitative statement at the time. In a subsequent experiment on antiprotonic X-rays of P, Cl, K, of Sn, I, Pr, and of W, the same group[41] explored more fully the effect of the strong interaction on the deepest lying bound states, and in addition to determining the reduction of some transitions, also observed the effect of the strong interaction via the line width and energy shift in one of the target nuclei, sulfur. These X-ray studies — as in the case of pionic atoms — also allowed probing isotopic effects, this time with antiprotons: the first measurements (still at

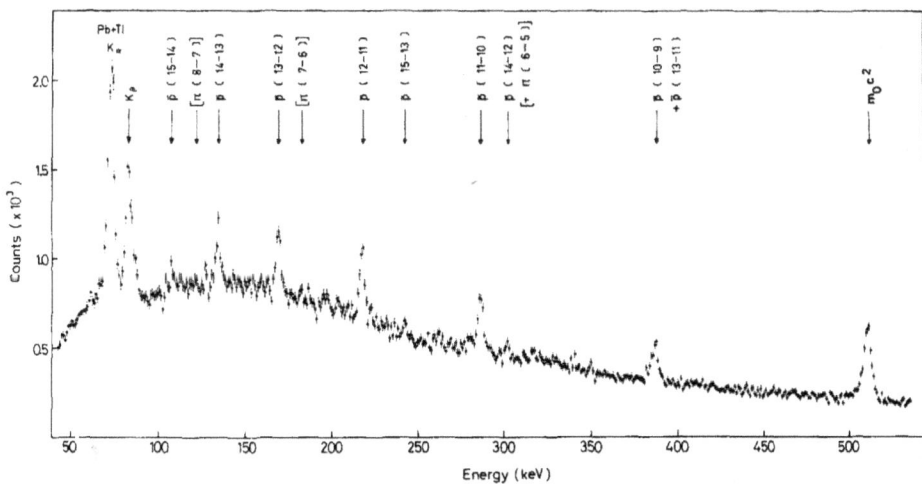

Fig. 3. X-ray spectrum of the first antiprotonic atom to be observed: $\bar{p}-^{81}$Tl obtained from 14×10^6 stopped antiprotons measured with a 10 cm^3 Ge (Li)-detector.[8]

Table 1   Antiproton mass and magnetic moment measurements.

Quantity	Year	Value	Rel. precision	Method	Ref.
$\|m_p - m_{\bar{p}}\|$	1970	< 0.5 MeV	5×10^{-4}	$\bar{p}$-Tl	[8]
$\|m_p - m_{\bar{p}}\|$	1977	< 0.05 MeV	5×10^{-5}	$\bar{p}$-Zr and $\bar{p}$-Y	[25]
$m_{\bar{p}}/m_p$	1990	0.999999977(42)	4×10^{-8}	trapped $\bar{p}$	[74]
$m_{\bar{p}}/m_p$	1995	0.9999999995(11)	1×10^{-9}	trapped $\bar{p}$	[66]
$(q/m)_{\bar{p}}/(q/m)_p$	1999	$-0.99999999991(9)$	5×10^{-11}	trapped $\bar{p}$	[63]
$(\mu_p - \|\mu_{\bar{p}}\|)/\mu_p$	1972	(-0.04 ± 0.1)	3×10^{-2}	$\bar{p}$-Pb	[79]
$(\mu_p - \|\mu_{\bar{p}}\|)/\mu_p$	2009	$(2.4 \pm 2.9) \times 10^{-3}$	10^{-3}	$\bar{p}$-He	[70]
$\mu_{\bar{p}}/\mu_p$	2013	$-1.000000(5)$	5×10^{-6}	trapped $\bar{p}$	[78]

the PS) of these isotope effects with antiprotonic atoms (Fig. 2) in $\bar{p}$–^{16}O/^{18}O provided important new information on the $\bar{p}$–n interaction,[26] allowing to separate the $\bar{p}$–n and $\bar{p}$–p scattering lengths, although some assumptions about the neutron distributions in the nuclear tail needed to be made.

It subsequently took several more years for more precise measurements of the transitions in antiprotonic atoms ($\bar{p}$Pb and $\bar{p}$U) to result in improved measurements of antiproton parameters: the anomalous magnetic moment in 1972,[79] subsequently improved in 1975,[9] a measurement that also yielded a 10-fold improvement over the first spectroscopic antiproton mass measurement at CERN. Further measurements, still at Brookhaven,[25] of the fine structure splitting continued to improve the knowledge of the antiproton mass ($m_{\bar{p}} = 938.229 \pm 0.049$ MeV, Table 1) but were still hampered by theoretical uncertainties and — more importantly — detector resolution.

The sensitivity of early measurements of antiprotonic He at the PS[44] was not sufficient to detect the small fraction of long-lived ($\sim \mu$s) meta-stable states discovered in 1991.[43] These states opened the door to high precision laser-spectroscopic studies of the energy levels of the antiprotonic atoms by the PS205 (at LEAR, until 1996[45]) and the ASACUSA (at the AD) experiments. This methodological advance over resolution-limited X-ray detectors consequently improved the knowledge of the mass, charge (Section 4.1) and magnetic moment (Section 4.2) of the antiproton by many orders of magnitude beyond what measurements of the de-excitation cascade could provide.

3.1.   *Protonium*

This simplest antiprotonic atom provides a particularly pure measurement of the $\bar{p}$–p interaction via the usual energy level broadening and shift studies. Furthermore, it offers the very attractive possibility of tagging the initial state of $\bar{p}$–p annihilation via detection of the corresponding X-ray transition. Knowing the quantum numbers of the initial state consequently had a significant impact on meson spectroscopy in the 1980s at CERN, by limiting the possible contributing waves to the Dalitz plot fits that were the workhorse of many experiments at LEAR.[46] This in turn made it possible to better identify the quantum numbers of short-lived resonances produced in p̄–p annihilations. It was only in 1978 that the first observation of (Balmer Series) X-rays from antiprotonic hydrogen took place[47] at the low-momentum antiproton line of the PS. Subsequent experiments on protonium benefitted greatly from CERN's dedicated antiproton program from 1980 onwards (Section 4), in particular from the background-free low energy antiproton beam provided by the LEAR facility.

A series of spectroscopic measurements of transitions between low-lying protonium bound states using different techniques (with different sensitivities and resolutions) took place in the first and second generation of LEAR experiments. PS171 (ASTERIX collaboration) identified the Lyman-α line,[48] measured the strong interaction shift and broadening of the ground state of the p̄–p atoms,[49] and obtained the cascade time of antiprotons in gaseous hydrogen using a drift chamber as X-ray detector,[16] a measurement extended by PS201 (Obelix collaboration) to a large range of H_2 densities.[17] Improved resolution was provided by a dedicated experiment (PS174) using a Si(Li) detector,[50] which subsequent experiments using a much higher resolution spectrometers (PS175, using a cyclotron trap for formation of protonium at very low pressures[51] or PS207, using CCD's coupled to a crystal spectrometer[52]) improved by several orders of magnitude. The results on the spin triplet and spin singlet strong interaction shifts and widths were instrumental in finessing the N̄N potential models and even uncovered problems (resolved since) with QED calculations for both hydrogen and other isotopes.

4. Antiprotons

Experiments on and with antiprotons at CERN started almost immediately after the required energy for producing them was available, and have carried on uninterruptedly since then, spawning new experimental facilities and opening new experimental areas on the way. The first facility capable of producing antiprotons at CERN was the Proton Synchrotron; completed in 1959, the first antiprotons were produced shortly afterwards.[1] Experiments using the newly accessible antiproton-proton annihilations to investigate the time-like structure of the proton[53] or — in tandem with developments of large bubble chambers — as a testbed for measurements of cross sections and meson spectroscopy (from 1965) quickly started up at CERN[54] and at Brookhaven's AGS.[55] However, precision experiments on antiprotons and exotic atoms incorporating them could start in earnest only once a pure beam of antiprotons was available, requiring a dedicated facility for their production, storage and extraction.

This became possible through the invention of stochastic cooling (proposed in 1968 by S. van der Meer, published in 1972),[56] and its successful test in the Initial Cooling Experiment (ICE) in 1978[57] which demonstrated an increased beam density, as well as an extension of the beam lifetime from hours (without cooling to counteract Coulomb scattering on residual gas) to days. This test also allowed a dramatic improvement in a test of the CPT symmetry: the equality of lifetimes of protons and antiprotons. Prior to this test, the experimental lower limit on the lifetime of antiprotons (derived from bubble chamber tracks) was 120 μs; by inserting a simple antiproton production target upstream of the ICE set-up and cooling and storing a minute number of antiprotons in ICE, a 9-order of magnitude improved lower limit on the lifetime of antiprotons could be obtained.[58]

This fundamental advance allowed CERN's antiproton complex (Figs. 4 and 5) consisting of the antiproton production target, the Antiproton Accumulator (AA) and the Antiproton Collector (AC) and the low energy antiproton ring (LEAR) — and relying on the Proton Synchrotron which produces 26 GeV/c protons, to be proposed[59] and rapidly built (AA start-up in 1980, LEAR began operation in 1982, AC from 1987 onwards).

The development of antiproton trapping and electron cooling techniques by Gabrielse *et al.*[60] in 1986 at CERN's dedicated antiproton experimental facility LEAR finally allowed carrying out precision experiments on trapped and cooled antiprotons and working towards the study of antihydrogen atoms first at LEAR

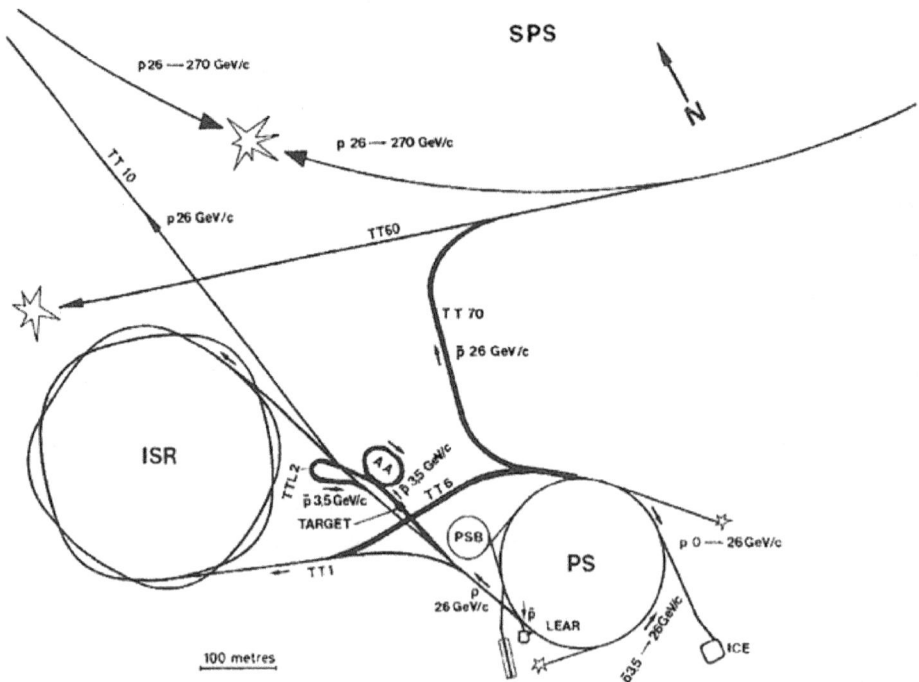

Fig. 4.   Layout of the CERN accelerators in 1981. LEAR, still under construction in 1981, is also shown.[58] From 2000 onwards, the AA was transformed into the Antiproton Decelerator, and now houses at CERN all experiments worldwide relying on low energy antiprotons.

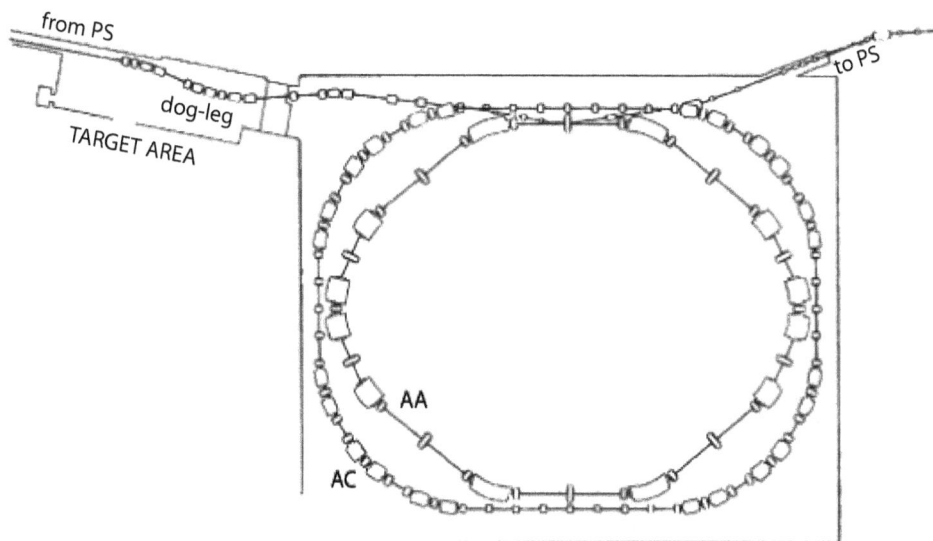

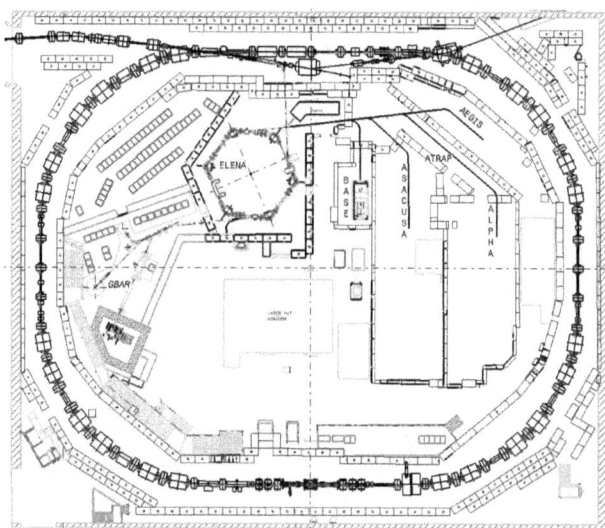

Fig. 5. Top: In 1986/87, the AC was built around the AA and a dog-leg incorporated into the injection line to diminish the flux of electrons and π^- reaching the hall.[61] Bottom: AD layout with experimental areas (status at the end of 2014) with the future "Extra Low Energy Antiproton ring" (ELENA) indicated.

from 1986 to 1996, and — since 2000 — at the unique Antiproton Decelerator (AD) facility (transformed from the AC), which hosts all existing experiments requiring trapped antiprotons. Confinement of antiprotons in ion traps for seconds (or days) opened up major improvements in the determination of the antiproton's mass, charge, but also of its magnetic moment (although it took until 2012 to surpass the precision achievable in antiprotonic helium transitions). Furthermore, the formation and trapping of antihydrogen atoms, precision spectroscopy of antihydrogen, or precise measurements of the protonium energy levels without collisional broadening could also be envisaged, prepared and — by 2014 — partly achieved.

4.1. *Charge and mass: TRAP, ATRAP and ASACUSA*

A comparative measurement of the charge and mass of antiprotons and protons allows a direct test of CPT invariance. While exotic antiprotonic atoms had allowed reaching a relative precision of 10^{-5}, further improvements were limited by the precision with which the energies of the transition could be measured as well as calculated. With the techniques (Fig. 6) for slowing, trapping, cooling and stacking of antiprotons now established by his PS196 collaboration,[60, 64] Gabrielse's group rapidly carried out a first determination of the q/m ratio of antiprotons[74] with 100 simultaneously trapped antiprotons. For this, he measured their cyclotron frequencies in the Penning trap in which the antiprotons were confined and with this method achieved a factor of 1000 improvement in precision relative to the previous best values coming from exotic antiprotonic atoms.[25, 65] In rapid succession, they further refined and improved their sensitivity, reaching first a precision of 1 ppb[66] with a single trapped antiproton before extending the technique to a 90 ppt precision (Table 1). The main systematics limiting the previous measurements were addressed in this final measurement at the LEAR complex: the use of an H^- ion instead of a proton, and simultaneous confinement of the antiproton and the H^- ion in the same trap (Fig. 6). This avoids non-reproducible electric trapping potentials, which had led to differences in the magnetic field experienced by the antiproton and proton of up to 1 ppb. As well, due to the ease with which $\bar{p}$ and H^- could be switched and probed, magnetic drifts affected the (alternating) measurements far less than earlier measurements.

At the same time, the PS205 experiment focused on precision measurements of transitions in antiprotonic helium ($\bar{p}He^+$), benefiting from the co-temporaneous discovery[43] that these could form meta-stable states ($\tau \sim 3-4\,\mu s$), thus opening the window to inducing (via laser pulses) transitions between these meta-stable states and unstable states which would rapidly decay and produce a clear annihilation signal. The technique developed by PS205[67] is based on injecting a large number of antiprotons into gaseous helium. While short-lived states will annihilate within few ns, some long-lived meta-stable states may be populated and can be probed via a delayed laser pulse which — if it corresponds to a transition energy from a populated meta-stable state into a short-lived state — will lead to a short increase in the annihilation rate (Fig. 7). The significant advance over direct measurements of the emitted X-rays is that the precision is now limited mainly by the external laser system applied to stimulate the transitions. In a series of improvements of their apparatus and technique (frequency combs, doppler-free transitions, low density

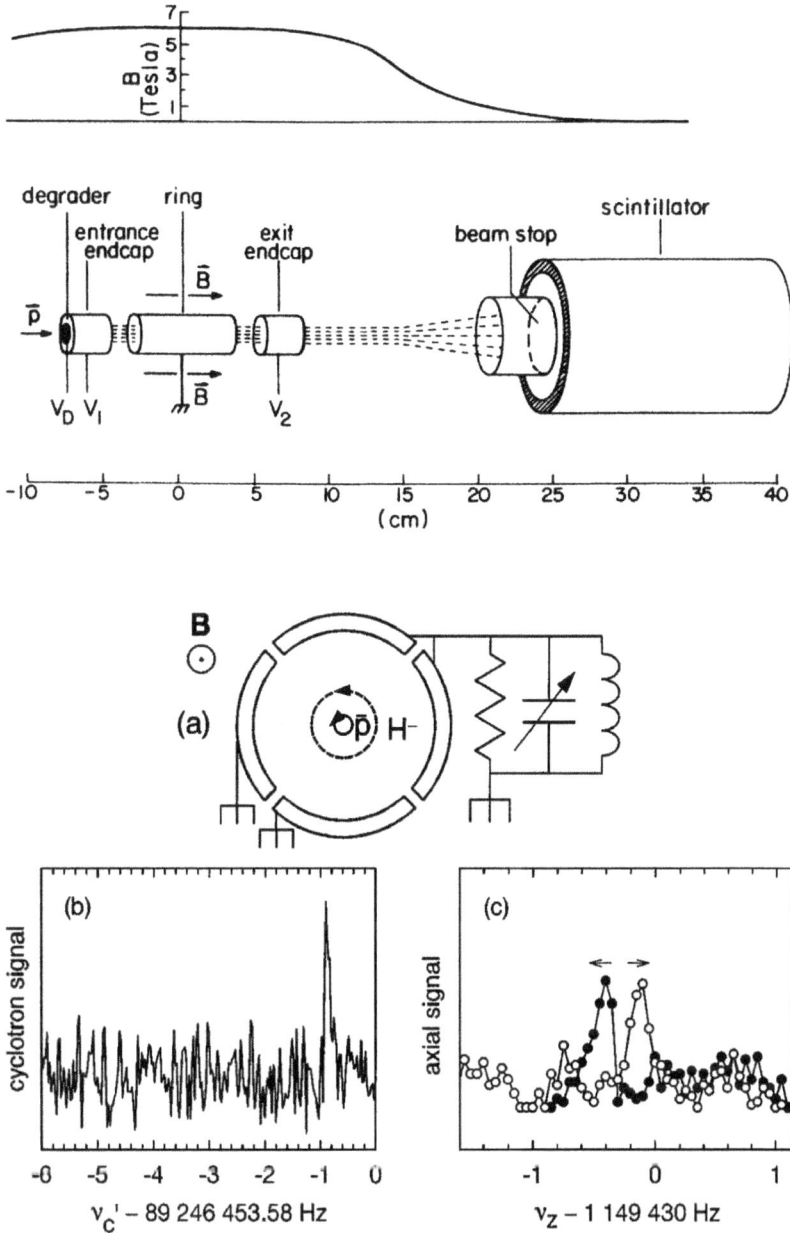

Fig. 6. Top: Outline of the $\bar{p}$ trap electrodes and the scintillator. The direction of the homogeneous magnetic field is indicated by the arrow, its magnitude along the center axis is plotted above and important field lines are indicated by dashed lines.[60] Bottom: (a) Central trap electrode, viewed along B, and the LCR detection circuit used to observe the signal (b) induced by free cyclotron motion. The driven axial signal (c) induced across a similar circuit, as the drive is stepped up or down in frequency every 4 s, is delayed by a detector time constant. Thus ν_z (needed to ± 0.7 Hz to determine it to 600 ppb in order to determine the cyclotron frequency ν_c to 90 ppt) is midway between the peaks.[63]

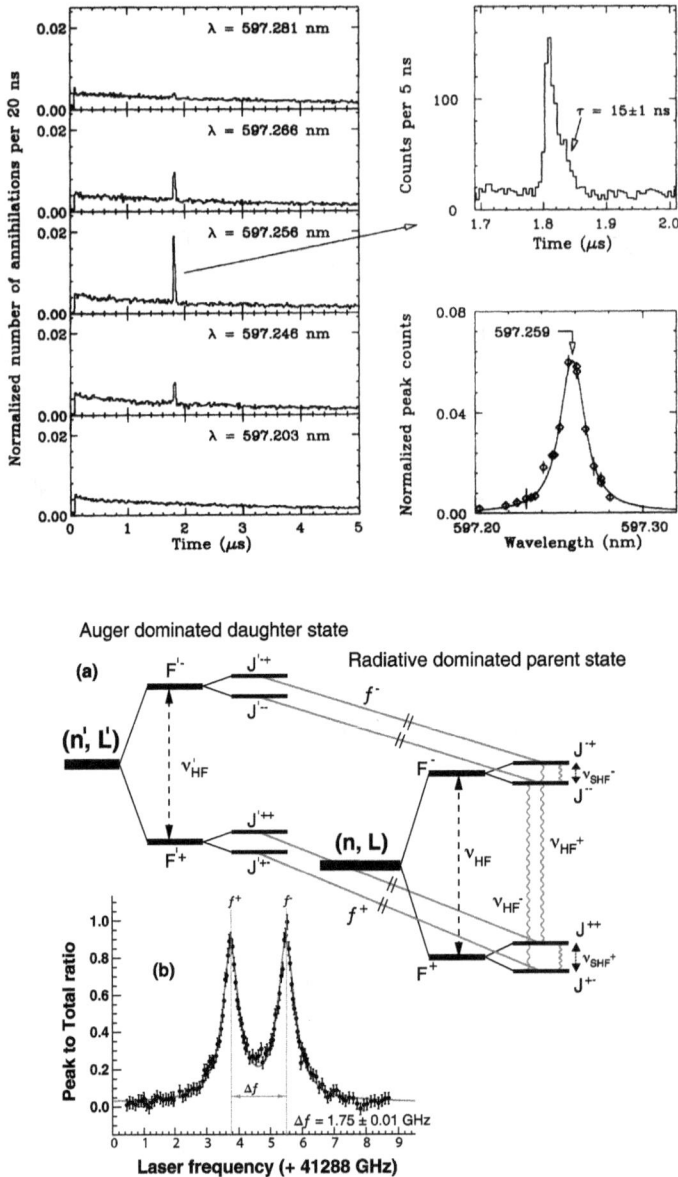

Fig. 7. Top: Observed time spectra of delayed annihilation of antiprotons with laser irradiation of various vacuum wavelengths near 597.2 nm, normalised to the total delayed component.[12] Spikes due to forced annihilation through the resonance transitions are seen at 1.8 ps. Upper right: Enlarged time profile of the resonance spike. A damping shape with a time constant of 15 ± 1 ns is observed. Middle right: Normalised peak count versus vacuum wavelength in the resonance region, showing a central wavelength 597.259 ± 0.002 and a FWHM 0.018 nm. Bottom: (a) Schematic view of the level splitting of $\bar{p}He^+$ for the $(n, l) \to (n-1, l+1)$ electric dipole transitions.[70] The laser transitions f^+ and f^-, from the parent to daughter states, are indicated by straight lines and the microwave transitions, between the quadruplets of the parent, by wavy ones. For this experiment $(n, L) = (37, 35)$ and $(n', L') = (38, 34)$. (b) Laser resonance profile demonstrating the two sharp peaks and HF laser splitting $\Delta f = f^- - f^+$. Although there are four SHF laser transitions only the HF transitions were resolved in this experiment (see Section 4.2).

targets), the PS205 experiment (and the successor experiment AD-3/ASACUSA) have improved the knowledge of the transitions by several orders of magnitude, now reaching ppb precisions. In parallel, the crucial theoretical treatment of the $\bar{p}He$ system has kept pace.[68, 69]

This antiprotonic atomic system also allows a precision comparison of the charges of a particle and its antiparticle: different functional dependencies of the observables (i.e. the Rydberg constant of the exotic atom, and the cyclotron frequency of trapped negative particles) allow factorising charge and mass. A test to 10 ppm of the equality of the charges and masses of antiprotons and protons[24] combines measurements of transitions in antiprotonic atoms[25] and of the cyclotron frequencies of trapped (anti)protons,[74] and is limited only by the precision achievable in the exotic atoms' transitions.

With recent calculations of the theoretical transition frequencies ($O(\alpha^7)$) at the level of 0.1 ppb accuracy in H_2^+, HD^+ and antiprotonic helium,[71, 72] the bar has been raised once more, and further even more precise measurements planned by the ATRAP and ASACUSA collaborations should thus allow improving the knowledge of $m_{\bar{p}}$ and of $q_{\bar{p}}$ by another order of magnitude (or more).

4.2. *Magnetic moment: ATRAP, ASACUSA*

A first observation of (hyperfine) level splitting in antiprotonic helium (caused by the coupling of the orbital angular momentum of the antiproton to the spin of the remaining electron) was reported in 1997 by PS205.[73] This splitting is of the order of $10 \sim 15$ GHz. The interaction between the antiproton spin and that of the electron causes a further *super-hyperfine* and much smaller splitting of $150 \sim 300$ MHz.[75] In order to see this minute splitting induced by the antiproton spin a far more complex microwave resonance experiment was required. A first detection of this hyperfine structure via a laser-microwave-laser resonance method was reported in 2002[76] but only in 2009[70] was the resolution of the transition sufficient (Fig. 7) to be able to extract the antiproton magnetic moment at the level of 10^{-3}, establishing equality with the proton magnetic moment to the same precision.

While further improvements allowed detecting even the nuclear-spin induced splitting in antiprotonic $^3He^+$,[77] the precision that is achievable cannot rival with very recent measurements of the antiproton magnetic moment in traps. The first ever measurement of spin flips on a single trapped antiproton were carried out by the ATRAP collaboration[78] in a specially-prepared Penning trap that adds a finely tuned magnetic bottle gradient to the trap's axial B field. High resolution measurements of the resulting axial frequency shifts (stemming from the interactions of the cyclotron, magnetron and spin moments with ΔB) allowed the ATRAP collaboration to improve the sensitivity on the comparison of the proton and antiproton magnetic moment by three orders of magnitude from the best exotic atom measurement to 4.4 ppm (Table 1).

## 5.	Antihydrogen

Antihydrogen formation in traps was proposed already in 1986.[60] Several production processes mixing antiprotons ($\bar{p}$) with positrons (e^+) or positronium (Ps) are possible:

$$\bar{p} + e^+ + e^+ \rightarrow \bar{H} + e^+, \tag{4}$$

$$\bar{p} + Ps \rightarrow \bar{H} + e^-, \tag{5}$$

$$\bar{p} + e^+ \rightarrow \bar{H} + \gamma. \tag{6}$$

Unfortunately, these processes require high positron densities (three-body formation), production and transport of positronium towards trapped antiprotons, or have a very low cross-section (radiative formation). In 1994, before the positron accumulation technique based on radio-isotope decays developed in 1989 by the group of C. Surko[80] had become advanced enough that the required numbers of positrons were routinely available for antihydrogen production in Penning traps to be attempted, an alternative route was proposed by members of the PS202 experiment[81] to produce antihydrogen atoms in flight, using the interaction between the antiprotons stored in the LEAR ring and a jet-gas target consisting of Xe atoms. The production process of antihydrogen is then:

$$\bar{p}Z \rightarrow \bar{p}\gamma\gamma Z \rightarrow \bar{p}e^+e^-Z \rightarrow \bar{H}e^-Z \tag{7}$$

where the requisite positrons are formed via the space-like interaction between photons formed by the antiprotons in the Coulomb field of the nucleus with charge Z. The experiment (called PS210) was carried out during 15 hours in the course of 1995. With an integrated luminosity (based on the number of antiprotons and the gas-target thickness) of $\mathcal{L} = 5 \times 10^{33}\,\mathrm{cm}^{-2}$, a total of 11 antihydrogen atom candidate events[82] were detected (with an estimated background of 2 events). Although the high momentum of the antiproton beam ($1.94\,\mathrm{GeV}/c$) meant that the resulting antihydrogen atoms could not be studied, this proof-of-principle experiment gave great support to the subsequent modification of the AC into a full-fledged antiproton deceleration facility at which antihydrogen production in Penning traps, trapping of the produced antihydrogen atoms, and spectroscopy of these atoms could be attempted.

### 5.1.	*Low energy antihydrogen: ATHENA, ATRAP*

Several proposals to carry out experiments at this new antiproton facility, the Antiproton Decelerator, planned for 1999, were submitted as soon as its construction had been decided. Two of them (P302: ATHENA and P306: ATRAP) specifically focused on the production and study of cold antihydrogen atoms, and were approved in 1997 with the names of AD1 and AD2.

Both ATHENA and ATRAP were based on a similar experimental design: a multi-well Penning trap that would hold antiprotons and positrons simultaneously, and allow bringing them into contact in a controlled manner. In both experiments, the produced antihydrogen atoms — being neutral — would leave the formation region. Detection in the ATHENA case was via reconstruction of the annihilation vertex of the antiproton (through a two-layer double-sided silicon micro-strip detector that detected the annihilation pion trajectories) and of the opening angle (from the vertex) of the two 511 keV photons produced in the positron annihilation: an excess at an opening angle of π is the signal for antihydrogen annihilation. In the case of the ATRAP experiment, detection of $\bar{H}$ was performed by field-ionising the produced atoms, and subsequently detecting the resulting, trapped, antiprotons. This scheme is well matched to the 3-body production process (expected to be dominant) of Eq. (4) which produces mostly highly excited states of antihydrogen.

The first observation of the production of antihydrogen via mixing of antiprotons and positrons was provided by ATHENA,[83] followed only a few weeks later by a confirmation of the process by ATRAP[84] (Fig. 8). This second paper however went further, by giving a first indication of the production process, since field-ionisation of antihydrogen would not have allowed detecting deeply bound antihydrogen atoms, such as would have been produced in the competing radiative production process of Eq. (6).

Further information on the distribution of populated antihydrogen states could also be immediately obtained via the same field-ionisation detection scheme by varying the ionising field strength,[85] confirming that antihydrogen formed by mixing

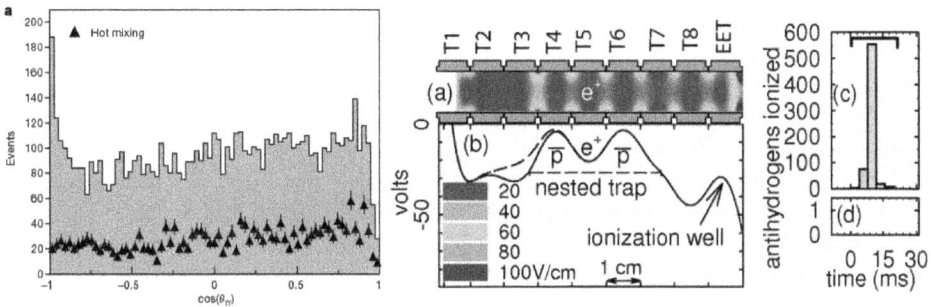

Fig. 8. Left: ATHENA experiment: Angular distribution between two detected photons from e^+e^- annihilation as determined from an antiproton annihilation vertex. The peak at $\cos(\theta) = -1$, but also the bulk of the distribution, corresponds to antihydrogen annihilations. The 'hot mixing' data correspond to mixing of cold antiprotons with RF-heated positrons, where no antihydrogen production can take place.[83] Right: Electrodes for the ATRAP nested Penning trap, upon which a representation of the magnitude of the electric field that strips $\bar{H}$ atoms is superimposed. (b) Potential on axis for positron cooling of antiprotons (solid line) during which $\bar{H}$ formation takes place, with the (dashed line) modification used to launch $\bar{p}$ into the well. (c) Antiprotons from $\bar{H}$ ionisation are released from the ionisation well during a 20 ms time window. (d) No $\bar{p}$ are counted when no e^+ are in the nested Penning trap.[84]

antiprotons and positrons in a nested-well Penning trap is mainly produced in Rydberg states. Subsequent analyses of plasma physics processes and simulations of the interaction of Rydberg antihydrogen atoms with the dense positron plasma confirmed and finessed this picture.[86]

With trapping of antihydrogen atoms the next goal after formation, measurements of the velocity distributions of the produced atoms unfortunately revealed that the antihydrogen atoms formed in the ATHENA and ATRAP experiments were far too energetic to be trapped. By employing a doublet of field-ionising electrodes and superimposing a temporal modulation on the first doublet, a velocity-dependent transmission probability could be imposed on the continuous flux of antihydrogen atoms. The corresponding measurement by ATRAP[87] confirmed the underlying problem that because the formation rate is determined by the relative velocity of antiprotons and positrons in the nested trap, even quite energetic antiprotons will have a velocity comparable to those of cold (but still fast, since light) positrons, consequently producing antihydrogen atoms with (relatively) high kinetic energy. The production of high temperature antihydrogen atoms in spite of the cryogenic environment in which they are formed was confirmed by a measurement by ATHENA of the axial distribution of antihydrogen annihilation vertices.[88]

One alternative to the nested well technique of producing antihydrogen, and that potentially could lead to much colder atoms being produced, as long as the antiprotons are far colder, is the charge exchange reaction of Eq. (5) whose cross-section scales with the Ps principal quantum number $n_{P_s}^4$. By producing and exciting positronium, it is thus possible to produce large amounts of Rydberg antihydrogen, with the additional benefit of having control of its Rydberg state. The large mass difference between the antiproton and the Ps entails that the kinetic energy of the formed antihydrogen is again dominated by that of the antiprotons. The scheme used by the ATRAP collaboration in 2004[89] is explained in Fig. 9.

5.2. *Trapping: ALPHA, ATRAP*

Precision experiments on antihydrogen atoms benefit greatly from trapping them; if they are produced in nested electric potential wells (required to mix antiprotons with positrons), they will only survive for a few μs after being formed before impacting on the walls of the electrodes forming the potential wells and annihilating. It is currently not possible to slow and cool energetic antihydrogen atoms formed randomly. Instead, trapping antihydrogen atoms relies on forming them inside a (neutral atom) trap, and at energies lower than the trap's potential. A magnetic minimum trap, which relies on the coupling of (anti)hydrogen atoms to magnetic fields via their small magnetic dipole moment can be formed by overlaying a transverse multipole (quadrupole, octupole) and two axial Helmholz-like coils. In its ground state, the magnetic moment of the antihydrogen atom is minimal (Rydberg states have a much larger dipole moment) but these are the states that need to be

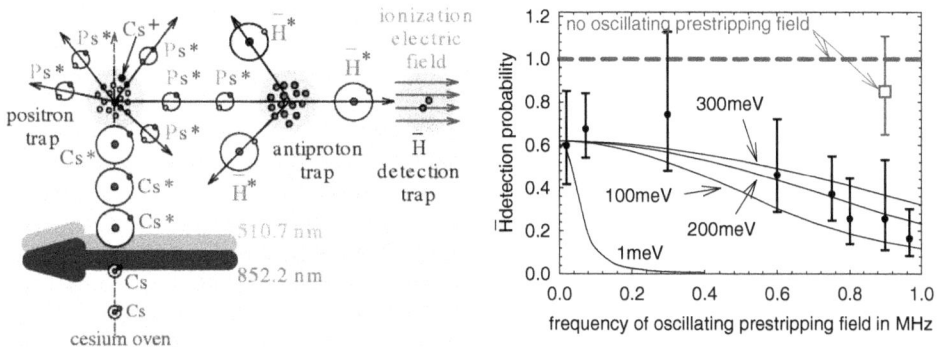

Fig. 9. Left: schematic for laser-controlled $\bar{H}$ production: Cs atoms in a gas jet are excited into Rydberg states via two laser pulses (infrared: 852.2 nm; $6S_{1/2} \rightarrow 6P_{3/2}$; green: 510.7 nm, $6P_{3/2} \rightarrow 37D$); these Rydberg Cs atoms undergo a charge-exchange reaction with trapped cold positrons to form positronium (also in a Rydberg state); these neutral positronium atoms diffuse, and if they encounter antiprotons trapped nearby, will undergo a second charge exchange reaction to form Rydberg antihydrogen.[89] Right: Transmission probability through a pre-field-ionising oscillating potential for high frequencies, only high velocity $\bar{H}$ are transmitted: 300 meV corresponds to $\approx$1000 K.

trapped; the depth for these is $0.76\,\mathrm{K T}^{-1}$. State-of-the-art systems achieve a trap depth of about 1 Tesla, and thus of 0.76 K, corresponding to an antihydrogen kinetic energy of 65 μeV.

Even forming antihydrogen atoms in this challenging (magnetic) environment is a recent development. It is only in 2008 that the first antihydrogen atoms were formed in a quadrupolar[90] or octupolar magnetic trap. Furthermore, antihydrogen atoms need to be formed at the lowest possible temperature, in order to trap even a fraction of the formed atoms. In 2010, the ALPHA collaboration reported on the first trapping of such ultra-cold antihydrogen atoms[91] which correspond to a minute fraction of all atoms produced in their trap (Figs. 10(a)–10(c)); in a second paper,[92] they furthermore showed that the trapped atoms had decayed into the ground state. Comparable results were also obtained by the ATRAP collaboration[93] in 2012, in spite of much slower trap release time constants (Fig. 10(d)).

5.3. *Spectroscopy: ALPHA*

The ultimate goal of antihydrogen experiments is to carry out precision spectroscopy. For this, two transitions are attractive: the transition between the ground state (1s) and the first excited states (2s) which can only decay via a two-photon transition, and has been measured in hydrogen[94] to one part in 10^{14}; and the hyperfine transition (HFS) in ground state antihydrogen, which — in hydrogen — has been measured with a relative precision of 10^{-12}, and can be measured — in antihydrogen — with a relative precision of 10^{-7} or better via microwave-cavity based methods pursued by the ASACUSA collaboration.

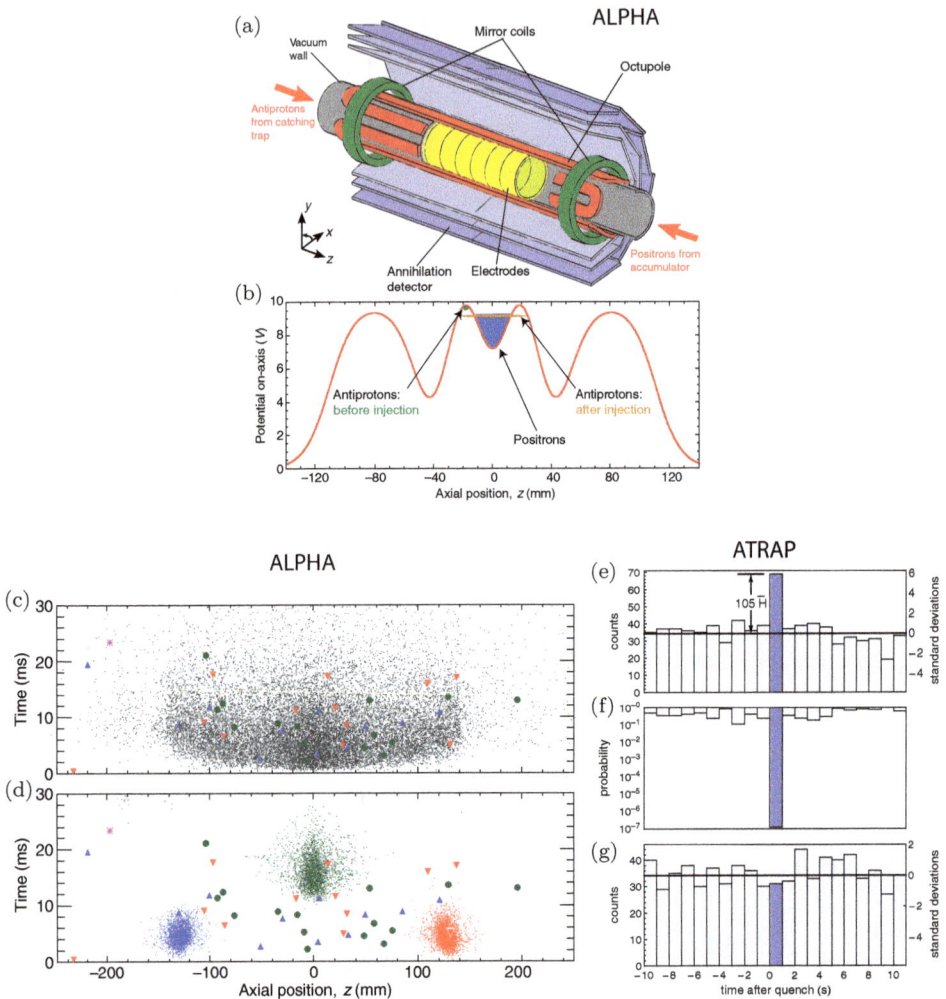

Fig. 10. Trapping of antihydrogen by ALPHA and ATRAP. Top: Antihydrogen synthesis and trapping region of the ALPHA apparatus (a) and nested-well potential used to mix antiprotons and positrons (b).

Bottom left: Measured $t - z$ distribution of annihilations obtained for three (red, green, blue) different experimental conditions (to differentiate trapped antihydrogen atoms from trapped antiprotons) during the opening of the ALPHA magnetic trap. Colored symbols are data, the grey dots are simulations for antihydrogen atoms (c); the same data are shown in figure (d), this time in comparison to expected distributions for antiprotons (tiny coloured dots) that could have been trapped instead of antihydrogen atoms, for three different experimental conditions.[91] The colour codes are the same for data and simulations.

Bottom right: Detected antihydrogen annihilations after trap release at $t = 0$ in the ATRAP experiment. The solid line at 35 counts corresponds to the average cosmic ray counting rate. (e) detected annihilation rate as a function of time; (f) probability that cosmic rays produce the observed counts or more; (g) control sample showing no signal during the trap quench.[93]

To date, a single spectroscopic measurement of antihydrogen has been carried out. The ALPHA collaboration focused first on the (HFS) microwave transition.[95] Although the energy levels, and thus their splitting, of the trapped antihydrogen atoms depend on the position of the atom within the magnetic potential well, the field minimum of 1 Tesla ensures that no hyperfine splitting below that of the value at 1 T can take place. By exposing trapped antihydrogen atoms to a broad-band microwave radiation (15 MHZ around this minimum), the ALPHA collaboration was able to induce spin-flips in the trapped atoms, which then — because they were now in an un-trapped configuration — could be detected through annihilation. Figure 11 shows the Zeeman splitting, the spin-flip transition line shapes and microwave scan windows (two possible transitions), and the rate of detected antihydrogen atoms for 15-s scans over each window.

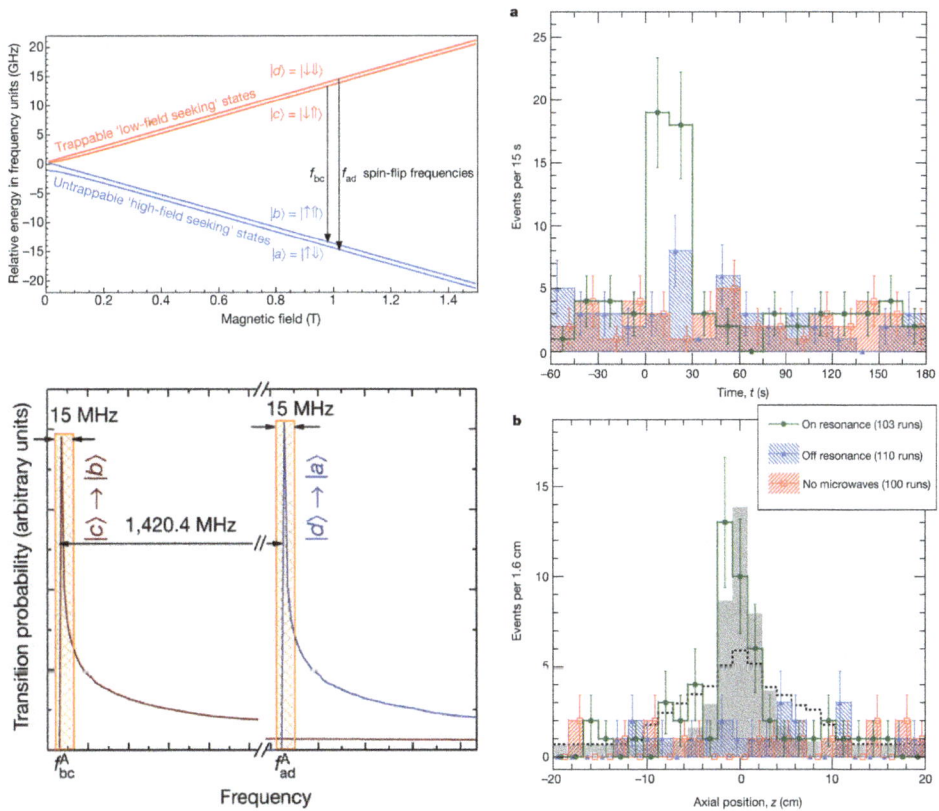

Fig. 11.   Left: Breit–Rabi diagram of the hyperfine energy levels of ground state antihydrogen. below: Spin-flip transition line shapes and microwave scan windows for the $|c\rangle \rightarrow |b\rangle$ and the $|d\rangle \rightarrow |a\rangle$ transitions. Right: Number of detected antihydrogen annihilations before, during and after application of the microwaves (top) and axial position of the detected annihilations for $0 < t < 30$ s (bottom) in the ALPHA experiment.[95]

5.4. *Gravity: AEgIS, GBAR*

The latest antihydrogen experiments at CERN aim at testing another fundamental symmetry: the weak equivalence principle. While experiments with the goal of measuring the behavior of matter and antimatter in the Earth's gravitational field have been contemplated before, the weakness of the gravitational interaction, and the impossibility of sufficiently shielding remnant electric and magnetic interactions for charged (anti)particles has hindered their realisation. Several groups have proposed experiments using neutral antihydrogen atoms as gravitational probes. The first such experiment (currently undergoing commissioning) is the AEgIS/AD-6 experiment.[96] It aims to produce a moderately focused pulsed horizontal beam of antihydrogen atoms, whose parabolic trajectory can be measured via a high resolution annihilation detector (a silicon-photographic emulsion hybrid). Two periodic gratings (a classical moiré deflectometer)[97] are employed to produce a spatially modulated distribution of transmitted atoms. Since the vertical shift of the periodic pattern depends on the time during which a set of monoenergetic atoms fall, each atom's velocity is determined by forming all atoms simultaneously (through charge exchange between laser-excited positronium and ultra-cold antiprotons) and measuring the arrival time of the atom (in the annihilation detector) together with its vertical position.

A second method has been proposed by the GBAR collaboration:[98] by interacting a low energy beam of antiprotons with a high-density cloud of positronium atoms, a dual charge exchange process (as described in Section 5.1), first forming ground state antihydrogen, and subsequently the bound $\bar{H}^+$ ion, a stable and positively charged antihydrogen ion is formed. Trapping and interacting this positive anti-ion with other laser cooled cations (e.g. Cs^+) allows a first pre-cooling before a final cooling step with laser-cooled Be^+ to μK is carried out. Finally, the $\bar{H}^+$ is laser-ionised; the neutral $\bar{H}$'s free fall time from the trap to a detector measures its gravitational behavior.

Lastly, trapped antihydrogen atoms can also be released, and — if they are sufficiently cold — their subsequent free-fall behavior can be investigated. While first attempts[99] by the ALPHA experiment do not yet have the necessary sensitivity — at several $100\,mK$, the temperatures of the (neutral, uncooled) trapped atoms are still far too high — this method may well become competitive, should laser cooling of antihydrogen atoms succeed in the coming years.

References

1. G. von Dardel *et al.*, *Phys. Rev. Lett.* **5**, 333 (1960).
2. A. Wightman, *Phys. Rev.* **77**, 521 (1950).
3. M. Leon and H. Bethe, *Phys. Rev.* **127**, 636 (1962).
4. P. Brix *et al.*, *Phys. Lett.* **1**, 56 (1962).

5. G. Backenstoss *et al.*, *Phys. Lett. B* **25**, 365 (1967).
6. G. Backenstoss *et al.*, *Phys. Lett. B* **32**, 399 (1970).
7. G. Backenstoss *et al.*, *Phys. Lett. B* **33**, 230 (1970).
8. A. Bamberger *et al.*, *Phys. Lett. B* **33**, 233 (1970).
9. E. Hu *et al.*, *Nucl. Phys. A* **254**, 403 (1975).
10. E. Fermi and E. Teller, *Phys. Rev.* **72**, 399 (1947).
11. L. I. Ponomarev, *Ann. Rev. Nucl. Sci.* **23**, 395 (1975).
12. N. Morita *et al.*, *Phys. Rev. Lett.* **72**, 1180 (1994).
13. T. B. Day, G. A. Snow and J. Sucher, *Phys. Rev. Lett.* **3**, 61 (1959).
14. R. Knop *et al.*, *Phys. Rev. Lett.* **14**, 767 (1965).
15. R. A. Burnstein *et al.*, *Phys. Rev. Lett.* **15**, 639 (1965).
16. G. Reifenröther *et al.*, *Phys. Lett. B* **214**, 325 (1988).
17. A. Bianconi *et al.*, *Phys. Lett. B* **487**, 224 (2000).
18. R. Schafer, *Phys. Rev.* **163**, 1451 (1967).
19. R. Kunselman, *Phys. Lett. B* **34**, 485 (1971).
20. S. C. Cheng *et al.*, *Nucl. Phys. A* **254**, 381 (1975).
21. G. Dugan *et al.*, *Nucl. Phys. A* **254**, 396 (1975).
22. G. Backenstoss *et al.*, *Phys. Lett. B* **36**, 403 (1971).
23. G. Backenstoss *et al.*, *Phys. Lett. B* **43**, 431 (1973).
24. R. Hughes and B. Deutsch, *Phys. Rev. Lett.* **69**, 578 (1992).
25. P. Robertson *et al.*, *Phys. Rev. C* **16**, 1945 (1977).
26. H. Poth *et al.*, *Nucl. Phys. A* **294**, 435 (1978).
27. G. Backenstoss *et al.*, *Phys. Lett. B* **38**, 181 (1972).
28. W. Weise and L. Tauscher, *Phys. Lett. B* **64**, 424 (1976).
29. B. Adeva *et al.*, *Phys. Lett. B* **704**, 24 (2011).
30. B. Adeva *et al.*, *Phys. Lett. B* **674**, 11 (2009).
31. J. Schweizer, *Phys. Lett. B* **587**, 33 (2004).
32. H. Acker *et al.*, *Nucl. Phys.* **87**, 1 (1966).
33. G. Backenstoss *et al.*, *Nucl. Phys.* **62**, 449 (1965).
34. S. A. De Wit *et al.*, *Nucl. Phys.* **87**, 657 (1967).
35. G. Backenstoss *et al.*, *Phys. Lett. B* **25**, 547 (1967).
36. A. Trzcinska *et al.*, *Phys. Rev. Lett.* **87**, 082501 (2001).
37. R. Pohl *et al.*, *Annu. Rev. Nucl. Part. Sci.* **63**, 175204 (2013).
38. R. Pohl *et al.* (CREMA Collaboration), *Nature* **466**, 213 (2010).
39. A. Antognini *et al.* (CREMA Collaboration), *Science* **339**, 417 (2013).
40. B. Desai, *Phys. Rev.* **119**, 1385 (1960).
41. G. Backenstoss *et al.*, *Phys. Lett. B* **41**, 552 (1972).
42. J. Fox *et al.*, *Phys. Rev. Lett.* **29**, 193 (1972).
43. M. Iwasaki *et al.*, *Phys. Rev. Lett.* **67**, 1246 (1991).
44. H. Poth *et al.*, *Phys. Lett. B* **76**, 523 (1978).
45. F. Hartmann, *Hyperfine Int.* **119**, 175 (1999).
46. H. Koch, Hadron physics with antiprotons, in *Proc. of the Int. School of Physics "Enrico Fermi"*, Course CLVIII, eds. T. Bressani, A. Filippi and U. Wiedner (IOS Press, 2005), p. 305 ff., DOI: 10.3254/1-58603-526-6-305.
47. E. G. Auld *et al.*, *Phys. Lett. B* **77**, 454 (1978).
48. S. Ahmad *et al.*, *Phys. Lett. B* **157**, 333 (1985).

49. M. Ziegler *et al.*, *Phys. Lett. B* **206**, 151 (1988).
50. T. P. Gorringe *et al.*, *Phys. Lett. B* **162**, 71 (1985).
51. M. Augsburger *et al.*, *Nucl. Phys. A* **658**, 149 (1999).
52. D. Gotta *et al.*, *Nucl. Phys. A* **660**, 283 (1999).
53. M. Conversi *et al.*, *Phys. Lett.* **5**, 195 (1963).
54. U. Amaldi *et al.*, *Nuovo Cimento*, **34**, 825 (1964).
55. N. Barash *et al.*, *Phys. Rev.* **139**, B1659 (1965).
56. S. van der Meer, CERN Int. Report ISR- PO/72-31 (1972).
57. G. Carron *et al.*, *Phys. Lett. B* **77**, 353 (1978).
58. S. Gilardoni and D. Manglunki (eds.) *Fifty Years of the CERN Proton Synchroton*, Volume II, CERN-2013-005 (CERN, Geneva, 2013).
59. K. Killian, U. Gastaldi and D. Möhl, CERN/PS/DL 77-19.
60. G. Gabrielse *et al.*, *Phys. Rev. Lett.* **57**, 2504 (1986).
61. CERN report CERN/PS/86-30.
62. V. Chohan (ed.), CERN report CERN-2014-002.
63. G. Gabrielse *et al.*, *Phys. Rev. Lett.* **82**, 3198 (1999).
64. G. Gabrielse *et al.*, *Phys. Rev. Lett.* **63**, 1360 (1989).
65. B. Roberts, *Phys. Rev. D* **17**, 358 (1978).
66. G. Gabrielse *et al.*, *Phys. Rev. Lett.* **74**, 3544 (1995).
67. H. Torii *et al.*, *Nucl. Inst. Meth. A* **396**, 257 (1997).
68. V. I. Korobov, *Phys. Rev. A* **54**, 1749 (1996).
69. V. I. Korobov *et al.*, *Hyperfine Int.* **194**, 15 (2009).
70. T. Pask *et al.* (ASACUSA Collaboration), *Phys. Lett. B* **678**, 55 (2009).
71. V. I. Korobov *et al.*, *Phys. Rev. Lett.* **112**, 103003 (2014).
72. V. I. Korobov *et al.*, *Phys. Rev. A* **89**, 032511 (2014).
73. E. Widmann *et al.*, *Phys. Lett. B* **404**, 15 (1997).
74. G. Gabrielse *et al.*, *Phys. Rev. Lett.* **65**, 1317 (1990).
75. D. Bakalov and V. I. Korobov, *Phys. Rev. A* **57**, 1662 (1998).
76. E. Widmann *et al.*, *Phys. Rev. Lett.* **89**, 243402 (2002).
77. S. Friedreich *et al.* (ASACUSA Collaboration), *J. Phys. B* **46**, 125003 (2013).
78. G. Gabrielse *et al.*, *Phys. Rev. Lett.* **110**, 130801 (2013).
79. J. D. Fox *et al.*, *Phys. Rev. Lett.* **29**, 193 (1972).
80. C. Surko *et al.*, *Phys. Rev. Lett.* **62**, 901 (1989).
81. G. Baur *et al.*, ERN/SPSLC 94-29, P283 (1994).
82. G. Baur *et al.*, *Phys. Lett. B* **368**, 251 (1996).
83. M. Amoretti *et al.* (ATHENA Collaboration), *Nature* **419**, 456–459 (2002).
84. G. Gabrielse *et al.*, *Phys. Rev. Lett.* **89**, 213401 (2002).
85. G. Gabrielse *et al.*, *Phys. Rev. Lett.* **89**, 233401 (2002).
86. S. Jonsell *et al.*, *J. Phys. B: At. Mol. Opt. Phys.* **42**, 215002 (2009).
87. G. Gabrielse *et al.*, *Phys. Rev. Lett.* **93**, 073401 (2004).
88. N. Madsen *et al.* (ATHENA Collaboration), *Phys. Rev. Lett.* **94**, 033403 (2005).
89. C. H. Storry *et al.* (ATRAP Collaboration), *Phys. Rev. Lett.* **93**, 263401 (2004).
90. G. Gabrielse *et al.* (ATRAP Collaboration), *Phys. Rev. Lett.* **100**, 113001 (2008).
91. G. B. Andresen *et al.* (ALPHA Collaboration), *Nature* **468**, 673676 (2010).
92. G. B. Andresen *et al.* (ALPHA Collaboration), *Nature Physics*, **7**, 558 (2011).
93. G. Gabrielse *et al.*, *Phys. Rev. Lett.* **108**, 113002 (2012).

94. M. Niering *et al.*, *Phys. Rev. Lett.* **84**, 54965499 (2000).
95. C. Amole *et al.*, *Nature* **483**, 439 (2012), doi:10.1038/nature10942.
96. CERN-SPSC-2007-017, http://cds.cern.ch/record/1037532/files/spsc-2007-017.pdf.
97. S. Aghion *et al.*, *Nature Commun.* **5**, 4538 (2014), doi: 10.1038/ncomms5538.
98. CERN-SPSC-2011-029, http://cds.cern.ch/record/1386684/files/SPSC-P-342.pdf.
99. The ALPHA Collaboration and A. E. Charman *Nature Commun.* **4**, 1785 (2013), doi: 10.1038/ncomms2787.

Muon $g - 2$ and Tests of Relativity

Francis J. M. Farley[*]

*Energy and Climate Change Division, Engineering and the Environment,
Southampton University, Highfield, Southampton, SO17 1BJ, England, UK
f.farley@soton.ac.uk*

After a brief introduction to the muon anomalous moment $a \equiv (g-2)/2$, the pioneering measurements at CERN are described. This includes the CERN cyclotron experiment, the first Muon Storage Ring, the invention of the "magic energy", the second Muon Storage Ring and stringent tests of special relativity.

1. Introduction

Creative imagination. That is what science is all about. Not the slow collection of data, followed by a generalisation, as the philosophers like to say. There is as much imagination in science as in art and literature. But it is grounded in reality; the well tested edifice of verified concepts, built up over centuries, brick by brick. All this is well illustrated by the muon $(g-2)$ theory and measurements at CERN.

It also illustrates the reciprocal challenges. Theorists come up with a prediction, for example that light should be bent by gravity: how can you measure it? Eddington found a way. Conversely experiments show that the gyromagnetic ratio of the electron is not 2, but slightly larger: then the theorists are challenged to explain it, and they come up with quantum electrodynamics and a cloud of virtual photons milling around the particle. How can we check this? And so on. By reciprocal challenges the subject advances; step by step. And of course, some of the ideas turn out to be wrong; they are quietly dropped.

Over the years the muon $(g-2)$ has proved to be a marker, a lighthouse, a fixed reference that theories must accommodate; and many zany speculations have come to grief on this rock.

In this review I will not recap the detail which is given in the published papers and the many reviews.[1, 2] Instead I try to highlight the main creative steps, how they were reached, plus the many precautions needed to make the experiments work: and to give the correct answer.

---

[*]Address for correspondence: 8 Chemin de Saint Pierre, 06620 le Bar sur Loup, France.

The muon $(g-2)$ at CERN has a unique record. The number published at an early stage always turned out to be correct; it was verified by the next experiment; the new number always fell inside the one sigma error bars of the previous. The final CERN measurement was confirmed by the later experiment at Brookhaven. At one stage our number disagreed with the theory, but we published anyway. The theorists then revised their calculations, and they agreed with us.

The gyromagnetic ratio g is the ratio of the magnetic moment of a system to the value obtained by multiplying its angular momentum by the Larmor ratio $(e/2mc)$. For an orbiting electron $g = 1$. When Goudschmit and Uhlenbeck[3] postulated the spinning electron with angular momentum $(h/4\pi)$ to explain the anomalous Zeeman effect, it was surprising that its magnetic moment, one Bohr magneton, was twice the expected value: the gyromagnetic ratio for the electron was apparently 2. Later, Dirac[4] found that this value came out as a natural consequence of his relativistic equation for the electron.

Another surprise was to come. Experimentally[5] the magnetic moment of the electron was in fact slightly larger, so $g = 2(1 + a)$ with a being defined as the "anomalous magnetic moment". In its turn the anomaly was understood[6] as arising from the quantum fluctuations of the electromagnetic field around the particle. The calculation of this quantity,[7] in parallel with measurements of increasing accuracy, has been the main stimulus to the development of quantum electrodynamics. For the electron, astonishingly, theory and experiment agree for this pure quantum effect to 0.02 ppm (parts per million) in a, the limit being set by our independent knowledge of the fine structure constant α.

For the muon, the $(g-2)$ value has played a central role in establishing that it behaves like a heavy electron and obeys the rules of quantum electrodynamics (QED). The experimental value of $(g-2)$ has been determined by three progressively more precise measurements at CERN and a recent experiment at Brookhaven,[8] now achieving a precision of 0.7 ppm (parts per million) in the anomaly $a \equiv (g-2)/2$. In parallel, the theoretical value for $(g-2)$ has improved steadily as higher order QED contributions have been evaluated, and as knowledge of the virtual hadronic contributions to $(g-2)$ has been refined.

At CERN, theorists and experimenters work in close proximity and interact. So CERN theorists have made important contributions to the calculation of the muon $(g-2)$, starting with Peterman[9] who corrected an error in the $(\alpha/\pi)^2$ term, and continuing with Kinoshita, Lautrup and de Rafael.[10] Kinoshita[7] in particular was alerted to the problem during a tour of the experiments arranged by John Bell in 1962 and spent the rest of his career calculating higher and higher orders for the electron and the muon.

The story starts in 1956 when the magnetic anomaly $a \equiv (g-2)/2$ of the electron was already well measured by Crane *et al.*[11] Berestetskii *et al.*[12] pointed out that the postulated Feynman cutoff in QED at 4-momentum transfer $q^2 = \Lambda^2$, would reduce the anomaly for a particle of mass m by

$$\delta a/a = (2m^2/3\Lambda^2). \tag{1}$$

Therefore, a measurement for the muon with its 206 times larger mass would be a far better test of the theory at short distances (large momentum transfers). (At present the comparison with theory for the electron is 35 times better than for the muon; but to be competitive it needs to be 40,000 times better! The muon is by far the better probe for new physics).

In 1956, parity was conserved and muons were unpolarised, so there was no possibility of doing the experiment proposed by Berestetskii. But in 1957 parity was violated in the weak interaction and it was immediately realised that muons coming from pion decay should be longitudinally polarised. Garwin, Lederman and Weinrich,[13] in a footnote to their classic first paper confirming this prediction, used the $(g − 2)$ precession principle (see below) to establish that its gyromagnetic ratio g must be equal to 2.0 to an accuracy of 10%. This was the first observation of muon $(g − 2)$ 57 years ago.

In 1958, the Rochester conference took place at CERN; Panofsky[14] reviewing electromagnetic effects said that three independent laboratories, two in the USA and one in Russia, were planning to measure $(g − 2)$ for the muon. In the subsequent discussion, it was clear that leading theorists expected a major departure from the predicted QED value, either due to a natural cutoff (needed to avoid the well known infinities in the theory) or to a new interaction which would explain the mass of the muon. Feynman in 1959 told me that he expected QED to breakdown at about 1 GeV momentum transfer. At that time renormalisation was regarded as a quick fix to deal with infinite integrals, not a real theory.

2. Principle

The orbit frequency ω_c for a particle turning in a magnetic field B is

$$\omega_c = (e/mc)B/\gamma. \tag{2}$$

While for a particle at rest or moving slowly, the frequency at which the spin turns is

$$\omega_s = g\,(e/2mc)\,B \tag{3}$$

At low energy $(\gamma \sim 1)$ if $g = 2$ these two frequencies are equal, so polarised particles injected into a magnetic field would keep their polarisation unchanged. But if $g = 2(1 + a)$, then the spin turns faster than the momentum and the angle between them increases at frequency ω_a given by

$$\omega_a = \omega_s - \omega_c = a(e/mc)B. \tag{4}$$

This is the $(g − 2)$ principle discovered by Tolhoeck and DeGroot[15] and used so successfully by Crane for the electron.

Equation (4) for the $g - 2$ precession is true even at relativistic velocities.[15, 16] Significantly, there is no factor γ in this equation so at high energies the muon lifetime is dilated but the precession is not slowed down. With relativistic muons many $(g - 2)$ cycles can be recorded and the measurement becomes more accurate.

The magnetic field is measured by the proton NMR frequency ω_p and the experiment gives the ratio $R = \omega_a/\omega_p$. The ratio $\lambda = \omega_s/\omega_p$ in the same field is known from other experiments: careful studies of muon precession at rest and the hyperfine splitting in muonium.[17] Combining (3) and (4) a is calculated from

$$a = \frac{\omega_a}{\omega_s - \omega_a} = \frac{R}{\lambda - R}. \tag{5}$$

The $(g - 2)$ experiments are essentially measurements of the frequency ratio $R = \omega_a/\omega_p$. If the value of λ changes a should be recalculated.

3. 6 m Magnet with the CERN Cyclotron 1958–1962

In 1957 parity violation was discovered, muon beams were found to be highly polarised and, better still, the angular distribution of the decay electrons could indicate the muon spin direction as a function of time. The possibility of a $(g - 2)$ experiment for muons was envisaged, and groups at Berkeley, Chicago, Columbia, and Dubna started to study the problem.[14] Compared with the electron, the muon $(g - 2)$ experiment was much more difficult because of the low intensity, diffuse nature, and high momentum of available muon sources. The lower value of (e/mc) made all precession frequencies 200 times smaller, but the time available for an experiment was limited by the decay lifetime, 2.2 μs. Therefore, large volumes of high magnetic fields would be needed to give a reasonable number of precession cycles.

The main problem was how to inject muons into a magnetic field so that they made many turns. For the electron, Crane used a thermionic source already inside the solenoidal field and the spin was measured by scattering on a foil also inside the field. At CERN the muons were born inside the cyclotron (paradoxically already in a strong magnetic field); they came out and we needed to get them back in. To inject into a static field requires some kind of perturbation, usually a pulsed magnet which kicks the particle into a new direction (as used in most accelerators); otherwise the particle will exit the field after less than one turn.

Another option is a degrader in which the particle loses energy and so turns more sharply in the field. In a uniform field, the particle will then make one turn and return to the degrader. To inject successfully requires a horizontal field gradient, so that the orbit turns more sharply on one side than the other. The orbit then "walks" at right angles to the gradient and misses the degrader after one turn. We used a beryllium block about 10 cm thick to minimise multiple scattering and the edge was curved to fit the expected orbit.

In 1958 CERN acquired its first digital computer, the Ferranti Mercury with a programming language rather similar to Fortran, called Mercury Autocode. This was soon put to use[18] for tracking pions and muons coming out of the CERN cyclotron with a view to installing optimised beam pipes through the shielding. The program followed the tracks step by step in the horizontal plane and also included vertical focusing effects due to field gradients. It was put to work to follow muons turning in the horizontal plane of a long bending magnet with specially designed transverse gradients. Using a degrader, it was fairly easy to get the muons into the field. But could they be ejected? This was the key question, answered eventually by the computer.

To measure the spin angle one has to stop the muons in some block, wait for them to decay and record the distribution of the emitted electrons. But if the block is inside the magnet, the muons at rest will continue to rotate so the new spin direction will be scrambled. One must get the muons into the field, let them make many turns, and then get them out before stopping them in a field free region.

The problem is complicated by a fundamental theorem for particles turning in a magnetic field. In slowly varying fields, the flux through the orbit is an invariant of the motion. So the experts argued that once the muons were trapped in the field, it would be impossible to get them out. The experiment would fail.

At the end of the magnet the field decreases: inevitably there is a longitudinal gradient. When the particle reaches this point it feels the longitudinal gradient and moves sideways, to the side of the magnet where it either hits something or walks back along the fringing field to the beginning. It is not ejected.

What about using a very large transverse gradient? Then with a large step size, the particle will arrive suddenly at the end of the magnet and come out without moving sideways... as it does in a normal beam line. Ah yes, said the experts, but in a large gradient there will be strong alternating gradient focusing, the beam will blow up vertically and the particles will be lost.

The computer could address this question. It followed the muons for many turns, from injection in a medium gradient, through a transition to a very weak gradient where they made many turns, then gradually into a strong gradient with a very large step size, all the way to the end of the magnet where, it turned out, they were ejected successfully without any excessive vertical focusing!

Successful storage requires vertical focusing. Otherwise the particles will spiral up or down into the poles. A muon turning in a linear gradient is focused on one side of the orbit and defocused on the other. This gives a net focusing effect, but far too small to be useful. One needs to add a parabolic term so that the field decreases outwards on both sides of the orbit. Storage, Fig. 2, for up to 18 turns was demonstrated in a small magnet, Fig. 1, borrowed from the University of Liverpool. This result and the ejection calculation gave the lab confidence to order a special magnet 6 m long which could store the muons for many turns.

Fig. 1. First experimental magnet in which muons were stored at CERN for up to 30 turns. Left-to-right: Georges Charpak, Francis Farley, Bruno Nicolai, Hans Sens, Antonino Zichichi, Carl York, Richard Garwin.

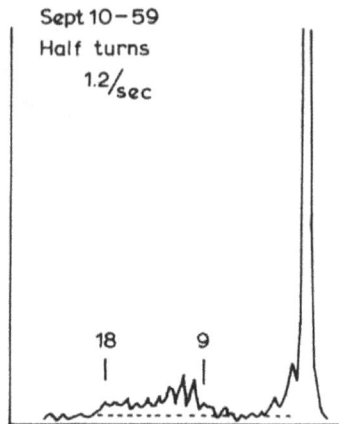

Sept 10 – 59
Half turns
$1.2/\text{sec}$

18   9

Fig. 2. First evidence of muons making several turns in the experimental magnet, shown in Fig. 1. The time of arrival of the particles at a scintillator fixed inside the magnet is plotted horizontally (time increases to the left). The first peak (right side) coincides with the moment of injection. The equally-spaced later peaks correspond to successive turns. Owing to the spread in orbit diameters and injection angles, some muons hit the counter after nine turns, while others take 18 turns to reach the same point (Charpak *et al.*, unpublished).

The 6 m magnet had removal poles 5 cm thick, which could be rolled out and shimmed. Hundreds of thin layers of iron were held in place by aluminium covers and specially shaped by trial and error to give the required field shape, a titanic task executed by Zichichi and Nicolai. At injection the step size was 1.2 cm to give reasonable clearance from the degrader. Moving along the magnet, the gradient was gradually reduced so that the muons advanced only 4 mm per turn and spent longer in the field. At the far end a very large gradient increased the step to 11 cm per turn.

The theorem mentioned above, that the flux through the orbit is an invariant of the motion, was used to good effect. If the average field varies along the magnet, the orbit will move sideways in the gradient, to keep the flux constant, so the particles can be lost. This was checked with a flux coil 40 cm diameter (the size of the orbit) which could be moved along the magnet. The coil was connected to a fluxmeter and any deviation from constancy was corrected with a special set of "longitudinal" shims. This was particularly important in the transition regions where the gradient was changing. Moving the flux coil sideways measured the lateral gradient. The theorem also implied that we could calibrate the field with NMR at the centre of the magnet; and the result would be valid everywhere.

An overall view of the final storage system[19] is shown in Fig. 3. The magnet pole was 6 m long and 52 cm wide, with a gap of 14 cm. Muons entered on the left through a magnetically shielded iron channel and hit a beryllium degrader in the injection part of the field. Here the step size s was 1.2 cm. Then there was a transition to the long storage region, where $s = 0.4$ cm. Finally, a smooth transition

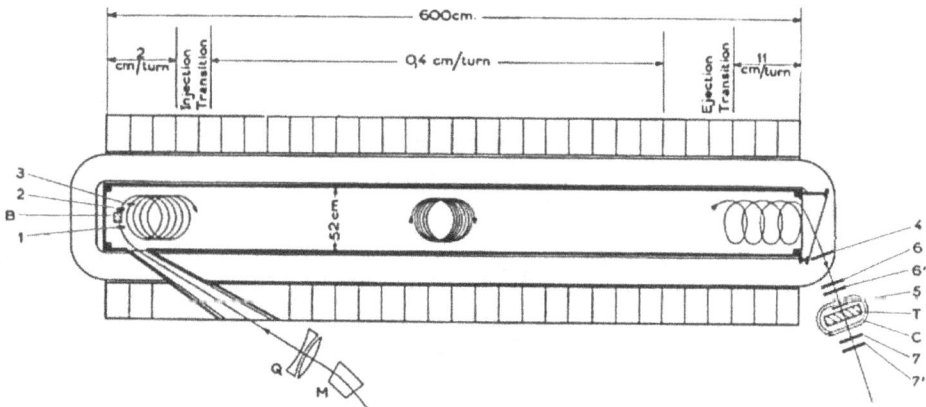

Fig. 3. The 6 m bending magnet used for storing of muons for up to 2000 turns. A transverse field gradient makes the orbit walk to the right. At the end a very large gradient is used to eject the muons which stop in the polarisation analyser. Coincidences 123 and 466′57, signal an injected and ejected muon respectively. The coordinates used in the text are x (the long axis of the magnet), y (the transverse axis in the plane of the paper), and z (the axis perpendicular to the paper).

was made to the ejection gradient, with $s = 11$ cm per turn. The ejected muons fell onto the polarisation analyser Fig. 4, where they decayed to e^+.

The muons were trapped in the magnet for 2–8 μs depending on the location of the orbit centre on the varying parabolic gradient. About one muon per second was stopped finally in the polarisation analyser, and the decay electron counting rate was 0.25 per second.

To obtain the anomalous moment a from Eq. (4) one must measure the time a muon has spent in the field and the spin angle before and after storage. Time was measured with a 10 MHz clock, started when a muon came out of the magnet and stopped by a delayed signal from a muon at the entrance. An elaborate veto system rejected events with two signals close to each other at either end, so there was no chance of confusion leading to incorrect times.

The spin angle was measured by the polarisation analyser, Fig. 4. The same counters were used to signal a muon stopping in the central absorber E and to record the subsequent decay electron emitted either backwards or forwards. The ratio of backward (B) to forward (F) counts measures the asymmetry, but this is not sensitive to the transverse angle. Therefore the muon spin was flipped through

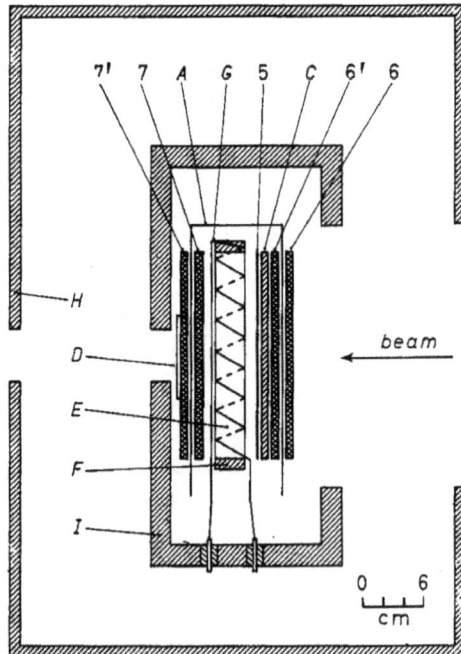

Fig. 4. Polarisation analyser. When a muon stops in the liquid methylene iodide E a pulse of current in coil G is used to flip the spin through $\pm 90°$. Backward or forward decay electrons are detected in counter telescopes 66' and 77'. The static magnetic field is kept small by the double iron shield H, I and the mu-metal shield A. The muon must pass the thin scintillator 5, backed by plexiglass C. D is a mirror used for alignment.

±90 degrees by a short pulse of vertical field applied to the absorber every time a muon stopped. The ratio

$$A = \frac{F_+ - F_-}{F_+ + F_-} \tag{6}$$

for forward counts with +90 and −90 flipping was then a measure of the transverse spin component. Similar data was obtained from the backward telescope. The flipping angle should be consistent, but its exact value is not important.

For this to work, the absorber in which the muons stopped had to be non-conducting (no metals) and not depolarising, which ruled out most plastics. Luckily liquid methylene iodide had the right properties. A double iron shield plus an inner mumetal shield was used to reduce the magnetic field in the absorber.

The direction of the arriving muons was measured with a venetian blind made of parallel slats of scintillator used to veto the event. The only particles recorded were those that got through the spaces between the slats without touching any of them.

When the polarisation analyser was used to study the muons coming out of the cyclotron the transverse angle was found to vary rapidly with muon momentum (range). This could create an error because the band of momentum selected by the storage magnet could be very different. The effect was eliminated by passing the muon beam through a long solenoid with field parallel to the beam. This rotated all transverse spin components through 90°, horizontal into vertical and vice versa. Because of vertical symmetry inside the cyclotron the result was no spin-momentum correlation in the horizontal plane.

For muons that had been through the magnet, the analyser recorded the asymmetry A as a function of the time t the particle had spent in the field. This showed a sinusoidal variation due to the $(g - 2)$ precession in the magnet.

$$A = A_0 \sin \theta_s = A_0 \sin\{a(e/mc)Bt + \phi\} \tag{7}$$

where ϕ is an initial phase determined by measuring the initial polarisation direction and the orientation of the analyser relative to the muon beam.

The experimental data are given in Fig. 5, together with the fitted line obtained by varying A_0 and a in Eq. (7). Full discussion of the precautions needed to determine the mean field B seen by the muons, and to avoid systematic errors in the initial phase ϕ, are given in Ref. 19. The first experiment gave ±2% accuracy in a and this was later improved to ±0.4%. The figures agreed with theory within experimental errors. The corresponding 95% confidence limit for the photon propagator cut-off, Eq. (1), was $\Lambda > 1.0\,\text{GeV}$.

This was the first real evidence that the muon behaved so precisely like a heavy electron. The result was a surprise to many, because it was confidently expected that g would be perturbed by an extra interaction associated with the muon to account for its larger mass. When nothing was observed at the 0.4% level, the muon became accepted as a structureless point-like QED particle, and

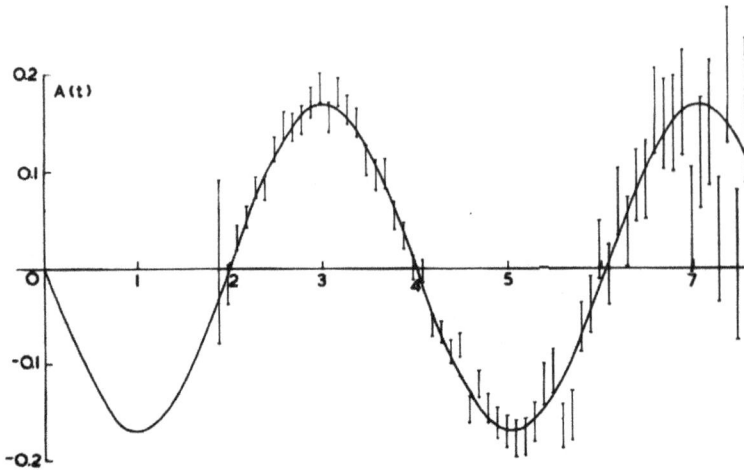

Fig. 5.   Asymmetry A of observed decay electron counts as a function of the storage time t. The time t spent in the magnet depended on the transverse position of the orbit on the parabolic magnetic field. The muons that were stored for 7.5 μs made 1600 turns in the magnet and then emerged spontaneously at the far end. The sinusoidal variation results from the $(g-2)$ precession; the frequency is measured to ±0.4%.

the possibility of finding a clue to the $\mu - e$ mass difference now appeared more remote.

In retrospect this experiment was quite remarkable. We poured muons into the magnet at one end, they were trapped inside for almost 2000 turns (2.5 km) and then came out at the other end, all of their own accord: no pulsed fields, no kickers. Nothing like this has ever been done, before nor since.

4.   First Muon Storage Ring 1962–1968

4.1.   *Overview*

The muon $(g - 2)$ experiment was now the best test of QED at short distances. To go further and to search again for a new interaction, it was desirable to press the experiment to new levels. Relativistic particles with dilated lifetimes were available from the CERN PS and there is no factor γ in Eq. (4) so in principle high energy muons would give more precession cycles and greater accuracy. Storing muons of GeV energy in a magnetic field and measuring their polarisation required totally new techniques. Farley[20] proposed to measure the anomalous moment using a muon storage ring. Simon van der Meer designed the magnet and participated in the whole experiment.

Time dilation in a straight path was well established. But no one had proved it for a two way journey, out and return or a circular orbit. The twin paradox (clock paradox) was still a puzzle and some people did not believe it. Notably

Herbert Dingle,[21] who had written a short but excellent textbook on relativity, lost faith and carried on a campaign against it. The predictions of special relativity were clear: the twin who suffers acceleration ends up younger. But perhaps this was not the whole story, acceleration was said to be equivalent to gravity and the gravitational redshift could change time. So perhaps there was no time dilation in a circular orbit and the experiment would fail. It was a leap in the dark. Luckily there were no Dingles on the committee.

The experiment is made possible by four miracles of Nature. (First, identify your miracle, then put it to work for what you wish to do.) The first miracle is that it is easy to inject muons into a storage ring. One simply injects pions for a few turns; they decay in flight and some of the muons will fall onto permanently stored orbits. The easy way to inject pions is to put the primary target of the accelerator inside the storage magnet and hit it with high energy protons, thus producing the pions inside the ring. The second miracle is that stored muons come from forward decay, so they are strongly polarised. The third miracle is that when the muons decay the electrons have less energy; bent by the field they come out on the inside of the ring and hit the detectors. The higher energy electrons must come from forward decay: so as the spin rotates, the electron counting rate is modulated by the $(g − 2)$ frequency ($\sim 270\,\text{kHz}$). One simply reads it off.

An advantage of this method is that it works equally well for μ^+ and μ^-. Most muon precession experiments can only be done with μ^+, because stopped μ^- are captured by nuclei and largely depolarised. $g − 2$ can be measured for μ^- as well as μ^+.

It was later realised that the injected muons would be localised in azimuth (injection time 10 ns, rotation time 52 ns), so the counting rate would also be modulated at the much faster rotation frequency ($\sim 20\,\text{MHz}$). This would enable the mean radius of the stored muons to be calculated, leading to a precise knowledge of the corresponding magnetic field.

With the primary target of the accelerator inside the storage ring there would be a huge background in the counters. Would this swamp the observations? A test inside the PS tunnel revealed radiation lasting for many milliseconds and decaying roughly as $1/t$. This could only come from neutrons banging around inside the building from wall to wall. A theory of neutron slowing down[22] gave a reasonable fit to the data. The typical neutron velocity after a time t is obtained by dividing the width of the room by t. This paper is widely used where short lived radioactive isotopes are studied, e.g. at ISOLDE.

We later discovered that the main background in the counters came from neutrons trapped inside the plexy light pipes, creating Cherenkov light after an (n,γ) process. Adopting air filled light pipes with white walls reduced the effect.

The first Muon Storage Ring[23] was a weak-focusing ring (Fig. 6) with $n = 0.13$, orbit diameter 5 m, and a useful aperture of $4\,\text{cm} \times 8\,\text{cm}$ (height × width); the

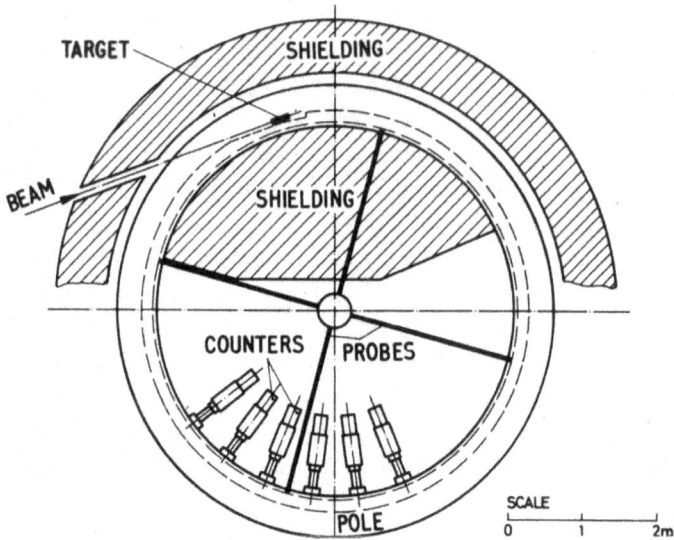

Fig. 6. First Muon Storage Ring: diameter 5 m, muon momentum 1.3 GeV/c, time dilation factor 12. The injected pulse of 10.5 GeV protons produces pions at the target, which decay in flight to give muons.

muon momentum was 1.28 GeV/c corresponding to $\gamma = 12$ and a dilated lifetime of 27 μs. The mean field at the central orbit was $\bar{B} = 1.711$ T. The injection of polarised muons was accomplished by the forward decay of pions produced when a target inside the magnet was struck by 10.5 GeV protons from the PS. The proton beam consisted of one to three radio-frequency bunches (fast ejection), each ~10 ns wide, spaced 105 ns. As the rotation time in the ring was chosen to be 52.5 ns, these bunches overlapped exactly inside the ring. Approximately 70% of the protons interacted, creating, among other things, pions of 1.3 GeV/c that started to turn around the ring. The pions made about four turns before again hitting the target, and in each turn about 20% decayed.

Typically the pions go round the magnet with momentum 1–2% above the nominal central momentum. Muons with the top energy follow the same orbit as the pions and will eventually hit something and be lost. But muons with 1–3% lower momentum fall onto permanently stored trajectories. Because they come from almost forward decay the polarisation is of order 97%.

This was the theory. But in practice the muon polarisation was found to be much lower, around 30%. A high energy pion only has a short track inside the storage region but it can decay at a large angle and inject a stored muon with small polarisation. It is a rare process, but there were very many higher energy pions and a majority of the stored muons were born in this way... low average polarisation.

4.2. *Muon decay in flight*

There was no need to get the muons out of the field to study their spin. Just observe their decay in flight. The highest energy electrons in the lab have the same momentum as the muons, and are trapped in the field. But those with lower energy are bent more and exit the ring on the inside. Here they hit one of the lead-scintillator detectors in which they produce a shower and the light output is proportional to the electron energy. By selecting pulse height in the detector, one selects a band of decay electron energies. By recording the high energy particles, one selects forward decays: as the spin rotates the number is modulated by the $(g − 2)$ frequency.

When the muon decays the electron energy is boosted by the Lorentz transformation. The broad rest-frame spectrum becomes a falling triangle with a large number at low momentum dropping to zero at the end point which is equal to the stored muon momentum, Fig. 7. To have this maximum momentum in the lab, the electron must be emitted exactly forward and have the top energy in the muon frame; so the asymmetry for these particles in the lab is $A = 1$. These particles carry the maximum information about the muon spin, but there are none of them. At lower energy a mixture of rest frame electron energies and decay angles can contribute, the number rises and the asymmetry falls, Fig. 7. To have high energy in the lab, the electron must be emitted forwards in the muon frame.

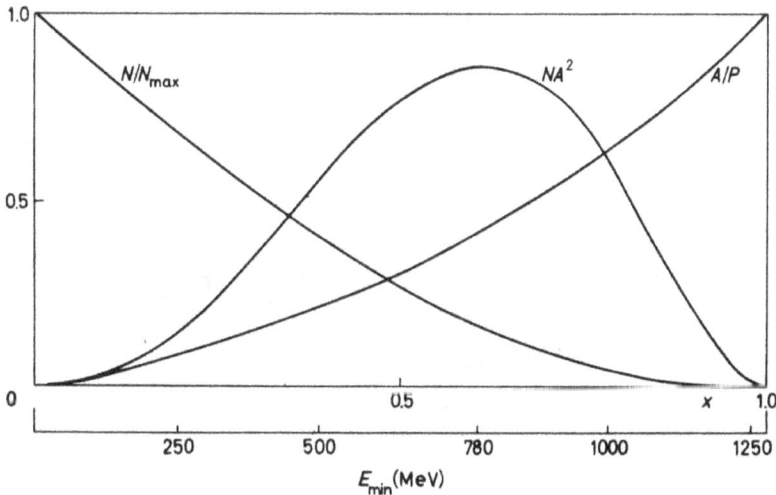

Fig. 7. Integral energy spectrum N of decay electrons hitting a detector: asymmetry coefficient A and NA^2 versus electron energy threshold E_{min}. The maximum of NA^2 occurs when E_{min} is about 0.65 times the stored muon energy.

4.3. *Experimental details and results*

The dilated muon lifetime was now $27\,\mu s$ so the muon precession could be followed out to storage time $t = 130\,\mu s$ as shown in Fig. 8. Data for t less than $20\,\mu s$ could not

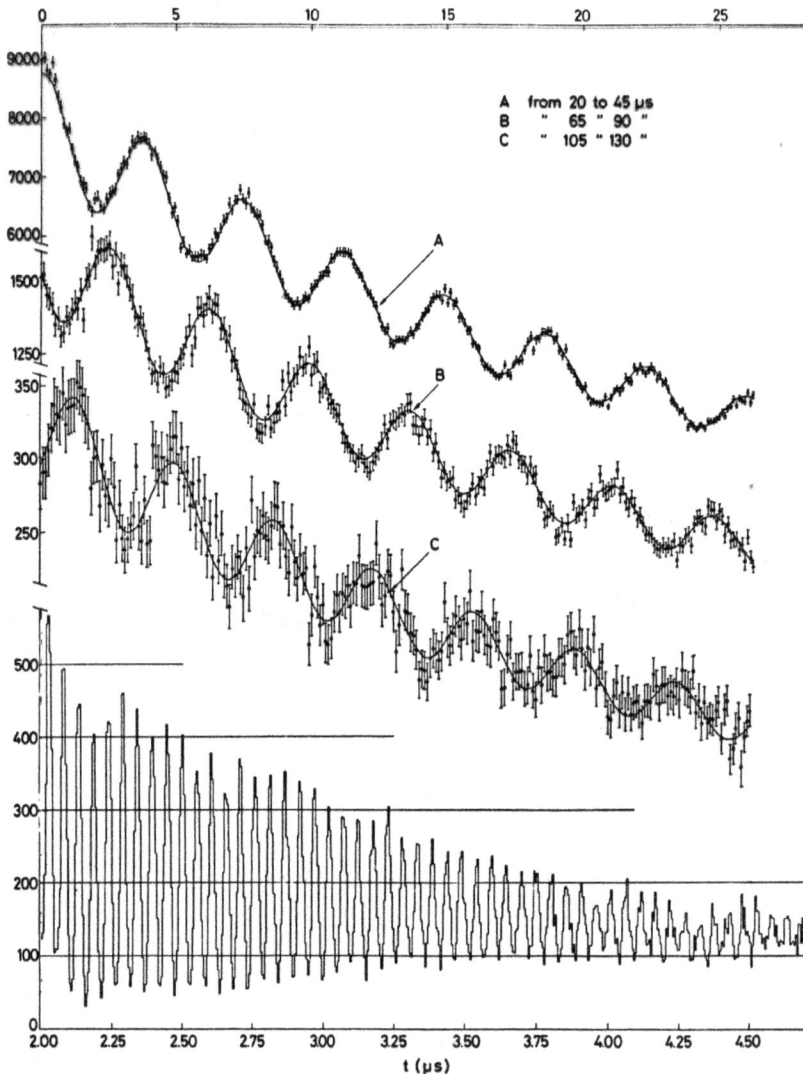

Fig. 8. First Muon Storage Ring: decay electron counts as a function of time after the injected pulse. The lower curve 1.5–4.5 μs (lower time scale) shows the 19 MHz modulation due to the rotation of the bunch of muons around the ring. As it spreads out the modulation dies away. This is used to determine the radial distribution of muon orbits. Curves A, B, and C are defined by the legend (upper time scale); they show various sections of the experimental decay (lifetime $27\,\mu s$) modulated by the $(g-2)$ precession. The frequency is determined to 215 ppm, $\bar{B}$ to 160 ppm leading to 270 ppm in a.

be used because of background due to neutrons and other effects created when the protons hit the target in the ring. The initial polarisation angle of the muons is not needed for the measurement: one just fits the oscillations that are seen. With thirty $(g-2)$ cycles to fit the accuracy in ω_a was now much better. Fitting a frequency ω to exponentially decaying oscillations the error is

$$\delta\omega/\omega = \frac{\sqrt{2}}{\omega\tau A\sqrt{N}} \tag{8}$$

where N is the total counts, τ the dilated lifetime and A the amplitude of the oscillations (asymmetry). To get good accuracy one should increase the number of cycles per lifetime by using high magnetic field and high energy, and maximise the product NA^2. The best value of NA^2 was obtained by accepting decay electrons above 780 MeV.

The magnetic field was measured between runs with the vacuum chamber removed at 288 positions in the azimuth and ten radii. During the runs it was monitored by four plunging NMR probes which could be driven into the centre of the ring. The radial magnetic gradient needed for vertical focusing implied a field variation of ±0.2% over the horizontal aperture of the storage ring (8 cm), so a major problem was to know the mean radius of the ensemble of muons that contributed to the data.

The muons are bunched at injection so there is a strong modulation of the counts at the rotation frequency, as seen in the lower curve of Fig. 8. Because of their various radii and rotation periods, they gradually spread around the ring, and the modulation dies away. The envelope of the modulation is the Fourier transform of the frequency spectrum, or equivalently of the radial distribution. By making the inverse transform one recovers the radial distribution of the muon equilibrium orbits. Using this and the map of the magnetic field, the mean field for the muon population is readily calculated. A conservatively assigned error of ±3 mm in radius implied an error of 160 ppm in the field.

This method of finding the muon radius has an elegant advantage: it uses the same electron data that are used for fitting the $(g-2)$ frequency. Muons at larger radii have less chance of sending an electron into the counters than muons on the inside of the ring; so there can be a bias. Here the same detectors are used for both measurements, so there is no bias. Further details, together with checks to ensure that the measurement at early times was representative of the muon population at later times when the $(g-2)$ precession was measured, are given in Ref. 23 and the review article Ref. 2.

To calculate a from ω_a using (5) one needs the value of λ. At that time the best measurement was the measurement by Hutchinson of μ^+ precession in water. The result[23] was

$$a = (116\ 616\ \pm 31) \times\ 10^{-8}\ (270\,\text{ppm}). \tag{9}$$

Initially, this was 1.7 standard deviations higher than the theoretical value, suggesting that there was more to be discovered about the muon. In fact the discrepancy came from a defect in the theory. Theorists had originally speculated that the contribution of the six $(\alpha/\pi)^3$ diagrams involving photon–photon scattering in the QED expansion[7] for a would be small, and perhaps these terms would cancel exactly; but they had never been computed. The experimental result stimulated Aldins, Kinoshita, Brodsky and Dufner[24] to make the calculation and they obtained the surprisingly large coefficient of 18.4! The theory then agreed with the measurement, to the great satisfaction of the experimental team,

$$a_{\text{exp}} - a_{\text{th}} = 240 \pm 270 \text{ ppm}. \tag{10}$$

The limit for the Feynman cutoff (1) was now $\Lambda > 5\,\text{GeV}$.

Time dilation in a circular orbit was spectacularly confirmed. After this there was no serious doubt about the twin (clock) paradox: it was an uncomfortable fact. The measured lifetime was just $1.2 \pm 0.2\%$ shorter than the expected value of $26.69\ \mu\text{s}$, probably due to imperfections in the magnetic field and a slow loss of muons. A more precise verification of the Einstein time dilation was obtained with the second muon storage ring.

5. Second Muon Storage Ring 1969–1976

The success of the muon storage ring and the apparent difference from theory justified a larger ring to achieve better accuracy. This project was master-minded by Emilio Picasso, aided by John Bailey. Higher energy would increase the muon lifetime and a larger aperture would improve the statistics. But there was a fundamental limitation: the magnetic gradient needed for vertical focusing was 50 ppm per millimetre and it would be impossible to locate the muons more precisely.

5.1. *Electric focusing*

After much discussion between Bailey, Farley and Picasso[25] it was decided to use a uniform magnetic field with no gradient and focus the particles vertically with an electric quadrupole field spread all around the ring. The vertical field focuses the particles while the horizontal component defocuses, slightly offsetting the semicircular focusing effect of the magnet. Overall it has the same effect as a magnetic gradient. A voltage of 10–20 kV would be required.

The horizontal electric field would bend the orbit; but in the muon rest frame it would transform to a vertical magnetic field, which would turn the spin. How would this affect the $(g-2)$ precession? Stray electric fields had been a major worry for the electron $(g - 2)$ measurement. The change in $(g - 2)$ frequency[16] for an electric field E is

$$\Delta f / f = (\beta - 1/a\beta\gamma^2)(E/B). \tag{11}$$

One observes that at a particular energy given by $\beta^2\gamma^2 = 1/a$, or equivalently when $\gamma = \sqrt{1 + 1/a}$, the electric field has no effect. This is the so-called "magic" energy[25] which is 3.1 GeV for muons. Here electric quadrupoles do not change the spin motion: one can use them with impunity.

The fourth miracle of Nature, mentioned above, is that the magic energy was conveniently accessible with the CERN PS and a reasonable step up from the previous storage ring. The muon lifetime was increased to 64 μs.

What about the spread in momentum? At the centre of the aperture the muons would have the magic energy exactly, but in any case the electric field there would be zero. At smaller radii the field would be inwards and the energy less than magic, the $(g - 2)$ frequency would be reduced. At larger radii both effects would be opposite, so the frequency again reduced. The frequency change would be parabolic with a maximum at the centre of the aperture: the average correction was only 1.7 ppm. The pitch correction for muons oscillating vertically was re-evaluated by Farley, Field and Fiorentini[26] and extended to focusing by electric fields.

5.2. *Electric quadrupoles and scraping*

It turned out that operating the electric quadrupoles in the strong magnetic field was not easy. The configuration is similar to a Penning gauge for measuring small pressures. Electrons are trapped and oscillate up and down, gradually increasing the ionisation of the residual gas. This happened in the ring and led to sparks, flashover, electric breakdown. The effect was worse when μ^- were studied. But Frank Krienen discovered that several milliseconds were required for the ionisation to build up, and we only needed the voltage for less than a millisecond while the muons were stored. By turning off the quadrupoles between fills the problem was solved.

Muon losses during the storage time can change the mean spin angle, if those that are lost started with a different spin angle from those that remain. This was not a serious error for the $(g - 2)$ measurement, but for the measurement of the lifetime it was essential to reduce the late-time muon losses to a minimum. This was done by shifting the muon orbits at early times both vertically and horizontally in order to "scrape off" the muons which passed near the edge of the aperture and were most likely to be lost.

The orbits were shifted by applying asymmetric voltages to opposite quadrupole plates at injection time, and then gradually bringing them back to normal. The result was that the aperture of the ring was reduced both vertically and horizontally during scraping, then gradually restored to normal with a time constant of about 60 turns, slow enough not to excite extra oscillations. The net result was to leave a clear space of a few millimetres around the stored muons. Any slow growth of oscillation amplitudes, would not cause muons to be lost.

A lost muon would hit something, lose energy and come out on the inside of the ring. A muon telescope sampled the lost muons. It was calibrated with no scraping when the losses were large enough to change the lifetime and then used to measure the losses when they were small.

5.3. *Ring magnet*

The major component of the new experiment was the 14 m diameter ring magnet. We needed to know the field on the muon orbit to a few ppm; but there was no way to measure it with NMR while the muons were there. One needed to stop the run, turn off the magnet, extract the vacuum chamber, then turn the magnet back on and survey the field. This process would have to be repeated many times. Guido Petrucci brilliantly designed a ring magnet that could be turned on and off and always came back to the same field.[27] This could only be achieved with some very special precautions, including:

- Temperature controlled room
- Independent temperature controlled concrete base with internal water pipes
- Coils not touching the iron, independently supported from the floor and able to deflect elastically to accommodate thermal expansion
- 40 separate iron yokes close to each other but not mechanically connected supporting quasi continuous poles
- 40 individual NMR probes with feedback loops to 40 compensating coils.

Usually the coils of a large magnet are strapped to the iron. The strong magnetic forces and thermal expansion makes the coils move, sliding and slipping whenever the magnet is turned on. Magnets always squeak and creak. The movement implies change: the field never repeats exactly. Petrucci's design avoided this. His ring made no noise. After a warmup period of two days, during which the field changed by about 5 ppm, the field averaged over the muon orbit reached a steady value, always the same to ±1 ppm.

The 40 pole pieces were touching but because of the gaps in the yokes the field was 400 ppm less at the junctions. This did not significantly perturb the orbits nor the measurement of the average field seen by the muons. With the field stabilised at 40 points no azimuthal harmonics could develop. Overall, this magnet was mechanically far more stable than the BNL ring built later with superconducting coils.

5.4. *Pion injection*

Instead of injecting protons which gave a large background, a beam of momentum selected pions was brought into the ring just outside the muon storage region. This required a pulsed inflector to kick them onto a tangential orbit. As the inflector was a closed concentric line, the leakage of the pulsed field into the muon storage region was very small. It was measured with pick up coils to compute a small correction.

The pions had slightly higher momentum and after half a turn they passed through the centre of the aperture. $\pi−\mu$ decay in this region launched the stored muons. They came from forward decay so the polarisation was high. With the pions matched to the ring acceptance, this gave many more muons, and the background in the counters was far less. Detectors for the decay electrons could be positioned all the way around the ring.

With zero magnetic gradient, the average value of magnetic field did not depend on the assumed radial distributions of muons. Even in extreme cases the average magnetic field was the same within less than 2 ppm, compared with the 160 ppm uncertainty in the previous experiment and the new statistical accuracy of ∼7 ppm. The $(g − 2)$ frequency was essentially independent of the distribution of muons within the storage region. However, an accurate value for the mean radius (and momentum) was needed for checking the Einstein time dilation (see below).

5.5. *Radial distribution*

As before, the radial distribution of the muons was obtained by analysing the pattern of counts at early times when the data is modulated by the rotating bunch. Now in Fig. 9 the rotation signal and the $(g − 2)$ modulation can be seen together!

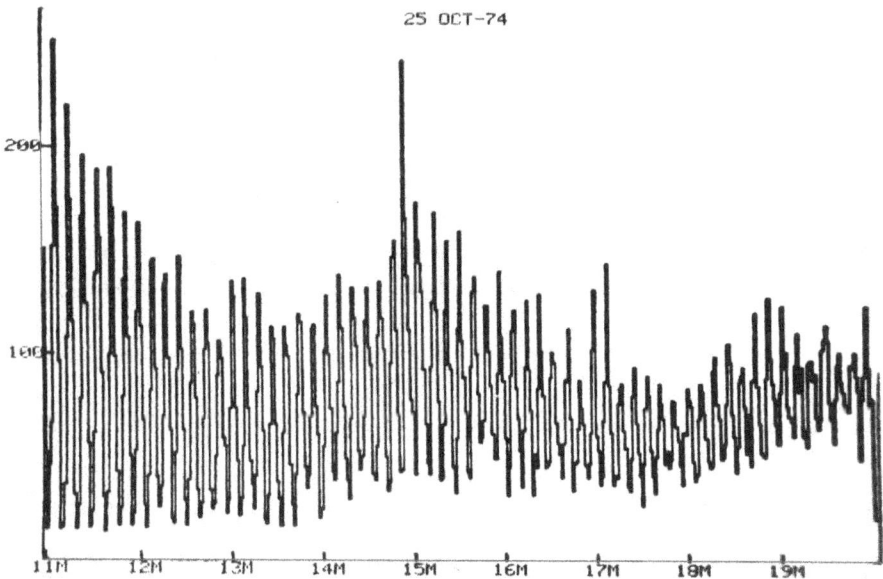

Fig. 9.   Counting rate vs. time (11 to 20 μs) showing both the rotation frequency and the $(g − 2)$ modulation, (online computer output for one run). The rotation signal dies away as the bunch spreads around the ring. The Fourier transform of the rotation data gives the radial distribution of the muons.

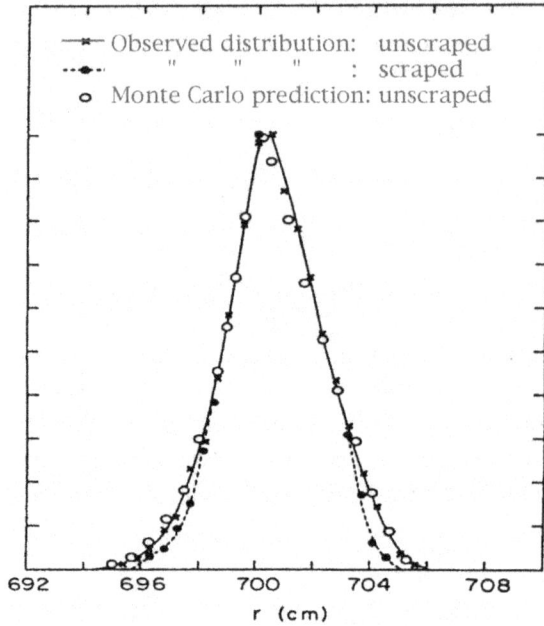

Fig. 10. Fourier transform of rotation data scraped (black dots) and unscraped (crosses), compared to prediction (open circles).

The computed radial distributions are in Fig. 10. The unscraped data agrees well with the prediction and the narrowing of the distribution by scraping is clearly seen. The mean rotation frequency ω_{rot} gives the relativistic γ factor:

$$\gamma = 2\lambda\omega_{\text{rot}}/g\omega_p \tag{12}$$

in which ω_p is the proton frequency corresponding to the magnetic field, $\lambda = \omega_s/\omega_p$ is known from mu precession at rest and muonium[17] and g is of course known from this experiment to better than 1 in 10^8. Equation (12) is used in checking the time dilation (see below).

The radial distribution is used to calculate the electric field correction (1.7 ppm) and pitch correction.[26] For $n = 0.135$, $v = 4$ cm, $r = 700$ cm, the pitch correction was 0.5 ppm. The statistical error in the mean radius was typically 0.1–0.2 mm.

5.6. *Results*

Figure 11 gives the combined decay electron counts versus storage time for the whole experiment, now showing the $(g - 2)$ precession out to $534\,\mu s$ with a strictly exponential decay. As the muon lifetime at rest is $2.2\,\mu s$ that was quite remarkable. A maximum likelihood fit was made to the data to obtain the $(g - 2)$ frequency ω_a.

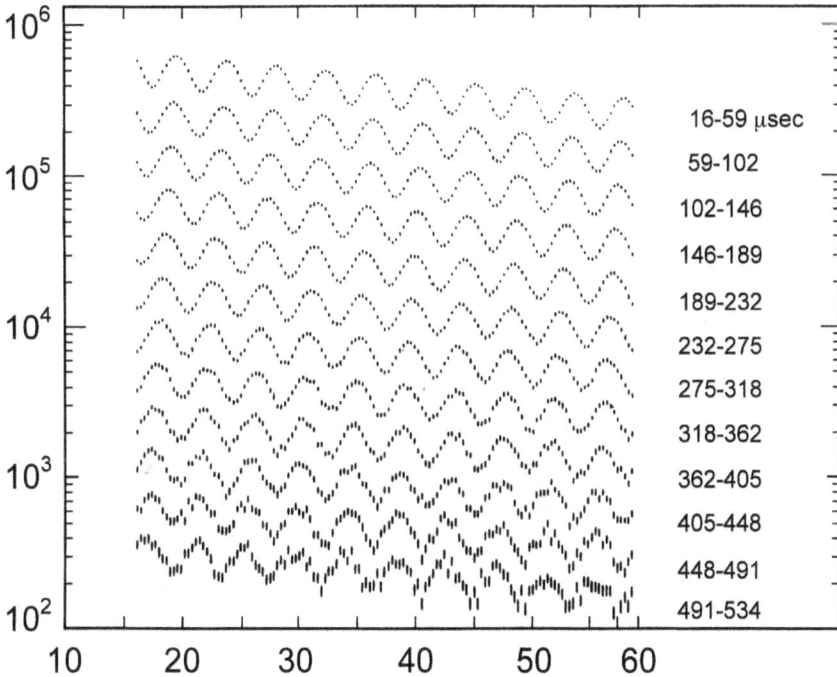

Fig. 11. Second Muon Storage Ring: decay electron counts versus time after injection. Range of time for each line is shown on the right (in microseconds).

Nine separate runs were made over a period of two years and fitted separately. As the field was determined in terms of the proton resonance frequency ω_p, the measurement of the $(g − 2)$ precession frequency ω_a is expressed as the ratio $R = \omega_a/\omega_p$. The nine R values, six for μ^+ and three for μ^- were consistent ($\chi^2 = 7.3$ for eight degrees of freedom). The overall mean value was the essential result of the experiment:

$$R = \omega_a/\omega_p = 3.707\,213\,(27) \times 10^{-3}(7\,\text{ppm}). \tag{13}$$

The error was 7.0 ppm statistical from ω_a plus 1.5 ppm from ω_p.

The corresponding value of the anomaly is given by Eq. (5) using the current result for λ.[17] The result is slightly different from that published in Ref. 28 because the value of λ has changed. Combining the data for μ^+ and μ^-,

$$a = 1\,165\,923\,(8.5) \times 10^{-9}\,(7\,\text{ppm}) \tag{14}$$

in agreement with the theory. The 95% confidence limit for the Feynman cutoff (1) was increased to $\Lambda = 23\,\text{GeV}$.

6.   Summary

In summary, the cyclotron measurement confirmed QED and established the muon as a heavy electron. The first storage ring discovered the $(\alpha/\pi)^3$ term in the QED expansion (scattering of light by light). The second verified the hadronic loops in the cloud of virtual particles around the muon, which contribute about 50 ppm to the anomalous moment.

In $(g-2)$ two worlds collide. The theorist is surrounded by esoteric concepts, wave functions, amplitudes, complex formulae many pages long. He evaluates endless integrals and after painstaking calculation comes up with a number. The experimenter deals with nanoseconds, huge magnets, racks of electronics, mazes of cables, and flashing lights. After years of effort he comes up with a number. These two worlds have nothing in common. And yet they agree on the same answer, accurate to parts per million. How is this possible? This is the deep enduring mystery of $(g-2)$.

7.   Tests of Relativity

7.1.   *Einstein's second postulate*

CERN's direct test of the second postulate of special relativity,[29] that the velocity of light is independent of the motion of the source is not widely known. Gamma rays from the PS target have been shown to come from the decay of π^0 in flight. Gammas of 6 GeV were selected with a lead glass Cherenkov counter. They must come from the forward decay of π^0 with energy at least 6 GeV, so the source velocity was greater than $0.99975c$. Would the velocity of these gammas be greater than normal?

Gamma ray time of flight is normally impossible, because they only interact once. But the PS beam is bunched in time by the RF driver, so the π^0 are bunched and the gamma rays also. Bunches of gammas are sweeping across the lab: they can be timed relative to the phase of the RF. When the detector is moved, the relative phase changes. If the displacement corresponds to one RF time period, then the relative phase should again be the same. This provides a sensitive test that the gamma ray velocity is the same as the velocity of light, independent of the calibration of the timing circuits.

The data are shown in Fig. 12. Position B for the detector is one RF wavelength further away from the accelerator than position A, and the timing curves look the same. The velocity of gammas from the moving source was found to be the same as the standard velocity of light to 1 part in 10^4. This confirms the second postulate to high accuracy at very high velocities. It is also the best measurement of the velocity of any gamma rays.

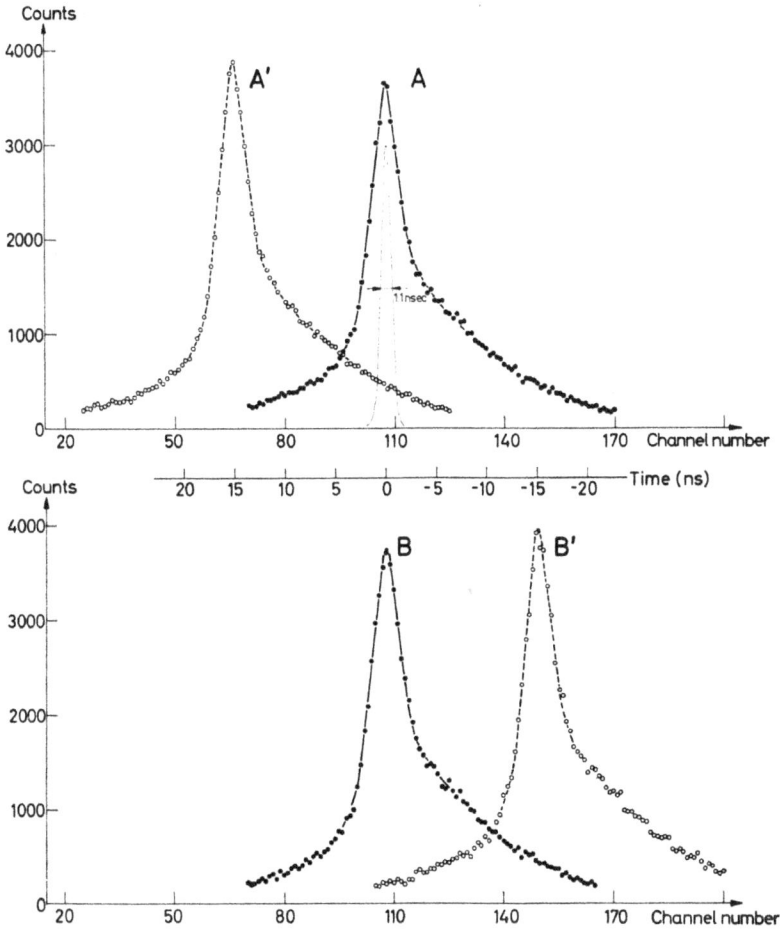

Fig. 12. Gamma arrival time vs PS radiofrequency. Positions A and B are one RF wavelength apart. A′ and B′ are offset by 4.5 m: the peaks move correspondingly 15 ns. Comparing A with B gives the velocity of gammas from the moving π^0.

7.2. *Muon lifetime in flight*

The muon lifetime in a circular orbit is a stringent test of relativity. It can also measure the life of μ^- which cannot be measured at rest and therefore tests the CPT invariance of the weak interaction.

The twin paradox was discussed in Einstein's first paper.[30] It is a paradox because, if only relative motion is important, one can ask which twin moves and which remains at rest? The difference is that to return to the same point, one twin must have suffered some acceleration which the other (older) twin did not. It seems that, according to relativity, the one with a history of acceleration finishes younger than the sessile partner; a result which is hard for the human mind to grasp, though people driving fast sports cars do seem to be younger than the average.

Time dilation was established by the first muon storage ring. With the second we measured it accurately.

The scraping system described above minimised the losses. A correction was made for the residual loss rate ($\sim 0.1\%$ per lifetime) measured with the calibrated loss detector. The rotation frequency gave the radial distribution as shown in Fig. 10 and Eq. (12) gave the mean value of $\gamma = 29.327\,(4)$. Multiplying by the lifetime[31] at rest $2.19711\,(8)$ μs gave a predicted lifetime of $64.435\,(9)$ μs to be compared to the experimental value $64.378\,(26)$ μs. So the Einstein time dilation was verified to 0.9 ± 0.4 parts per thousand. Further details are given in Bailey *et al.*[32]

This is the best reported measurement of time dilation in a circular orbit. The lifetime of negative muons was the same as μ^+.

In Memoriam

This review is dedicated in warm appreciation to the memory of my esteemed colleagues Emilio Picasso, Simon van der Meer and Frank Krienen.

References

1. F. J. M. Farley Y. K. Semertzidis, *Prog. Part. Nucl. Phys.* **52**, 1 (2004).
   F. J. M. Farley and E. Picasso, *Annu. Rev. Nucl. Part. Sci.* **29**, 243 (1979).
   F. Combley, F. J. M. Farley and E. Picasso, *Phys. Rep.* **68**, 2 (1981).
   F. J. M. Farley, *Cargese Lectures in Physics* Vol 2, ed. M. Levy (1968), pp. 55–117.
2. F. J. M. Farley and E. Picasso, "The muon (g-2) experiments" in *Quantum Electrodynamics*, ed. T. Kinoshita (World Scientific, 1990), pp. 479–559.
3. S. Goudsmit and G. Uhlenbeck, *Zeit. Phys.* **35**, 618 (1926).
4. P. A. M. Dirac, *Proc. Roy. Soc.* **117**, 610 (1928); *ibid.* **118**, 351 (1928).
5. P. Kush and H. M. Foley, *Phys. Rev.* **72**, 1256 (1947); *ibid* **74**, 250 (1948).
6. J. S. Schwinger, *Phys. Rev.* **73**, 416 (1948); **76**, 790 (1949).
7. T. Kinoshita, "Theory of the Anomalous Magnetic Moment of the Electron" in *Quantum Electrodynamics*, ed. T. Kinoshita (World Scientific, 1990), pp. 218–321.
8. G. W. Bennett *et al.*, *Phys. Rev. D* **73**, 072003 (2006).
9. A. Petermann, *Helv. Phys. Acta* **30** 407 (1957); *Phys. Rev.* **105**, 1931 (1957).
10. B. E. Lautrup, A. Petermann and E. de Rafael, *Phys. Rep.* **3C**, N4 (1972).
11. W. H. Louisell, R. W. Pidd and H. R. Crane, *Phys. Rev.* **91**, 475 (1953).
    A. A. Schupp, R. W. Pidd and H. R. Crane, *Phys. Rev.* **121**, 1 (1961).
12. V. B. Berestetskii, O. N. Krokhin and A. X. Klebnikov, *Zh. Eksp. Teor. Fiz.* **30**, 788 (1956), Transl. *JETP* **3**, 761 (1956).
    W. S. Cowland, *Nucl. Phys* **8**, 397 (1958).

13. R. L. Garwin, L. Lederman and M. Weinrich, *Phys. Rev.* **105**, 1415 (1957).
    J. I. Friedman and V. L. Telegdi, *Phys. Rev.* **105**, 1681 (1957).
14. W. K. H. Panofsky, in *Proc. 8th Int. Conf. on High Energy Physics*, CERN, Geneva ed. B. Ferretti (1958), p. 3.
15. H. A. Tolhoek and S. R. DeGroot, *Physica* **17**, 17 (1951); H. Mendlowitz and K. M. Case, *Phys. Rev.* **97**, 33 (1955).
    M. Carrassi, *Nuovo Cimento* **7**, 524 (1958).
16. V. Bargmann, L. Michel and V. L. Telegdi, *Phys. Rev. Lett.* **2**, 435 (1959).
17. W. Liu *et al.*, *Phys. Rev. Lett.* **82**, 711 (1999).
    D. E. Groom *et al.*, *Eur. Phys. J. C* **15**, 1 (2000).
18. F. J. M. Farley, *Computation of particle trajectories in the CERN cyclotron*, CERN yellow report 59–12 (1959).
19. G. Charpak, F. J. M. Farley, R. L. Garwin, T. Muller, J. C. Sens, V. L. Telegdi and A. Zichichi, *Phys. Rev. Lett.* **6**, 128 (1961).
    G. Charpak, F. J. M. Fariey, R. L. Garwin, T. Muller, J. C. Sens and A. Zichichi, *Phys. Lett.* **1**, 16 (1962).
    G. Charpak, F. J. M. Farley, R. L. Garwin, T. Muller, J. C. Sens and A. Zichichi, *Nuovo Cimento* **37**, 1241 (1965).
20. F. J. M. Farley, *Proposed high precision (g − 2) experiment*, CERN Intern. Rep. NP 4733 (1962).
21. H. Dingle, *Special theory of relativity*, Methuen (1940); *Science at the Crossroads* (Martin Brian & O'Keeffe, London, 1972).
22. F. J. M. Farley, *Nucl. Inst. Methods* **28**, 279 (1964).
23. F. J. M. Farley, J. Bailey, R. C. A. Brown, M. Giesch, H. Jöstlein, S. van der Meer, E. Picasso and M. Tannenbaum, *Nuovo Cimento* **45**, 281 (1966).
    J. Bailey, G. von Bochmann, R. C. A. Brown, F. J. M. Farley, H. Jöstlein, E. Picasso and R. W. Williams, *Phys. Lett. B* **28**, 287 (1968).
    J. Bailey, W. Bartl, G. von Bochmann, R. C. A. Brown, F. J. M. Farley, M. Giesch, H. Jöstlein, S. van der Meer, E. Picasso and R. W. Williams, *Nuovo Cimento A* **9**, 369 (1972).
24. J. Aldins, T. Kinoshita, S. J. Brodsky and A. J. Dufner, *Phys. Rev. Lett.* **23**, 441 (1969).
    J. Aldins, S. J. Brodsky, A. J. Dufner and T. Kinoshita, *Phys. Rev. D* **1**, 2378 (1970).
25. See F. J. M. Farley, *G minus TWO plus Emilio*, Colloquium in honour of Emilio Picasso (1992), CERN-OPEN-2002-006.
26. F. J. M. Farley, *Phys. Lett.*, **42B**, 66 (1972).
    J. H. Field and G. Fiorentini, *Nuovo Cimento A* **21**, 297 (1074).
27. H. Drumm, C. Eck, G. Petrucci and O. Rúnolfsson, *Nucl, Inst. Methods* **158**, 347 (1979).
28. J. Bailey, K. Borer, F. Combley, H. Drumm, C. Eck, F. J. M. Farley, J. H. Field, W. Flegel, P. M. Hattersley, F. Krienen, F. Lange, G. Lebee, E. McMillan, G. Petrucci, E. Picasso, O. Rúnolfsson, W. von Rüden, R. W. Williams and S. Wojcicki, *Nucl. Phys. B* **150**, 1 (1979).
29. T. Alväger, J. M. Bailey, F. J. M. Farley, J. Kjellman and I. Wallin, *Phys. Lett.* **12**, 260 (1964); *Arkiv för Fysik* **31**, 145 (1966).
30. A. Einstein, *Ann. der Phys.* **17**, 891 (1905).

31. M. P. Balandin, V. M. Grebenyuk, V. G. Zinov, A. D. Konin and A. N. Ponomarev, *J. Exp. Theor. Phys.* **40**, 811 (1974).

32. J. Bailey, K. Borer, F. Combley, H. Drumm, F. J. M. Farley, J. H. Field, W. Flegel, P. M. Hattersley, F. Krienen, F. Lange, E. Picasso and W. von Rüden, *Nature* **268**, 301 (1977); see also F. Combley, F. J. M. Farley, J. H. Field, and E. Picasso, *Phys. Rev. Lett.* **42**, 1383 (1979).

The Discoveries of Rare Pion Decays
at the CERN Synchrocyclotron

Giuseppe Fidecaro

CERN, CH-1211 Geneva 23, Switzerland

giuseppe.fidecaro@cern.ch

In 1957 the CERN 600 MeV Synchrocyclotron started to operate and could detect for the first time already in 1958, and later in 1962, the two still missing β-decays of the charged pion, providing crucial verifications of the universal V–A coupling.

1. Introduction

In October 1955, one year after its birth, CERN was exclusively a building site, where the foundations of the two CERN accelerators, the 25 GeV Proton Synchrotron (PS) and the 600 MeV Synchrocyclotron (SC) were laid down and the first buildings were still under construction. In the middle of 1956 the construction of the synchrocyclotron was fairly advanced and experimental physicists started gathering in the SC Division. In 1957, when the CERN synchrocyclotron started accelerating protons, and the first buildings were ready to offer office space to physicists still living in the wooden barracks at the Geneva airport site, the SC Division was already an active and lively point of attraction for people from Europe and the U.S. interested in working at the SC, and for numerous visiting physicists from other countries in the World.

Benefitting from experience at other laboratories, physicists and engineers at the SC began developing and building the first elements of experimental equipment, such as scintillation counters, all kinds of electronic circuits, and even a "synchrocyclotron simulator" to test electronic circuits in conditions similar to those encountered at the SC.[1] A very helpful electronic instrument was a pulse generator based on a mercury switch relay driven by the local A.C. 50 Hz network, widely used in U.S. laboratories but unknown in Europe. Built by Oreste Piccioni, a visiting professor at CERN from Brookhaven in 1957, this pulse generator turned out to be invaluable for nanosecond work.

In June 1957 the engineers were getting ready to start acceleration. With the agreement of the SC engineers, in collaboration with Tito Fazzini, I installed a scintillation counter in the synchrocyclotron hall on July 16 to detect background radiation, and a rate meter with a pen recorder in the nearby experimental hall. A steady counting rate was obtained on August 1, 1957, the official date of the first CERN SC operation. The event was recorded in a sheet signed by Wolfgang Gentner, Head of the Division, and by all people present including Tito Fazzini, Alec Merrison and myself. We were the first physicists to see particles accelerated at CERN. A number of very interesting results were obtained in the following years, among which the two most important ones are reported in this article.

2.   Universal Fermi Interaction and Pion Decay: Two Parallel Tales

2.1.   *The weak interaction before the π-meson discovery*

After Fermi proposed in 1934 his β-decay theory,[2] which involved the direct emission of a pair of light particles, an electron and a neutrino, Yukawa put forward in 1935 his theory[3] of exchange nuclear forces induced by emission and absorption of an intermediate particle, charged or neutral, that Yukawa named meson. There was, however, a disagreement between Fermi's and Yukawa's theories. In Yukawa's theory the electron–neutrino pair originated from the decay of the intermediate meson, while Fermi's theory was based on the direct emission of an electron–neutrino pair, with no intermediate particle at all.

In 1936, the discovery in cosmic rays of a charged particle having the mass predicted by Yukawa, between that of the electron and that of the proton, brought full support to his theory. The new cosmic-ray particle, named mesotron, was identified with Yukawa's meson, and the mesotron interaction became known as Yukawa interaction. The mesotron was recognised as the strong interaction carrier.

In 1940 it was known, as suggested by Møller, Rosenfeld and Rozenthal,[4] that there were two kinds of cosmic-ray mesons, one with a mean life of the order of 10^{-8} s, the other with a mean life of the order of 10^{-6} s. In the same year Sakata,[5] in an attempt to understand the mean lives of these mesons, described the meson decay as a compound of the Yukawa interaction $N \leftrightarrow P + \Pi^-$ with the original Fermi β-decay interaction $N \leftrightarrow P + e^- + \bar{\nu}$, namely $\Pi^- \rightarrow \bar{P} + N \rightarrow e^- + \bar{\nu}$, and the charge-conjugate reaction for Π^+. Here the Fermi notation Π has been used for the Yukawa meson[6] (the π-meson had not been discovered yet).

While discussions on the Fermi's and Yukawa's theories were going on, a wealth of information was being accumulated from experimental work on nuclear β-decay and cosmic rays. In 1947, Conversi, Pancini and Piccioni[7] discovered in Rome that negative mesotrons in the cosmic radiation coming to rest in carbon were not captured by nuclei, but they decayed into an electron and a neutral particle, either a neutrino or a γ-ray, just like positive mesotrons do.

In the same year, Pontecorvo[8] was the first to notice that the capture rate of a bound negative meson ($\sim 10^6$ s^{-1}) is of the order of the rate of ordinary K-capture processes, when allowance is made for the difference in the disintegration energy and the difference in the volumes of the K-shell and of the meson orbit. Thus he called attention to the possible equality of the coupling constants of electrons and mesons (or mesotrons) to nucleons, and essentially laid down the first two sides of the Puppi triangle,[9] a graphic representation of the universal weak interaction invented by J. Tiomno.

In the following years, Pontecorvo's idea developed through the work of other authors (Clementel–Puppi, O. Klein, Lee–Rosenbluth–Yang, Leite Lopes, Marty–Prentki, Puppi, Tiomno–Wheeler) into the more general idea of a Universal Fermi Interaction. Namely, the various weak processes are "universal" in the sense that they are different manifestations of a single fundamental interaction (Sakurai, 1964). The name "Universal Fermi Interaction" was coined in 1950 by Yang and Tiomno.[10]

2.2.   The weak interaction after the π-meson discovery

The true π-meson had been discovered in cosmic rays at Bristol in 1947[11] in nuclear emulsions and detected at the Berkeley synchrocyclotron in the following year by Gardner and Lattes.[12] Two examples of π-meson β-decay had been published by the Bristol group, but ... surprise! the decay products were not electrons but a new type of meson that was named μ. Electrons were neither found in π-decay at the Berkeley synchrocyclotron by Bishop, Burfening, Gardner and Lattes,[13] nor by Lattes,[14] in nuclear emulsion experiments.

The absence of electron decays became a mystery, since among physicists the idea was developing that the Fermi coupling constants between Dirac particles might all be equal. For example, in the case of the following three interactions discussed by Fermi in his lectures at Yale in April 1950:[6]

$$N \rightarrow P + e + \bar{\nu} \quad (\beta \text{ decay}), \tag{1}$$

$$\mu \rightarrow e + \nu + \bar{\nu} \quad (\mu \text{ decay}), \tag{2}$$

$$P + \mu^- \rightarrow N + \nu \quad (\mu^- \text{ capture}). \tag{3}$$

Tiomno and Wheeler,[15] Lee, Rosenbluth and Yang[16] and other authors had found close equality of the coupling constants, though the errors were not negligible.

However, because the decay π–eν had never been observed the equality of the coupling constants of the above interactions (1), (2), (3) was not sufficient proof of existence of a Universal Fermi Interaction.

In 1949, Ruderman and Finkelstein[17] computed the ratio of the π–eν to the π–$\mu\nu$ decay rate for various types of mesons and Fermi couplings and pointed out that, while no conclusion could be drawn on absolute rates, this ratio was independent of divergent integrals and was only a function of the pion, electron and muon masses:

$$R = \frac{\Gamma(\pi \rightarrow e\nu)}{\Gamma(\pi \rightarrow \mu\nu)} = \left(\frac{M_\pi^2 - M_e^2}{M_\pi^2 - M_\mu^2}\right)^2 \times \frac{M_e^2}{M_\mu^2} = 1.28 \times 10^{-4} \quad \text{(pseudovector coupling)},$$

$$R = \frac{\Gamma(\pi \rightarrow e\nu)}{\Gamma(\pi \rightarrow \mu\nu)} = \left(\frac{M_\pi^2 - M_e^2}{M_\pi^2 - M_\mu^2}\right)^2 = 5.49 \quad \text{(pseudoscalar coupling)}.$$

In the case of vector, scalar and tensor interaction both π–$\mu\nu$ and π–eν decays are forbidden.

In 1949, Steinberger[18] also made an attempt to compute the decay rates for various types of mesons and couplings using a subtraction method proposed by Pauli and Villars to deal with divergences. If one calculates R from the decay rates so obtained, one gets the same result as Ruderman and Finkelstein.

Anyhow, a definite prediction for the value of R existed and, according to Ruderman and Finkelstein, any theory which coupled π-mesons to nucleons also predicted π–eν decay. However, this decay had not been observed. Was it simply because it was a rare event in comparison with π–$\mu\nu$ decay?

On the one side, it was important to start a systematic search for π–eν decay. On the other side, in the years around late 1940s and early 1950s, Scalar (S) and Tensor (T) couplings were favoured to describe the weak interaction responsible for nuclear β-decay. In such a case, according to Wu and Muszowski,[19] the absence of π–eν decay did not concern physicists too much, because only two of the five possible couplings, namely A and P, can be formed out of the pseudoscalar pion field and one four-vector, representing the non-local nature of the intermediate state. Therefore, A and P couplings are the only ones that can induce the decay of pseudoscalar pion; the other three (S, V, T) are forbidden. If both A and P are lacking in π-meson decay, then π–eν is naturally forbidden.

Frota-Pessôa and Margem[20] were the first to search for π–eν decay in emulsions exposed at Berkeley and found none out of 200 π–$\mu\nu$ decays ($R < 0.5 \times 10^{-2}$). Then, in 1951 Smith[21] analysed emulsions that was also exposed at Berkeley and found $R = (0.3 \pm 0.4) \times 10^{-2}$ ("... less than 1% and probably zero ...").

The most sensitive emulsion experiment was done by Friedman and Rainwater who published their final results in 1951.[22] In the nuclear emulsions exposed at the Nevis synchrocyclotron, they found " ... one or zero π–e events compared to 1419 π–μ events ... ($R \leq 7 \times 10^{-4}$)".

After the experiment by Friedman and Rainwater, it became clear that the limit on R could only be lowered by counter experiments. The first experiment of this type was performed by Lokanathan and Steinberger[23] in 1954 using the apparatus shown in Fig. 1. A π^+ beam from the Nevis synchrocyclotron was brought to rest in a thin CH_2 target and decay positrons were detected by a scintillation counter hodoscope, which contained a variable thickness of CH_2 absorber. Most of the data were taken with an absorber of 23 cm in thickness, which corresponds to an energy loss of 55 MeV from ionisation alone, above the end-point of $\mu^+ \rightarrow e^+\nu\bar{\nu}$ decay. However, some of these positrons could still traverse the telescope through the conversion of their bremsstrahlung. As a consequence, even with a 23-cm thick absorber the telescope could see not only the $\sim$70 MeV positrons from $\pi^+ \rightarrow e^+\nu$ decay, but also the tail of the much more frequent positrons from $\mu^+ \rightarrow e^+\nu\bar{\nu}$ decay. They

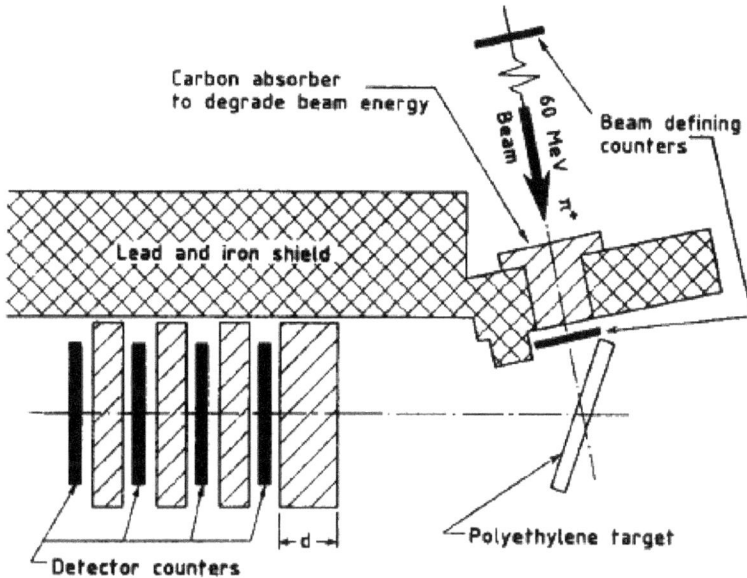

Fig. 1. Arrangement of counters and absorbers in the experiment of Lokanathan and Steinberger.[23]

measured $R = (-0.3 \pm 0.9) \times 10^{-4}$ and concluded: "It seems therefore improbable that the pion is coupled symmetrically to the muon."

I became interested in π–$e\nu$ decay in the Summer 1954 at the Varenna Physics School in Varenna, Lake Como, Italy, where I attended a lecture by Steinberger on the first results from his experiment. This interest became stronger three years later, after attending a seminar at CERN by Herbert Anderson on June 12, 1957 on his search for π–$e\nu$ decay together with Lattes. That experiment, published on December 1, 1957, ten years after the discovery of the pion,[24] produced the most striking negative result. They used a magnetic spectrometer to measure the momentum distribution of electrons from stopped π^+-mesons (see Fig. 2). This technique was expected to provide a much higher rejection against electrons from μ^+ decay, thus being sensitive to $\pi^+ \rightarrow e^+\nu$ decays with R values well below 10^{-4}. The spectrometer was calibrated with 5.15 MeV α-particles which have the same curvature as 98 MeV positrons. Once again, no evidence for $\pi^+ \rightarrow e^+\nu$ was found, giving $R = (-4.0 \pm 9.0) \times 10^{-6}$. The authors concluded: "This appears to be statistically significant and thereby allows only a 1% probability that R could be greater than 2.1×10^{-5}."

After hearing the Chicago result, I had more discussions on this subject with Fazzini and Merrison, who were also interested in the subject, but we considered the possibility of doing an experiment rather remote at that time.

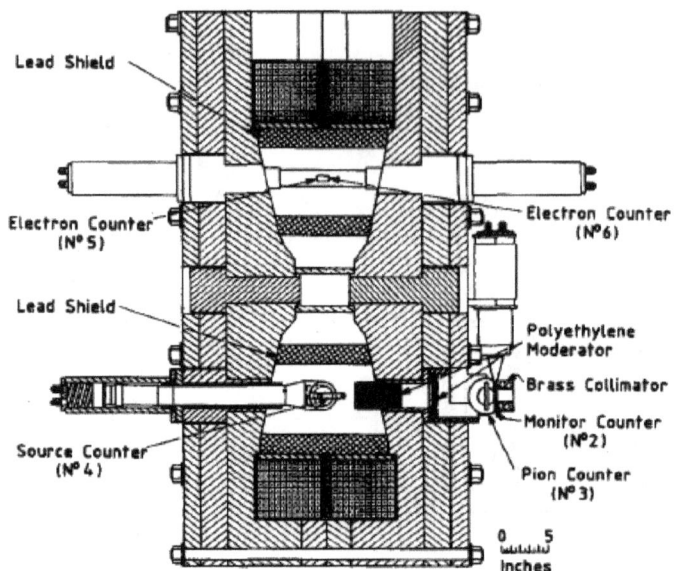

(a)

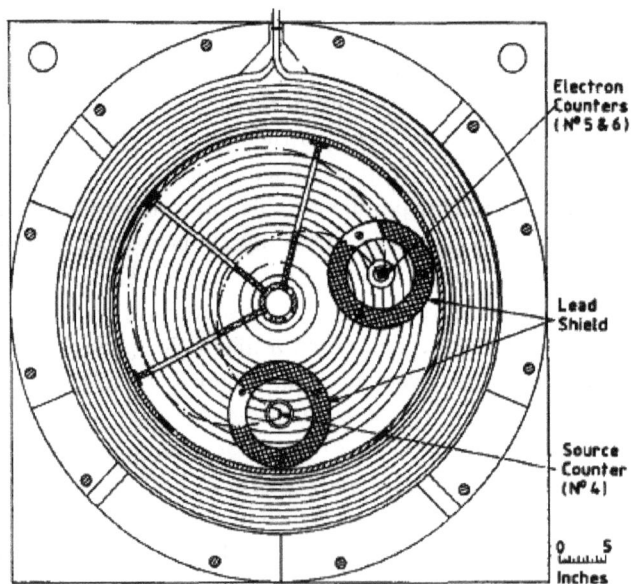

(b)

Fig. 2. Sectional view of the Chicago spectrometer.[24] (a) Section along the incident beam direction. Pions are injected from the right parallel to the magnetic field direction through the brass collimator and come to rest in Counter 4. (b) Section through the median plane normal to the magnetic field direction.

In the second half of 1957, three satisfactory formulations of the Universal Fermi Interaction were proposed almost at the same time:

- On July 16, Feynman and Gell-Mann submitted their famous paper "Theory of the Fermi interaction", proposing a universal V–A form.[25] In this paper one could read "Experimentally no $\pi-e\nu$ have been found, indicating that the ratio (to $\pi-\mu\nu$) is less than 10^{-5}. This is a very serious discrepancy. The authors have no idea on how it can be resolved." And then at the end of the paper they concluded: "These theoretical arguments seem to the authors to be strong enough to suggest that the disagreement with the ^{6}He recoil experiment and with some other less accurate experiments indicates that these experiments are wrong. The $\pi-e\nu$ problem may have a more subtle solution."
- On September 22–28, the International Conference on Mesons and Recently Discovered Particles took place in Padova and Venice. At this conference, Sudarshan and Marshak presented the paper "The Nature of the Four-Fermion Interaction"[26] in which they expressed doubts on the validity of the results from some experiments, on the searches for $\pi-e\nu$ decay, in particular.
- On October 31, Sakurai submitted a paper "Mass Reversal and Weak Interactions"[27] along the same lines.

Sudarshan and Marshak stressed that although a mixture of vector and axial vector was the only universal four-fermion interaction that was possible and at the same time possessed many elegant features, it appeared that several published and unpublished experiments could not be reconciled with that hypothesis. They listed four experiments that had to be redone. Should any of the four experiments be confirmed, it would be necessary to abandon the hypothesis of a universal V–A four-fermion interaction or at least one of the assumptions of a two-component neutrino and the conservation of leptons.

However, the suggestion of redoing the quoted experiments did not go too far because the Proceedings of the Padova–Venice Conference were only published in mid-1958 due to technical difficulties. Also, despite our interest in pion experiments, all of us (Fazzini, Merrison and myself) missed Sudarshan's talk in Venice because it was delivered in a session devoted to strange particles chaired by Heisenberg. We only knew of Sudarshan's suggestions several years later.

Nevertheless, the interest raised by the above papers called the attention to the experimental difficulties. The ^{6}He recoil experiment, namely the measurement of the electron-neutrino angular correlation in the decay ^{6}He $\rightarrow^6$ Li $+ e^- + \bar{\nu}$,[28] which required Tensor (T) coupling, appeared as a very serious difficulty. Even more serious appeared the absence of $\pi-e\nu$ decay. There were perhaps already encouraging signals in the air that the ^{6}He recoil experiment had to be redone anyhow, as it looked like from a post-deadline paper presented at the New York meeting of the American Physical Society (APS) in January 1958, but that was certainly not the case for the absence of the $\pi-e\nu$ decay.

These problems were solved in 1958 by the first experiment performed at CERN with particles from a CERN accelerator. It was a search for $\pi^+ \to e^+\nu$ decay.[29] That experiment, which "put CERN on the map of high energy physics",[30] started the great CERN tradition of experimental work in weak interaction physics. It is remembered as a European success, but also as a success of the 600 MeV CERN SC, a machine rightly conceived to start research in Europe as early as possible before the 26 GeV proton synchrotron was ready.

3.　π-Meson Decay to Electron and Neutrino: A CERN Discovery

The paper by Feynman and Gell-Mann[25] proposing a universal V–A form for the weak interaction was published on January 1, 1958. Unaware of that paper, I was attending the January 1958 meeting of the APS in New York, where I heard about the V–A theory directly from Feynman who was giving an invited talk at that meeting. In the first part of that talk Feynman brilliantly described the V–A theory and its successes. Then, in the second part he presented some ideas on how to strongly suppress $\pi \to e\nu$ with respect to $\pi \to \mu\nu$ decay, probably an anticipation of the invited talk "Forbidding of π–β decay" that he gave in the Summer of the same year at the International Conference on High-Energy Physics at CERN.[31] In his talk Feynman was arguing that $\pi \to e\nu$ decay might be strongly suppressed by the effect of large radiative corrections canceling the leading-order weak decay amplitude.

Feynman's arguments were not convincing, because radiative corrections were known to produce only effects at the few percent level in other processes. It was on that occasion that I decided — rightly or wrongly — to start a search for $\pi^+ \to e^+\nu$ decay at the CERN SC with a detector as simple as possible, without a spectrometer magnet, by stopping π^+-mesons in a scintillation counter and displaying the signals from this counter on the screen of an oscilloscope. This electronic system could be applied to an arrangement of counters and absorbers similar to that used by Lokanathan and Steinberger,[23] where, however, π^+-mesons were stopped in an inert CH_2 target, a crucial difference between the two experiments.

Such an experiment could be set up fairly quickly and also produce results quickly, but there would be a risk of failure should the $\pi \to e\nu$ decay not exist at all. Or else it might not be approved because of the absence of a magnetic spectrometer. The general belief that the $\pi \to e\nu$ decay either did not exist at all or its rate was much smaller than the theoretical prediction was so strong that very few people in a position of responsibility would have approved such an experiment (no committee would approve today an experimental search which, at least on paper, has a lower sensitivity than previous experiments that gave a null result).

In previous experiments, separation between electrons from $\pi^+ \to e^+\nu$ decays and from the $\pi^+ \to \mu^+ \to e^+$ decay chain had been achieved by measuring the decay electron energy, either by total absorption counters or by magnetic deflection. The main background of electrons from the $\pi^+ \to \mu^+ \to e^+$ decay chain at

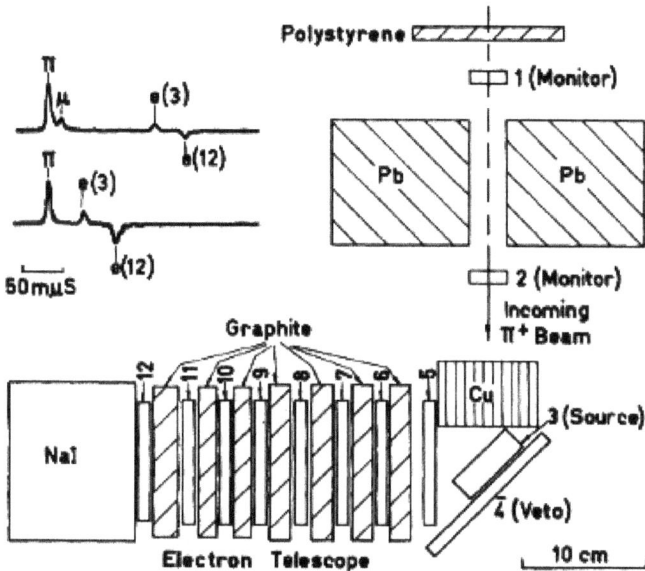

Fig. 3. Layout of the SC experiment[29] together with typical $\pi^+ \to \mu^+ \to e^+$ and $\pi^+ \to e^+$ signals, as recorded on a fast oscilloscope (the time scale unit, "milli-micro-second" (mμs) is called "nanosecond" (ns) today). Counter 3 is the active target where incident π^+ mesons stop. The NaI counter information was not used in the final analysis.

rest has a maximum energy of ~53 MeV while electrons from $\pi^+ \to e^+\nu$ decay have an energy of ~70 MeV. In the experiment of Lokanathan and Steinberger,[25] the energy measurement method was adopted using a variable thickness counter telescope.

The SC experiment was performed in 1958 by Fazzini, Fidecaro, Merrison, Paul and Tollestrup.[29] Work to set up the detector started in February, its construction was completed in May and the first run with beam took place on June 23.

The detector layout is shown in Fig. 3. Although the arrangement of counters and absorbers was similar to that of the experiment by Lokanathan and Steinberger[23] (Fig. 1), the methods to recognise $\pi^+ \to e^+\nu$ decays from the $\pi^+ \to \mu^+ \to e^+$ decay chain in the SC experiment were completely different because, as mentioned earlier, in the SC experiment the π^+-mesons were stopped in an active target, namely in a plastic scintillator optically coupled to a photomultiplier.

The decay electron following a π^+ stop, no matter from π^+ or μ^+ decay, opened a gate whose length was equal to the length of the oscilloscope trace. The gate opened the door to the π^+ stop signal (properly delayed) that in turn started the oscilloscope trace. The latter was always started at the same time with respect to the π^+ stop signal, which, however, was shifted forward by an appropriate delay, so as to allow the inspection of the trace also before the arrival of the primary π^+. Figure 3 displays two typical traces. The upper one corresponds to a $\pi^+ \to \mu^+ \to e^+$ decay event, with the second signal after the π^+ stop associated with the muon from

$\pi^+ \to \mu^+$ decay, and the third one, labelled e(3) associated with the positron from $\mu^+ \to e^+$ decay. In the lower trace, there are only two signals from Counter 3, with no intermediate μ^+ signal between the two, as expected from a $\pi^+ \to e^+$ decay event. In both traces, the signal labelled e(12) is obtained from Counter 12, after a suitable delay, and is present on the trace together with signal e(3) only if the electron has traversed all the graphite absorbers and reached Counter 12.

The trace length covered a few π^+ lifetimes after the π^+ stop signal but only a fraction of the μ^+ lifetime (the π^+ lifetime is $\sim$26 ns, while the μ^+ lifetime is $\sim$80 times longer). Under normal running conditions, the rate at which the oscilloscope trace started was only a few per hour.

In the experiment of Lokanathan and Steinberger,[23] the signal from the incoming π^+ could still be used, but the presence of the intermediate μ^+ in the $\pi^+ \to \mu^+ \to e^+$ decay chain could not be detected because the μ^+ remained invisible in the inert CH_2 target (the μ^+ range from $\pi^+ \to \mu^+$ decay at rest is less than 1 mm in CH_2).

Figure 4 is a photograph of the target region of the SC experiment, while Fig. 5 shows the fast oscilloscope and associated camera that were used to record the signals from the scintillator where the incident π^+-mesons were stopped. Finally, the racks containing the electronics in the counting room of the of the SC experiment are shown in Fig. 6.

In the choice of the method used in the SC experiment[29] to distinguish $\pi^+ \to e^+\nu$ decays from the $\pi^+ \to \mu^+ \to e^+$ decay chain, I was influenced by the single photomultiplier experiment of Janes and Kraushaar[32] performed in 1953 at the M.I.T. 300 MeV electron synchrotron to measure the photoproduction cross-section

Fig. 4. Counter arrangement in the target region of the SC experiment.[29]

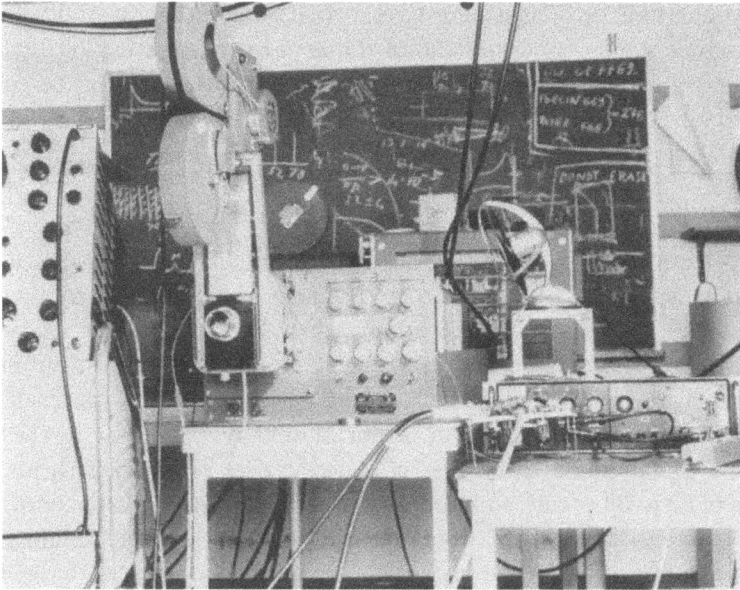

Fig. 5. The travelling-wave fast oscilloscope equipped with a photographic camera used in the SC experiment to record the target signals.[29] On the blackboard one can see hand-written notes from a discussion on the first results, including a preliminary lower limit on R, suggesting that this photograph was taken at the end of August 1958.

Fig. 6. The main electronic racks in the counting room of the SC experiment.[29]

of π^+-mesons from hydrogen and carbon at $90°$, down to 10 MeV. To identify the π^+-mesons against a background of stable particles, they exploited the unique property of π^+-decays at rest in a scintillator by measuring the two consecutive signals from $\pi^+ \rightarrow \mu^+$ decay on the trace of a fast oscilloscope.

3.1. Results

Figure 7 shows the first results of this experiment in which the electron rate is presented as a function of the absorber thickness for both $\pi^+ \rightarrow \mu^+ \rightarrow e^+$ events and $\pi^+ \rightarrow e^+$ candidates. The contamination of false $\pi^+ \rightarrow e^+\nu$ decays, i.e., $\pi^+ \rightarrow \mu^+ \rightarrow e^+$ events with the μ^+ signal in the target too near in time to the π^+ signal to be resolved from it, was directly measured with small absorber thickness, where the electron rate is dominated by $\pi^+ \rightarrow \mu^+ \rightarrow e^+$ decays. A total of 40 candidates of $\pi^+ \rightarrow e^+\nu$ decay were observed with an absorber thickness of 30 to 34 g/cm^2, to be compared with an expected number of four false $\pi^+ \rightarrow e^+\nu$ decays. The time distribution of the positrons in the 40 candidates had an exponential form with a decay constant $\tau = 22\pm4$ ns, consistent with the known π^+ lifetime.

Here it should be stressed that, contrary to the experiment by Lokanathan and Steinberger,[23] the selection of $\pi^+ \rightarrow e^+\nu$ events was based on the presence of only two signals on the oscilloscope trace, and not on the absorber thickness traversed by the positron. On hindsight, the experiment could have been done with a simpler electron telescope consisting of only two counters with a variable absorber in between.

These results became available only after the 1958 International Conference on High-Energy Physics at CERN. I presented them for the first time on Thursday, September 4, 1958, at an informal session on Fundamental and High Energy Physics of the 2nd United Nations International Conference on the Peaceful Uses of Atomic Energy, which took place in Geneva on September 1–13, 1958. The session was chaired by Weisskopf and the Scientific Secretaries were I. Ulehla and A. Salam. The audience was rather small, only two or three dozens of people, contrary to other sessions with gigantic audience. Only three speakers were on the session programme. Feynman, particularly interested in the subject of my presentation, was in the audience. The results of the CERN experiment were quoted by several invited speakers and in the closing talk of the conference.[33]

The results shown in Fig. 7 represent the first experimental evidence for the decay $\pi^+ \rightarrow e^+\nu$. In order to derive a value of R, it was necessary to estimate the positron detection efficiency (this was not trivial at that time because electronic computers were still in their infancy). Lacking a precise knowledge of this efficiency, the observation gave the lower limit $R > 4 \times 10^{-5}$,[29] consistent with the V–A expectation.

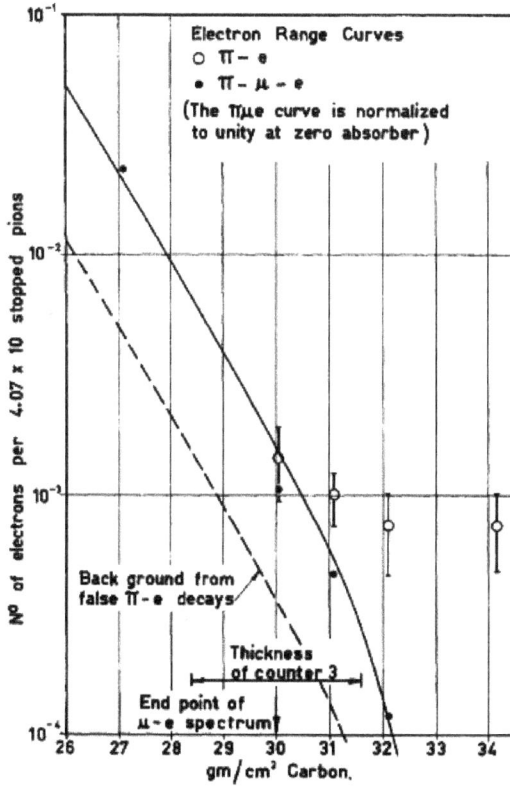

Fig. 7. Range curves for $\pi-\mu-e$ (full circles) and $\pi-e$ (open circles), as measured in the SC experiment.[29] The full curve is a smooth line through the $\pi-\mu-e$ points. The dashed line is the range curve for unresolved $\pi-\mu-e$ events, as obtained from runs with no absorber in the electron telescope. The fraction of unresolved $\pi-\mu-e$ events was 0.23 of the total detected number of $\pi-\mu-e$ events.

At the end of 1958, Julius Ashkin from Carnegie-Mellon joined the group and gave important contributions to the calculation of the positron detection efficiency by developing a Monte Carlo program to this purpose. These calculations were first done in Rome using the computer of the National Research Council (C.N.R.) and then at CERN, when the first electronic computer (a British-made Ferranti Mercury computer) was installed. The final paper, published in 1959, includes the results from these calculations, giving the result $R = (1.22 \pm 0.30) \times 10^{-4}$,[34] which is in excellent agreement with the electron–muon universality of the A coupling and in disagreement with the result of the Chicago experiment.[24]

This result was soon confirmed by the Columbia group,[35] who found evidence for $\pi^+ \to e^+\nu$ decay from a re-analysis of 65,000 π^+ stops in a liquid hydrogen bubble chamber operating in a magnetic field of 0.88 T.

It is amusing to compare the 1958 measurement of R at the SC with the present world average,[36] $R = (1.230 \pm 0.004) \times 10^{-4}$.

4.   First Observation of the Decay $\pi^+ \to \pi^0 e^+ \nu$

A further important experiment performed at the SC in 1962 achieved the first measurement of the pion beta decay mode $\pi^+ \to \pi^0 e^+ \nu$. This provided an excellent confirmation of the theory since the rate of this decay can be reliably predicted. It is a $0^- - 0^-$ transition between two levels of an isotopic triplet and thus a "superallowed" pure Fermi transition. The strength of such transitions is known from nuclear beta decays and, after correcting for the different phase space, the decay rate could be predicted as $\Gamma(\pi^+ \to \pi^0 e^+ \nu) = (0.393 \pm 0.002)s^{-1}$, which corresponds to a very small branching ratio, $B(\pi^0 e^+ \nu) = 1.02 \times 10^{-8}$.

The first observation of this rare decay mode was made at the SC[37] by Depommier, Heintze, Mukhin, Rubbia, Sörgel and Winter using the apparatus shown in Fig. 8 (in the group, Mukhin was a visiting scientist from JINR, Dubna, USSR). A π^+-beam was brought to rest in a scintillation counter that served to

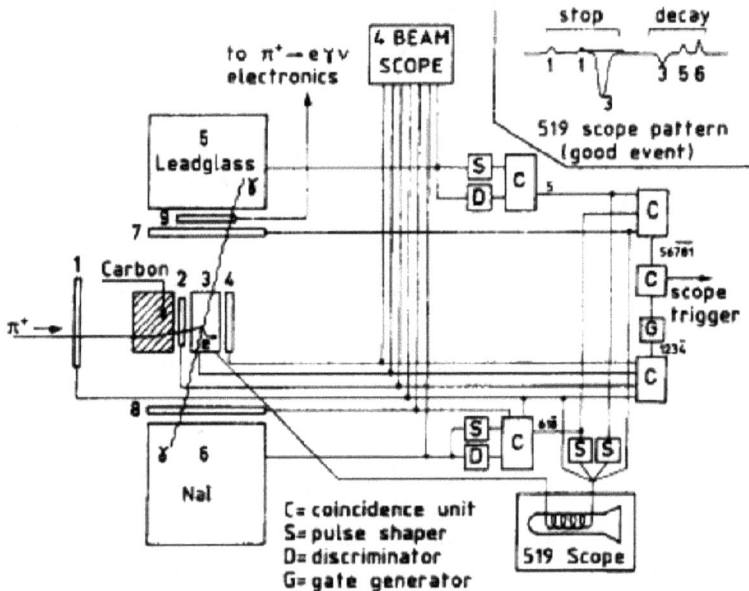

Fig. 8.   Counter arrangement and electronics diagram in the first SC experiment which measured the $\pi^+ \to \pi^0 e^+ \nu$ decay rate.[37, 38] The pattern of signals recorded on a fast oscilloscope is shown in the upper right corner.

Fig. 9.   The SC experiment to measure the $\pi^+ \to \pi^0 e^+ \nu$ decay rate[37, 38] during installation.

detect the decay e^+ and to measure its energy from pulse height (the maximum e^+ energy from $\pi^+ \to \pi^0 e^+ \nu$ decay is 4.5 MeV, including the contribution from $e^+ e^-$ annihilation in the counter). The two photons from π^0 decay, which are emitted with an opening angle always greater than $176°$, were detected by a NaI crystal and a lead-glass Cherenkov counter in coincidence. A photograph of the apparatus during installation is shown in Fig. 9.

In a first run 16 candidates of $\pi^+ \to \pi^0 e^+ \nu$ decay were observed with an estimated background of 2.0 ± 1.3 events,[37] giving a branching ratio $B(\pi^0 e^+ \nu) = (1.7 \pm 0.5) \times 10^{-8}$. Additional data-taking increased the event sample to 44 candidates with an estimated background of 6 ± 2 events,[38] corresponding to a branching ratio $B(\pi^0 e^+ \nu) = (1.15 \pm 0.22) \times 10^{-8}$, which is in good agreement with the theoretical prediction.

A second experiment performed a few years later[39] with a lead-glass photon spectrometer having a much larger angular coverage (see Fig. 10) provided a sample of 411 candidates with an estimated background of 79 ± 10 events. This gave the branching ratio $B(\pi^0 e^+ \nu) = (1.00^{+0.08}_{-0.10}) \times 10^{-8}$, confirming the theoretical prediction at the 10% level.

The present world average is[36] $B(\pi^0 e^+ \nu) = (1.036 \pm 0.006) \times 10^{-8}$.

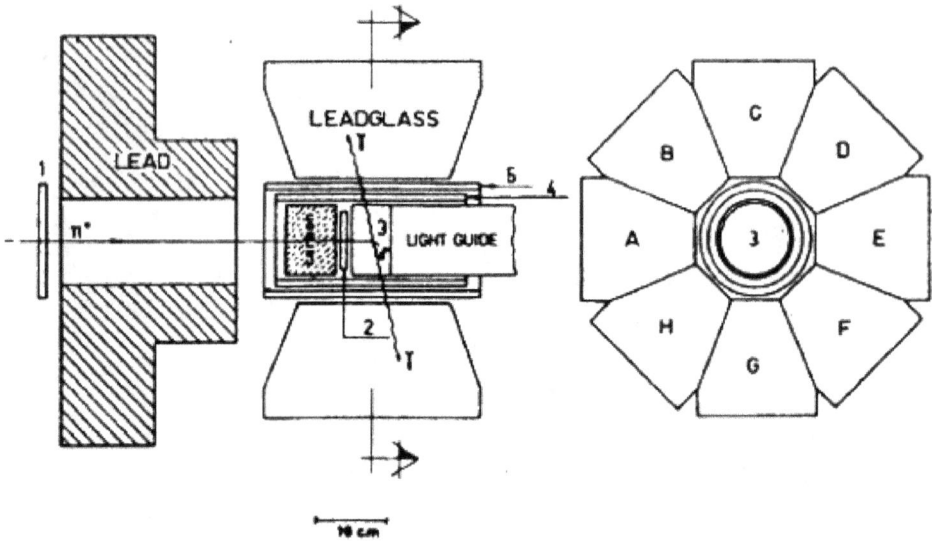

Fig. 10.   Counter arrangement of the second SC measurement[39] of the $\pi^+ \rightarrow \pi^0 e^+ \nu$ decayrate.

5.   Conclusions

The SC was the first accelerator to be built at CERN, with the main purpose of providing an opportunity for European physicists to learn how to do high energy physics. It began operation several years after other machines of similar energy and intensity, such as the synchrocyclotrons at Berkeley, Dubna, Chicago, Liverpool or Nevis. Nevertheless, it made remarkable contributions to particle physics, among which are the results of historical importance on $\pi^+ \rightarrow e^+ \nu$ and $\pi^+ \rightarrow \pi^0 e^+ \nu$ decay described in this article.

Other important particle physics experiments at the SC include:[40]

- Searches for $\mu \rightarrow e\gamma$ decay and neutrinoless μ^--capture, whose negative results pointed to the existence of a second neutrino;
- The measurement of the positron helicity from μ^+-decay;
- Measurements of the μ^- capture rate in hydrogen, both in liquid and gaseous form;
- Last but not least, the first measurement of the muon anomalous magnetic moment, described in Farley's article of this book.[41]

From its spectacular start to its closing down in 1990, the SC has considerably contributed to the scientific reputation of CERN.

References

1. T. Fazzini, G. Fidecaro and H. Paul, *Nucl. Instr.* **3**, 156 (1959).
2. E. Fermi, *Z. Phys.* **88**, 161 (1934) (in German); Nuovo Cim. **11**, 1 (1934) (in Italian).
3. H. Yukawa, *Proc. Math. Soc. Japan* **17**, 48 (1935).
4. C. Møller, L. Rosenfeld and S. Rozenthal, *Nature* **144**, 629 (1939).
5. S. Sakata, *Phys. Rev.* **58**, 576 (1940).
6. E. Fermi, *Elementary Particles,* p. 110, Yale University Press, New Haven (1951).
7. M. Conversi, E. Pancini and O. Piccioni, *Phys. Rev.* **71**, 209 (1947).
8. B. Pontecorvo, *Phys. Rev.* **72**, 246 (1947).
9. G. Puppi, *Nuovo Cim.* **5**, 587 (1948).
10. C. N. Yang and J. Tiomno, *Phys. Rev.* **79**, 495 (1950).
11. C. M. G. Lattes, G. P. S. Occhialini and C. F. Powell, *Nature* **160**, 453 (1947).
12. E. Gardner and C. M. G. Lattes, *Science* **107**, 270 (1948).
13. A. S. Bishop, J. Burfening, E. Gardner and C. M. G. Lattes, *Phys. Rev.* **74**, 1558 (1948).
14. C. M. G. Lattes, *Phys. Rev.* **75**, 1468 (1949).
15. J. Tiomno and J. A. Wheeler, *Rev. Mod. Phys.* **21**, 153 (1949).
16. T. D. Lee, M. Rosenbluth and C. N. Yang, *Phys. Rev.* **75**, 905 (1949).
17. M. Ruderman and R. Finkelstein, *Phys. Rev.* **76**, 1458 (1949).
18. J. Steinberger, *Phys. Rev.* **76**, 1180 (1949).
19. C. S. Wu and S. A. Moszkowski, *Beta Decay*, Interscience, New York, (1966), p. 239.
20. E. F. Pessôa and N. Margem, *An. Acad. Brasil. Ciênc.* **22**, 371 (1950) (in Portuguese).
21. F. M. Smith, *Phys. Rev.* **81**, 897 (1951).
22. H. L. Friedman and J. Rainwater, *Phys. Rev.* **81**, 644 (1951); *Phys. Rev.* **84**, 684 (1951).
23. S. Lokanathan and J. Steinberger, *Suppl. Nuovo Cim.* **2**, 151 (1955).
24. H. L. Anderson and C. M. G. Lattes, *Nuovo Cim.* **6**, 1356 (1957).
25. R. P. Feynman and M. Gell-Mann, *Phys. Rev.* **109**, 193 (1958).
26. E. C. G. Sudarshan and R. E. Marshak, The nature of the four-fermion interaction, in *Proc. Int. Conf. on Mesons and Recently Discovered Particles*, Padova-Venezia (Italy), Sept. 22–28, 1957 (Borghero, Padova, 1958), p. V–14.
27. J. J. Sakurai, *Nuovo Cim.* **7**, 649 (1958).
28. B. M. Rustad and S. L. Ruby, *Phys. Rev.* **97**, 991 (1955).
29. T. Fazzini, G. Fidecaro, A. W. Merrison, H. Paul and A. V. Tollestrup, *Phys. Rev. Lett.* **1**, 247 (1958).
30. J. J. Sakurai, The structure of charged currents, in *Proc. Int. Conf. on Neutrino Physics and Astrophysics*, Dept. of Physics and Astronomy, Honolulu (1981), Vol. 2, p. 457.
31. R. P. Feynman, Forbidding of $\pi-\beta$ decay, in *Proc. Int. Conf. on High Energy Physics*, CERN, Geneva, ed. B. Ferretti (1958) p. 216.
32. G. S. Janes and W. L. Kraushaar, *Phys. Rev.* **93**, 900 (1954).
33. *Proceedings of the 2nd United Nations Int. Conf. on the Peaceful Uses of Atomic Energy* (United Nations, Geneva, 1958), Vol. 1, p. 389; Vol. 30, pp. 42, 57, 136, 327–328.

34. J. Ashkin, T. Fazzini, G. Fidecaro, A. W. Merrison, H. Paul and A. V. Tollestrup, *Nuovo Cim.* **13**, 1240 (1959).

35. G. Impeduglia, R. Plano, A. Prodell, N. Samios, M. Schwartz and J. Steinberger, *Phys. Rev. Lett.* **1**, 249 (1958).

36. K. A. Olive *et al.* (Particle Data Group), *Chinese Physics C* **38**, 090001 (2014), see page 34.

37. P. Depommier, J. Heintze, A. Mukhin, C. Rubbia, V. Sörgel and K. Winter, *Phys. Lett.* **2**, 23 (1962).

38. P. Depommier, J. Heintze, C. Rubbia and V. Sörgel, *Phys. Lett.* **5**, 61 (1963).

39. P. Depommier, J. Duclos, J. Heintze, K. Kleinknecht, H. Rieseberg and V. Sörgel, *Nucl. Phys.* **B4**, 189 (1968); *Nucl. Phys.* **B4**, 432 (1968).

40. L. Di Lella, Elementary particle physics at the SC, *Phys. Repo.* **225**, 45 (1993).

41. F. J. M. Farley, *Muon g − 2 and Tests of Relativity*, in this book, pp. 371–396.

Highlights at ISOLDE

K. Blaum[1], M. J. G. Borge[2,3], B. Jonson[4] and P. Van Duppen[5]

[1] *Max-Planck-Institut für Kernphysik, D-69117 Heidelberg, Germany*
[2] *ISOLDE-PH, CERN, CH-1211 Geneva-23, Switzerland*
[3] *Instituto de Estructura de la Materia, CSIC, Serrano 113 bis,*
E-28006 Madrid, Spain
[4] *Fundamental Physics, Chalmers University of Technology,*
SE-41296 Göteborg, Sweden
[5] *KU Leuven, Instituut voor Kern- en Stralingsfysica,*
B-3001 Leuven, Belgium
[2] *mgb@cern.ch*

The ISOLDE Radioactive Ion Beam Facility at CERN started fifty years ago as an interesting attempt to widen the palette of nuclear species for experimental investigations. During this half century, one has witnessed a continuous development and refinement of the experimental programme. On the road towards today's installation many scientific breakthroughs have been achieved. We present some of them here.

1. Introduction

The ISOLDE Radioactive Beam Facility is the dedicated CERN installation for the production and acceleration of radioactive nuclei. Isotopes from a variety of elements are produced in a target directly connected to the ion source of an isotope separator, which results in a very short time-delay between production of a nucleus and its arrival at the experimental set-up. Thus, the possibility to study isotopes with extreme neutron-to-proton ratios and with very short half-life, is provided. The radioactive isotopes produced at ISOLDE are used in experiments in nuclear-, atomic-, solid-state- and biophysics, as well as in applications, particularly in medicine. The study of properties of nuclei all over the nuclear landscape gives not only clues to a detailed understanding of the structure of the nucleus but also about reactions in the Cosmos, where the chemical elements building up the Nature around us are born (Fig. 1).

The pioneering experiment using an isotope separator directly linked to an accelerator was carried out in Copenhagen already in 1951.[1] Inspired by this achievement, the European nuclear-physics community proposed to build a general-purpose experiment for the production of short-lived isotopes connected to the synchrocyclotron (SC) at CERN. The project was approved on December 17, 1964 by the CERN director Victor Weisskopf. An underground laboratory was built and

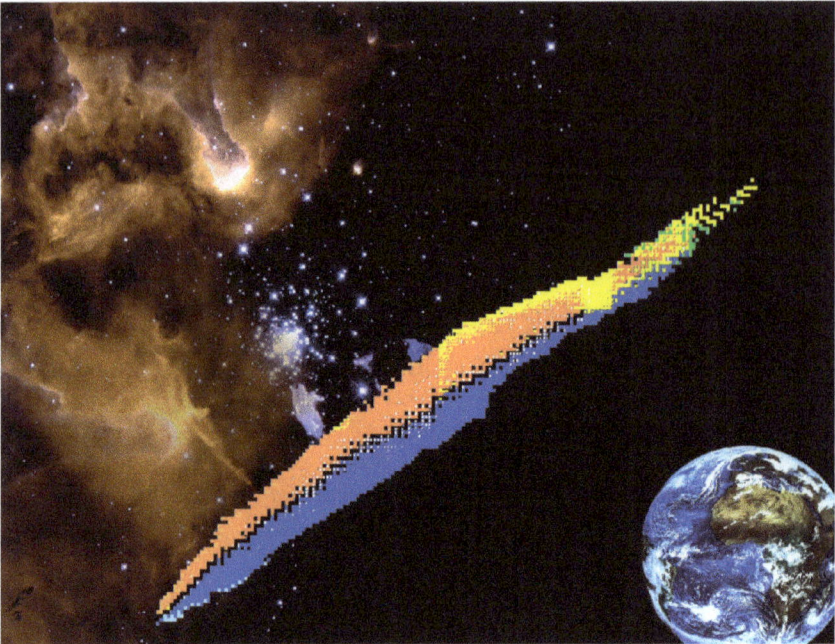

Fig. 1.   Atomic nuclei are organised in a grid of squares, each of which represents a certain number of protons (vertically) and neutrons (horizontally), together forming the chart of nuclides, as shown here. The black squares are the stable nuclei and indicate the valley of stability. This chart, or nuclear landscape, is the working field of ISOLDE, where the main emphasis is on the most exotic nuclei. The nuclei studied at ISOLDE give important new insight in the complex nuclear many-body system. They give clues to the simplicity hidden in the complexity, they tell us about the elements that build up the Nature around us and about their cosmic origin and plays a prominent role in our understanding of the formation of the chemical elements.

protons from the SC were brought via a tunnel to hit a production target. The first experiment at this on-line isotope separator, named ISOLDE, was performed on September 17, 1967 (Fig. 2). ISOL (acronym for Isotope Separator On Line) has since then been the standard name for this type of radioactive isotope production method — the ISOL technique.

Just at the time of the first successful experiments at the new underground hall, CERN decided for a major upgrade of the SC. This SC improvement programme (SCIP) aimed at an increase of the internal beam intensity from 1 to 10 μA together with an improved extraction efficiency giving a proton beam intensity increase of more than a factor 100 at the ISOLDE target. An essential part of the upgrade was to change the frequency system at the SC, which had been based on a tuning fork, to a rotating condenser. In order to cope with the higher proton current offered to ISOLDE, an advanced technical development programme was launched. A new design of the target–ion source systems was proposed and, as it turned out, gave access to more and more isotopes of different chemical elements. The SCIP

Fig. 2.   The ISOLDE experimental hall in 1967. Note that ISOLDE at that time was part of the CERN Nuclear Chemistry Group, which meant white lab coats.

programme took place in the years 1972–1974 and the new layout of the separator and its target–ion source became referred to as ISOLDE 2. The high intensity of produced isotopes and the large variety of different elements meant that ISOLDE had become a major international facility to perform experiments on radioactive isotopes.

The SC machine had been in operation since 1957 and it became clear in the middle of the 1980s that this accelerator had to be closed. To maximise the use of the last years of the SC, the ISOLDE Collaboration proposed to build a second isotope separator. This new separator, ISOLDE 3, was constructed with a two-stage separation (one 90° magnet followed by a 60° one) to achieve a very high mass resolution. A new target was placed in the SC vault and the produced radioactive isotopes were brought into the proton hall. The new separator gave a mass resolution of M/ΔM of 7000 and was a pre-runner for the design of the present High-Resolution Separator (HRS) at the PS-Booster.

The future of the ISOLDE programme after the SC shutdown was discussed and the general consensus was that the most attractive option would be to move ISOLDE closer to the PS complex and to place its targets in an extracted 1 GeV proton beam from the PS Booster. The ISOLDE Collaboration set up a Technical Committee that helped CERN to find the optimal design of the new facility. A suitable layout was found and on May 4, 1990 the CERN Directorate approved the proposal to move ISOLDE to the PS Booster. The ground work for the new ISOLDE building started in October. At noon on December 19, 1990 the last shift of protons was delivered to ISOLDE from the SC leaving a legacy of more than a quarter of a century of pioneering experiments that benchmarked the future of the ISOL facilities in the world. The ISOLDE-PS Booster Facility was built in the usual CERN spirit and already in May 1992 the new installation could be inaugurated. The first experiment, a study of the beta decay of the two-proton halo nucleus ^{17}Ne,[2] was successfully completed on June 26.

The ISOLDE programme was traditionally mainly dedicated to study nuclear ground-state properties and excited nuclear states populated in radioactive decays. With the large palette of different isotopes, some of them produced with high intensity, it was an attractive possibility to build a post-accelerator at ISOLDE. In 1994, such a proposal was presented to CERN asking for permission of the ISOLDE community to build a suitable accelerator to get exotic nuclear beams in the energy range of 2–3 MeV/u. The project was approved and the REX-ISOLDE accelerator was built in an extension to the experimental hall (see Fig. 3 and next section for details). The first beams were accelerated on October 31, 2001, and this addition to the ISOLDE programme has turned out to be both successful and very prolific.

Fig. 3. The ISOLDE experimental hall in 2007.

2. Production, Manipulation and Acceleration of Radioactive Ion Beams

The success of the ISOLDE facility is based on intertwined developments of radioactive ion beams (RIB) and instrumentation for physics experiments. The cross fertilisation leads to a broad spectrum of beams available with masses varying from ^{6}He to ^{232}Ra, with half-lives down to the ms range (e.g. ^{14}Be $T_{1/2} = 4.45$ ms), intensities up to the nA level (e.g. ^{213}Fr with $\sim 8 \times 10^9$ particles per second) and energies from rest to a few MeV/u.[3, 4] A continuous development programme implementing new techniques, like e.g. the use of nano-structured target material, laser resonance ionisation, ion cooling and charge state breeding, keeps the facility at the forefront of RIB science ever since it was constructed. The RIB production and manipulation process adapts the beam properties to the different experimental setups. As one mainly deals with short-lived radioactive isotopes that are produced in minute quantities compared to the vast amount of unwanted species produced (ratios over 10^{12} between the production rate of the unwanted versus wanted isotopes are routinely reached) the overall RIB production process has to be efficient, fast and selective.

2.1. *The target–ion source system — The heart of the matter*

ISOLDE's radioactive isotopes are produced in high-energy proton induced reactions impinging on different target materials. The primary proton beams from the CERN-PS Booster induce spallation, fragmentation and fission reactions which allow, by a proper choice of the target material, to produce a range of isotopes that covers a substantial part, 80%, of the chart of nuclei below uranium ($Z = 92$). As the reaction mechanisms are barely selective, the target–ion source system at the origin of the low-energy ion beam combined with the mass analysing magnet and other ion manipulation devices are used to reduce the unwanted contaminants and/or to identify the isotopes of interest. Pioneering work was necessary to integrate the target and the ion source into one compact system that is kept at high temperature to speed up the diffusion and effusion of the radioactive atoms from the target container.[3] This led to a successful design that today is still competitive and that allows using different atomic and chemical processes to purify the beam. A simple but effective approach is cooling the transfer line between target and ion source, allowing only the gaseous elements (noble gases) or most volatile molecules to reach the ion source. The suppression of elements that make a chemical bonding with the surface of a quartz line installed between target and ion source represents another approach. Recently, new developments including the use of nano-structured target materials are explored to reduce the delay time and obtain more ruggedised systems.

ISOLDE's successful laser spectroscopy programme and the fact that powerful pulsed laser systems became available led to the implementation of laser resonance ionisation for the production of RIB in the mid-1980s.[5] This element selective and efficient ionisation process, that is based on the use of different laser beams to invoke multi-step atomic excitations into the continuum, results in clean beams. The first on-line production of photo-ionised radioactive Yb beams was soon followed by isotopes from a wide range of different elements.[6, 7] Now the laser ion source is routinely used for over 50% of ISOLDE's beam time. A recent improvement of the selectivity of the laser ionisation is the Laser Ion Source Trap (LIST)[8] approach that integrates a standard target–ion source system, laser ionisation and ion manipulation. It is based on the photo-ionisation of the plume of atoms escaping from the high temperature ISOLDE target–ion source system, subsequent capturing of the ions in a radio-frequency trap and transporting them to the extraction region. While losses in overall efficiencies are encountered, LIST improves the selectivity by about four orders of magnitude.[9]

2.2.   *Cooled beams, isomeric beams and in-source laser spectroscopy*

Adapting the longitudinal and transverse RIB emittance or the beam pulse characteristics to the needs of the experiments was pioneered at ISOLDE's high-precision mass spectrometry set-up ISOLTRAP. The potential of buffer-gas cooling in radio-frequency or in Penning traps to produce cooled, bunched radioactive ion beams with good efficiency could be demonstrated. Larger versions of both the radio-frequency quadrupole ion trap and of the Penning trap were developed to deliver cooled and bunched beams to other ISOLDE users like the collinear laser spectroscopy set-up (see Section 5) where it increased the signal-to-background ratio up to four orders of magnitude, and to the REX-ISOLDE post-accelerator (see Section 2.3).

Soon after the first laser ionised RIB, beams of long-lived states, called isomeric, were produced and separated using the hyperfine splitting of the atomic transition[11] as it depends on the nuclear properties of the isomer. By changing the laser frequency of the first atomic transition, the specific hyperfine structure of the different nuclear states can be probed. Combining this isomer selectivity with β-decay and mass spectroscopy studies led to the discovery of three β-decaying states in ^{70}Cu: the ground state and two isomeric states (see Fig. 4). Their existence could be explained as due to the coupling of one proton and one neutron to a ^{68}Ni core.[12] With ISOLDE's post-accelerator (see Section 2.3), isomeric beams were post-accelerated and used for Coulomb excitation measurements probing the strength of the $Z = 28$ shell and $N = 40$ sub-shell closures.[13] This pioneering experiment moreover demonstrated that Coulomb excitation could trigger the depopulation of

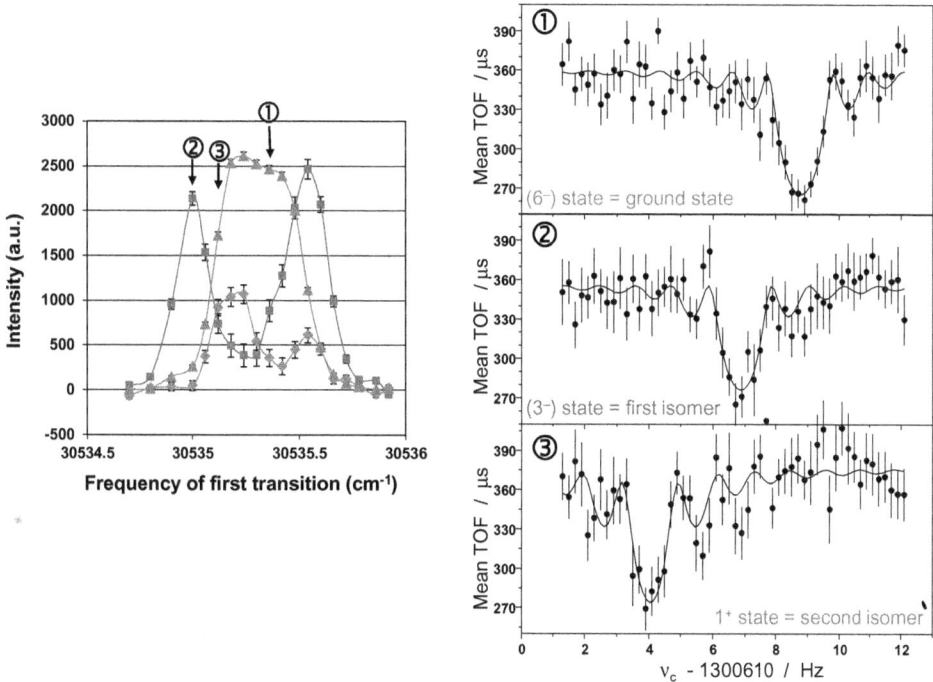

Fig. 4. Isomer selection is performed using the hyperfine splitting of the atomic levels probed in the resonant laser ionisation process used in the ISOLDE laser ion source. The yield of the different isomers in ^{70}Cu as a function of the laser frequency of the first atomic transition are shown for the ground state (6^-) (triangles), (3^-) (squares) and 1^+ (circles) β-decaying isomers of ^{70}Cu (left). The modified cyclotron frequency resonance spectra obtained at ISOLTRAP are shown for three different laser frequencies (1, 2 and 3). The line represents a fit through the data and the resonance frequency is inversely proportional to the mass of the nuclear state. The ISOLTRAP measurements demonstrated the presence of three different long-lived states in ^{70}Cu and to obtain their mass whose difference is in perfect agreement with β-spectroscopy studies (right). The line is a fit through the data points. The spectrum on the bottom right (3) was obtained after an extra purification step in the Penning trap.

an isomer towards an excited state that subsequently decays to the ground state, which called for a detailed study of other spin-multiplets in odd–odd nuclei.

Because of the high sensitivity of the laser ion source, laser ionisation spectroscopy measurements became possible with very weak beams (intensities down to less than one atom per second). However, this so-called in-source laser spectroscopy method was mainly limited to heavy mass nuclei because of the limited spectral resolution of the method. Charge radii and electromagnetic moments of a number of neutron deficient nuclei around the lead isotopes $(Z = 82)$ were obtained extending the pioneering work on optical spectroscopy of the mercury isotopes using samples from ISOLDE[14] (see Section 4). This technique allowed for the determination of the unknown ionisation potential of astatine, the only element in the table of Mendeleev below uranium for which this fundamental atomic property was not known experimentally.[15] The result benchmarks quantum chemistry calculations and has moreover an impact on the field of innovative medical radioisotope production. For example the isotope ^{211}At, because of its decay properties, is an interesting pharmaceutical radioisotope for targeted alpha therapy in cancer treatment provided its chemistry is well understood.

The large variety of RIBs with different decay properties and from various elements make them tailored probes for condensed matter and biophysics studies. The radioactive atoms act as spies and their emitted radiation provides information on their lattice position or on the magnetic and electrical properties of the surrounding atoms. Because of the high radiation detection efficiency, only very low concentrations of radioactive impurity atoms are necessary to provide unique nano-scale information in materials, surfaces or interfaces (see Fig. 5).

2.3.   *REX ISOLDE — A new concept for post-acceleration of radioactive ion beams*

In order to broaden its physics scope and triggered by the successful post-acceleration of light RIB at the Louvain-le-Neuve (Belgium) project,[16] new ways to accelerate the singly charged RIB in a universal, fast, efficient and cost-effective way were explored. This resulted in a novel concept based on ion beam cooling and bunching in the buffer gas of a Penning trap, charge-state breeding in an Electron Beam Ion Source (EBIS) and post-acceleration in a room-temperature linear accelerator. Ion beam cooling and bunching modulates the RIB from ISOLDE into bunches suited for injection in EBIS and was based on the ISOLTRAP experience. The efficient injection of singly charged ions and extraction of highly charged

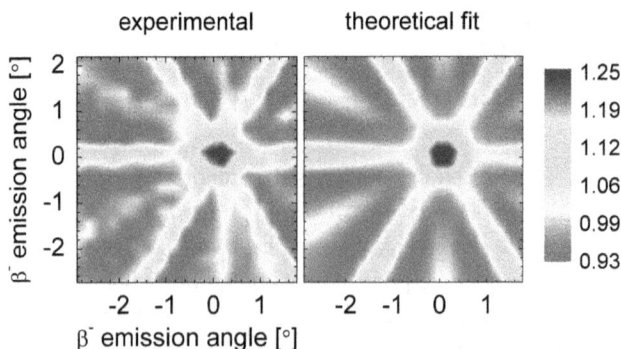

Fig. 5.   Emission channeling data recorded using a position sensitive silicon detector obtained after the implantation of ^{56}Mn nuclei into a GaAs (semiconductor) sample.[10] The detector views the GaAs sample from the implantation site. The colour scale (arbitrary units) corresponds to the angular dependent rate of channeled beta particles emitted from the radioactive ^{56}Mn nuclei sitting in the GaAs matrix. The data (left) are compared to simulations (right) from which the site location of Mn in the GaAs sample can be determined. The latter enhances the understanding of electrical, optical and magnetic influence of dopants in semiconductors.

Fig. 6. The REX-ISOLDE post-accelerator delivering radioactive ion beams from ^{6}He to ^{224}Ra with energies up to 3 MeV/u.

ions from EBIS was based on a concept from the Manne Siegbahn Laboratory (Stockholm, Sweden). Finally, the room temperature accelerator cavities were based on designs from the Max-Planck-Institute for Nuclear Physics (Heidelberg, Germany), the GSI HLI-IH-structure (Darmstadt, Germany) and the lead LINAC at CERN.

While the original goals of this Radioactive beam EXperiment at ISOLDE — REX-ISOLDE project[17] (Fig. 6) were limited to energies up to 2 MeV/u and masses below $A = 50$, the concept proved to be very successful and meanwhile beams with A/Q ratio <4.5 and with masses up to 220 have been accelerated up to 3 MeV/u, with efficiencies reaching 10%. Most of the beams have been used for Coulomb excitation measurements or few-nucleon transfer reactions using a dedicated particle and gamma-ray detector array for low-intensity low-multiplicity RIB experiments (see Fig. 7).

3. Shell Structure: The Decline of the Magic Numbers

The nucleus presents typical characteristics of few-body and many-body quantum systems at the same time. Its microscopic and mesoscopic manifestation are governed by effective 2- and 3-body interactions of great complexity that depends not only on the distance between nucleons but also on its spins and moments.

In its macroscopic behaviour one observes properties equivalent to those of a liquid drop such as energy surface deformation, vibrations, rotations and shapes.

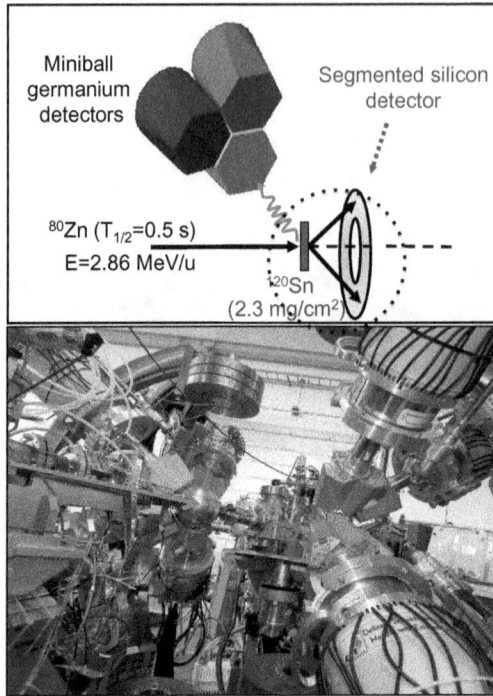

Fig. 7. Top: A schematic drawing of the set-up used for Coulomb excitation experiments. An inelastic collision takes place between a post-accelerated ^{80}Zn beam when hitting a thin ^{120}Sn target. After the collision, the scattered beam and target particles are detected in a segmented silicon detector while the de-exciting gamma rays are recorded in the Miniball germanium array. The beam diagnostics (not shown in the drawing), target and silicon detector array are situated in the spherical reaction chamber. Bottom: The picture shows Miniball germanium array, developed for low-intensity RIB experiments. Eight clusters containing three hexagonal shapes germanium crystals each are positioned around the spherical reaction chamber.

Experimental achievements related with the latter are described in Section 4. Understanding the manifestation of these semiclassical behaviours in terms of the quantum dynamics of the nuclear constituents, protons and neutrons, is one of the main challenges of nuclear theory.

Experiments done with stable and near stable nuclei have shown, that nuclei with N or Z equal to the so-called magic numbers 8, 20, 28, 50, 82 are more difficult to excite than their neighbours. This fact has supported the use of the well-established nuclear shell model developed independently by Maria Goeppert Mayer and J. Hans Jensen in 1949 giving them the Nobel Prize in 1963. This model, based on the interactions between nucleons and their arrangement in orbits, has been a success for the understanding of nuclear properties for stable or near stable atomic nuclei. The nuclei with magic numbers exhibit highly symmetric spherical configurations and are since then the milestones of the nuclear landscape. Although this model emerged from a pure phenomenological approach, modern nuclear theory can trace the magic numbers down to nucleon–nucleon forces derived from low-energy QCD.

The advent of experimental facilities for the production of radioactive nuclei permitted to reach nuclei with a completely different balance between protons and neutrons, i.e. different isospin. It turned out that the traditional shell structure changed in some regions of the nuclear chart, showing dramatic effects on the neutron-rich side near the neutron binding limit, the neutron drip-line, where the magic character appears at different, N or Z values. Many current studies of nuclear structure with exotic radioactive nuclei focus on the question of whether these magic numbers persist or are altered in going away from the 'valley of stability'. These studies challenge the predictive power of nuclear theory and will eventually lead the way towards a universal description of nuclear structure.

In this section we will describe CERN's contribution to the discovery of the decline of the classic magic numbers. This discovery has changed the perception of nuclear systems and questioned established knowledge. Although the first anomaly in the expected order of nuclear orbits was observed in 1960 in ^{11}Be,[18] it was the measurement of the masses of the exotic sodium isotopes, ^{31,32}Na performed at CERN PS that revealed that these nuclei were tighter bound than expected.[19] The excess of binding energy was associated with deformation and explained due to the excitation of a neutron across the $N = 20$ gap. This was soon further supported by measurements of other ground-state properties: spin, magnetic moments and mean charged radii.[20] Further, the sodium beta-decay studies allowed for the determination of the first excited states of the magnesium isobars,[21] while the highly sensitive collinear laser techniques allowed for the mapping of the ground state properties of the neutron rich magnesium isotopes.[22] The data revealed that the $N = 20$ isotopes undergo a sudden onset of deformation and thus the expected magicity vanishes. The shell gap in these neutron-rich isotopes is not robust enough to avoid excitations to a higher shell. This gives rise to quadrupole correlations that favour deformation. A schematic representation of this effect of inversion of orbits is shown in Fig. 8. Many studies have been dedicated to map and define the so-called *island of inversion* region where ^{32}Mg is situated in the centre. The ^{30}Mg ($N = 18$) isotope has a spherical 0^+ ground state with an excited 0^+ state at 1788.2 keV. Based on the measured monopole strength it could be shown within a two-level model approach that the 0_2^+ state is strongly deformed.[23] Moreover theory predicts that its wave function contains a strong intruder configuration, i.e. shape coexistence (see next section). The ground-state of ^{32}Mg, a semi-magic nucleus with $N = 20$, is strongly deformed as shown from the large B(E2; $0_{gs}^+ \rightarrow 2_1^+$) value. Thus it appears that in ^{32}Mg an inversion takes place: the ground state being deformed and composed of intruder configurations, while a so far unidentified excited 0^+ state being spherical. The best proof would be to identify this near-spherical excited 0^+ state, the analogue of the 0^+ ground state in ^{30}Mg and to characterise the underlying neutron particle-hole structure of the ground and excited 0^+ states. In spite of several attempts this state could not be identified before the work at ISOLDE.

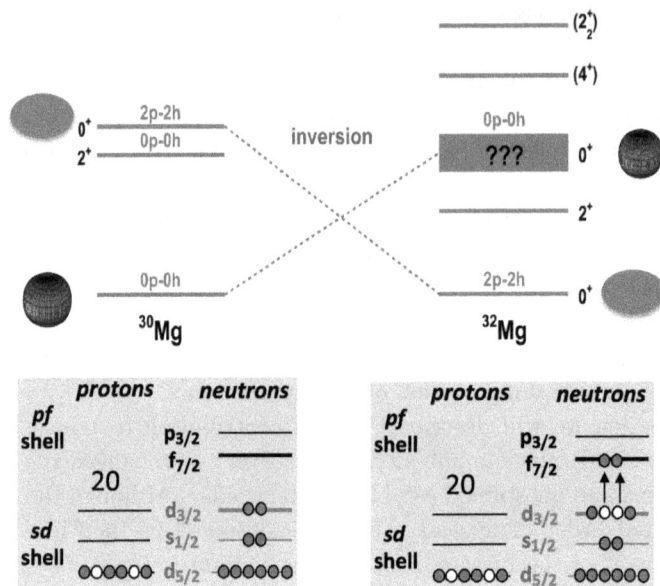

Fig. 8.   The fascinating phenomenon of different nuclear shapes coexisting at similar energies —
the difference is less than 1% of the total binding energy. It is the reduction of the $N = 20$ shell gap
that enables neutron-pair excitations across the $N = 20$ shell gap leading to quadrupole correlations
and giving rise to low-lying deformed two-particle two-hole $2p$–$2h$ states. The so-called intruder
states with two neutrons in the pf shell (right: ^{32}Mg ground state) coexist at low excitation energy
with the normal spherical zero-particle zero-hole ($0p$–$0h$) neutron states in the sd shell (left: ^{30}Mg
ground state).

A key idea was to study the addition of two neutrons to the spherical
ground state of ^{30}Mg populating either the deformed ground state in ^{32}Mg or
the excited presumed spherical 0^+ state in ^{32}Mg. The proof for the correctness
of this assumption was achieved at ISOLDE when a 1.8 MeV/u beam of ^{30}Mg
($T_{1/2} = 335$ ms) was used to populate the ground-state and excited states in ^{32}Mg
by a two-neutron (t, p) transfer reaction in inverse kinematics, so called when the
heavy element is the projectile. The detection of protons and gamma rays were
done with the T-REX array and the MINIBALL Ge detector array, see Fig. 7.
The T-REX is a 4π array consisting of a barrel of silicon-strip detectors together
with an annular double-sided segmented silicon-strip detector of CD shape. Energy
and direction of emission were measured for protons, deuterons and tritons. By
studying angular distributions for the emitted protons, one can determine the
angular momentum transferred and from that the spin and parities of the states
populated can be deduced. In the experiment, an excited 0^+ state at 1058 keV was
identified that appears to be an excellent candidate for the spherical state shape
co-existing with the strongly deformed 0^+ ground state.[24] From the (t, p) cross-
sections neutron occupancies across the $N = 20$ shell gap for the states in ^{32}Mg
were inferred, confirming the inversion and occurrence of shape coexistence.

The disappearance of magic numbers far from stability is accompanied by the emergence of new ones. Its proper prediction is closely related to our understanding of the different components of the strong force that acts between protons and neutrons. Very recently the predicted magic number far from stability, $N = 32$, was confirmed in the study of calcium ($Z = 20$) isotopes at the verge of existence. The $^{51-52}$Ca masses were determined using the ISOLTRAP Penning-trap mass spectrometer described in Section 6 and for the extremely rarely produced and short-lived species, $^{53-54}$Ca, a multi-reflection time-of-flight spectrometer was used. The latter was designed for isobar separation and used for the first time for mass determination. The measured masses confirmed the existence of a prominent shell closure at $N = 32$ and provided a formidable benchmark for nuclear theory.[25]

4. Nuclear Shapes — Shape Coexistence and Quadrupole Deformations

Atomic nuclei exhibit single-particle and collective degrees of freedom. Understanding the delicate balance between these two extremes underpinning the structure of atomic nuclei is a challenge for theory. In general single-particle effects dominate the structure of nuclei at and around closed proton and neutron shells, while deformation is observed in nuclei situated on the nuclear chart in between doubly closed shell nuclei. Shapes have been studied at ISOLDE using laser spectroscopy and Coulomb excitation measurements. While the former method results in charge radii, magnetic dipole and electrical quadrupole moments of ground states and long-lived isomeric states, the latter allows for the determination of quadrupole moments and quadrupole or higher order transition strengths of excited states. Throughout the nuclear chart experimental evidence has been accumulated for a phenomenon called shape coexistence whereby quantum states with different deformation but similar binding energy appear at low energies in the nucleus. In the heavy nuclei shape coexistence was discovered at ISOLDE serendipitously in the light mercury isotopes in optical spectroscopy measurements.[14] The strong staggering in the charge radii for the lightest mercury isotopes (see Fig. 9) was interpreted as due to the appearance in the odd-mass mercury isotopes of a strongly deformed ground state co-existing with a more spherical isomer. The in-source laser spectroscopy data obtained in the neighbouring lead and polonium isotopes[26, 28] combined with Coulomb excitation measurements in this region using the post-accelerated REX beams allowed to deduce the oblate nature of the ground-state deformation and supported the proposed interpretation of shape coexistence induced by particle–hole excitation across the closed proton shell.

The prevalent shape of nuclei is quadrupole deformation, symmetric against reflection, but in some heavy unstable nuclei circumstantial evidence for octupole deformation has been reported. This type of reflection-asymmetric or pear shaped deformation is not only important to test nuclear models, but isotopes exhibiting

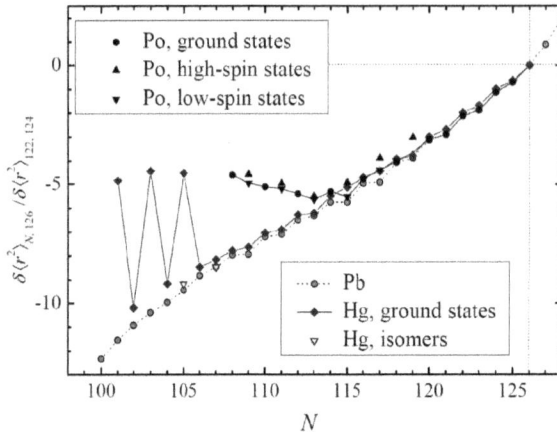

Fig. 9. Relative change in mean-square nuclear charge radii, $\delta\langle r^2\rangle$, for the even-Z ${}_{80}$Hg (blue), ${}_{82}$Pb (red) and ${}_{84}$Po (black) isotopes. While the relative change in charge radii of one isotope compared to its neighbour for the heaviest isotopes are very similar for these three elements, large differences are observed further away from the $N = 126$ neutron shell closure. The large staggering observed in the Hg data is interpreted as shape coexistence caused by the occupation of specific single-particle states. The deviation observed for the Po isotopes is linked to an onset of collective behaviour possibly caused by the same mechanism. Adapted from Ref. 26.

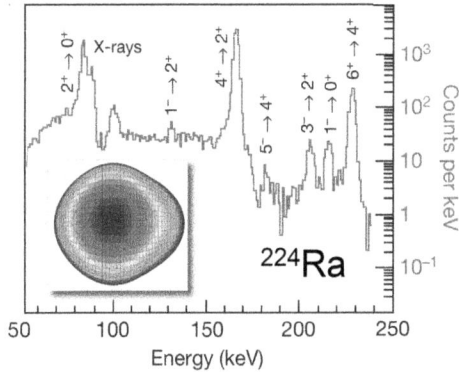

Fig. 10. Part of the gamma-ray energy spectrum obtained after Coulomb excitation of ^{224}Ra. From the intensities of the gamma-rays (indicated with the spin and parity of initial and final state), especially the ones de-exciting negative parity states, information on the octupole transition strength was deduced evidencing enhanced octupole deformation.[27] The inset shows the static octupole deformation of ^{224}Ra in the intrinsic frame as deduced from the experiment.

this type of deformation are ideal probes to look for physics beyond the Standard Model. For the search of an atomic electric-dipole moment in odd-mass isotopes, static electric octupole deformation of the atomic nucleus amplifies the sensitivity by several orders of magnitude. In a recent Coulomb excitation experiment using energetic beams of ^{224}Ra, enhanced octupole deformation was evidenced through the measurement of octupole transition strengths (see Fig. 10).[27] This constrains the region of suitable isotopes for studies of the atomic electric-dipole moments.

5. Nuclear Halos

A considerable fraction of the experimental programme at the ISOLDE Facility concerns studies of beta decay, which is a well-proven probe of nuclear structure as well as of weak interactions (see Section 6). There is an important difference between beta-decay of near stable nuclei and those in the drip-line regions. Close to stability, the transitions occur between discrete bound levels while the decay closer to the drip-lines also involves states in the continuum. In the neighbourhood of the drip-lines one also encounters beta-delayed particle emission processes, i.e. the particle emission, mediated by the strong force, is delayed as the de-exciting state is slowly populated in the beta decay process. In near drip-line nuclei this decay mode dominates over decays to bound states.[29] A quite spectacular example of a beta-delayed particle emitter nucleus is provided by the last particle-bound lithium isotope, ^{11}Li. This nucleus has a beta-decay Q value, i.e. the mass difference between mother and daughter nuclei, of 20.623 MeV, while the neutron separation energy of its daughter, ^{11}Be, is as low as 504 keV. This energy unbalance opens up several possible beta-delayed particle emission channels, as illustrated in Fig. 11. In a

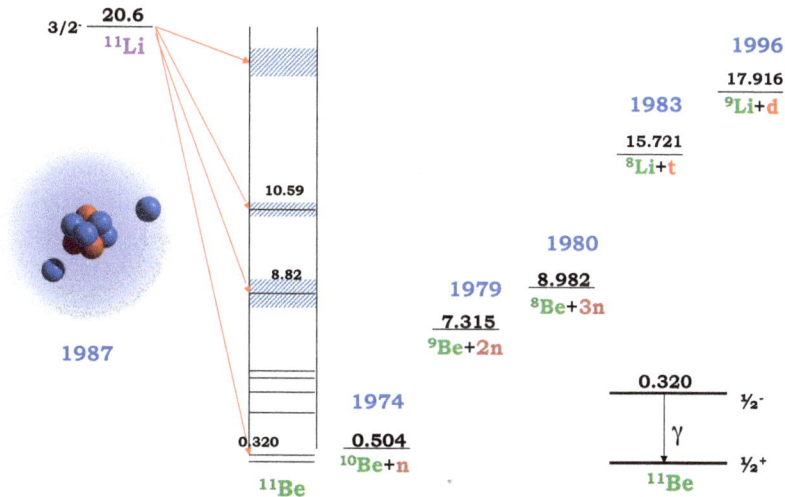

Fig. 11. The two-neutron halo nucleus ^{11}Li. The very high Q-value of 20.6 MeV for beta-decay (red arrows) combined with the comparably low separation energies for various particles in the daughter nucleus ^{11}Be results in a multitude of different beta-delayed particle emission decay modes in this nucleus. The ^{11}Be excitation energies and threshold energies for the different beta-delayed decay modes in MeV (in black) are given relative to the ^{11}Be ground state. The years (in blue) are given when these decay modes were first observed and when the halo structure of ^{11}Li was suggested.

number of experiments at ISOLDE at the end of the seventies and the beginning of eighties the decay modes of beta-delayed two-neutron, three-neutron and triton emission were observed for the first time.

Just at the time when the study of ^{11}Li was in focus at ISOLDE, a group at Berkeley led by I. Tanihata[30] studied interaction cross-sections of Li isotopes. The remarkable result of this experiment was a large and abrupt increase in the matter radius for ^{11}Li. Scientists working with ^{11}Li at ISOLDE came with an explanation for the increase of the ^{11}Li radius based on the low two-neutron separation energy.[31] A model, where ^{11}Li should have a novel type of structure, a halo, was proposed. The basic idea is that the ^{11}Li nucleus is built up by a ^{9}Li core surrounded by two loosely bound neutrons forming a veil of neutron matter around the core.

The realisation of the occurrence of halo structure at the drip-lines sparked off an intense experimental activity and many nuclei having neutron and proton halos are known today.[32] Important early ingredients for the understanding of the halo structure came from ISOLDE where the spin, magnetic moment and the electric quadrupole moments were measured for the chain of bound Li isotopes in a combination of optical and beta-decay measurements,[33, 34] results that later got confirmed and improved.[35] The results showed that the magnetic dipole and the electric quadrupole moments of the two isotopes ^{9}Li and ^{11}Li were very similar. This proves that the increase in the radius arises from the neutron tail, while the charged core is little affected.

Another consequence of the two-neutron halo structure is the occurrence of beta-delayed deuteron emission. The Q value for this process is $Q(\beta^- d) = (3.007 - S_{2n})$ MeV,[36] where S_{2n} is the separation energy for the last two neutrons. The occurrence of this decay mode was shown at ISOLDE for the first time for the two-neutron halo nucleus ^{6}He[37] and later also for ^{11}Li.[38]

The beta-decay daughter of ^{11}Li is ^{11}Be, which is an example of a one-neutron halo nucleus. Its magnetic moment was measured at ISOLDE in a very beautiful experiment.[39] The Be isotopes are produced in a UC$_2$ target matrix irradiated by 1 GeV protons from the PS Booster (see Section 2.1). The produced Be evaporates into a tungsten cavity, where two laser beams excite the atoms from the $2s^2$ $^1S_0$ atomic ground state to an auto-ionising state via the atomic $2s2p^1P_1$ state. The nuclei of the ^{11}Be$^+$ beam are then optically polarised by a collinear frequency-doubled CW dye laser beam. The polarised ions are implanted in a Be crystal placed in the centre of an NMR magnet. The first-forbidden beta decay to ^{11}B of the polarised nuclei are detected with two scintillators and the beta-decay asymmetry is measured. From the observed Larmour frequency the magnetic moment is determined as $\mu(^{11}\text{Be}) = -1.6816(8)\,\mu_N$. This value is confirming a 16% core polarisation admixture in the ^{11}Be ground-state wave function.[40]

A major experimental success was the use of the collinear laser technique to determine the nuclear charge radii for Be isotopes. In collinear laser spectroscopy

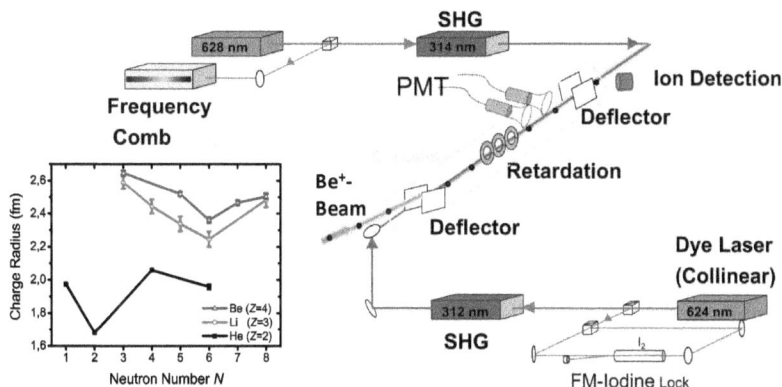

Fig. 12. Set-up for collinear laser spectroscopy with parallel and antiparallel excitation and a frequency comb as reference for the determination of the charge radius for Be isotopes. Key components of the experimental set-up are shown (SHG: second harmonic generator, PMT: photomultiplier tube). The inset shows the state-of-the-art of charge radii measurements[41] for light drip-line nuclei.

the laser beam is superimposed with a beam of fast (typically 30–60 keV) ions or atoms and the resonance fluorescence is detected with a photomultiplier perpendicular to the flight direction. Since the atoms are propagating in a parallel or antiparallel manner to the laser beam, the resonance frequency of the atom is shifted in the laboratory system by the relativistic Doppler effect.[42] Collinear laser spectroscopy has the advantage that the acceleration of an ion ensemble with a static electric potential compresses the longitudinal velocity distribution. Thus, the Doppler width is considerably reduced. With a beam of Be^+ ions, a frequency comb and measuring the absolute transition frequencies for parallel and antiparallel geometry of the ion and laser beams, see Fig. 12, a hitherto impossible precision was obtained. The beauty of the technique is that the rest frame frequency, ν_0, is obtained independent of the acceleration voltage by combining the measured absolute transition frequencies for parallel (ν_p) and antiparallel (ν_a) laser beams so that $\nu_p \nu_a = \nu_0^2 \gamma^2 (1 + \beta)(1 - \beta) = \nu_0^2$. The required accuracy in the isotope shift measurement of 1 MHz was obtained. The charge radii were observed to decrease for the isotopes ^{7}Be to ^{10}Be and then increase for ^{11}Be, see Fig. 12.[43] This increase of the charge radius is expected since the centre of mass and the centre of charge do not coincide in a one-neutron halo nucleus like ^{11}Be.

6.   Fundamental Interaction Studies

Radioactive beams are ideal probes for fundamental studies of weak interaction and of the Standard Model (SM) in general. Among others, precision measurements of masses, half-lives and branching ratios of superallowed β-emitters allow in combination with nuclear theory a precise determination of the first (V_{ud}) element of the Cabibbo–Kobayashi–Maskawa (CKM) quark-mixing matrix, which relates the quark weak-interaction eigenstates to the quark mass eigenstates assuming three quark generations. Taking V_{us} and V_{ub} from the Particle Physics Data Group (PDG), a stringent top-row unitarity test of the CKM matrix can be performed:[44]

$$\sum_j |V_{uj}|^2 = |V_{ud}|^2 + |V_{us}|^2 + |V_{ub}|^2 = 1. \tag{1}$$

Any deviation from 1 can be related to concepts beyond the SM such as the existence of an additional Z-boson or the existence of right-handed currents in the weak interaction. V_{ud} can be determined from the fundamental vector coupling constant G_V and the well-known weak-interaction constant G_F of purely leptonic muon decay: $V_{ud} = G_V/G_F$, where G_V in turns can be derived from the corrected strengths (Ft-value) of superallowed β-transitions, which are a function of the experimental parameters: β-decay Q-value, half-life $T_{1/2}$, and branching ratio b, as well as of different correction terms including isospin-symmetry-breaking and radiative correction. The uncorrected ft-values can be derived purely from nuclear physics experiments, namely from mass, half-life and branching ratio measurements of superallowed β-transitions. Many ISOLDE experiments, especially high-precision mass measurements with the Penning-trap mass spectrometer ISOLTRAP (see Fig. 13), being the first of its type installed at a radioactive ion beam facility in 1986, have contributed to this kind of research, providing the most accurate V_{ud} value to date of $|V_{ud}| = 0.97417(21)$.[44] Taking V_{us} and V_{ub} from PDG, one obtains the result:

$$|V_{ud}|^2 + |V_{us}|^2 + |V_{ub}|^2 = 0.99978(55), \tag{2}$$

i.e. the unitarity is fully satisfied to a precision of 0.06%.

Another pillar of the SM is the conserved-vector-current (CVC) hypothesis, stating that the vector part of the weak interaction is not influenced by the strong interaction.[44] Thus, Ft should be constant for all superallowed transitions. Taking all presently available data result in an amazing consistency at the 0.03% precision level.[44]

Radioactive nuclei are also ideal systems to put constraints on scalar currents in weak interaction which is supposed to be a pure A-V interaction. One highlight example of ISOLDE is the measurements on ^{32}Ar ($T_{1/2} = 98$ ms) an isotopic spin $T_z = -2$ nucleus, which was discovered at ISOLDE in 1977.[45] One of the best known values for the positron-neutrino angular correlation coefficient a was determined in the $0^+ \rightarrow 0^+$ β-decay of the ^{32}Ar experiment by Adelberger *et al.* in 1999.[46] The effect of lepton recoil on the shape of the narrow proton peak in the particle spectrum following the superallowed decay was analysed. Since the mass of ^{32}Ar was only known to an uncertainty of 50 keV at that time, the accuracy in the determination of a was limited to 6%. Thus, the mass prediction by the isobaric multiplet mass equation (IMME) was used instead and

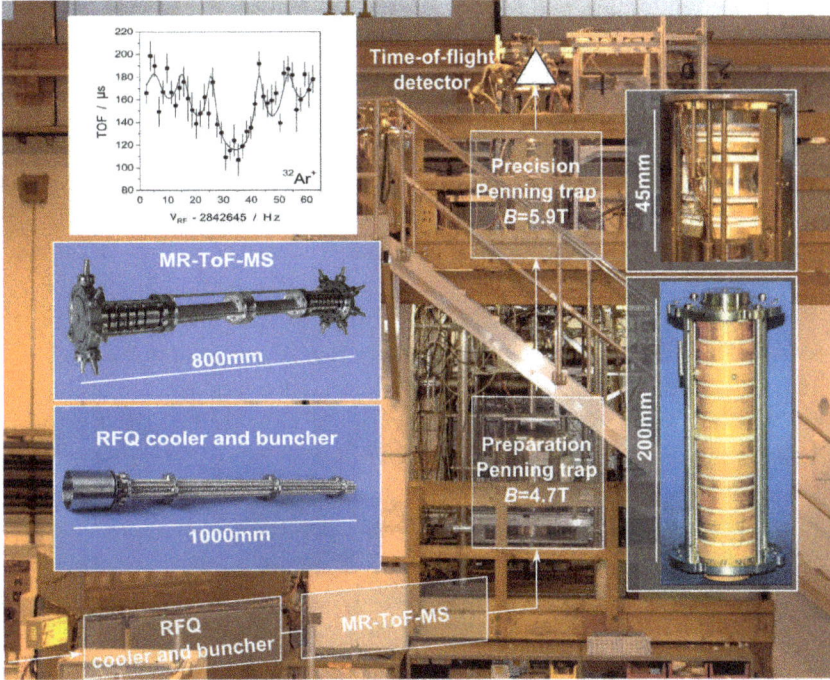

Fig. 13. Present set-up of the ISOLTRAP Penning-trap mass spectrometer at ISOLDE for high-precision mass measurements on short-lived nuclides. The inset shows on the top left the cyclotron resonance of ^{32}Ar$^+$ with the fit of the theoretically expected line-shape. The other insets show the different trapping devices at ISOLTRAP. At the time of the Ar measurement in 2001 the multi-reflection time-of-flight (MR-ToF-MS) spectrometer was not installed at ISOLTRAP.

resulted in the beta-neutrino correlation coefficient for vanishing Fierz interference of $a = 0.9989 \pm 0.0052(\text{stat}) \pm 0.0039(\text{syst})$ at the 68% confidence level, thus being fully consistent with the SM prediction. Furthermore, a new limit on the masses of scalar particles with gauge coupling strength could be derived being at that time $M_S \geq 4.1\,M_W$. A few years later the mass of ^{32}Ar was measured for the first time directly by Penning-trap mass spectrometry at ISOLTRAP[47] (see inset of Fig. 13) with a mass uncertainty of only $\delta m = 1.8\,\text{keV}/c^2$, thus allowing for an improved value of the positron-neutrino angular correlation coefficient a and relying no longer on the IMME prediction.

7. ISOLDE at the Doorstep to the Next Half-Century

In the period of writing this historic review of some of the landmark experiments, we witness an intense period building up new projects for the future:

The new post-accelerator structure HIE-ISOLDE. The HIE-ISOLDE (High Intensity and Energy) project will provide major improvements in energy range, beam intensity and beam quality. An important element of the project will be an increase of the final energy of the post-accelerated beams to 10 MeV/u throughout the periodic table. The first stage will boost the energy of the current REX-LINAC to 5.5 MeV/u where the multistep Coulomb excitation cross-sections are strongly increased with respect to the previous 3 MeV/u and many transfer reaction channels will be opened.[48] The construction of the Linac is underway and the full physics programme with post accelerated beams up to 5.5 MeV/u will start in 2016.

The TSR storage ring. Recently, a low-energy storage ring was proposed to be installed at HIE-ISOLDE using the existing ring TSR presently in operation at the Max-Planck-Institute for Nuclear Physics in Heidelberg.[49] The addition of a storage ring to an ISOL facility opens up an extremely rich scientific programme in nuclear physics, nuclear astrophysics and atomic physics. Reaction and decay studies can benefit from the "recycling" of the rare exotic nuclei stored in the ring and from low background conditions. Studies of the evolution of the atomic structure can be extended to isotopes outside the valley of stability. In addition to experiments performed using beams recirculating within the ring, cooled beams can be extracted and exploited by external spectrometers for high-precision measurements.

The CERN-MEDICIS project. Building further on ISOLDE's know-how in RIB production and on CERN's proton beam capabilities, the CERN-MEDICIS (Medical Isotopes Collected from ISOLDE) project was initiated. MEDICIS will exploit targets installed at ISOLDEs beam dump position and produce long-lived radioisotopes for fundamental studies in cancer research, for new imaging and therapy protocols and for pre-clinical trials.

These projects will create new opportunities for radioactive beam research and bring ISOLDE at the doorstep of the next half-century.

Acknowledgments

It has been a privilege for us to write this chapter in the book celebrating CERN's 60 years. ISOLDE has been around at CERN for 50 years and we want to take this opportunity to thank, on behalf of all our experimental and theoretical colleagues that have worked at ISOLDE, CERN for its generous support and especially the ISOLDE technical team has always been supporting the users and made this story possible. We also want to convey our sincere thanks to the national funding bodies for their continued support, which made this 50 years journey possible.

References

1. O. Kofoed-Hansen and K. O. Nielsen, *Phys. Rev.* **82**, 96 (1951); *Kgl. Dan. Vidensk. Selsk. Mat. Fys. Medd.* **26**, No. 7 (1951).
2. M. J. G. Borge *et al.*, *Phys. Lett. B* **317**, 25 (1993).

3. H. L. Ravn and B. W. Allardyce, On-line Mass separators, in *Treatise on Heavy Ion Science*, Vol. 8 (Springer, 1989), pp. 363–439.
4. P. Van Duppen and K. Riisager, *J. Phys. G* **38**, 1 (2011).
5. H.-J. Kluge *et al.*, Laser Ion Sources, in *Proc. Accelerated Radioactive Beams Workshop*, Parksville, Canada (1985).
6. F. Scheerer *et al.*, *Rev. Sci. Instr* **63**, 2831 (1992).
7. V. Mishin *et al.*, *NIM B* **73**, 550 (1993).
8. K. Blaum *et al.*, *NIM B* **204**, 331 (2003).
9. D. Fink *et al.*, *NIM B* **344**, 83 (2015); *NIM B* **317**, 661 (2013).
10. L. M. C. Pereira *et al.*, *Appl. Phys. Lett.* **98**, 201905 (2011).
11. U. Köster *et al.*, *NIM B* **160**, 528 (2000).
12. J. Van Roosbroeck *et al.*, *Phys. Rev. Lett.* **92**, 112501 (2004).
13. I. Stefanescu *et al.*, *Phys. Rev. Lett.* **98**, 122701 (2007).
14. J. Bonn *et al.*, *Phys. Lett. B* **38**, 308 (1972).
15. S. Rothe *et al.*, *Nature Comm.* **4**, 1835 (2013).
16. P. Delrock *et al.*, *Phys. Rev.* **67**, 808 (1991).
17. O. Kester *et al.*, *NIM B* **204**, 20 (2003).
18. I. Talmi and I. Unna, *Phys. Rev. Lett.* **4**, 469 (2006).
19. C. Thibault *et al.*, *Phys. Rev. C* **12**, 644 (1975).
20. G. Huber *et al.*, *Phys. Rev. C* **18**, 2342 (1978).
21. D. Guillemaud-Mueller *et al.*, *Nucl. Phys. A* **426**, 37 (1984).
22. G. Neyens *et al.*, *Phys. Rev. Lett.* **94**, 022501 (2005).
23. W. Schwerdtfeger *et al.*, *Phys. Rev. Lett.* **103**, 012501 (2009).
24. K. Wimmer *et al.*, *Phys. Rev. Lett.* **105**, 252501 (2010).
25. F. Wienholtz *et al.*, *Nature* **498**, 346 (2013).
26. M. D. Seliverstov *et al.*, *Phys. Lett. B* **719**, 362 (2013).
27. L. Gaffney *et al.*, *Nature* **497**, 199 (2013).
28. H. De Witte *et al.*, *Phys. Rev. Lett.* **98**, 112502 (2007).
29. M. J. G. Borge, *Phys. Scr. T* **152**, 014013 (2013).
30. I. Tanihata *et al.*, *Phys. Rev. Lett.* **55**, 2676 (1985).
31. P. G. Hansen and B. Jonson, *Europhys. Lett.* **4**, 409 (1987).
32. K. Riisager, *Phys. Scr. T* **152**, 014001 (2013).
33. E. Arnold *et al.*, *Phys. Lett. B* **197**, 311 (1987).
34. E. Arnold *et al.*, *Phys. Lett. B* **281**, 16 (1992).
35. R. Neugart *et al.*, *Phys. Rev. Lett.* **101**, 132502 (2008).
36. B. Jonson and K. Riisager, *Nucl. Phys. A* **693**, 77 (2001).
37. K. Riisager *et al.*, *Phys. Lett. B* **235**, 30 (1990).
38. I. Mukha *et al.*, *Phys. Lett. B* **367**, 65 (1996).
39. W. Geitner *et al.*, *Phys. Rev. Lett.* **83**, 3792 (1999).
40. T. Suzuki, T. Otsuka and A. Muta, *Phys. Lett. B* **364**, 69 (1995).
41. K. Blaum, J. Dilling and W. Nörtershäuser, *Phys. Scr. T* **152**, 014017 (2013).
42. S. L. Kaufman, *Opt. Commun.* **17**, 309 (1976).
43. W. Nörtershäuser *et al.*, *Phys. Rev. Lett.* **102**, 062503 (2009).
44. J. C. Hardy and I. S. Towner, *Phys. Rev. C* **91**, 025501 (2015).
45. E. Hagberg *et al.*, *Phys. Rev. Lett.* **39**, 792 (1977).

46. E. G. Adelberger *et al.*, *Phys. Rev. Lett.* **83**, 1299 (1999) and E. G. Adelberger *et al.*, *Phys. Rev. Lett.* **83**, 3101 (1999).
47. K. Blaum *et al.*, *Phys. Rev. Lett.* **91**, 260801 (2003).
48. K. Riisager *et al.* (eds.), *HIE-ISOLDE: The scientific opportunities.* CERN Report, CERN-006-013.
49. M. Grieser *et al.*, *Eur. Phys. J. Special Topics* **207**, 1 (2012).

Index